FIRE ECOLOGY

FIRE ECOLOGY

UNITED STATES AND SOUTHERN CANADA

HENRY A. WRIGHT

Texas Tech University

and

ARTHUR W. BAILEY

University of Alberta

A Wiley-Interscience Publication

JOHN WILEY & SONS

New York   •   Chichester   •   Brisbane   •   Toronto   •   Singapore

Copyright © 1982 by John Wiley & Sons, Inc.

All rights reserved. Published simultaneously in Canada.

Reproduction or translation of any part of this work
beyond that permitted by Section 107 or 108 of the
1976 United States Copyright Act without the permission
of the copyright owner is unlawful. Requests for
permission or further information should be addressed to
the Permissions Department, John Wiley & Sons, Inc.

Library of Congress Cataloging in Publication Data:

Wright, Henry A.
  Fire ecology, United States and southern Canada.

  "A Wiley-Interscience publication."
  Includes bibliographies and index.
  1. Fire ecology—United States. 2. Fire ecology
—Canada. 3. Botany—United States—Ecology.
4. Botany—Canada—Ecology. 5. Prescribed burning
—United States. 6. Prescribed burning—Canada.
I. Bailey, Arthur W. II. Title.

QH1O4.W74   581.5′222   81-14770
ISBN 0-471-09033-6   AACR2

Printed in the United States of America

10 9 8 7 6 5 4 3 2

To

Janet and Ruth
with our affection

and

to the pioneers of prescribed fire
whose foresight was not swayed
by ridicule and paved the way for
use of prescribed fire in a
responsible, stewardly manner

FOREWORD

Forest and range managers today have an unparalleled opportunity to use fire as a resource management tool. Increased awareness of the benefits of fire and new policies of land management agencies now encourage a more flexible approach to using and controlling fire than was previously permitted. I am sure that many of the fire research pioneers, such as Ed and Roy Komarek of Tall Timbers Research Station, Tallahassee, Florida, are pleased. Many ecologists have long advocated that fire be recognized as an ecological force and as a management tool.

The traditional policy of suppressing all forest and rangeland fires at any cost served a useful purpose during the period of custodial management of wildlands, especially on federal and state lands. The policy strengthened the vital campaign against careless burning of wildlands. Prevention of careless burning is as valid today as ever before. Nevertheless, custodial management has ended. We must now manage our wildlands in a cost-effective manner, with emphasis on producing goods and services for a multitude of purposes. Fire is a cost-effective tool for altering forest and range ecosystems to meet management goals. This concept includes wilderness and other reserved areas where fire, carefully monitored, is allowed to play its natural role.

The new opportunity to use fire presents several serious responsibilities. Prescribed fire must be conducted safely, with good control, and must attain stated land management objectives. Successful prescribed fire requires training and experience, careful planning, and skillful execution, all pursued in a patient, deliberate manner.

Skillful application of fire to meet resource management objectives requires definitive and quantified knowledge of the physical, biological, and ecological effects of fire on the specific ecosystem involved. In this book, Drs. Wright and Bailey wisely emphasize the impact of fire on plants and plant communities. Long ago I concluded understanding the effects of fire on plants is the key to predicting the effects on resource production whether goals are for timber, range management, wildlife, water, or recreational values. This book provides knowledge on the role and use of fire in many different plant communities over a broad geographic area.

The effects of fire have been studied since the beginning of organized forest and range management research. Unfortunately, the research has often been conducted and published outside the fire research community or has been scattered over a wide range of outlets. For example, for years I conducted research in lodgepole pine, a noted fire species, but the information was not identified as fire research.

The authors have provided us with a prodigious work collected from many sources, infused it with their considerable experience, and have presented us with a valuable guide to using fire over an extremely complex geographic area and for a wide variety of plant communities. It should prove invaluable to fire researchers, wildland planners, and land and resource managers. I believe that *Fire Ecology* will be a standard reference for years to come.

JAMES E. LOTAN
Supervisory Research Forester
Intermountain Forest and Range Experiment Station
Missoula, Montana

PREFACE

Fire ecology is of interest to many, and there are various ways to write about fire and its impact on ecosystems. We have chosen to emphasize the impact of fire on plants, particularly the native plant communities, giving a historical perspective on the role of fire in major ecosystems in the United States and southern Canada. The past impact of fire on an ecosystem gives the manager some idea of the potential role that fire may play in a managed ecosystem for both plants and animals. Most of the natural ecosystems in the United States and southern Canada have only come under some form of management within this century. In most cases there is still time to define the role that fire should play in the management of these ecosystems for the future.

The principles and use of prescribed fire to achieve management objectives is another important aspect of this text. Both authors have had considerable experience burning low- and high-volatile fuels. The benefit of their experience and that of many other land managers is included in the last chapter.

Fire ecology is experiencing a new wave of popularity as a general outgrowth of work by hardy pioneers in research and management who disagreed with the feeling that all fire was bad. They started a base of knowledge against a groundswell of criticism. From their courage came a tremendous flush of knowledge that has been developing in the 1960s, 1970s, and 1980s. Knowledge is currently being added to existing information at a rapid rate at both the research and management levels.

This book should be considered a progress report on the present state of knowledge on fire ecology in the United States and southern Canada. We hope it will help to disseminate current information about fire ecology and serve as a catalyst to improve the general knowledge base.

HENRY A. WRIGHT
ARTHUR W. BAILEY

Lubbock, Texas
Edmonton, Alberta
January 1982

ix

ACKNOWLEDGMENTS

We wish to acknowledge help from many colleagues and practitioners who provided criticism, suggestions, and photos for this text. Contributors included Martin E. Alexander, Robert R. Alexander, Stephen E. Arno, Robert G. Barse, Wilbert H. Blackburn, Charles E. Boldt, Eric G. Bolen, Donald E. Boyer, Carlton M. Britton, Stephen C. Bunting, David L. Caraher, Warren P. Clary, John F. Corliss, William F. Davis, Norbert V. DeByle, John H. Dieterich, D. Lynn Drawe, C. Theodore Dyrness, George R. Fahnestock, Robert M. Frank, Jerry F. Franklin, Neil C. Frischknecht, Robert F. Gartner, Lisle R. Green, Harold E . Grelen, George E. Gruell, Fred S. Guthery, James R. Habeck, Frederick C. Hall, Brad C. Hawkes, Carlton H. Herbel, Min Hironaka, Ragnar W. Johansen, Alex Johnston, Bruce M. Kilgore, Donald A. Klebenow, Arnold D. Kruse, Warren J. Kubler, John L. Launchbaugh, Bruce Lawson, Clifford E. Lewis, Robert M. Loomis, Robert E. Martin, S. Clark Martin, Phillip M. McDonald, Alstair McLean, William H. McNab, Don Minore, Leon F. Neuenschwander, David R. Patton, Stephen S. Sackett, Edward F. Schlatterer, Gilbert H. Schubert, Charles J. Scifres, Forest A. Sneva, H. W. Springfield, Peter F. Stickney, Roy M. Strang, Robert C. Szaro, Rita P. Thompson, Arthur R. Tiedeman, John F. Vallentine, Charles E. Van Wagner, Leslie A. Viereck, Stephen D. Viers, Jr., Leonard A. Volland, Dale D. Wade, Robert J. Warren, Neil E. West, and James A. Young. We also wish to express our appreciation to Cynthia L. Fulton and C. Ellen Stanley for their time to draw the vegetation maps.

The USDA Forest Service, Intermountain Forest and Range Experiment Station, Ogden, Utah, provided partial support to write the chapters titled Grasslands, Semidesert Grass-Shrub, Sagebrush-Grass, and Pinyon-Juniper in three USDA Forest Service General Technical Reports. These reports were written in a somewhat different form with fewer illustrations than the chapters in this text. The Intermountain Forest and Range Experiment Station also provided support for portions of the Ponderosa Pine chapter that was published as Texas Tech University Range Science Series No. 2. We gratefully acknowledge the support of the Intermountain Forest and Range Experiment Station for providing financial support and for providing a driving force to help get this text written.

H. A. W.
A. W. B.

CONTENTS

Large Mammals, 64

    Deer, 67
    Elk, 67
    Pronghorn Antelope, 68
    Caribou and Moose, 69
    Other Mammals, 69

       General, 69
       Fisher and Marten, 70
       Beaver, 70

Ants and Acacia, 71
Stream Fauna, 71
References, 80

CHAPTER 5   GRASSLANDS   **81**
Fire History of the Grasslands, 81
Ecological Characteristics and Effects of Fire, 83

    Southern Great Plains, 84

       Distribution, Climate, Soils, and Vegetation, 84
       Fire Effects—Shortgrass Prairie, 88
       Fire Effects—Mixed Prairie, 91
       Fire Effects—Mixed Tallgrass-Forest, 95

    Central Great Plains, 96

       Distribution, Climate, Soils, and Vegetation, 96
       Fire Effects—Shortgrass and Mixed Grass Prairie, 99
       Fire Effects—Tallgrass Prairie, 100

    Northern Great Plains, 103

       Distribution, Climate, Soils, and Vegetation, 103
       Fire Effects—Semiarid Mixed Prairie, 108
       Fire Effects—Mesic Mixed Prairie, 109
       Fire Effects—Tallgrass Prairie, 111
       Fire Effects—Fescue Prairie, 113
       Fire Effects—Aspen Forest, 121

    Rio Grande Plains and Coastal Prairie, 121

       Distribution, Climate, Soils, and Vegetation, 121
       Fire Effects, 123

    California Prairie, 125

       Distribution, Climate, Soils, and Vegetation, 125
       Fire Effects, 126

    Palouse Prairie, 127

       Distribution, Climate, Soils, and Vegetation, 127
       Fire Effects, 128

    Mountain Grasslands, 128

FIRE ECOLOGY

CHAPTER

1

INTRODUCTION

Fire has been a part of natural ecosystems since the origin of climate on earth. Lightning has always provided a natural ignition source and has shown no partiality as to where it will strike, although it is more prevalent in some areas than others (Komarek 1966). The potential of a lightning strike igniting material depends on the fuel source, the quantity and continuity of fuel, the fuel moisture, and the existing weather. Dry lightning storms in continuous fuels with gentle to rolling topography are most apt to cause large fires if winds and temperatures are high and relative humidity is low. Historically, a range of fire intensities that included headfires and backfires have burned under a wide range of fuel and weather conditions and have resulted in varied vegetative responses.

Thus for most ecosystems, fire was the natural catalyst for diversity that provided stability in those ecosystems (Vogl 1971b). Without fires forested communities become monocultures and are plagued with overstocking, excessive fuel accumulation, stagnation, and inadequate reproduction, which encourage diseases and insects (Vogl 1971a). Similarly, grasslands become stagnant and are invaded by shrubs and trees, whereas shrublands become decadent or impenetrable thickets (Wright 1974). Wildlife, in most cases, decline in numbers and diversity with a lack of fire, except in shrubby grasslands. Only deserts with less than 17 cm (7 in.) of precipitation probably escaped the influence of fire on their ecosystems because of a lack of fine fuel.

Man has probably used and "kept" fire for more than 500,000 years, but has learned to use it only during the last 20,000 years (Johnston 1970). Man's use of fire, along with lightning, intensified the effect of fire in many ecosystems, particularly in grasslands and those communities bordering grasslands. This is where man could

1

live year-round (Komarek 1965). It was easy to grow crops in grasslands and it was easy to harvest game on freshly burned areas. Moreover, dead wood, resulting from burns near grasslands, provided an important source of woody fuel to burn in the winter.

Fire continued to have a major impact on North American plant communities after the arrival of European man because of lightning, purposeful use of fire, and careless-ness. At this time European-trained foresters in the forest industry and government had a major impact upon use of fire in the United States and Canada. Fire was bad be-cause it killed trees. Having grown trees in Europe under cultivated systems and pro-tected them until harvest time, this philosophy was understandable. Fire suppression started in earnest about the time of Teddy Roosevelt's presidency. It is ironic that the champion conservationist was so misled about a force that is a part of our heritage: fire. Man has a natural tendency to fear fire because it is difficult to control. The ex-clusion of natural fires in grasslands and forests had no place in North America, how-ever, because the ecosystems evolved with fire.

European fire protection philosophy in North American forests spilled over into every plant community in North America. Not until the Leopold Report of 1963 (Leopold et al. 1963) was the general public in the United States informed that pro-tecting all plant communities from fire could be bad—excessive fuel buildups, stag-nant young pine trees, dense understories of shrubs and trees in forests (e.g., Sequoia National Park) that could lead to catastrophic stand-replacing fires, decadent shrub and grassland communities, encroachment of shrubs and trees into grasslands, mon-ocultures of trees that lead to increased disease and insect damage [e.g., lodgepole pine *(Pinus contorta)* in the Rocky Mountains] as well as less diversity in numbers and species of wildlife, and ultimately, devastating fires (Dodge 1972) that cannot be controlled with any amount of manpower.

Fire is a natural force in most of our plant communities and should be permitted to play a greater natural role where possible. Where natural fires are not possible, ap-plied fire ecology should play a greater role to achieve management objectives such as fuel reduction, seedbed preparation, disease control, thinning, suppression of shrubs, removal of litter, increased herbage yields, increased availability of forage, and increased wildlife. These objectives can be looked upon as the spokes of a wheel that radiate from the hub of diversity, which was historically created by fire.

A CHANGE IN PHILOSOPHY

Since the early 1960s biologists of all disciplines began taking a constructive view of fire in North America. Fire's reintroduction as a natural force began in the southeast and northwest and has quickly spread on a limited scale to all areas of North America, although the Flint Hills of Kansas have been burned since 1880. Interest has acceler-ated in the past few years because alternative management tools (e.g., chemicals and mechanical treatments) are environmentally unacceptable to the public, are not effec-tive, or are too expensive. In the case of insect and disease outbreaks in forests, alternatives to fire have been unsatisfactory (Loope 1972). Further, with 80 years of fire protection in our western forests, we can no longer tell the ''good'' fire years from the ''bad'' fire years if we look at costs for fire prevention. There is so much dead fuel in chaparral communities and in the understory of protected forests that man cannot control wildfires. The leading front of a fire generally stops because of natural barriers or a change in weather.

Today, the national parks are letting fire take a more natural role in vegetation management. With discretion, fires are used to restore the original condition of parks. Fire research during the 1970s and 1980s throughout the United States and Canada is providing the basis for using fire as a management tool for renewable natural resources.

Use of fire in forest communities of the western United States and Canada is being accepted more slowly and with more wariness than in shrub and grassland ecosystems. This is natural because personnel of the various forest service agencies have seen many devastating wildfires and are not convinced that the benefits of prescribed fire outweigh the risks. Unfortunately, the risks are becoming more serious with time (Dodge 1972), and the western forests will need a massive infusion of hazard-reduction burning to solve the problems of fuel buildups, stagnation, monocultures, and disease and insect damage, which create the potential for massive catastrophic fires. The hazards associated with prescribed burning are now much greater because of dense, overstocked forests and massive fuel buildups. These fuels must be removed with fire by highly experienced personnel before maintenance burning can be used with safety, reduced cost, and reduced risk of the fire getting away.

It is unfortunate that the governments in the United States and Canada do not understand the benefits from massive infusions of money to be spent on hazard reduction, forest and range restoration, and wildlife habitat improvement through the use of prescribed burning on our public lands. Instead, they waste massive sums of money either fighting fires that they cannot stop or putting fires out that would go out naturally. True, spending money to put fires out appeals to the public, but often it is not good stewardship over our natural resources or our money.

LEARNING TO USE PRESCRIBED FIRE

Prescribed fire is a reasonable way to reintroduce fire into many ecosystems, but too much emphasis has been placed on the risks of prescribed burns. Based on our combined experience of more than 200 shrub and grass fires, including volatile fuels such as juniper (*Juniperus* spp.) and chaparral, prescribed burns are relatively safe if personnel are experienced and cautious. Each of us started conducting prescribed burns with no experience. We had one escape of 250 ha (500 acres), one 100 ha (200 acres), and three that were less than 4 ha (10 acres). All escapes occurred within the first three years of research, except for one holdover fire that started last year 15 hours after a prescribed burn. We had numerous spot fires from our burns but these are controlled with either wide firelines or a few select teams of fire suppression crews.

During this time we were learning how to use fire, testing the limits for prescribed burning, and testing limits of plant response to fire.

We never had more than one D-7 caterpillar tractor, one or two slip-on pumpers, or 15 people to control a fire on any burn. Moreover, we did not have access to backup fire crews or air-tankers. Therefore, the present concern about risk is exaggerated in our opinion, even though we must be cautious during and after fires. Current state-of-the-art techniques on prescribed burning reduce the risk of fires getting away, but several years of prescribed burning experience is necessary before one should be given this responsibility as crew boss. There is a definite need for the training of knowledgeable, skilled users of prescribed fire throughout North America.

The use of fire can be frightening, however, and many burns do not get started because of fear in setting a fire. Even people who have been trained but not forced to make decisions to light a fire, based on existing weather, are afraid to start a fire. As long as someone else makes the decision whether to proceed, they can conduct the burn safely. As a consequence, judgment, common sense, experience, and the willingness to make a decision are of utmost importance in conducting a prescribed burn.

In the hands of confident prescribed burners (trained for two or three years in the field), fire is a versatile tool that can achieve many objectives simultaneously in many plant communities (Wright 1974). It is natural and frequently the only management tool available to achieve particular objectives. Often fire is the cheapest management alternative. Risk is still a factor in prescribed burns, but generally prescribed burns are safe in the hands of skilled personnel. Fire in the hands of unskilled and untrained personnel often leads to disaster.

REALITY AND THE FUTURE

Proponents of fire have a tendency to show all the good things about fire and few of the undesirable effects. Our experience has shown that desirable effects are sold most easily when the whole story of fire, good and bad, is told. The secret to convincing the public of the place of prescribed fire and wildfire is to tell how to achieve desirable effects and how to avoid undesirable effects. For example, burning with low subsoil moisture in the Great Plains usually leads to low grass yields, whereas burning with high subsoil moisture leads to high grass yields. Therefore, we recommend burning during wet years. Wildfires in some parks and wilderness areas are permitted to burn under certain weather conditions, but if there is a chance of a massive stand-replacing fire that has unacceptable risks in terms of lives and property, usually because of extremely dry conditions, the fire is suppressed as quickly as possible.

Safety is uppermost in everyone's minds. Fires must be conducted so that they do not go beyond the planned firelines, park, or wilderness. There is generally adequate data on how to conduct prescribed burns, but the techniques have not been taught to many people. There are excellent practitioners who can conduct prescribed burns, but their rules are often general, sometimes regional in nature, and very little information is documented. Procedures on burning grasslands have been handed down from father to son in the Flint Hills of Kansas. Most users have no specific prescriptions for relative humidity, wind speed, fuel moisture, and so on, yet they are very skilled.

To conduct burns safely, the fire boss needs precise research data on which to base a fire plan. Many researchers have accumulated such data and are refining their prescriptions throughout North America. We can give precise prescriptions to burn safely in many communities and hope researchers continue to refine prescriptions and collect new data where it is needed.

Despite our efforts to get good field-tested research on prescribed burns, we must still acknowledge that conducting wildland fires is dangerous. People must be informed about these dangers—what they are and how to make plans to cope with them. The unexpected behavior of a fire is always a threat and only the person with years of experience can attempt to forecast most of the dangers.

The disadvantages to burning are common knowledge. The news media are willing to play up the dramatic and the negative: wildfires destroy forests, watersheds, wildlife, homes, and so on. On a limited scale, there is a place for wildfires in national parks, wilderness areas, and forested areas with limited commercial timber values. Wildfires are most common during dry years following a fuel buildup and can be extensive where fuel loads are high. Prescribed fire can be a useful tool to lower fuel loads and limit the damage and extent of wildfires. Vegetative cover heals quickly on prescribed burns if soil is moist, and soil loss is minimal unless followed by an unusual, intense rain storm.

Many people have a misconception that many animals are killed by fires. However, most vertebrate animals manage to escape the heat of fires by flying or running away, going below the ground several centimeters, hiding in rock outcrops, or seeking islands missed by the fire. Fire studies in California by Howard et al. (1959) showed no harm to cottontails (*Sylvilagus* spp.), wood rats (*Neotoma* spp.), mourning doves *(Zenaida macroura)*, quail (*Colinus* spp.), and several species of birds. In fact, the numbers of all animals increased immediately after the fire. Phillips (1965) lists a large number of ungulates associated with forests and woodland-savannahs in east Africa; the only report of any mature animals being killed by fire was after a fire in 1869 in which a dead elephant *(Laxodonta africana)* and buffalo *(Syncerus caffer)* were found. Phillips occasionally noted that a few species of very young animals were killed by fires.

The principal way that fires adversely affect population densities of animals is by altering the habitat, not by killing. Habitat, more than anything else, determines the species and their densities. In general, fires greatly enrich the wildlife species and numbers on all vegetation types (R. Komarek 1963; Marshall 1963).

Airborne particulates are the primary pollutant of fires, although they are also short-lived (Dieterich 1971). Hydrocarbons are another combustion product, but few, if any, appear in the combustion of wood products that are important in photochemical reactions (Fritchen et al. 1970). Carbon monoxide is a pollutant from fires, but it seems to oxidize quite readily (Fritchen et al. 1970) and does not pose an immediate threat to people, plants, or animals (Dieterich 1971).

Management after a burn is important to the success of a burn. Grazing animals will frequently concentrate on a burn because the herbage or browse is more accessible, palatable, and nutritious. For this reason prescribed burning is generally done on a management-unit basis. If small areas within large areas are burned, animals will concentrate on the burn, even two years afterwards. In Texas, burns have greatly improved ranges and all of them have been grazed within three to seven months after the burns. There was full control over the areas because they were burned on a field basis.

We must use imagination on how to help the users of our resources with fire without making them pay penalties such as reduced game or livestock numbers. Reducing livestock numbers has always been the standby answer to improve depleted forage-producing ranges. Frequently our ranges need more than reduced livestock numbers and we should think of ways to improve ranges without cutting livestock or wildlife numbers. We have practiced this philosophy in our burning programs and have been successful in improving ranges and keeping the users satisfied with the results.

In summary, there are uses for prescribed burning and some wildfire in the management and conservation of forests, chaparral, grasslands, watersheds, and wildlife. There are dangers in using fire, both in its application and in its results. To minimize harmful effects, fire should rarely be used during extended dry periods. Moreover, the user should be an experienced professional with a thorough knowledge of ecosystems, weather, and fire behavior. The following chapters in this text provide an ecological basis for making decisions as to whether fire is desirable to achieve specific management objectives in the United States and southern Canada. Moreover, where data is available, we show how to conduct prescribed fires to achieve many management objectives.

REFERENCES

Dieterich, J. H. 1971. Air-quality aspects of prescribed burning, p. 139–151. In Proc. Prescribed Burning Symp. USDA For. Serv. Southeastern For. Exp. Stn., Asheville, N.C., pp.139–151.

Dodge, M. 1972. Forest fuel accumulation—a growing problem. *Science* **177**:139–142.

Fritchen, L. J., H. Bovee, K. Buettner, R. Charlson, L. Monteith, S. Pickford, and J. Murphy. 1970. Slash fire atmospheric pollution. USDA For. Serv. Res. Paper PNW-97. Northwest For. and Range Exp. Stn., Portland, Ore.

Howard, W. E., R. L. Fenner, and H. E. Childs, Jr. 1959. Wildlife survival in brush burns. *J. Range Manage.* **12**:230–234.

Johnston, V. R. 1970. Nature before us successfully managed the forest. *Audubon,* Sept. 1970, pp. 78–81, 85, 88, 90, 92, 99, 100, 102, 104, 106, 108, 110, 112, 114, 116, 117, 118, 119.

Komarek, E. V., Sr. 1965. Fire ecology—grasslands and man. *Proc. Tall Timbers Fire Ecol. Conf.* **4**:169–220.

Komarek, E. V., Sr. 1966. The meteorological basis for fire ecology. *Proc. Tall Timbers Fire Ecol. Conf.* **5**:85–125.

Komarek, R. 1963. Fire and the changing wildlife habitat. *Proc. Tall Timbers Fire Ecol. Conf.* **2**:35–43.

Leopold, A. J., S. A. Cain, C. M. Cottam, I. N. Gabrielson, and T. L. Kimball. 1963. Wildlife management in the national parks. *Amer. For.* **69**(4):32–35, 61–63.

Loope, L. 1972. The fires next time. *Time* **100**(6):48, 49.

Marshall, J. T. 1963. Fire and birds in the mountains of southern Arizona. *Proc. Tall Timber Fire Ecol. Conf.* **2:**135–141.

Phillips, J. 1965. Fire—as Master and Servant: Its influence in the bioclimatic regions of Trans-Saharan Africa. *Proc. Tall Timbers Fire Ecol. Conf.* **4:**7–109.

Vogl, R. J. 1971a. The future of our forest. *Ecol. Today* **1**(1):6–8, 58.

Vogl, R. J. 1971b. Smokey: A bearfaced lie. *Ecol. Today* **1**(3):14–17.

Wright, H. A. 1974. Range burning. *J. Range Manage.* **27:**5–11.

2

TEMPERATURE AND HEAT EFFECTS

Temperatures and their durations are often measured to determine heat effects on plant parts at specific locations. These measurements are not intended to measure fire intensity. Moreover, surface fire intensity will not have any bearing on the various types of burning in duff or a plant crown sometime after a fire passes a given point. Some researchers (e.g., Van Wagner and Methven 1978) try to tie fire behavior to fire effects. This can be done for above ground-effects (Van Wagner 1973) but not below the humus layer. Here moisture and quantity of organic matter become overriding factors. Therefore, temperatures and their durations can be used to determine what is going on near a rootstock or in a plant, and this result can be similar for a variety of fire intensities (Wright 1971). This chapter will cover temperature regimes for different fuels above and below the soil surface as well as heat effects on living material.

TEMPERATURE REGIMES

Grasslands and Other Fine Fuels

SOIL SURFACE

Mineral soil surface temperatures of grassland headfires are a linear function of the amount of uncompacted fine fuel available for burning (Fig. 2.1). For fuels that varied from 1685 to 7865 kg/ha (1500 to 7000 lb/acre), our research (Stinson and

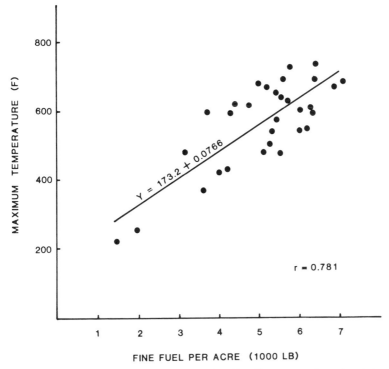

Fig. 2.1. Average maximum temperatures of grassland headfires in west Texas in relation to total yield of herbage. Each point represents an average of six thermocouples.

Wright 1969) has shown that average soil surface temperatures of grass fires in the Great Plains normally varied from 102° to 388°C (215° to 730°F). Temperature data from other grassland studies by Bentley and Fenner (1958), Ito and Iizumi (1960), Smith and Sparling (1966), Britton and Wright (1971), and Bailey and Anderson (1980) also fall within this range. Temperature extremes, however, for the 1685 to 7865 kg/ha (1500 to 7000 lb/acre) range of fuels will vary from 83° to 682°C (182° to 1260°F) (Stinson and Wright 1969). Similarly, McKell et al. (1962) have reported a soil surface temperature range of 168° to 593°C (335° to 1100°F) for medusahead *(Taeniatherum asperum)* fuels that averaged 5840 kg/ha (5200 lb/acre). The highest soil surface temperatures are probably associated with local accumulations of loosely arranged litter and intense winds created by the fire.

Air temperature, relative humidity, and soil moisture do not seem to affect soil surface temperatures in grassland fires (Britton and Wright 1971). The effect of wind, however, can vary from slight (Sparling and Smith 1966; Britton and Wright 1971) to very significant (Hare 1961; Whittaker 1961). Whittaker observed that an increase from a slight wind to a moderate one and the subsequent fanning of the flames caused a temperature rise of 78° to 156°C (172° to 312°F) at ground level.

High fuel moisture slows down the combustion rate (Byram 1958) and rate of fire spread (Batchelder and Hirt 1966), but within a range of 6.4 to 33.1 percent fine fuel moisture, Britton and Wright (1971) did not find that it had any effect on mineral soil

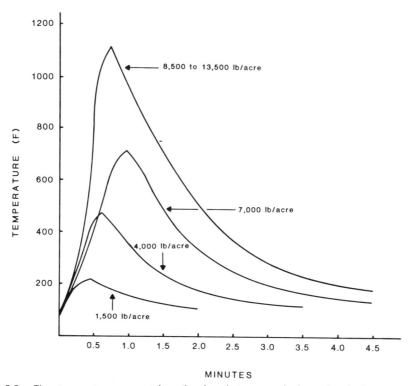

Fig. 2.2.   Time-temperature curves at the soil surface during prescribed grass fires for four amounts of fine fuel. At the time of burning air temperature, relative humidity, and wind were 21° to 27°C (70° to 80°F), 20 to 40 percent, and 13 to 24 km/h (8 to 15 mi/h). (From Wright et al. 1976.)

surface temperatures. Fires were equally effective in consuming dead honey mesquite *(Prosopis glandulosa* var. *glandulosa)* stems, as long as fine fuel moisture was below 20 percent. Similar data on consumption of woody materials were reported for pilot ignition times by Stockstad (1979) in which he showed that rotten wood sections of ponderosa pine *(Pinus ponderosa)* and Douglas fir *(Pseudotsuga menzesii)* ignited equally well over a fuel moisture content range of 5 to 24 percent. A fine fuel moisture of approximately 33 percent appears to be the threshold value for ignition of dead fuels.

Time-temperature curves for various fine fuel loadings are illustrated in Fig. 2.2. For fuels up to 7865 kg/ha (7000 lb/acre), soil surface temperatures above 66°C (150°F) generally last only from 0.9 to 5.4 min (Stinson and Wright 1969), which indicates that seeds of most plants can survive grass fires (Daubenmire 1968). Similarly, Smith and Sparling (1966), working in vegetation dominated by grasses and shrubs in northeastern Ontario, Canada, found that temperatures above 66°C (150°F) lasted from 1.4 to 6.0 min at the soil surface. The time-temperature curves as shown in Fig. 2.2 can be simulated with a propane gas burner (Britton and Wright 1980), which is very useful for individual plant studies.

BELOW SOIL SURFACE

Below the mineral soil surface, temperatures decrease sharply (Heyward 1938; Bentley and Fenner 1958; Ito and Iizumi 1960; Norton and McGarity 1965; Trabaud 1979). Under longleaf pines *(Pinus palustris)*, where the principal fuel was grass, the majority of soil temperatures at depths of 0.3 to 0.64 cm (0.12 to 0.25 in.) ranged from 66° to 79°C (150° to 175°F) and generally persisted for 2 to 4 min after which they declined rapidly (Heyward 1938). Soil temperature increases were negligible below a depth of 0.64 cm (0.25 in.), regardless of soil texture, even when flames were 3.7 m (12 ft) high. Similarly, when burning pastures with grass 38 cm (15 in.) high [5280 to 10,786 kg/ha (4700 to 9600 lb/acre)] in northern New South Wales, Norton and McGarity (1965) recorded a maximum temperature of 75°C (168°F) at 1 mm (0.04 in.) below the soil surface. Marked temperature rises were recorded only in the upper 10 mm (0.4 in.) of the soil. Thus, temperatures of a grass fire should have little direct effect on soil organic matter, microbial populations, or buried seeds (Norton and McGarity 1965). Heat penetration to the mineral soil below fine fuels in forests is dependent on prefire fuel depth, moisture regime of the duff layer, and fire persistence or duration.

ABOVE SOIL SURFACE

Above the soil surface, up to a height of 6 to 15 cm (2 to 6 in.), temperatures of headfires rise very rapidly. At a height of 6 to 15 cm (2 to 6 in.), temperatures are about twice as high as those at the soil surface (Lindenmuth and Byram 1948; Ito and Iizumi 1960; McKell et al. 1962; Smith and Sparling 1966), although the differences may range from being 0.2 to 0.5 times higher (Smith and Sparling 1966; Bailey and Anderson 1980) to 3 times higher (Ito and Iizumi 1960; McKell et al. 1962).

The magnitude of temperatures above the 5- to 15-cm (2- to 6-in.) height depends primarily on the quantity of fine fuel being burned and wind velocity (Trabaud 1979). For moderate fuels [3933 kg/ha (3500 lb/acre)], Ito and Iizumi (1960) recorded temperatures of 90°, 310°, and 135°C (194°, 590°, and 275°F) for heights of 0, 5, and 30 cm (0, 2, and 12 in.) respectively. For a wide variety of moderate lichen-shrub or grass-shrub fuels [2310 kg/ha (2055 lb/acre)] in Ontario, Canada, Smith and Sparling (1966) recorded temperatures of 266°, 401°, 203°, and 121°C (510°, 754°, 398°, and 250°F), for heights of 0, 5, 10, 20, and 51 cm (0, 2, 4, 8, and 20 in.), respectively. At the same heights in light fuels [544 kg/ha (484 lb/acre)] Smith and Sparling (1966) recorded temperatures of 142°, 158°, 140°, 93°, and 66°C (288°, 316°, 284°, 200°, and 151°F). Similarly, under cooler burning conditions Bailey and Anderson (1980) recorded temperatures of 165°, 185°, 200°, 180°, and 160°C (329°, 365°, 392°, 356°, and 320°F) at heights of 0, 5, 15, 30, and 45 cm (0, 2, 6, 12, and 18 in.) in heavy [5085 kg/ha (4523 lb/acre)] grassland fuels.

Shrublands

BELOW SOIL SURFACE

Using data from several sources, DeBano et al. (1977) reported that maximum soil surface temperatures for intense, moderate, and light burns in California chaparral were 685°C (1265°F), 430°C (806°F), and 260°C (500°F), respectively. These temperatures generally did not cool down to 100°C (212°F) at the soil surface for 15 min. At 2.5 cm (1 in.) below the soil surface, maximum temperatures were 195°C (383°F), 175°C (347°F), and 90°C (194°F), respectively, for the same fire intensities. At 5.0 cm (2 in.), maximum temperatures for all fire intensities were about 50°C (122°F).

ABOVE SOIL SURFACE

Burning a scrub oak *(Quercus coccifera)* community in France, Trabaud (1979) found that the highest temperatures were always near the top [1.3 m (4 ft)] of the vegetation during autumn burnings. Temperatures were 300°, 900°, 1000°, 1000°+, 450°, and 280°C (572°, 1652°, 1832°, 1832°+, 842°, and 536°F) at heights of 0, 0.5, 1.0, 1.3, 2.3, and 3.0 m (0, 1.5, 3.2, 4.3, 7.5, and 9.8 ft), respectively. Whittaker (1961), working in heath lands in Scotland, found that some temperatures were 500°C (932°F) higher at 0.2 m (0.7 ft) above than at ground level.

In western snowberry *(Symphoricarpos occidentalis)* brushland, Bailey and Anderson (1980) found cooler temperatures in spring burns, although the trends of maximum temperatures were similar to those of Trabaud's slightly below the top of the shrubs. Maximum temperatures for headfires were 510°, 590°, 530°, and 490°C (950°, 1094°, 1085°, 986°, and 914°F), at heights of 0, 0.07, 0.2, 0.45, and 0.75 m (0,3,8,18, and 30 in.) respectively. Thus, neither the gradient nor the maximum temperatures are as steep for spring fires as for fall fires, but maximum temperatures occur near the top of the vegetation.

In 8-year-old gallberry *(Ilex glabra)*-saw-palmetto *Serenoa repens)* roughs where the fuel was probably 11.2 to 22.4 metric tons/ha (5 to 10 tons/acre), Davis and Martin (1960) measured temperatures in headfires as high as 870°C (1600°F) at 30 cm (1 ft) above the surface. The maximum temperature dropped to 290°C (550°F) at 1.2 m (4 ft) above the soil surface.

Heavy Slash

Where heavy logs and deep slash are piled, duff surface temperatures vary from 620° to 1005°C (1150° to 1841°F) (Nelson and Sims 1934; Isaac and Hopkins 1937; Bentley and Fenner 1958). Below the soil surface Neal et al. (1965) recorded temperatures of 432°, 182°, 83°, and 62°C (810°, 360°, 182°, and 143°F) at soil depths of 0.6, 2.5, 7.6, and 12.5 cm (0.25, 1, 3, and 5 in.), respectively. These data are comparable to those for mean maximum temperatures found beneath burning piles of dead trees by Beadle (1940) in Australia.

Above the soil surface Countryman (1964), burning 90 to 112 metric tons/ha (40 to 50 tons/acre) of pinyon (*Pinus* spp.) and juniper (*Juniperus* spp.) slash found that temperatures rose from the presumed range of 649° to 982°C (1200° to 1800°F) at the soil surface to 1100°C (2000°F) at the middle [0.76 m (2.5 ft)] and top [1.5 m (5 ft)] of the fuel bed. Temperature maximums gradually rose to 1260°C (2300°F) at 6 m (20 ft) with an occasional maximum at 1430°C (2600°F). Temperatures at the middle of the fuel bed were above 820°C (1500°F) for at least 40 min, and above 40°C (1000°F) for at least 60 min.

Tree Trunk Temperatures

Temperatures on tree trunks are generally twice as high on the leeward side as the windward side (Fahnestock and Hare 1964; Tunstall et al. 1976). The magnitude, however, depends on quantity of fine fuel and wind speed. Fahnestock and Hare (1964) have recorded temperatures of 340° to 380°C (650° to 720°F) on the windward side of pine trees and 700° to 790°C (1300° to 1450°F) on the leeward side (Table 2.1). These data were taken where the fuel was almost entirely pine litter [16 to 31 metric ton/ha (7 to 14 ton/acre)] and the trees varied from 20 to 34 cm (8 to 13 in.) diameter at breast height (dbh).

Table 2.1. Maximum Temperatures Recorded at the Bark Surface and in Adjacent Air, in Relation to Fire Type and Thermocouple Height

Height of Thermocouple		Windward		Leeward		Ambient	
(m)	(ft)	(°C)	(°F)	(°C)	(°F)	(°C)	(°F)
Headfires							
0	0	549	1020	666	1230	577	1070
0.3	1	354	670	767	1412	349	660
0.6	2	361	682	714	1318	293	560
0.9	3	380	717	700	1292	204	400
Backfires							
0	0	428	802	614	1137	704	1300
0.3	1	402	762	468	874	177	350
0.6	2	200	393	498	928	71	160
0.9	3	133	272	224	435	146	295

(From Fahnestock and Hare 1964. Reproduced by permission of Journal of Forestry and USDA Forest Service.)

In west Texas we have recorded similar temperatures on leeward and windward sides of honey mesquite trees. Temperatures averaged 340°C (640°F) on the windward and 660°C (1220°F) on the leeward sides of trees that were 5 to 15 cm (2 to 6 in.) in diameter. Tree diameter had no effect on maximum temperatures. Fine fuel was 6180 kg/ha (5500 lb/acre) and wind speeds varied from 13 to 19 km/h (8 to 12 mi/h).

Fahnestock and Hare (1964) also compared temperatures on bark plates with those in bark fissures during a headfire. They found that the surface plates were hotter and stayed hot longer than the adjacent fissures. As a result, the total heat effect (time and temperature) was 10 to 25 percent higher on the bark plates than in fissures.

Ignition Temperature

The temperature of fuel must reach a threshold value of 346° ± 40°C (655° ± 72°F) for combustion to occur. Combustion occurs when volatile gases are emitted from the fuel and react with oxygen of the air to ignite (Anderson 1970), although pilot ignition of rotten wood can take place at temperatures as low as 250°C (482°F) (Stockstad 1979). When a fuel is exposed to heat, it warms up to where the water vapor boils off and steam distillation takes place. At this point organic chemicals with a boiling range similar to water vapor are carried along with the steam. Continued heating after removal of the water vapor results in volatilization of additional organic compounds, which, upon reaction with oxygen of the air, burst into flame, and a selfsustaining combustion process is established (Batchelder and Hirt 1966).

The time required to reach ignition temperature will depend on moisture content, material density, specific heat, material thickness, and heat source intensity. Using a constant moisture, density, and specific heat, Anderson (1970) has shown the relationship between ignition delay time, material thickness, and heat source intensity (Fig. 2.3). Ignition is defined by Anderson as the time when fuel begins to generate heat or when flaming or glowing combustion is evident. Notice in Figure 2.3 that thickness of wood material has less effect on ignition time as the heat source intensity increases.

Moisture delays ignition time considerably, depending on the amount of moisture. Fons (1950) showed that for 0.6-cm (0.25-in.) wood cylinders exposed to 620°C (1150°F), ignition times were 6.83, 8.90, and 15.47 sec when the moisture content was 2.2, 7.8, and 19.3 percent, respectively. Using a constant time, Batchelder and Hirt (1966) found that dry wood ignited at 290° to 330°C (554° to 626°F), whereas wood with sap did not ignite until the surface temperature reached 500°C (932°F) or higher.

Type of Burn

Relatively little research has been done to compare temperatures in headfires versus backfires. On forest floors in South Carolina, Lindenmuth and Byram (1948) found that headfires were consistently 54°C (130°F) hotter than backfires up to a height of

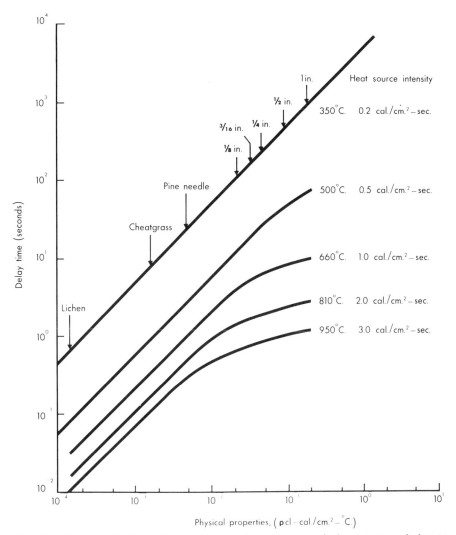

Fig. 2.3. Relation of ignition delay time to physical properties [sample density(p), specific heat(c), and thickness(l)]. Sample density, specific heat, and moisture content were held constant. (From Anderson 1970. Reproduced from the November 1970 *Fire Technology*. Copyright © 1970 National Fire Protection Association, Quincy, Massachusetts.)

12.5 cm (5 in.) and then the difference declined to −18°C (0°F) at a height of 46 cm (18 in.). This is because backfires tend to burn more slowly and deeply than headfires in pine needle beds (Beaufait 1965). Above 46 cm (18 in.) headfires are consistently hotter than backfires (Lindenmuth and Byram 1948) and they do more damage to brush and trees than slow-moving backfires (Fahnestock and Hare 1964).

By contrast, Bailey and Anderson (1980) found that headfires were consistently hotter than backfires at all heights measured in grasslands and shrublands. The difference was 60° to 90°C (140° to 194°F) in grasslands and 105° to 180°C (221° to 356°F) in shrublands. In similar quantities of grassland fuel [5843 kg/ha (5200 lb/acre)], McKell et al. (1962) found that slow-moving fires had a soil surface temperature of 293°C (560°F) compared to 193°C (380°F) for fast-moving fires. Thus, the literature is contradictory relative to soil surface temperatures. Above 46 cm (18 in.), however, headfires are 80° to 200°C (175° to 390°F) hotter than backfires (Fahnestock and Hare 1964; Bailey and Anderson 1980).

HEAT EFFECTS

Seeds

Seeds are very tolerant to heat (Daubenmire 1968). Sampson (1944) showed that grass seeds could tolerate temperatures of 82° to 116°C (180° to 240°F) for 5 min. He found that chaparral species were even more tolerant to fire. Most species could easily tolerate temperatures of 115° to 127°C (240° to 260°F) for 5 min, and these temperature exposures usually increased the percentage of germination, as has also been shown by Went et al. (1952). Chaparral species with thick or hard seed coats can tolerate temperatures of 125° to 140°C (260° to 280°F), and a few can tolerate temperatures of 140° to 150°C (280° to 300°F) (Wright 1931; Sampson 1944). Thus, if seeds are slightly covered by soil, they can survive a relatively intense fire. Grass fires would probably have only a slight effect on the mortality of dormant seeds, even if they were lying on the soil surface. In some cases, seeds harvested from recently burned grassland were found to have a higher percentage of germination than those on nearby unburned grassland (Ehrenreich and Aikman 1963; Grant et al. 1963).

Vascular Plant Tissue

Vascular plant tissue is easily killed by heat, and can be killed throughout a wide range of temperatures if any given temperature is maintained for the appropriate length of time (Hare 1961; Yarwood 1961; Wright 1970). Thus, death of plant tissue is an exponential function between temperature and time if the moisture content is constant (Fig. 2.4). Plant moisture, which is closely tied to the phenological stage of herbaceous plants, increases the susceptibility of plants to heat. This is why we see such a wide range of ''lethal'' plant temperatures reported in the literature.

A temperature of 60°C (140°F) is usually given as the thermal death point [the lowest temperature that results in no survival after a fixed period of exposure, usually 10 min (Schmidt 1954)] for most plant tissues (Hare 1961; Kayll 1966). However, Baker (1929), reporting on a compilation of data from several researchers, stated that the thermal death points for 20 plants ranged from 47° to 59°C (117° to 139°F). Only cactus (*Opuntia* spp.) survived temperatures over 62°C (144°F). Baker concluded that the thermal death point at the cellular level for herbage of average mesophytic plants lies between 50° and 55°C (122° and 131°F).

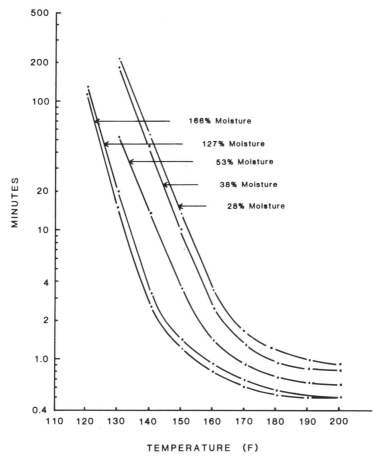

Fig. 2.4. Time required to kill tissue of needle-and-thread *(Stipa comata)* for five moisture contents in relation to temperature.

Jameson (1961) reported lethal temperatures of culms for four grass species to vary between 60° and 75°C (140° and 165°F). Shirley (1936) found that pine needles can withstand 50°C (122°F) for 2 hr. For a shorter period of time (24 min), Nelson (1952) found that southern pine needles had a thermal death point of 52°C (126°F) and could take 63°C (145°F) for 1 min. Lorenz (1939) found that lethal temperatures for eastern white pine *(Pinus strobus)* ranged from 57° to 59°C (135° to 138°F) for 30 min and 65° to 69°C (149° to 156°F) for 1 min. Thus, one can readily see that death of plant tissue depends primarily on moisture content and is an exponential function of temperature and time, as shown by Wright (1970). The widely quoted temperature of 60°C (140°F) for 1 min to kill plant tissue appears to be erroneous. Based on data in Figure 2.4, the time to kill plant tissue at 60°C (140°F) may vary from 2 to 60 min, depending on moisture content of the tissue (Wright 1970). Salts, sugars, lignin, and

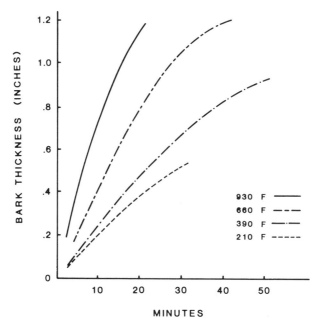

Fig. 2.5. Time required to reach 60°C (140°F) at the cambium. (From Kayll 1966. Reproduced by permission of the Minister of Supply and Services Canada.)

pectin are other variables that might be involved in the tolerance of plant tissue to heat.

Using 60°C (140°F) as a thermal death point, Kayll (1966) showed how long it would take for cambial tissues of different sized trees to reach a lethal temperature when various bark thicknesses are exposed to different temperatures (Fig. 2.5). However, type of bark also affects heat conductivity (Hare 1965). For example, Hare showed that longleaf and slash pines *(Pinus elliottii)* were about twice as fire resistant as sweetgum *(Liquidambar styraciflua)* even though all three species had the same bark thickness.

Morphological Adaptations

BARK INSULATION

Bark is a good insulator for trees. Fahnestock and Hare (1964) found that where bark surface temperatures varied from 290° to 800°C (550° to 1470°F) on longleaf pine, cambial temperatures varied from 38° to 82°C (100° to 180°F). They also found that the external temperature must be near 95°C (200°F) to raise the internal temperature appreciably. Very little heat damage occurred on trees if the bark was 1.0 to 1.3 cm (0.4 to 0.5 in.) thick. Heat lesions were mostly above the groundline on the leeward side and at or very near the groundline on the windward side. Tunstall et al. (1976) found that the maximum height of heat lesions on the leeward side in cool fires was

40 cm (1.6 ft). The maximum heat lesion height on mesquite trees in our field-tested trials in tobosagrass *(Hilaria mutica)* communities was 30 cm (1.2 ft).

Heat resistance is dependent on density, moisture content, and thickness of bark (Spalt and Reifsnyder 1962). In western North America trees with thick corky bark such as western larch *(Larix occidentalis)* are most resistant to surface fires. Ponderosa pine and Douglas fir usually have very thick bark and are very resistant to heat damage. Grand fir *(Abies grandis)* with thick bark and lodgepole pine *(Pinus contorta)* with thin bark have medium resistance to fire. Young trees of both species, however, are easily killed by fire. Western hemlock *(Tsuga heterophylla)* with a medium bark thickness and thin-barked species such as Engelmann spruce *(Picea engelmanii)* and subalpine fir *(Abies lasiocarpa)* have low resistance to fire. Thus, depending on surface fire intensity and age and species of trees, composition of a forest can be altered greatly with one fire.

CONE SEROTINY

Cone serotiny in lodgepole pine, jack pine *(Pinus banksiana)*, and to some extent in black spruce *(Picea mariana)* is an extremely important fire-adaptive feature for these tree species. Seeds in the cones are very resistant to heat, and the cones may remain high in the crowns attached to living branches for 25 years. Jack pine seeds can tolerate 150°C (300°F) for 30 to 45 sec and 370°C (700°F) for 10 to 15 sec (Beaufait 1960). These temperatures are rarely exceeded in slash burns above a height of 5.2 m (17 ft) (Beaufait 1961). As a consequence, natural regeneration from cones remaining on the trees that have received enough heat to melt the resinous bonds [temperatures in excess of 60°C (140°F)], but not enough to kill the seed is a very effective method of reestablishment on bare mineral soils (McRae 1979).

GROWTH FORM AND GROWING POINTS

Bunchgrass with large accumulations of dead plant material [e.g., needle-and-thread *(Stipa comata)*] often generate high temperatures for a long period of time after the main fire has passed (Wright 1971). The passing fire ignites the outer edge of the plant and then fuel within the plant generates temperatures at the soil surface up to 540°C (1000°F) within 1 hr (Fig. 2.6). Temperatures above 95°C (200°F) for over 2 hr are common in large bunchgrasses, but not in small bunchgrasses or rhizomatous species. This is why small bunchgrasses and most rhizomatous grasses recover very quickly after a fire compared to large, decadent bunchgrasses.

Rhizomatous grasses are usually resistant to fire because the rhizomes are 2.5 cm (1 in.) below the soil surface and there is no mass of dead plant material that can burn down to the growing point. Bunchgrasses with a growing point close to the soil surface, such as threeawns *(Aristida* spp.), are more susceptible to fire than other grasses, especially if they contain much dead material. Small plants containing only a few living culms generally survive fire regardless of depth of growing point unless they are burned at a susceptible phenological stage of growth (usually when flowering).

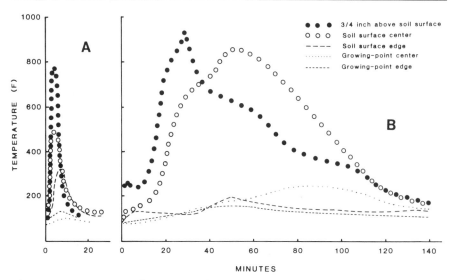

Fig. 2.6. Temperature in relation to time for five thermocouples in bunchgrass plants burned at 204°C (400°F) for 30 sec: *A, Sitanion hystrix* (a loose stemmy plant); *B, Stipa comata* (a dense leafy plant). (From Wright 1971.)

Trees also show varied resistance to fire depending on depth of bud zone. For example, velvet mesquite *(Prosopis glandulosa* var. *velutina)* is more susceptible to fire than honey mesquite because its crown bud zone is closer to the soil surface (Wright 1980). Similar data have been shown when comparing age of tree susceptibility to depth of crown bud zone (Wright et al. 1976; Steuter and Wright 1979).

Measurement

PLOT SIZE

Where the chief fuel is fine fuel, the plot size for measuring temperatures is of minor importance. Soil temperatures were no higher for a fire which burned 16 ha (40 acres) than for a fire which burned 0.001 ha (0.0023 acre) having the same type of vegetation (Heyward 1938). Temperatures up to 650°C (1200°F) as measured by McKell et al. (1962) on plots 9 by 9 m (30 by 30 ft) also indicate that small plots are probably just as adequate as large plots for measuring temperatures. Where wind is a prominent factor in collecting headfire data, White (1969) suggests that the plots be at least 15 to 18 m (50 to 60 ft) long.

SAMPLING TECHNIQUES AND EQUIPMENT

Maximum temperatures and durations are the most informative data and should be recorded with a multipoint recorder powered by a 12-volt car battery and Inverter if possible (Stinson and Wright 1969). Iron-constantan thermocouples attached to a double glass-wrapped silicone-impregnated wire (24 AWG) are satisfactory for collecting most field temperature data. Generally, soil surface temperatures are commonly collected, but temperatures above and below the soil surface are also useful for specific studies.

"Tempil cards," which contain commercial temperature pellets that melt upon reaching specific temperatures, can be embedded between asbestos and mica. They work well for recording maximum temperatures up to 425°C (800°F), but do not melt easily above this temperature; they simply char. The appearance of tempils at different temperatures in a muffle furnace should be checked before high temperature data are read in the field. The disadvantage of this data is that one cannot get time-temperature data which might be of value to simulate burning temperatures with a portable burner (Wright et al. 1976; Britton and Wright 1980).

REFERENCES

Anderson, H.E. 1970. Forest fuel ignitability. *Fire Technol.* **6:**312–319, 322.

Bailey, A.W., and M.L. Anderson. 1980. Fire temperatures in grass, shrub, and aspen forest communities of Central Alberta. *J. Range Manage.*3:37–40.

Baker, F.S. 1929. Effect of excessively high temperatures on coniferous reproduction. *J. For.* 27:949–975.

Batchelder, R.B., and H.F. Hirt. 1966. Fire in tropical forests and grasslands. Tech. Rep. U.S. Army Natick Lab. Natick, Mass.

Beadle, N.C.W. 1940. Soil temperatures during forest fires and their effect on the survival of vegetation. *J. Ecol.* **28:**180–192.

Beaufait, W.R. 1960. Some effects of high temperatures on the cones and seeds of jack pine. *For. Sci.* 6:194–199.

Beaufait, W.R. 1961. Crown temperatures during prescribed burning in jack pine. *Mich. Acad. Sci., Arts, and Letters* **46:**251–257.

Beaufait, W.R. 1965. Characteristics of backfires and headfires in a pine needle fuel bed. USDA For. Serv. Res. Note INT-39. Intermt. For. and Range Exp. Stn., Ogden, Utah.

Bentley, J.R., and R. L. Fenner. 1958. Soil temperatures during burning related to postfire seedbeds on woodland range. *J. For.* **56:**737–774.

Britton, C. M., and H. A. Wright. 1971. Correlation of weather variables to mesquite damage by fire. *J. Range Manage.* 24:136–141.

Britton, C. M., and H. A. Wright. 1980. A portable burner for evaluating effects of fire on plants. *J. Range Manage.* 32:475–476.

Byram, G. M. 1958. Some basic thermal processes controlling the effects of heat on living vegetation. USDA For. Serv. Res. Note 114. Southeastern For. Exp. Stn., Asheville, N.C.

Countryman, C. M. 1964. Mass fires and fire behavior. USDA For. Serv. Res. Paper PSW-19. Pac. Southwest For. and Range Exp. Stn., Berkeley, Calif.

Daubenmire, R. 1968. Ecology of fire in grasslands. *Adv. Ecol. Res.* **5:**209–266.

Davis, L. S., and R. E. Martin. 1960. Time-temperature relationships of test headfires and backfires. USDA For. Serv. Res. Note 148. Southeastern For. Exp. Stn., Asheville, N.C.

DeBano, L. F., P.H. Dunn, and C. E. Conrad. 1977. Fire's effect on physical and chemical properties of chaparral soils. In Symposium on Environmental Consequences of Fire and Fuel Management in Mediterranean Ecosystems. USDA For. Serv. Gen. Tech. Rep. WO-3. Washington, D.C., pp. 65–74.

Ehrenreich, J. H., and J. M. Aikman. 1963. An ecological study of certain management practices on native prairie in Iowa. *Ecol. Monogr.* **33**:113–130.

Fahnestock, G. R., and R. C. Hare. 1964. Heating of tree trunks in surface fires. *J. For.* **62**:799–805.

Fons, W. L. 1950. Heating and ignition of small wood cylinders. *Ind. and Eng. Chem.* **24**:2130–2133.

Grant, S. A., R. F. Hunter, and C. Cross. 1963. The effects of muirburning *Molina*-dominant communities. *J. British Grassland Soc.* **18**:249–257.

Hare, R.C. 1961. Heat effects on living plants. USDA For. Serv. Occas. Paper S-183. Southern For. Exp. Stn., New Orleans, La.

Hare, R. C. 1965. Contribution of bark to fire resistance in southern trees. *J. For.* **63**:248–251.

Heyward, F. 1938. Soil temperatures during forest fires in the longleaf pine region. *J. For.* **36:** 478–491.

Ito, M., and S. Iizumi. 1960. Temperatures during grassland fires and their effect on some species in Kawatabi, Miyagi Prefecture. *Tokyo Univ. Sci. Rep. Res. Inst.* **D-11**(2):109–114.

Isaac, L. A., and H. G. Hopkins. 1937. The forest soil of the Douglas fir region and the changes wrought upon it by logging and slash burning. *Ecology* **18**:264–279.

Jameson, D. A. 1961. Heat and desiccation resistance of important trees and grasses of the pinyon-juniper type. *Bot. Gaz.* **122**:174–179.

Kayll, A. J. 1966. A technique for studying the fire tolerance of living tree trunks. Can. Dept. For. Pub. No. 1012. Ottawa, Ont.

Lindenmuth, A. W., Jr., and G. M. Byram. 1948. Headfires are cooler near the ground than back-fires. *Fire Contr. Notes* **9**:8, 9.

Lorenz, R. W. 1939. High temperature tolerance of forest trees. Minn. Agric. Exp. Stn. Tech. Bull. 141. Minneapolis, Minn.

McKell, C. M., A. M. Wilson, and B. L. Kay. 1962. Effective burning of rangelands infested with medusahead. *Weeds.* **10**:125–131.

McRae, D. J. 1979. Prescribed burning in jack pine logging slash: A review. Can. For. Serv. Rep. O-X-289. Great Lakes For. Res. Cent., Sault Ste. Marie, Ont.

Neal, J. L., E. Wright, and W. B. Bollen. 1965. Burning Douglas-fir slash: Physical, chemical, and microbial effects on the soil. For. Res. Lab., Res. Paper. Oregon State Univ., Corvallis.

Nelson, R. M. 1952. Observations on heat tolerance of southern pines. USDA For. Serv. Res. Paper SE-14. Southeastern For. Exp. Stn., Asheville, N. C.

Nelson, R. M., and I. H. Sims. 1934. A method of measuring experimental forest fire temperatures. *J. For.* **32**:488–490.

Norton, B. E., and J. W. McGarity. 1965. The effect of burning of native pasture on soil temperature in northern New South Wales. *J. British Grassland Soc.* **29**:101–105.

Sampson, A. W. 1944. Plant succession on burned chaparral lands in northern California. Calif. Agric. Exp. Stn. Bull. 685. Berkeley.

Schmidt, C. F. 1954. Thermal resistance of micro-organisms. In G. F. Reddish (ed.) *Antiseptics, Disinfectants, Fungicides, and Sterilization.* Lea and Febiger, Philadelphia, pp. 720–759.

Shirley, H. L. 1936. Lethal high temperatures for conifers, and the cooling effect of transpiration. *J. Agric. Res.* **53**:239–258.

Smith, D. W., and J. H. Sparling. 1966. The temperatures of surface fires in jack pine barron. I. The variation in temperature with time. *Can. J. Bot.* **44**:1285–1292.

Spalt, K. W., and W. E. Reifsnyder. 1962. Bark characteristics and fire resistance: A literature survey. USDA For. Serv. Occas. Paper S-193. Southern For. Exp. Stn., New Orleans, La.

Sparling, J. H., and D. W. Smith. 1966. The temperatures of surface fires in jack pine barren. II. The effects of vegetation cover, wind speed, and relative humidity on fire temperatures. *Can. J. Bot.* **44**:1293–1298.

Steuter, A. A., and H. A. Wright. 1979. Redberry juniper mortality following prescribed burning. *Research Highlights: Noxious Brush and Weed Control; Range Wildlife Manage.* **10:**14. Texas Tech Univ., Lubbock.

Stinson, K. J., and H. A. Wright. 1969. Temperature of headfires in the southern mixed prairie of Texas. *J. Range Manage.* **22:**169–174.

Stockstad, D. S. 1979. Spontaneous and piloted ignition of rotten wood. USDA For. Serv. Res. Note INT-267. Intermt. For. and Range Exp. Stn., Ogden, Utah.

Trabaud, L. 1979. Etude du comportement du feu dans la Garrigue de Chene kermes à partir des températures et des vitesses de propgagtion. *Ann. Sci. For.* **36:**13–38.

Tunstall, B. R., J. Walker, and A. M. Gill. 1976. Temperature distribution around synthetic trees during grass fires. *For. Sci.* **22:**269–276.

Van Wagner, C. E. 1973. Height of crown scorch in forest fires. *Can. J. For. Res.* **3:**373–378.

Van Wagner, C. E., and I. R. Methven. 1978. Discussion: Two recent articles on fire ecology. *Can. J. For. Res.* **8:**491–492.

Went, F. W., G. Juhren, and M. C. Juhren. 1952. Fire and biotic factors effecting germination. *Ecology* **33:**351–363.

White, R. S. 1969. Fire temperatures and the effect of burning on South Texas brush communities. M. S. Thesis, Texas Tech Univ., Lubbock.

Whittaker, E. 1961. Temperature in heath fires. *J. Ecol.* **49:**709–715.

Wright, E. 1931. The effect of high temperatures on seed germination. *J. For.* **29:**679–687.

Wright, H. A. 1970. A method to determine heat-caused mortality in bunchgrass. *Ecology* **51:**582–587.

Wright, H. A. 1971. Why squirreltail is more tolerant to burning than needle-and-thread. *J. Range Manage.* **24:**277–284.

Wright, H. A. 1980. The role and use of fire in the semidesert grass-shrub type. USDA For. Serv. Gen. Tech. Rep. INT-85. Intermt. For. and Range Exp. Stn., Ogden, Utah.

Wright, H. A., S. C. Bunting, and L. F. Neuenschwander. 1976. Effect of fire on honey mesquite. *J. Range Manage.* **29:**467–471.

Yarwood, C. E. 1961. Translocated heat injury. *Plant Physiol.* **36:**721–726.

3

SOIL AND WATER PROPERTIES

Fire has a variety of effects on soil and water properties depending on intensity of the burn, fuel type, soil, climate, and topography. Thus fire can be good or bad, depending on objectives and where and how it is used. In this chapter we try to delineate the ramifications of fire effects on soil and water properties, and why the effects reported in the literature are highly variable. Using wisdom, prescribed burning can be a beneficial and versatile management tool without damage to soil productivity and water quality.

ORGANIC MATTER

Litter and soil organic matter have several benefits. They help to develop an aggregated, granular soil structure that increases infiltration compared to bare soils (Beutner and Anderson 1943; Weaver and Rowland 1952). Moreover, soil organic matter increases cation exchange capacity (Wells et al. 1979) and reduces soil erosion (Boyer and Dell 1980). Litter also stabilizes soil surface temperatures (Barkley et al. 1965).

On semidesert range in Arizona, infiltration of water was two to three times greater with 4.5 metric ton/ha (2 ton/acre) of litter than on either bare soil or soil where 4.5 metric ton/ha of litter had been mixed in the top 7.5 cm (3 in.) of soil

(Beutner and Anderson 1943). Weaver and Rowland (1952) determined that the infiltration rate under big bluestem *(Andropogon gerardi)* was 46 percent more rapid than under Kentucky bluegrass *(Poa pratensis)*. The big bluestem was in heavy litter [13.5 to 20 metric ton/ha (6 to 9 ton/acre)] whereas the bluegrass was growing in only a thin layer of litter.

Litter retards the evaporation of soil moisture. Weaver and Rowland (1952) found that the loss of water from mulched soil during the first day after wetting averaged 75 percent less than from unmulched soil. Similarly, Russell (1939) reported that a mulch of 3.8 cm (1.5 in.) reduced evaporation 91 percent, which was no more effective than 1.2 cm (0.5 in.) of mulch. Fallen and standing debris also act favorably to retain snow and its accumulation when blown by wind (Weaver and Rowland 1952) as well as reduce wind and water erosion (Dyksterhuis and Schmutz 1947).

Litter reduces soil erosion by protecting soil aggregates from dispersal by rainfall impact (Boyer and Dell 1980). Therefore, macropore space, infiltration, and soil aeration remain optimum (Wells et al. 1979).

A thick cover of litter is capable of intercepting rainfall and causing considerable loss of water to the soil. For example, Clark (1940) showed that a thick stand of big bluestem withheld about two-thirds of the precipitation during heavy rains of 3·to 4.5 cm (1.2 to 1.8 in.) and as much as 97 percent of very light showers. Weaver and Rowland (1952) found that when water was added to litter by means of a spray syringe at a rate of 0.6 surface-cm (0.25 surface-in.) in 30 min, more than half of this 0.6-cm (0.25-in.) rain would not reach the soil. Litter absorbed a lot of the water, but most of the absorbed water was eventually lost to evaporation. The absorption capacity of litter has been illustrated by Flory (1936) where 225 g of litter on a square meter of grassland absorbed 650 g of water. Results of the work of Weaver and Rowland (1952) indicated that saturated litter could hold above the soil as much as 0.8 cm (0.3 in.) of rain which in turn would be lost to evaporation.

Decomposition and Accumulation

Several factors affect the decomposition of organic matter (Kelly 1971; Bollen 1974). Interaction between soil microbes and forest residues are most important and are controlled by water, temperature, aeration, pH, food supply, and biological relationships (Bollen 1974). Rapid decomposition of residues can best be achieved by incorporation of rotted wood, leaves, and other material into the soil. Reducing particle sizes and assuring good contact of residue particles with the soil will improve aeration, temperature stability, drainage, and moisture retention properties (Bollen 1974).

Bacterial populations decline immediately after a burn but increase 3- to 10-fold within a month (Fig. 3.1) (Miller et al. 1955; Jurgensen et al. 1979) because soil temperatures and nutrients for their growth are more favorable (Greene 1935; Ahlgren and Ahlgren 1965; Renbuss et al. 1973; Harvey et al. 1976).

Thus, even though nitrifying bacteria have thin walls and are easily killed between 53° and 56°C (127° and 132°F) (Gibbs 1919), they are generally protected by soil layers above them and populations appear to be able to recover quickly after a fire and

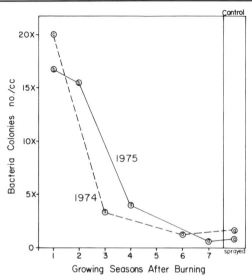

Fig. 3.1. Bacteria colonies per cc taken in two successive years responded similarly through seven growing seasons after control burning. Means for the same year with the same letter are not statistically different ($P \leq .01$). (From Neuenschwander 1976.)

produce nitrates from organic matter (Ahlgren and Ahlgren 1965; Sharrow and Wright 1977a). Only in the cases of very intense fires (areas beneath piles of burning debris) would large reservoirs of nitrogen be volatilized and restoration of nitrogen through symbiotic fixation be insufficient to balance losses from fires for many years (Jones and Richards 1977).

Coleman (1916) has reported that the optimum range for bacterial growth is 30° to 38°C (86° to 100°F); Black (1957) has reported on optimum range of 35° to 45°C (95° to 113°F); and Piene (1974) mentioned that bacterial growth declined after temperatures reached 37°C (100°F). Maximum spring soil temperatures on unburned grasslands in the southern United States, when precipitation is adequate, usually averages 22°C (72°F) (Greene 1935; Sharrow and Wright 1977a). Therefore, the usual average increase of 10°C (18°F) after a fire can have a dramatic effect on the mineralization of nitrogen as long as nitrogen is present in an available form such as ammonium, nitrate, or organic nitrogenous compounds (Neal et al. 1965). Neuenschwander (1976) reported a 10-fold increase in soil bacteria the first year after a burn in west Texas. As the organic nitrogen supply becomes depleted, microbial activity slows to an equilibrium minimum, because the nitrogen becomes extensively immobilized in microbial protein (Waksmann 1922) until fresh litter comes into contact with the soil. Nitrogen in residues is gradually increased by the death of microbial cells.

The main ecological effect of fire on organic matter above the soil surface is to compress the oxidative processes of decay into a very short time span (Harvey et al. 1976). Most of the products from either biological decomposition or heat oxidation are similar in chemical composition and in quantity (Komarek 1970), although volatilization of nitrogen (Klemmedson 1976; Alban 1977) and sulfur (Tiedemann and Anderson 1980) are significant losses above the soil surface in forests for some period of time after a fire. Moreover, hot spots are changed physically and chemically for many years (Neal et al. 1965), and charcoal residues are highly resistant to decomposition (Harvey et al. 1976).

Grasslands

Excessive litter accumulations in the Great Plains will inhibit grass and seed yields during normal to wet years (Weaver and Rowland 1952; Old 1969). Primarily, litter lowers soil temperatures (Hensel 1923; Weaver and Rowland 1952; Kucera and Ehrenreich 1962; Sharrow and Wright 1977a), which reduces bacterial activity (Neuenschwander et al. 1974). As a consequence, the general nutrient cycling process is slowed, particularly during cool, wet years. In dry years, however, when wildfires are prevalent, litter is important for insulation and protection of the soil from flash floods (Wright 1972; 1974). Removal of litter during dry years tends to increase drought stress on plants and lower herbage yields.

When soil moisture is adequate, increased soil temperatures following fires will enhance nitrification of organic matter and increase forage yields on grasslands compared to unburned grasslands (Sharrow and Wright 1977a). Depending on the initial quantity of litter, the temperature differential between burned and unburned sites ranges from 3° to 16°C (5° to 28°F) (Hensel 1923; Greene 1935; Weaver and Rowland 1952; Kucera and Ehrenreich 1962; Scotter 1963; Peet et al. 1975). Generally, the temperature differential averages 10°C (18°F) (Wells et al. 1979; Boyer and Dell 1980).

Litter decomposition increases steadily from 10° to 37°C (50° to 99°F) and then decreases from 37° to 50°C (99° to 122°F) (Piene 1974), indicating that temperatures from 30° to 37°C (86° to 99°F) are optimum for organic matter decomposition when soil moisture is optimum (Greene 1935; Bartholomew and Norman 1946; Black 1957). Increased soil temperatures following spring fires can greatly enhance breakdown of organic matter that is in the soil (Fig. 3.2) (Whigham 1976; Sharrow and Wright 1977a).

On mesic climax sites in Oklahoma, Rice and Pancholy (1973) found that nitrifiers were inhibited by organic products from litter so that ammonium nitrogen was not

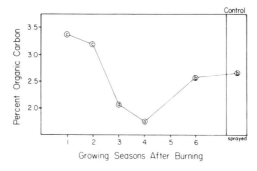

Fig. 3.2.   Percent organic carbon for six growing seasons after burning. Data was taken in 1974. (From Whigham 1976.)

oxidized to nitrate as readily as on the drier, lower stage plant successional sites. This inhibition of nitrification was related to tannins and tannin derivatives which occur in grasses, forbs, and trees and which accumulate in the soil. Fire probably oxidizes many of these inhibitors, but nitrification, nitrogen-fixation, and ammonification are increased by pH and increased concentration of electrolytes (mainly calcium) (Isaac and Hopkins 1937; Rode 1955) and the presence of charcoal (Tyron 1948).

Following a fire in medium or tall grasses, about 650 kg/ha (579 lb/acre) of humic mulch remains for a few months (Dix 1960), new shoots stand stiffly erect for two to three years, and decomposition of these erect stems is slow (McCalla 1943; Sharrow and Wright 1977b). After two to three growing seasons fresh mulch begins to accumulate on the soil surface and the normal organic matter recycling process is renewed. Once on the soil surface, litter requires about 3 years to decompose (Hopkins 1954; Old 1969). Fresh and humic mulch generally return to normal on grasslands within five years after a burn (Lemon 1949; Dix 1960; Ehrenreich and Aikman 1963; Sharrow and Wright 1977a; Neuenschwander et al. 1978), depending on annual precipitation and presence or absence of snow. Equilibrium might be reached more quickly with grazing since it tends to increase the weight of humic mulch at the expense of fresh mulch (Dix 1960) and thus increases the rate of microbial decomposition (McCalla 1943).

Frequent burning of grass beneath pines in the coastal plain of the southeastern United States has increased organic matter content in the upper 10 to 15 cm (4 to 6 in.) of the soil profile (Heyward and Barnette 1934; Greene 1935; Heyward 1936; Wahlenberg et al. 1939). In Mississippi this amounted to a 60 percent increase (Greene 1935). Even when charred material is separated from uncharred material, the humus content is still higher (Metz et al. 1961). However, if needle litter has crowded the grass out, burning does not generally increase soil organic matter and may cause a decrease (J. F. Corliss, pers. comm.).

After six consecutive years of burning in Kansas, humus did not decline (Aldous 1934). Also, in west Texas humus depth was the same as the control immediately after burning. The rapid mineralization of nitrogen after fire caused it to decline for two to three years before it began to increase and reach a normal level five years after burning (Odum 1960; Sharrow and Wright 1977b; Neuenschwander et al. 1978). Humus content in the soil of grasslands will recover in one or two years after a burn in a 100-cm (40-in.) precipitation zone, but will take five to eight years in a 50-cm (20-in.) precipitation zone (Sharrow and Wright 1977b).

In shortgrass communities, such as those that support buffalograss *(Buchloe dactyloides)*, undisturbed plant communities do not accumulate much litter [0.6 to 1.2 cm (0.25 to 0.5 in.)], which does not inhibit herbage production (Dix 1960). Therefore, fire in these communities seldom has beneficial effects.

Shrublands

Shrubs can occur as seral communities in forests, as postclimax communities in grasslands, or as climax communities such as the salt-desert shrub or chaparral communities. This section refers to stands of continuous shrubs capable of supporting a fire. Pure stands of big sagebrush *(Artemisia tridentata)* or chaparral usually burn extremely hot (Blaisdell 1953; DeBano et al. 1977). There is rarely a middle ground for cool or hot fires as in grassland and forest communities. Thus, the net effect of fire is to remove a significant portion of the organic matter [36 kg/ha (32 lb/acre) of nitrogen] from the upper 2 cm (0.8 in.) of soil (Blaisdell 1953; DeBano et al. 1977) and to distill organic compounds downward on moderate burns lasting 5 to 25 min, leaving many areas nonwettable for several years (DeBano et al. 1977). The distillation of organic compounds downward into the soil will be discussed under "Distillation of Organic Matter by Fire."

Forests

Forest litter delays freezing of the soil and minimizes the formation of concrete frost (MacKinney 1929), although unthinned tree canopies have a similar effect (Piene 1974). The intergranular spaces do not fill so noticeably with ice as in bare soil. Snow accumulations on open areas will insulate the soil and also minimize the formation of concrete frost and the depth of freezing (Weitzman and Bay 1963). As a consequence, litter enables water from winter rains and thaws to penetrate the soil, instead of running off the more compactly frozen soil (MacKinney 1929). MacKinney found that the depth of frost penetration in soil was reduced 40 percent by litter cover. However, Weitzman and Bay reported that frost lasted longer in organic material and that coarse-textured soils have less concrete freezing than fine-textured soils and freezing under hardwoods is less severe than under conifers.

Litter and duff varying from 22 to 224 metric ton/ha (10 to 100 ton/acre) (Isaac and Hopkins 1937) on forest soils may be reduced 3 to 70 percent by light burning (Beaton 1959; Sweeney and Biswell 1961). The remaining litter and humus, and presumably the unchanged organic matter content of the soil profile, had no harmful effects on infiltration and percolation (Biswell and Schultz 1957). Severe burns remove all litter and humus (Blaisdell 1953; Tarrant 1956) and destroy as much as 65 percent of the surface inch of soil organic matter (Fowells and Stephenson 1934). Such burns accelerate oxidation of soil organic matter, reduce infiltration and water storage capacity, and are detrimental to nutrient properties of the soil (Wells et al. 1979). Severe burns, however, rarely exceed 2 to 5 percent of slash areas that have been burned (Tarrant 1956). Forests produce about 4.5 metric ton/ha/year (2 ton/acre/year) of litter (Isaac and Hopkins 1937) of which one-half decays.

Distillation of Organic Matter by Fire

Humic acids of organic carbons are lost at temperatures below 100°C (212°F). At temperatures between 100° and 200°C (212° and 390°F), nondestructive distillation of volatile organic substances occurs, and at temperatures between 200° and 300°C (390° and 570°F) about 85 percent of organic substances are destroyed by destructive distillation (DeBano et al. 1977). Depending on the duration of a fire that is burning organic material, organic substances can distill downward into the soil and form a non-wettable hydrophobic layer in the soil (Wells et al. 1979). Fuels that burn quickly (e.g., grass) or that burn very hot (brush piles) do not generally form water repellent layers (DeBano and Krammes 1966).

Water repellent layers are most common in shrub communities where fires may last from 5 to 25 min (DeBano 1966; DeBano and Krammes 1966; DeBano et al. 1967; DeBano 1969; Adams et al. 1970; DeBano et al. 1976). Hydrophobic substances are also common occurrences in forest soils, particularly under ponderosa pine *(Pinus ponderosa)*, Douglas fir *(Pseudotsuga menziesii)*, and lodgepole pine *(P. Contorta)* (Zwolinski and Ehrenreich 1967; Dyrness 1976; Campbell et al. 1977). In chaparral communities that burn for 5 to 25 min, DeBano (1974) found that 90 percent of the decomposed organic matter was lost as smoke or ash. The remaining material distilled downward and condensed (DeBano 1966). The thickness and depth of this layer depended on the intensity and duration of the fire, soil water content, and soil physical properties. Thicker water repellent layers form in dry soils than in wet soils (DeBano et al. 1976) and coarse-textured soils are more likely to become water repellent than fine-textured soils (DeBano et al. 1967). Water repellent substances are believed to be long chain aliphatic hydrocarbons (Savage et al. 1972) that are destroyed when heated much over 280°C (536°F) (Savage 1974; Scholl 1975; DeBano and Krammes 1966).

In summary, nonwettable soils are most likely to form as a consequence of fire in chaparral and forest communities (Wells et al. 1979), but they do not occur extensively on most burned areas. Dry, sandy soils are most susceptible to forming water repellent soils under fires of moderate duration, whereas wet, fine-textured soils are the least susceptible to water repellency (DeBano et al. 1976).

SOIL FAUNA

Removal of soil litter by fire or other mechanisms causes a dramatic change in food supplies, water content, temperature, and pH of the soil, which cause a threefold reduction in numbers of soil animals as well as a loss in many genera (Pearse 1946). Litter not only serves to absorb and retain moisture, but it also tends to reduce evaporation from below, and this serves to retard drying of the soil surface (Hursh 1928). Therefore, litter helps the soil to retain high humidities and be reasonably thermostable so that the bodies of animals living in the soil do not lose moisture (Pearse 1946). Stunkard (1944) found that the optimum humidity for the development of an oribated mite *(Galumna* sp.) was 82 percent. Research by Gill (1969) has shown that microclimatological properties of the soil are far more important in controlling abundance and migration of soil organisms than nutritional properties of the litter. Loose organic material serves as a refuge for soil animals when rains may drive all air out of the soil and humus (Pearse 1946).

Vegetarians are the most common microorganisms and insects in the litter and upper soil surface (Heyward and Tissot 1936; Pearse 1946). The abundant fungi and bacteria play a key role in softening roots, leaves, stems, and feces to make them fit for consumption by termites, insect larvae, and other animals (Bornebusch 1930).

Among the most numerous vegetarian insects are mites *(Acarina* sp.) and springtails *(Collembola* sp.) (Heyward and Tissot 1936; Pearse 1946). Other vegetarians are many ants, termites, thrips, bugs, certain beetles, snails, millipedes, and symphylans (Pearse 1946). Those that feed largely on fungi are fungous gnats, pauropods, compodeans, certain collembolans (springtails), thrips, and beetles (erotylids) (Pearse 1946). Earthworms, which thrive best at a pH of 6 to 7 (Arrhenius 1921), mix organic matter with mineral soil when soil moisture is ideal.

Carnivores associated with the soil are primarily centipedes, chelonethids, spiders, pselaphid and staphylinid beetles, certain mites, ponerine and doryline ants, reduviid bugs, and certain collembolans (springtails) and thysanopterans (thrips) (Pearse 1946). Parasites consist chiefly of ants, parasitoid mites, and nematodes (Heyward and Tissot 1936; Pearse 1946). Together, all organisms in the soil decompose organic matter into H_2O, CO_2, and various mineral elements as well as make the soil more porous and penetrable. They are most effective in warm-wet ecosystems and least effective in cold or dry ecosystems (Harvey et al. 1979b).

Following fire, food supplies on the soil surface diminish to some degree (depending on fire intensity), moisture decreases, and temperature and pH increase. These changes cause at least 3- to 10-fold drops in numbers of most organisms (Heyward and Tissot 1936; Pearse 1946) and require three to five years to reach new population equilibria.

SOIL CHEMISTRY

Nitrogen and Mineral Elements

VOLATILIZATION OF ELEMENTS

Nitrogen and sulfur volatilize easily following burning (Bollen 1974; Klemmedson 1976; Tiedemann and Anderson 1980), but cations are much more difficult to volatilize and losses are minor (Lewis 1974; Alban 1977; Stark 1979). Volatilization temperatures of potassium [760°C (1400°F)], sodium [880°C (1616°F)], calcium [1240°C (2264°F)], and magnesium [1107°C (2025°F)] are quite high. Such temperatures are seldom reached at or below the soil surface (Wells et al. 1979), but can easily be higher in the upper portion of flames and in piles of debris that cause severe burns. Often nutrients move from the ashed litter into the soil (Alban 1977). Most losses of cations from burned sites can probably be attributed to surface erosion (Wells et al. 1979), movements of ions below the root zone because of increased soil water for transport of ions (Stark 1979), dilution effects of increased runoff (Tiedemann 1973; DeBano and Conrad 1978), and, to a large extent, losses in fly ash and volatilization (Grier 1975; DeByle 1976a).

Nitrogen volatilizes at 200°C (392°F) (White et al. 1973), which is easily exceeded in fires if fine fuels exceed 3370 kg/ha (3000 lb/acre) (Stinson and Wright 1969). Material burning above the ground has the greatest chance of being volatilized. Sharrow and Wright (1977a) found that over 90 percent of nitrogen in standing grass fuels volatilized, whereas in heavier fuels, Knight (1966) reported that 25 to 64 percent of the nitrogen was lost when forest floor temperatures varied from 300° to 700°C (572° to 1292°F). For chaparral fuels, DeBano et al. (1979) reported total nitrogen losses of 20 percent at 486°C (907°F), 40 percent at 600°C (1112°F), and 80 percent at 825°C (1517°F). Fire intensity, amount of green material, and fuel moisture all have a major effect on total nitrogen losses (Lewis 1974; Dunn and DeBano 1977). Burning over a moist soil should reduce nitrogen losses (DeBano et al. 1979).

Actual nitrogen losses have varied from 30 to 33 kg/ha (27 to 30 lb/acre) in grasslands with 4.5 to 6.7 metric ton/ha (2 to 3 ton/acre) of fine fuel (Elwell et al. 1941; Sharrow and Wright 1977b) to 907 kg/ha (807 lb/acre) on heavy slash burns (Grier 1975). Moderate fires in a ponderosa pine forest, where the average loss of litter and duff was 22 metric ton/ha (10 ton/acre) (40 percent of the litter and humus), yielded nitrogen losses of 75 to 139 kg/ha (67 to 124 lb/acre) (Klemmedson et al. 1962; Klemmedson 1976). Similar nitrogen losses [146 kg/ha (130 lb/acre)] have been reported for intense chaparral fires by DeBano and Conrad (1978).

Total nitrogen levels reported for mineral soil following light to moderate burns has been contradictory in the literature. It is either unaffected (Aldous 1934; Burns 1952; Scotter 1963; Sharrow and Wright 1977b), increases (Waksman 1927; Heyward and Barnette 1934; Greene 1935; Garren 1943; Reynolds and Bohning 1956; Metz et al. 1961), or decreases (Barnette and Hester 1930; Isaac and Hopkins 1937; Cook 1939; Austin and Baisinger 1955) after burning. For the most part, these variations reflect fire intensities, quantity of fuel burned, differences in soil moisture, and, to some extent, minor changes in total soil nitrogen in relation to the vast amount that is present in the soil. Severe fires in forests, slash burns, or chaparral communities always reduce total soil nitrogen (Austin and Baisinger 1955; Lutz 1956; Tarrant 1956; Neal et al. 1965; DeBano and Conrad 1978), whereas reductions of soil nitrogen in grasslands or cool forest fires apparently do not occur.

MINERAL ELEMENT CHANGES

Although mineral elements may increase severalfold in the upper 2.5 to 5.0 cm (1 to 2 in.) of soil on forest burns, the magnitude of increase is directly proportional to the amount of material burned. Ash in grasslands has a negligible effect on plant growth (Heyward 1936; Wahlenberg et al. 1939; Sampson 1944; Curtis and Partch 1950; Suman and Carter 1954; Reynolds and Bohning 1956; Hulbert 1969; Wright 1969).

In forests, however, the burning of organic matter in light to moderate fires yields significant ash which lowers nitrogen, phosphorus, potassium, calcium, and magnesium on the forest floor (Alban 1977). However all of these elements that are not volatilized are translocated downward into the mineral soil, resulting in a net gain (Grier and Cole 1971; Alban 1977). The addition of calcium has a favorable effect on growth of bacteria, which ultimately produce more nitrogen by mineralizing organic matter (Issac and Hopkins 1937).

Organic matter is translocated downward which increases cation exchange capacity as deep as 50 cm (20 in.) (Alban 1977). If soil temperatures exceed 300°C (570°F), net losses of calcium and magnesium occur below the root zone, but losses are insignificant when soil temperatures remain below 200° to 300°C (390° to 570°F) (Stark 1977). The hotter the fire, the less iron in soil water as a result of the alkaline pH. Loss of nutrients from overland flow and surface erosion were of little significance in a Douglas fir-western larch *(Larix occidentalis)* study in Montana (Stark 1977), although severe forest fires can result in substantial nutrient loss from overland flow and erosion (Tiedemann 1973). In a companion western larch-Douglas fir study, Stark (1979) found that less than 0.25 percent of the total content of eight biologically essential cations in the effective root zone were removed with wood and bark. Nutrient levels in an intermittent stream were essentially unchanged.

AVAILABILITY OF NUTRIENTS

Despite the rapid decomposition of organic material and nutrient losses that occur during a fire, large quantities of nitrogen, phosphorus, potassium, calcium, magnesium, sodium, and, to some extent, sulfur are readily soluble. Before the fire many of these elements in living and dead material on cold or dry sites were unavailable for plant growth (Lewis 1974; DeBano et al. 1977; Harvey et al. 1979b). In warm-wet ecosystems, however, decayed residue, in the form of brown, crumbly wood or in the soil, provides a site for continued nitrogen fixation (Harvey et al. 1979a). Fire is most beneficial for incorporation of woody materials into forest soils in cold or dry ecosystems (Harvey et al. 1979b).

The usual addition of some nitrogen to the soil surface after a fire (Sharrow and Wright 1977b; DeBano and Conrad 1978), a plentiful supply of mineral nutrients (Wells et al. 1979), increased soil temperatures and ashed minerals that stimulate nitrification and add nitrates to the soil via breakdown of organic matter (Fig. 3.1) (Fowells and Stephenson 1934; Sharrow and Wright 1977a; Raison 1979), and a readily available supply of soil moisture will greatly enhance soil fertility and plant growth (Sharrow and Wright 1977a). Moreover, heated soils [e.g. 2 hr at 400°C (750°F)] can increase NH^+-N from a few to 400 ppm, but nitrates are decomposed above 150°C (300°F) (Raison 1979). This increase of NH^+-N during a fire implies that plant growth could benefit immediately (before nitrification begins) when soil temperatures and moisture are favorable.

Free nitrate is highly mobile and can easily be lost from an ecosystem or quickly used (Vitousek and Melillo 1979). Rapidly growing plants in a warm environment will use nitrate as quickly as it is produced (Sharrow and Wright 1977a). Cool environments, however, will allow nitrate to accumulate and go unused (Sharrow and Wright 1977a). Destruction of a forest community that is followed with an ideal environment for nitrogen mineralization but insufficient plant growth will also allow free nitrate to accumulate (Vitousek and Melillo 1979). These excesses can be lost to groundwater or streamwater, depending on water for transport, volatilization, immobilization by decomposers, clay fixation of ammonium, nitrate reduction to ammonium, lags in nitrification, nitrate absorption on colloids, or plant nitrogen uptake (Vitousek and Melillo 1979). Therefore, losses of nitrate from ecosystems vary widely in plant communities, depending on plant growth and other factors after a burn.

REPLACEMENT OF NITROGEN

Replacement of nitrogen in the soil is largely through nitrogen fixation by leguminous and nonleguminous (e.g. *Ceanothus velutinus, Alnus rubra*) plants (Grier 1975) and partially combusted material (Raison 1979). Addition of nitrogen by rainfall is very slow—1.0 kg/ha/year (0.9 lb/acre/year) (Tiedemann and Helvey 1973), although there is evidence that the addition of nitrogen can be much higher from electrical storms or smoke pollution (J. F. Corliss, pers. comm.). Snowbrush *(Ceanothus velutinus)* has fixed from 715 kg/ha (636 lb/acre) to 1081 kg/ha (964 lb/acre) of nitrogen in two different plant communities over a 10- to 15-year period in the western United States (Youngberg and Wollum 1976). These quantities of nitrogen are equivalent to the greatest losses from severe forest fires of over 900 kg/ha (800 lb/acre) reported by Grier (1975).

pH

The hydrogen ion concentration decreases after burning, thereby raising the soil pH in the upper 1 to 10 cm (0.5 to 4 in.) (Lutz 1956; Beaton 1959; Moore 1960; Scotter 1963; Alban 1977; Raison 1979). In grasslands the increase in pH in the upper few centimeters is usually slight and persists for only a year or two. In southeastern Arizona Reynolds and Bohning (1956) reported no change in pH after burning, as did Blaisdell (1953) for sagebrush-bunchgrass in Idaho. Ehrenreich and Aikman (1963) showed a pH change from 5.8 to 6.1 in the upper 18 mm (0.7 in.) of soil in the Tall Grass Prairie. Similarly, Owensby and Wyrill (1973) showed a pH change from 5.87 to 6.07 on burned soils in the Tall Grass Prairie. In Nigeria, Moore (1960) showed a pH change from 6.0 to 6.2 after burning. The pH of soils after burning in California chaparral remains essentially unchanged (Sweeney 1956). Sampson (1944) reported pHs that ranged from 6.3 to 7.2 on burned and unburned chaparral soils, but no significant differences. In New Zealand Miller et al. (1955) reported pH values of 7 to 8 on scrub burns and pH values of 11 to 12 for ash slurry.

Forest soils usually show dramatic changes in soil pH. Beaton (1959), working on lodgepole pine sites two years after a fire in Canada, reported a pH of 4.6 in the unburned O horizon compared to 6.4 for the burned O horizon. On clear cuttings in Douglas fir of the Pacific Northwest, where annual precipitation exceeds 250 cm (100 in.) per year, Tarrant (1956) found that the pH increased from an average of 4.8 to 7.2 during the first year after burning. These changes are usually significant in only the upper 10 cm (4 in.) of soil (Alban 1977). Three years after burning, however, soils on light burns had returned to their original level of acidity. The pH on severe burns had a range from 5.0 to 6.0, which was well within the range found naturally in the Douglas fir region. Similar pH data for the Douglas fir region has been reported by Marshall and Averille (1928), Isaac and Hopkins (1937), and Austin and Baisinger (1955).

Cation Exchange Capacity

Cation exchange capacity usually decreases after burning on severe burns but is unaffected on light burns (Tarrant 1956; Scotter 1963). On severe burns this decrease in cation exchange capacity is probably due to a reduction in humus content (Edwards 1942). Coults (1945) found that destruction of humus colloids starts in the range of 100° to 250°C (212° to 482°F). Critical values vary with soil type. However, even at 500°C (932°F), exchange capacity is reduced only 20 percent. Since only the surface of the soil is heated to this level, Coults concluded that the effect of veld burning on base exchange capacity is very small and probably transitory.

Mild fires generally have no effect or may increase base saturation (Moore 1960). However, where ash is subject to horizontal displacement by wind or water, or perhaps to loss by leaching too quickly through the soil profile for it to be absorbed on colloids or taken up by soil organisms, no effect on base saturation may be noticed.

SOIL MOISTURE AND POROSITY

Rapid growth during the spring growing season after burning has been shown to reduce soil moisture in the upper 2.5 cm (1 in.) of prairie soil from 13.2 to 6.4 percent (Sharrow and Wright 1977a). At greater depths [2.5 to 12.5 cm (1 to 5 in.)] soil moisture was reduced from 18.8 to 10.2 percent. Evaporation accounted for 2 percent of the loss in soil moisture (Sharrow and Wright 1977b). Because plant yields were 2650 kg/ha (2360 lb/acre) on burned compared to 840 kg/ha (745 lb/acre) on unburned plots, plant growth was logically responsible for most of the reduction in soil moisture.

Nevertheless, burned grassland soils favor drought, especially during dry years. Soil temperatures increase. This factor alone decreases the wilting and field capacity moisture percentages (Sampson 1944) and increases evaporation and transpiration. Moisture-holding capacity is also affected by organic matter content. As the intensity of the burn increases on forest soils and thereby decreases the organic matter content of soil, especially in the upper 5 cm (2 in.), water-holding capacity is proportionately decreased (Neal et al. 1965). Therefore, because of the decrease in organic matter (particularly on intense fires) and the increase in soil temperature, plants on burned sites are most susceptible to drought injury during dry months (Phillips 1919; Scott 1934; Cook 1939; West 1965).

Generally, grass fires reduce infiltration and percolation rates slightly (Wahlenberg 1935; Hanks and Anderson 1957; McMurphy and Anderson 1965; Ueckert et al. 1978), or not at all (Greene 1935). Chaparral fires often reduce infiltration capacities on moderate burns by forming a water repellent layer, but cool or very hot fires have no effect on infiltration capacities (DeBano et al. 1976). In the Douglas fir region of the Pacific Northwest, Tarrant (1956) found that severe burning decreased percolation rates, but light burning increased percolation rates. Light burning increased macroscopic pore volume which was attributed to the burning of tree and shrub roots. The liberation of basic cations also may have improved soil aggregation.

Severe slash burns reduced macroscopic pore volume (presumably by burning soil organic matter which caused soil particles to disperse) and increased microscopic pore volume (Tarrant 1956). Since severe burns only occupied 2.8 percent of the total area, this effect was considered to be of minor importance. Scotter (1963) found that burning increased infiltration rates in the boreal forest, whereas in other areas it has been found that burning reduced infiltration rates (Kittredge 1938; Johnson 1940; Burns 1952; Beaton 1959). Severity of burning is probably the best indication of a fire's effect on infiltration and percolation. Biswell and Schultz (1957) mentioned that the partial decomposed duff and debris remaining in burned stands of ponderosa pine kept infiltration and percolation rates high. Generally, fires do not alter soil physical properties unless heating has been severe (Raison 1979).

SOIL LOSSES

Vegetation, dead stems and leaves, decaying organic matter, and humus all help to absorb the impact of rainfall, moderate the delivery of water to soil, obstruct the overland flow of water, and reduce the transport of soil material (Colman 1953). Without vegetation or other cover for mineral soil, raindrop splash action, surface runoff, soil creep, and debris flows can cause serious soil erosion losses. Therefore, prescribed burns need to be conducted carefully, especially on steep slopes, so as not to expose more than a minimum of mineral soil. Fuel moisture and fire intensity can be controlled to minimize exposure of mineral soil, although some exposure of mineral soil may be necessary for germination of shrub and tree seedlings (Shearer 1975).

Soil losses following fire are influenced by intensity of storms (Orr 1970), size and frequency of bare areas (Packer 1951), soil type, topography, and plant cover (Smith and Wischmeier 1962; Farmer and Van Haveren 1971). Among these factors, vegetative cover and slope are the most important under intense rainfall storms (Farmer and Van Haveren 1971; Meeuwig 1971). Meeuwig found that as cover decreases, slope becomes increasingly important for soils that vary from sandy loams to clay loams. For example, a 40 percent cover on a 5 percent slope is as effective as an 80 percent cover on a 35 percent slope. At less than 50 percent cover, erosion rates double for each 10 percent increase of slope within the range of 5 to 35 percent. Cover will account for as much as 76 percent of the variance of the logarithm of the weight of eroded soil (Meeuwig 1970). A cover of 60 to 70 percent is necessary for soil stability on slopes that range from 30 to 60 percent (Bailey and Copeland 1961; Packer and Laycock 1969; Orr 1970; Wright et al. 1976). However, as the size of the bare areas increases within a specific percentage cover category, the influence of ground cover decreases (Packer 1951; 1963). The influence of ground cover also decreases as the length of slope increases (Farmer and Van Haveren 1971). Fine-textured soils such as clay loams are far less erodible than coarse-textured soils such as loamy sand or sandy loam (Swanston and Dyrness 1973).

Following burning, unseeded chaparral or shrub covered watersheds with average slopes of 30 to 60 percent lose as much as 52 to 116 metric tons/ha (23 to 52 ton/acre) of soil during the first year, particularly on granitic, sandstone, and shale-derived soils (Pase and Ingebo 1965; Pase and Lindenmuth 1971; Pitt et al. 1978). On limestone-derived soils in central Texas, however, Wright et al. (1976) only recorded total soil losses of 16 to 22 metric ton/ha (7 to 10 ton/acre) over 2.5 years on 45 to 53 percent slopes, but a portion of these watersheds was covered with native grasses before burning. Four years are generally required for steep slopes to become stabilized with natural shrub vegetation (Pase and Ingebo 1965; Orr 1970; DeByle and Packer 1972; Anderson and Brooks 1975), although Rowe et al. (1954) reported that soil loss rates may not return to normal until 10 years after a fire.

Soil movement following burning in forest and chaparral communities is usually related to the intensity of the fire. Intense fires increase erosion and runoff (Connaughton 1935; Holland 1953; Rowe 1955; Hussian et al. 1969; Ursic 1969), whereas low intensity fires which leave some litter and a large portion of the humus on the soil surface have little or no effect on surface runoff and erosion (Biswell and Schultz 1957; Cooper 1961; Agee 1973). The most stable forest sites following logging and burning are on level terrain, have deep, well drained soils, have soils that are high in clay content, are on areas with low precipitation surplus, are seral forest types with a favorable growing season after treatment, and have a site history of periodic fires (DeByle 1976b).

Following intense fires on 40 to 80 percent slopes in chaparral and timbered areas of Arizona, erosion losses can vary from 72 to 272 metric ton/ha (32 to 165 ton/acre) during the first year after burning (Hendricks and Johnson 1944; Glendening et al. 1961). Where slopes exceed 80 percent, soil slips (mass erosion) are common, especially near stream channels, and soil losses can be as high as 795 metric ton/ha (354 ton/acre) (Rice et al. 1969; Pitt et al. 1978).

The principal element responsible for mass erosion in most areas is elimination of stability provided by anchoring roots of shrubs and trees as they gradually begin to decay (Rice and Krammes 1971). Decay progresses far enough in four to five years that mass movement either develops or is accelerated during periods of abnormally high rainfall (Swanston 1971; Tiedemann et al. 1979). High infiltration rates and high soil pore water pressures, particularly on areas seeded to grass, compound the mass erosion problem (Rice and Krammes 1971; Rice 1974). Therefore, the natural regeneration of shrubs and trees on steep slopes is very important for long-term stability.

Because of low annual precipitation [30 to 50 cm (12 to 20 in.)], removal of shrubs and trees from pinyon *(Pinus edulis)*-juniper *(Juniperus* spp.) communities by chaining or burning in the Rocky Mountain region does not generally affect erosion or runoff (Brown 1965; 1971; Collings and Myrick 1966; Gifford et al. 1970), except following 25-year storms (Baker et al. 1971) and occasionally following slash burnings or excessive soil disturbance treatments (Myrick 1971; Gifford 1973). However, on limestone-derived soils in central Texas where the annual precipitation is 66 to 71 cm (26 to 28 in.), Wright et al. (1976) found that juniper burns could significantly affect erosion and runoff. Until cover (live vegetation plus litter) reached about 70 percent, moderate slopes (15 to 20 percent) lost 0.2 to 2.5 metric ton/ha (0.1 to 1.1 ton/acre) and steep slopes (45 to 53 percent) lost 13.5 to 18 metric ton/ha (6 to 8 ton/acre), whereas only a trace was lost on the controls. Sampson (1944) and Burgy (1958) reported similar soil losses for California chaparral during the brush-to-grass conversion period after fire. The soil losses reported by Wright et al. (1976) occurred within 9 to 15 months on moderate slopes and within 30 months on steep slopes. After these periods of time, soil losses were minimal and appeared to be stabilized with healing by natural vegetation.

Soil losses on grasslands are very small after burning (Cook 1939; Wahlenberg et al. 1939; Edwards 1942; Rycroft 1947; Nye and Greenland 1960; Banks 1964; Ueckert et al. 1978), except on sandy soils (Blaisdell 1953). Long-term studies on the effects of regular burning of fire-maintained grass vegetation in Mississippi failed to reveal any evidence of accelerated erosion (Wahlenberg et al. 1939). Wright et al. (1976) and Ueckert et al. (1978) showed similar results. These results are most likely attributed to heavy textured soils, low slopes, and good vegetative cover. In the much drier and sandy *Artemisia* steppes, however, wind erosion is proportional to the intensity of the fire until after the second growing season (Blaisdell 1953).

Tolerable erosion limits have been discussed for various soils by Smith and Stamey (1965). They concluded that, in general, the earth's crust loses 0.74 to 2.24 metric ton/ha (0.33 to 1.0 ton/acre) of soil to the sea each year. This exceeds the annual weathering rate of 0.5 metric ton/ha (0.2 ton/acre), which could be used as a guide for tolerable erosion losses. Soil losses due to any management practice could be prorated over time for which the treatment is effective, with the 0.5-metric ton/ha (0.2-ton/acre) weathering rate as a guide. Nevertheless, materials such as limestone decompose at rates from 0.72 to 17.5 metric ton/ha (0.32 to 7.8 ton/acre) annually and areas with wind deposited soil may receive from 3.4 to 18.0 metric ton/ha/year (1.5 to 8.0 ton/acre/year). Thus erosion tolerances may vary with the soil and regional area. Moreover, for any burn, due consideration should be given to potential erosion losses by wind as well as water.

Seeding a burned watershed in the Black Hills of South Dakota reduced soil loss to 22 percent of that on unseeded areas (Orr 1970). Recovery of seeded watersheds (60 percent cover) took one to four years. In Oregon, Anderson and Brooks (1975) found that a burned forest had a satisfactory vegetative cover the first year after seeding, whereas natural vegetation did not provide a satisfactory cover in four years. Sometimes seedings are unsuccessful and soil losses continue (Corbett and Green 1965).

RUNOFF

Runoff is generally low in arid and semi arid zones (Slatyer and Mabbutt 1964; Branson et al. 1972). When annual precipitation is less than 40 cm (16 in.), water yield from brush-to-grass conversions is generally less than 5 cm (2 in.) (Branson et al. 1972), but will increase with rainfall at least up to 86 cm (34 in.) (Hibbert 1971).

As precipitation increased from 36 to 76 cm (14.3 to 30 in.) per year in the Rio Grande basin of New Mexico, Dortignac (1956) found that the percent of precipitation as runoff increased from 2.8 to 29.0 percent.

Brush-to-grass conversions following fire usually increase runoff in the California and Arizona chaparral zones (Rowe 1948; Pase and Ingebo 1965; Baldwin 1968; Brown 1970; Fox 1970; Pitt et al. 1978), but the increased water yield is highly variable and unpredictable depending on vegetation, intensity and duration of storms, antecedent soil moisture, and total precipitation (Pitt et al. 1978). Vegetation, and its associated soil properties (texture, soil aggregates, and organic material), before conversion to grass has been shown to be especially important in subsquent water yields. For example, conversions in chaparral are more likely to increase water yields than conversions in juniper vegetation. Baldwin (1968) and Hibbert et al. (1974) found that brush-to-grass conversions in a 38- to 64-cm (13.8- to 25-in.) precipitation zone in Arizona chaparral usually increased surface runoff from 1 to 12.5 surface-cm (0.3 to 5 in.) per year, whereas Gifford et al. (1970) reported no increased runoff following removal of pinyon and juniper trees in a 30- to 50-cm (12- to 20-in.) precipitation zone.

Natural shrub recovery after fire in Arizona reduced runoff close to prefire conditions by the end of the fourth year after burning (Pase and Ingebo 1965; Brown 1970). Similarly, in a 66-cm (26-in.) rainfall zone of Ashe juniper in central Texas, burning only increased average water yields from 0.05 to 2.87 surface-cm (0.02 to 1.13 in.) the first year after burning. This increase declined steadily to preburn water yields within three to five years. In general, conversion practices in pinyon-juniper communities with 30 to 50 cm (14 to 20 in.) of annual precipitation rarely increase water yields (Baker et al. 1971; Myrick 1971; Branson et al. 1972; Gifford 1973). Thus type of vegetation and its associated soil properties, before conversion to grass, are very influential on subsequent runoff yields.

In western Montana, DeByle and Packer (1972) observed that overland flow from snowmelt was eight times greater on logged and burned plots than on control plots. Vegetative cover was 98 percent on the control areas and 50 percent on the burned areas. Overland flow was highest during the first year and declined significantly during the second year (DeByle and Packer 1972), which is the trend for most burned and unseeded sites (Pase and Ingebo 1965; Brown 1970; Wright et al. 1976; Tiedemann et al. 1979).

When brush-to-grass conversions are practiced to increase water yield, it should be done where precipitation exceeds 40 cm (16 in.) per year (Branson et al. 1972) with slopes less than 30 percent to minimize large soil losses (Sampson 1944; Bentley 1967; Wright et al. 1976). A cover of 60 to 70 percent is considered necessary for soil stability (Bailey and Copeland 1961; Packer 1963; Orr 1970; Wright et al. 1976), which, if seeded, is reached in one to two years (Orr 1970; Anderson and Brooks 1975; Wright et al. 1981). If not seeded, stability will take one to two years on moderate slopes after burning (Wright et al. 1976) and two to four years on steep slopes (Glendening et al. 1961; Orr 1970; DeByle and Packer 1972; Anderson and Brooks 1975).

WATER QUALITY

Sediment and turbidity are the major pollutants that lower water quality following disturbance on agricultural lands (Grant 1971; Robinson 1971; Tiedemann et al. 1979), but total hardness (calcium and magnesium content) is also an important component of water quality (Hem 1970). Hardness is acceptable until it reaches 100 mg/liter (Hem 1970). Research by Wright et al. (1976) on limestone-derived soils showed that water quality, primarily turbidity and total hardness, was related to slope steepness. Following burning, water was soft (0 to 60 mg/liter) from level areas, moderately hard (61 to 120 mg/liter) for 6 months on 15 to 20 percent slopes, and moderately hard for more than 30 months on a 53 percent slope. Water from burned watersheds was more turbid than water from controls for 1.5 years on moderate slopes and for at least 2.5 years on steep slopes. Water quality increased significantly when plant cover reached 63 to 68 percent and soil loss rates were reduced to 0.5 metric ton/ha/year (0.2 ton/acre/year) (Wright et al. 1976). Sodium was usually less than 1.0 mg/liter and pH increased only slightly after burning moderate and steep watersheds.

Cation data (water hardness) summarized by Tiedemann et al. (1979) shows that fire in forest communities seldom has a pronounced effect on ionic composition of downstream water. Tiedemann et al. (1978) observed that concentration of cations was inversely proportional to flow. Therefore, high runoff levels reduced cation concentrations of streamwater (Tiedemann et al. 1979).

REFERENCES

Adams, S., B. R. Strain, and M. S. Adams. 1970. Water-repellent soils, fire, and annual plant cover in a desert scrub community of southeastern California. *Ecology* **51**:696–700.

Agee, J. K. 1973. Prescribed fire effects on physical and hydrologic properties of mixed-conifer forest floor and soil. Tech. Completion Rep. Water Res. Cent., Univ. Calif., Davis.

Ahlgren, I. F., and C. E. Ahlgren. 1965. Effects of prescribed burning on soil microorganisms in a Minnesota jack pine forest. *Ecology* **46**:304–310.

Alban, D. H. 1977. Influence on soil properties of prescribed burning under mature red pine. USDA For. Serv. Res. Paper NC-139. North Cent. For. Exp. Stn., St. Paul, Minn.

Aldous, A. E. 1934. Effect of burning on Kansas bluestem pastures. Kansas Agric. Exp. Stn. Tech. Bull. No. 38. Manhattan.

Anderson, E. W., and L. E. Brooks. 1975. Reducing erosion hazard on a burned forest in Oregon by seeding. *J. Range Manage.* **28**:394–398.

Arrhenius, O. 1921. Influence of soil reaction on earthworms. *Ecology* **2**:255–257.

Austin, R. C., and D. H. Baisinger. 1955. Some effects of burning on forest soils of western Oregon and Washington. *J. For.* **53**:275–280.

Bailey, R. W., and O. L. Copeland, Jr. 1961. Low flow discharges and plant cover relations on two mountain watersheds in Utah. *Intern. Assoc. Sci. Hydrol. Pub.* **51**:267–278.

Baker, M. C., Jr., H. E. Brown, and N. E. Champagne, Jr. 1971. Hydrologic performance of the Beaver Creek watersheds during a 100-year storm. Oral presentation. Amer. Geophys. Union Meeting, San Francisco, Dec. 7. (Mimeo.)

Baldwin, J. J. 1968. Chaparral conversion on the Tonto National Forest. *Proc. Tall Timbers Fire Ecol. Conf.* **8:**203–208.

Banks, C. H. 1964. Further notes on the effect of autumnal veld-burning on stormflow in the Abdolskloff catchment, Jonkershock. *Bosb. Suid-Afr.* **4:**79–84.

Barkley, D. G., R. E. Blaser, and R. E. Schmidt. 1965. Effect of mulches on microclimate and turf establishment. *Agron. J.* **57:**189–192.

Barnette, R. M., and J. B. Hester. 1930. Effect of burning upon the accumulation of organic matter in forest soils. *Soil Sci.* **29:**281–284.

Bartholomew, W. V., and A. G. Norman. 1946. The threshold moisture content for active decomposition of some mature plant materials. *Proc. Soil Sci. Soc. Amer.* **11:**270–279.

Beaton, J. D. 1959. The influence of burning on the soil in the timber range area of Lac le Jeune, British Columbia. II. Chemical properties. *Can. J. Soil Sci.* **39:**1–11.

Bentley, J. R. 1967. Conversion of chaparral areas to grassland: Techniques used in California. USDA Handb. No. 328. Washington, D. C.

Beutner, E. L., and D. Anderson. 1943. The effect of surface mulches on water conservation and forage production in some semi-desert grassland soils. *J. Amer. Soc. Agron.* **35:**393–400.

Biswell, H. H., and A. M. Schultz. 1957. Surface runoff and erosion as related to prescribed burning. *J. For.* **55:**372–375.

Black, C. A. 1957. *Soil-Plant Relationships.* Wiley, New York.

Blaisdell, J. P. 1953. Ecological effects of planned burning of sagebrush grass range on the upper Snake River Plains. USDA Tech. Bull. 1075. Washington, D. C.

Bollen, W. B. 1974. Soil microbes. In Environmental effects of forest residues management in the Pacific Northwest—a state of knowledge compendium. USDA For. Serv. Gen. Tech. Rep. PNW-24. Pac. Northwest For. and Range Exp. Stn., Portland, Ore. pp. B1–B41.

Bornebusch, C. H. 1930. The fauna of forest soil. *Proc. Int. Congr. For. Exp. Stn.* **1:**541–545.

Boyer, D. E., and J. D. Dell. 1980. Fire effects on Pacific Northwest soils. For. Serv. USDA Pac. Northwest Reg. Portland, Ore.

Branson, F. A., G. F. Gifford, and J. R. Owen. 1972. *Rangeland Hydrology.* Range Sci. Ser., No. 1. Soc. Range Manage., Denver.

Brown, H. E. 1965. Preliminary results of cabling Utah juniper, Beaver Creek watershed evaluation project. *Ariz. Watershed Symp. Proc.* **9:**16–22.

Brown, H. E. 1970. Status of pilot watershed studies in Arizona. Amer. Soc. Civil Eng. Proc., J. Irrigation and Drainage Div. (Paper 7129) **96:**11–23.

Brown, H. E. 1971. Evaluating watershed treatment alternatives. Amer. Soc. Civil Eng. Proc., J. Irrigation and Drainage Div. (Paper 7952) **97:**93–108.

Burgy, R. H. 1958. Hydrologic studies on California brush lands. Univ. Calif., Davis. (Mimeo.)

Burns, P. Y. 1952. Effect of fire on forest soils in the pine barren region of New Jersey. Yale Univ.: School of For. Bull. 57. New Haven, Conn.

Campbell, R. E., M. B. Baker, Jr., P. F. Folliott, F. R. Larson, and C. C. Avery. 1977. Wildfire effects on a ponderosa pine ecosytem: An Arizona case study. USDA For. Serv. Res. Paper RM-191. Rocky Mtn. For. and Range Exp. Stn., Fort Collins, Colo.

Clark, O. R. 1940. Interception of rainfall by prairie grasses, weeds, and certain crop plants. *Ecol. Monogr.* **10:**243–277.

Coleman, D. A. 1916. Environmental factors influencing the activity of soil fungi. *Soil Sci.* **2:**1–65.

Collings, M. R., and R. M. Myrick. 1966. Effects of juniper and pinyon eradication on streamflow from Corduroy Creek Basin, Arizona. U.S. Geol. Surv. Prof. Paper 491B.

Colman, E. A. 1953. Vegetation and hydrologic processes. In *Vegetation and Watershed Management.* Ronald Press, New York, pp. 68–109.

Connaughton, C. A. 1935. Forest fires and accelerated erosion. *J. For.* **33:**751, 752.

Cook, L. 1939. A contribution to our information on grass burning. *S. Afr. J. Sci.* **36:**270–282.

Cooper, C. F. 1961. Controlled burning and watershed conditions in the White Mountains of Arizona. *J. For.* **59:**438–442.

Corbett, E. S., and L. S. Green. 1965. Emergency revegetation to rehabilitate burned watersheds in southern California. USDA For. Serv. Res. Paper PSW-22. Pac. Southwest For. and Range Exp. Stn., Berkeley, Calif.

Coults, J. R. H. 1945. **Effects of veld burning on base exchange capacity of soils.** *S. Afr. J. Sci.* **41:**218–224.

Curtis, J. T., and M. L. Partch. 1950. Some factors affecting flower stalk production in *Agropyron gerardi. Ecology* **31:**488, 489.

DeBano, L. F. 1966. Formation of non-wettable soils...involves heat transfer mechanism. USDA For. Serv. Res. Note PSW-132. Pac. Southwest For. and Range Exp. Stn., Berkeley, Calif.

DeBano, L. F. 1969. Water repellent soils: A worldwide concern in management of soil and vegetation. *Agric. Sci. Rev.* **7:**11–18.

DeBano, L. F. 1974. Chaparral soils. In Proc. Symp. on Living with the Chaparral, March 1973, Univ. Calif., Riverside, pp. 19–26.

DeBano, L. F., and C. E. Conrad. 1978. The effect of fire on nutrients in a chaparral ecosytem. *Ecology* **59:**489–497.

DeBano, L. F., P. H. Dunn, and C. E. Conrad. 1977. Fire's effect on physical and chemical properties of chaparral soils. In Proc. Int. Symp. on the environmental consequences of fire and fuel management in Mediterranean ecosystems. USDA For. Serv. Gen. Tech. Rep. WO-3. Washington, D. C., pp.65–74.

DeBano, L. F., G. E. Eberlein, and P. H. Dunn. 1979. Effects of burning on chaparral soils: I. Soil nitrogen. *Soil Sci. Soc. Amer. J.* **43:**504–509.

DeBano, L. F., and J. S. Krammes. 1966. Water-repellent soils and their relation to wildfire temperatures. *Bull. Int. Assoc. Sci. Hydrol.* **11**(2):14–19.

DeBano, L. F., J. F. Osborn, J. S. Krammes, and J. Letey, Jr. 1967. Soil wettability and wetting agents...our current knowledge of the problem. USDA For. Serv. Res. Paper PSW-43. Pac. Southwest For. and Range Exp. Stn., Berkeley, Calif.

DeBano, L. F., S. M. Savage, and D. M. Hamilton. 1976. The transfer of heat and hydrophobic substances during burning. *Soil Sci. Soc. Amer. J.* **40:**779–782.

DeByle, N. V. 1976a. Soil fertility as affected by broadcast burning following clearcutting in Northern Rocky Mountain larch/fir forests. *Proc. Tall Timbers Fire Ecol. Conf.* **14:**447–464.

DeByle, N. V. 1976b. Fire, logging, and debris disposal effects on soil and water in northern conifer forests. Div. 1, Proc. XVI IUFRO World Congr., Oslo, Norway, pp. 201–212.

DeByle, N. V., and P. E. Packer. 1972. Plant nutrient and soil losses in overland flow from burned forest clearcuts. In Proc. Nat. Symp. on Watersheds in Transition, sponsored by the Amer. Water Resour. Assoc. and Colorado State Univ., Fort Collins, pp. 296–307.

Dix, R. 1960. The effects of burning on the mulch structure and species composition of grasslands in western North Dakota. *Ecology* **41:**49–56.

Dortgnac, E. J. 1956. Watershed resources and problems of the Upper Rio Grande Basin. USDA For. Serv. Res. Paper. Rocky Mtn. For. and Range Exp. Stn., Fort Collins, Colo.

Dunn, P. H., and L. F. DeBano. 1977. Fire's effect on the biological properties of chaparral soils. In Proc. Int. Symp. on the Environmental Consequences of Fire and Fuel Management in Mediterranean-Climate Ecosystems (Forests and Scrublands). USDA For. Serv. Gen. Tech. Rep. WO-3. Washington, D.C., pp. 75–84.

Dyksterhuis, E. J., and E. M. Schmutz. 1947. Natural mulches or "litter" of grasslands: With kinds and amounts on a southern prairie. *Ecology* **28:**163–179.

Dyrness, C. T. 1976. Effect of wildlife on soil wettability in the high Cascases of Oregon. USDA For. Serv. Res. Paper PNW-202. Pac. Northwest For. and Range Exp. Stn., Portland, Ore.

Edwards, D. C. 1942. Grass-burning. *Emp. J. Exp. Agric.* **10**:219–231.

Ehrenreich, J. H., and J. M. Aikman. 1963. An ecological study on certain management practices on native prairie in Iowa. *Ecol. Monogr.* **33**:113–130.

Elwell, H. M., H. A. Daniel, and F. A. Fenton. 1941. The effect of burning pasture and native woodland vegetation. Okla. Agric. Exp. Stn. Bull. B-247. Stillwater.

Farmer, E. E., and B. P. Haveren. 1971. Soil erosion by overland flow and raindrop splash on three mountain soils. USDA For. Serv. Res. Paper INT-100. Intermt. For. and Range Exp. Stn., Ogden, Utah.

Flory, E. L. 1936. Comparison of the environment and some physiological responses of prairie vegetation and cultured maze. *Ecology* **17**:67–103.

Fowells, H. A., and R. E. Stephenson. 1934. Effect of burning on forest soils. *Soil Sci.* **38**:175–181.

Fox, K. M. 1970. Prescribed fire as a tool for increasing water yield. In Proc. Symp. on Fire Ecology and the Control and Use of Fire in Wildland Management. Univ. Ariz., Tucson, pp. 66–68.

Garren, K. H. 1943. Effects of fire on vegetation of the southeastern United States. *Bot. Rev.* **9**:617–654.

Gibbs, W. M. 1919. The isolation and study of nitrifying bacteria. *Soil Sci.* **8**:427–471.

Gifford, G. F. 1973. Runoff and sediment yields from runoff plots on chained pinyon-juniper sites in Utah. *J. Range Manage.* **26**:440–443.

Gifford, G. F., G. Williams, and G. B. Coltharp. 1970. Infiltration and erosion studies on pinyon-juniper conversion sites in southern Utah. *J. Range Manage.* **23**:402–406.

Gill, R. W. 1969. Soil microarthropod abundance following old-field litter manipulation. *Ecology* **50**:805–816.

Glendening, G. E., C. P. Pase, and P. Ingebo. 1961. Preliminary hydrologic effects of wildfire in chaparral. In Modern Techniques in Water Management. *Ariz. Watershed Symp. Proc.* **5**:12–15. Phoenix.

Grant, K. E. 1971. Sediment: Everybody's pollution problem. In 33rd Nat. Farm Ins. Chamber of Commerce Proc. Des Moines, Iowa, pp. 23–28.

Greene, S. W. 1935. Effect of annual grass fires on organic matter and other constituents of virgin longleaf pine soils. *J. Agric. Res.* **50**:809–822.

Grier, C. C. 1975. Wildfire effects on nutrient distribution and leaching in a coniferous ecosystem. *Can. J. For. Res.* **5**:599–607.

Grier, C. C., and D. W. Cole. 1971. Influence of slash burning on ion transport in a forest soil. *Northwest Sci.* **5**:100–106.

Hanks, R. J., and K. L. Anderson. 1957. Pasture burning and moisture conservation. *J. Soil and Water Conserv.* **12**:288, 289.

Harvey, A. E., M. J. Larsen, and M. F. Jurgensen. 1979a. Interactive roles regulating wood accumulation and soil development in the northern Rocky Mountains. USDA For. Serv. Res. Note INT-263. Intermt. For. and Range Exp. Stn., Ogden, Utah.

Harvey, A. E., M. F. Jurgensen, and M. J. Larsen. 1979b. Role of forest fuels in the biology and management of soil. USDA For. Serv. Gen. Tech. Rep. INT-65. Intermt. For. and Range Exp. Stn., Ogden, Utah.

Harvey, A. E., M. F. Jurgensen, and M. J. Larsen. 1976. Intensive fiber utilization and prescribed fire: Effects on the microbial ecology of forests. USDA For. Serv. Gen. Tech. Rep. INT-28. Intermt. For. and Range Exp. Stn., Ogden, Utah.

Hem, J. D. 1970. Study and interpretation of the chemical characteristics of natural water. U. S. Geol. Surv. Water-Supply Paper 1473. Washington, D. C.

Hendricks, B. A., and J. M. Johnson. 1944. Effects of fire on steep mountain slopes in central Arizona. *J. For.* **24**:568–571.

Hensel, R. L. 1923. Recent studies on the effect of burning on grassland vegetation. *Ecology* **4**:183–188.

Heyward, F. 1936. Soil changes associated with forest fires in the longleaf pine region of the South. *Amer. Soil Surv. Assoc. Bull.* **17**:41, 42.

Heyward, F., and R. M. Barnette. 1934. Effect of frequent fires on chemical composition of forest soils in longleaf pine region. Univ. Fla. Agric. Exp. Stn. Tech. Bull. 265. Gainesville.

Heyward, F., and A. N. Tissot. 1936. Some changes in the soil fauna associated with forest fires in the longleaf pine region. *Ecology* **17**:659–666.

Hibbert, A. R. 1971. Increases in streamflow after converting chaparral to grass. *Water Resour. Res.* **7**:71–80.

Hibbert, A. R., E. A. Davis, and D. G. Scholl. 1974. Chaparral conversion in Arizona. Part 1: Water yield response and effects on other resources. USDA For. Serv. Res. Paper RM-126. Rocky Mtn. For. and Range Exp. Stn., Fort Collins, Colo.

Holland, J. 1953. Infiltration on a timber and a burn site in northern Idaho. USDA For. Serv. Res. Note RM-127. Rocky Mtn. For. and Range Exp. Stn., Ogden, Utah.

Hopkins, H. H. 1954. Effects of mulch upon certain factors of the grassland environment. *J. Range Manage.* **7**:255–258.

Hulbert, L. C. 1969. Fire and litter effects in undisturbed bluestem prairie in Kansas. *Ecology* **50**:874–877.

Hursh, C. R. 1928. Litter keeps forest soil productive. *Southern Lumberman* **134**:1–3.

Hussain, S. B., C. M. Skau, S. M. Bashir, and R. O. Meeuwig. 1969. Infiltrometer studies of water-repellent soils on the east slope of the Sierra Nevada. In L. F. DeBano and J. Letey (ed.) Water-Repellent Soils. Proc. Symp. Univ. Calif., Riverside.

Isaac, L. A., and H. G. Hopkins. 1937. The forest soil of the Douglas-fir region, and changes wrought upon it by logging and slash burning. *Ecology* **18**:264–279.

Johnson, W. M. 1940. Infiltration capacity of a forest soil as influenced by litter. *J. For.* **38**:520.

Jones, J. M., and B. N. Richards. 1977. Effects of reforestation on turnover of 15_N-labeled nitrate and ammonium in relation to changes in soil microfauna. *Soil Biol. Biochem.* **9**:383–392.

Jurgensen, M. F., M. J. Larsen, and A. E. Harvey. 1979. Forest soil biology-timber harvesting relationships. USDA For. Serv. Gen. Tech. Rep. INT-69. Intermt. For. and Range Exp. Stn., Ogden, Utah.

Kelly, J. M. 1971. Review of the literature on the effect of various levels and combinations of nitrogen and phosphorous fertilizer on forest litter decomposition. IBP Eastern Deciduous For. Biome Memo Rep. No. 71–68.

Kittredge, J. 1938. Comparative infiltration in the forest and open. *J. For.* **36**:1156, 1157.

Klemmedson, J. O. 1976. Effect of thinning and slash burning on nitrogen and carbon in ecosystems of young dense ponderosa pine. *For. Sci.* **22**:45–53.

Klemmedson, J. O., A. M. Schultz, H. Jenny, and H. H. Biswell. 1962. Effect of prescribed burning on forest litter on total soil nitrogen. *Soil Sci. Soc. Amer. Proc.* **26**:200–202.

Knight, H. 1966. Loss of nitrogen from the forest floor by burning. *For Chron.* **42**:149–152.

Komarek, E. V., Sr. 1970. Controlled burning and air pollution: an ecological review. *Proc. Tall Timbers Fire Ecol. Conf.* **10**:141–173.

Kucera, C. L., and J. H. Ehrenreich. 1962. Some effects of annual burning on central Missouri prairie. *Ecology* **43**:334–336.

Lemon, P. C. 1949. Successional responses of herbs in the longleaf-slash pine forest after fire. *Ecology* **30**:135–145.

Lewis, M. L., Jr. 1974. Effects of fire on nutrient movement in a South Carolina pine forest. *Ecology* **55**:1120–1127.

Lutz, H. J. 1956. Ecological effects of forest fires in the Interior of Alaska. USDA Tech. Bull. 1133. Washington, D.C.

MacKinney, A. L. 1929. Effect of forest litter on the soil temperature and soil freezing in autumn and winter. *Ecology* **10**:312–321.

Marshall, R., and C. Averill. 1928. Soil alkalinity on recent burns. *Ecology* **9**:533.

McCalla, T. M. 1943. Microbiological studies of the effect of straw used as a mulch. *Trans. Kansas Acad. Sci.* **43**:52–56.

McMurphy, W. E., and K. L. Anderson. 1965. Burning Flint Hills range. *J. Range Manage.* **18**:265–269.

Meeuwig, R. O. 1970. Infiltration and soil erosion as influenced by vegetation and soil in northern Utah. *J. Range Manage.* **23**:185–188.

Meeuwig, R. O. 1971. Soil stability on high-elevation rangeland in the Intermountain areas. USDA For. Serv. Res. Paper INT-94. Intermt. For. and Range Exp. Stn., Ogden, Utah.

Metz, L. J., T. Lotti, and R. A. Klawitter. 1961. Some effects of prescribed burning on Coastal Plain forest soils. USDA For. Serv. Res. Paper SE-133. Southeastern For. Exp. Stn., Asheville, N. C.

Miller, R. B., J. D. Stout, and K. E. Lee. 1955. Biological and chemical changes following scrub burning on a New Zealand hill soil. *New Zealand J. Sci. Technol.* **37**:290–313.

Moore, A. W. 1960. The influence of annual burning on a soil in the derived savanna zone of Nigeria. *Int. Congr. Soil Sci. Trans. 7th.* **4**:257–264.

Myrick, R. M. 1971. Cibecue ridge juniper project. In Proc. 15th Ariz. Watershed Symp. Phoenix, pp. 35–39.

Neal, J. L., E. Wright, and W. B. Bollen. 1965. Burning Douglas-fir slash: Physical, chemical, and microbial effects on the soil. For. Res. Lab. Res. Paper. Oregon State Univ., Corvallis.

Neuenschwander, L. F. 1976. The effects of fire in a sprayed tobosa-mesquite community on Stamford clay soils. Ph.D. Diss. Texas Tech Univ., Lubbock.

Neuenschwander, L. F., T. L. Whigham, D. N. Ueckert, and H. A. Wright. 1974. Effect of fire on organic carbon and bacterial growth in the mesquite-tobosa community. *Noxious Brush and Weed Control Res. Highlights* **5**:15. Texas Tech Univ., Lubbock.

Neuenschwander, L. F., H. A. Wright, and S. C. Bunting. 1978. The effect of fire on a tobosagrass-mesquite community in the Rolling Plains of Texas. *Southwestern Natur.* **23**:315–338.

Nye, P. H., and D. H. Greenland. 1960. The soil under shifting cultivation. *Commonw. Bur. Soils, Tech Commun.* **51**:156.

Odum, E. F. 1960. Organic production and turnover in old field succession. *Ecology* **41**:34–49.

Old, S. M. 1969. Microclimate, fire, and plant production in an Illinois prairie. *Ecol. Monogr.* **39**:355–384.

Orr, H. K. 1970. Runoff and erosion control by seeded and native vegetation on a forest burn: Black Hills, South Dakota. USDA For. Serv. Res. Paper RM-60. Rocky Mtn. For. and Range Exp. Stn., Fort Collins, Colo.

Owensby, C. E., and J. B. Wyrill, II. 1973. Effects of range burning on Kansas Flint Hills soil. *J. Range Manage.* **26**:185–188.

Packer, P. E. 1951. An approach to watershed protection criteria. *J. For.* **49**:639–644.

Packer, P. E. 1963. Soil stability requirements for the Gallatin elk winter range. *J. Wildl. Manage.* **27**:401–410.

Packer, P. E., and W. A. Laycock. 1969. Watershed management in the United States: Concepts and principles. In Lincoln Papers in Water Resources No. 8, New Zealand Agric. Eng. Inst., Lincoln College, Canterbury, New Zealand, pp. 1–22.

Pase, C. P., and P. A. Ingebo. 1965. Burned chaparral to grass: Early effects on water and sediment yields from two granitic soil watersheds in Arizona. In Proc. 9th Ariz. Watershed Symp., Phoenix, pp. 8–11.

Pase, C. P., and A. W. Lindenmuth, Jr. 1971. Effects of prescribed fire on vegetation and sediment in oak-mountain mahogany chaparral. *J. For.* **69**:800–805.

Pearse, A. S. 1946. Observations on the micro-fauna of the Duke Forest. *Ecol. Monogr.* **16**:127–150.

Peet, M., R. Anderson, and M. S. Adams. 1975. Effect of fire on big bluestem production. *Am. Midl. Natur.* **94**:15–26.

Phillips, E. P. 1919. A preliminary report on the veld-burning experiments at Groenkloof, Pretoria. *S. Afr. J. Sci.* **16**:285–299.

Piene, H. 1974. Factors influencing organic matter decomposition and nutrient turn-over in cleaned and spaced, young conifer stands on the Cape Breton Highlands, Nova Scotia. Can. For. Serv. Maritimes For. Res. Centre Information Rep. M-X-41. Fredericton, N. B.

Pitt, M. D., R. H. Burgy, and H. F. Heady. 1978. Influences of brush conversion and weather patterns from a northern California watershed. *J. Range Manage.* **31**:23–27.

Raison, R. J. 1979. Modification of the soil environment by vegetation fires, with particular reference to nitrogen transformations: A review. *Plant and Soil* **51**:73–108.

Renbuss, M. A., G. A. Chilvers, and L. D. Pryor. 1973. Microbiology of an ashbed. *Proc. Linn. Soc. N.S.W.* **97**:302–311.

Reynolds, G. H., and J. W. Bohning. 1956. Effects of burning on a desert grass-shrub range in southern Arizona. *Ecology* **37**:769–777.

Rice, E. L., and S. K. Pancholy. 1973. Inhibition of nitrification by climax ecosystems. II. Additional evidence and possible role of tannins. *Amer. J. Bot.* **60**:691–702.

Rice, R. M. 1974. The hydrology of chaparral watersheds. In Proc. Symp. on Living with the Chaparral, March 1973, Univ. Calif. Riverside, pp. 27–33.

Rice, R. M., and J. S. Krammes. 1971. Mass-wasting processes in watershed management. In Proc. Symp. Interdisciplinary Aspects of Watershed Management, Bozeman, Montana. Amer. Soc. Civil Eng., pp. 231–259.

Rice, R. M., J. S. Rothacher, and W. F. Megahan. 1969. Erosional consequences of timber harvesting: An appraisal. In Nat. Symp. on Watersheds in Transition, pp. 321–329.

Robinson, A. R. 1971. A primer on agricultural pollution: Sediment. *J. Soil Water Conserv.* **26**:61, 62.

Rode, A. A. 1955. *Soil Science*. (translated from Russian) Israel Progr. for Sci. transl., Jerusalem 1962.

Rowe, P. B. 1948. Influence of woodland chaparral on water and soil in central California. USDA and State of Calif. Dept. Natur. Res. Div. For. Unnumbered Pub. Sacramento, Calif.

Rowe, P. B. 1955. Effects of the forest floor upon disposition of rainfall in pine stands. *J. For.* **53**:342–348.

Rowe, P. B., C. M. Countryman, and H. C. Storey. 1954 Hydrologic analysis used to determine effects of fire on peak discharge and erosion rates in southern California watersheds. Calif. For. and Range Exp. Stn., USDA For. Serv. Berkeley.

Russell, J. C. 1939. The effect of surface cover on soil moisture losses by evaporation. *Proc. Soil Sci. Amer.* **4**:65–70.

Rycroft, H. B. 1947. A note on the immediate effects of veld burning on storm flow in a Jonkershock stream catchment. *J. S. Afr. For. Assoc.* **15**:80–88.

Sampson, A. W. 1944. Effect of chaparral burning on soil erosion and on soil-moisture relations. *Ecology* **25**:171–191.

Savage, S. M. 1974. Mechanism of fire-induced water repellency in soil. *Soil Sci. Soc. Amer. Proc.* **38**:652–657.

Savage, S. M., J. L. Osborn, and C. Heaton. 1972. Water repellency in soils induced by forest fires. *Soil Sci. Soc. Amer. Proc.* **26**:674–678.

Scholl, D. G. 1975. Soil wettability and fire in Arizona chapparral. *Soil Sci. Soc. Amer. Proc.* **39:**356–361.

Scott, J. D. 1934. Ecology of certain plant communities of the Central Province, Tanganyika Territory. *J. Ecol.* **22:**177–229.

Scotter, G. W. 1963. Effects of forest fires on soil properties in northern Saskatchewan. *For. Chron.* **39:**412–421.

Sharrow, S. H., and H. A. Wright. 1977a. Effects of fire, ash, and litter on soil nitrate, temperature, moisture and tobosagrass production in the Rolling Plains. *J. Range Manage.* **30:**266–270.

Sharrow, S. H., and H. A. Wright. 1977b. Proper burning intervals for tobosagrass in West Texas based on nitrogen dynamics. *J. Range Manage.* **30:**343–346.

Shearer, R. C. 1975. Seedbed characteristics in western larch forests after prescribed burning. USDA For. Serv. Res. Paper INT-167. Intermt. For. and Range Exp. Stn., Ogden, Utah.

Slatyer, R. O., and J. A. Mabbutt. 1964. Hydrology of arid and semi-arid regions. In Ven Te Chow (ed.) *Handbook of Applied Hydrology,* McGraw-Hill, New York, pp. 24-1 to 24-46.

Smith, D., and W. H. Wischmeier. 1962. Rainfall erosion. *Adv. Agron.* **14:**109–148.

Smith, R. M., and W. L. Stamey. 1965. Determining the range of tolerable erosion. *Soil Sci.* **100:**414–424.

Stark, N. M. 1977. Fire and nutrient cycling in a Douglas-fir/larch forest. *Ecology* **58:**16–30.

Stark, N. M. 1979. Nutrient losses from timber harvesting in a larch/Douglas-fir forest. USDA For. Serv. Res. Paper INT-231. Intermt. For. and Range Exp. Stn., Ogden, Utah.

Stinson, K. J., and H. A. Wright. 1969. Temperatures of headfires in the southern mixed prairie of Texas. *J. Range Manage.* **22:**169–174.

Stunkard, H. W. 1944. Studies on the oribated mite, *Galumna* sp., intermediate host of *Monozia expansa. Anat. Rec.* **89:**1–53.

Suman, R. F., and R. L. Carter. 1954. Burning and grazing have little effect on chemical properties of Coastal Plain forest soils. USDA For. Serv. Res. Note SE-56. Southeastern For. Exp. Stn., Asheville, N.C.

Swanston, D. N. 1971. Principal soil movement processes influenced by road-building, logging, and fire. In Proc. Symp. Forest Land Uses and Stream Environment. Oregon State Univ., Corvallis, pp. 28–40.

Swanston, D. N., and C. T. Dyrness. 1973. Stability of steep land. *J. For.* **71:**264–269.

Sweeney, J. R. 1956. Responses of vegetation to fire: A study of the herbaceous vegetation following chaparral fires. *Univ. Calif. Bot.* **28:**143–250.

Sweeney, J. R., and H. H. Biswell. 1961. Quantitative studies on the removal of litter and duff by fire under controlled conditions. *Ecology* **42:**572–575.

Tarrant, R. F. 1956. Effects of slash burning on some soils of the Douglas-fir region. *Soil Sci. Soc. Amer. Proc.* **20:**408–411.

Tiedemann, A. R. 1973. Stream chemistry following a forest fire and urea fertilization in north-central Washington. USDA For. Serv. Res. Note PNW-203. Pac. Northwest For. and Range Exp. Stn., Portland, Ore.

Tiedemann, A. R., and T. D. Anderson. 1980. Combustion losses of sulfur from native plant materials and forest litter. In Proc. 6th Conf. on Fire and Forest Meterology, April 22–24, 1980. Soc. Amer. For., Mills Bldg., Washington, D.C., pp. 220–227.

Tiedemann, A. R., C. E. Conrad, J. H. Dieterick, J. W. Hornbeck, W. F. Megahan, L. A. Viereck, and D. D. Wade. 1979. Effects of fire and water. USDA For. Serv. Gen. Tech. Rep. WO-10. Washington, D.C.

Tiedemann, A. R., and J. D. Helvey. 1973. Nutrient ion losses in streamflow after fire and fertilization in eastern Washington. *(Abstr.) Bull. Ecol. Soc. Amer.* **54:**20.

Tiedemann, A. R., J. D. Helvey, and T. D. Anderson. 1978. Stream chemistry and watershed nutrient economy following wildfire and fertilization in eastern Washington. *J. Environ. Qual.* **7:**580–588.

Tyron, E. H. 1948. Effect of charcoal on certain physical, chemical, and biological properties of forest soils. *Ecol. Monogr.* **18**:82–115.

Ueckert, D. N., T. L. Whigham, and B. M. Spears. 1978. Effect of burning on infiltration, sediment, and other soil properties in a mesquite-tobosagrass community. *J. Range Manage.* **31**:420–425.

Ursic, S. J. 1969. Hydrologic effects of prescribed burning on abandoned fields in northern Mississippi. USDA For. Ser. Res. Paper SO-46. Southern For. Exp. Stn., Pineville, La.

Vitousek, P. M., and J. M. Melillo. 1979. Nitrate losses from disturbed forests: Patterns and mechanisms. *For. Sci.* **25**:605–619.

Wahlenberg, W. G. 1935. Effect of fire and grazing on soil properties and the natural reproduction of longleaf pine. *J. For.* **33**:331–338.

Wahlenberg, W. G., S. W. Greene, and H. R. Reed. 1939. Effects of fire and cattle grazing on longleaf pine lands as studied at McNeill, Mississippi. USDA Tech. Bull. 683. Washington, D.C.

Waksman, S. A. 1922. Microbiological analysis of soil as an index of soil fertility. II. Methods of the study of numbers of micro-organisms in the soil. *Soil Sci.* **14**:283–298.

Waksman, S. A. 1927. *Principles of Soil Microbiology.* Williams and Wilkins, Baltimore, Md.

Weaver, J. E., and N. W. Rowland. 1952. Effects of excessive natural mulch on development, yield, and structure of native grassland. *Bot. Gaz.* **114**:1–19.

Weitzman, S., and R. R. Bay. 1963. Forest soil freezing and the influence of management practices, northern Minnesota. USDA For. Serv. Res. Paper LS-2. Lake States For. Exp. Stn., St. Paul, Minn.

Wells, C. G., R. E. Campbell, L. F. DeBano, C. E. Lewis, R. L. Frederickson, E. C. Franklin, R. C. Froelich, and P. H. Dunn. 1979. Effects of fire on soil. USDA For. Serv. Gen. Tech. Rep. WO-7. Washington, D.C.

West, O. 1965. Fire in vegetation and its use in pasture management, with special reference to tropical and subtropical Africa. Commonw. Bur. Pastures and Crops, Farnham Royal, Bucks, Engl.

Whigham, T. L. 1976. The effects of fire on selected physical and chemical properties of soil in a mesquite-tobosagrass community. M. S. Thesis. Texas Tech Univ., Lubbock.

White, E. M., W. W. Thompson, and F. R. Gartner. 1973. Heat effects on nutrient release from soils under ponderosa pine. *J. Range Manage.* **26**:22–24.

Wright, H. A. 1969. Effect of spring burning on tobosa grass. *J. Range Manage.* **22**:425–427.

Wright, H. A. 1972. Fire as a tool to manage tobosa grasslands. *Proc. Tall Timbers Fire Ecol. Conf.* **12**:153–167.

Wright, H. A. 1974. Range burning. *J. Range Manage.* **27**:5–11.

Wright, H. A., F. M. Churchill, and W. C. Stevens. 1976. Effects of prescribed burning on sediment, water yield, and water quality from dozed juniper lands in central Texas. *J. Range Manage.* **29**:294–298.

Wright, H. A., F. M. Churchill, and W. C. Stevens. 1982. Soil loss and runoff on seeded vs. non-seeded watersheds following prescribed burning. *J. Range Manage.* (in press)

Youngberg, C. T., and A. G. Wollum, II. 1976. Nitrogen accretion in developing *Ceanothus velutinus* stands. *Soil Sci. Soc. Amer. J.* **40**:109–112.

Zwolinski, M. J., and J. H. Ehrenreich. 1967. Prescribed burning on Arizona watersheds. *Proc. Tall Timbers Fire Ecol. Conf.* **7**:195–205.

CHAPTER

4

WILDLIFE

Habitat for wildlife is not a stable entity that can be preserved behind a fence, like a cliff dwelling or a petrified tree (Leopold et al. 1963). It is constantly changing and needs some form of maintenance to remain suitable for many animal populations. Historically, natural fires have always influenced wildlife habitats.

Fire in moist environments was infrequent because of marginal burning conditions. A coincidence of fuel buildup, extreme burning conditions, and ignition usually resulted in intense fires that perpetuated diverse wildlife habitats (Gruell 1980a; 1980b). Fires in drier environments (e.g., grasslands) were more frequent, but less intense. These fires inhibited shrub and tree development which depressed habitat diversity (Gruell 1980a; 1980b). Thus fire is not always beneficial to wildlife, nor is it beneficial to all species.

The value of prescribed burning for wildlife must take into account the total animal species complex, vegetation type, stages of plant succession, weather patterns, and intensity of planned burns. Generally, a mosaic of seral and climax stages of vegetation that may be created by a series of fires over time in shrub and forest communities is the preferred habitat for the highest diversity and number of wildlife species. Such habitat provides maximum "edge" for feeding, escape, loafing, and nesting (Leopold 1932). Fire in grasslands also creates variety (particularly feeding areas) for upland game birds but may be harmful to resident big game due to loss of tree and shrub cover.

Many people are concerned that fire is destructive to animals. However, on large fires, deaths of animals are rare (Vogl 1977), unless they happen to get cornered by a fire and lack escape routes. The beneficial effects of fire on fire-dependent habitats usually compensates for any potential losses (Vogl 1977). Animals are a product of the habitat, which is often the product of a fire. Therefore animals are generally well adapted to fire in their environment. In this chapter, we will discuss the role of fire in relation to the habitat of birds, mammals, and aquatic fauna.

SMALL MAMMALS

Survival of small mammals within a burn depends upon the uniformity, intensity, size, and duration of the burn (Buech et al. 1977), as well as the mobility and position of the animal relative tò the soil surface at the time of a passing fire. Since most rodents live in burrows beneath the soil surface, they are well insulated from heat that is released by fires (Howard et al. 1959). Nevertheless, some may die from suffocation (Tevis, 1956; Chew et al. 1959; Lawrence 1966), and slow moving mammals that happen to be above the ground such as wood rats *(Neotoma* spp.) may die directly. Mice can tolerate temperatures up to 63°C (145°F) for short periods if the relative humidity is below 22 percent (Howard et al. 1959), but can die quickly when exposed to 49°C (120°F) if the relative humidity is above 60 percent (Lawrence 1966). Thus, survival is greatest where the burn is incomplete in fuels of low density and high moisture content (Buech et al. 1977).

Voles, Mice, and Shrews

Regardless of the level of survival of small mammals in a fire, the resulting dramatic changes in habitat (temporary loss of shelter and food, exposure of surface runways and burrow openings, and increased predation) will decrease the number and diversity in small mammal populations for one to three years or longer after a fire (Cook 1959; Lawrence 1966; Klebenow and Beall 1977; Koehler and Hornocker 1977). Mesic habitat types support the greatest number of rodents, particularly red-backed voles *(Clethrionomys gapperi)* which are the most important food item for marten *(Martes americana)* (Koehler and Hornocker 1977). Burns destroy habitat of red-backed voles and the habitat for other species of voles for at least one year (Lensik et al. 1955; Cook 1959; Gashwiler 1970; Lowe 1975). Optimum habitat for red-backed voles will not occur in forests until the seventh year after a fire (Buech et al. 1977).

Deer mice *(Peromyscus maniculatus)* are very adaptable to fire and thrive in xeric environments (Koehler and Hornocker 1977). Immediately after the burn, their numbers may drop 50 to 84 percent (Buech et al. 1977; Crowner and Barrett 1979). However, after the first growing season, they remain constant on grassland burns and increase on shrub and forest burns (Cook 1959; Koehler and Hornocker 1977; Lowe et al. 1978). Gashwiler (1970) found that on a clearcut and burned area in west-central Washington deer mice increased to eight times that of the virgin forest two years after the burn.

Cook (1959) and Lawrence (1966) found that after burning in California chaparral, the composition of rodents shifted from chaparral dominants such as the chaparral mouse *(Peromyscus truei)* and the California mouse *(P. californicus)* to the grassland dominants such as the pocket mouse *(Perognathus californicus)*, the field mouse *(Microtus californicus)*, and western harvest mouse *(Reinthrodontomys megalotis)*. The chaparral mouse is associated with dominant shrub species such as wedgeleaf ceanothus *(Ceanothus cuneatus)*, the seeds and bark of which are important sources of food. Other, species except for field mice, eat primarily forb and grass seeds (James 1952; Cook 1959). Field mice are often restricted from an early recovery following burns because they need at least one year's accumulation of mulch to furnish runways (Cook 1959).

Shrews are usually rare on burned sites (Gashwiler 1970). Gashwiler found that Trowbridge's shrews *(Sorex trowbridgii)* and vagrant shrews *(S. vagrans)* were present on both the burned and virgin areas in relatively low numbers.

Squirrels

Holocaust fires are detrimental to tree squirrels. Gashwiler (1970) did not find Douglas's squirrels *(Tamiasciurus douglasii)* or northern flying squirrels *(Glaucomys sabriners)* on a clear-cut and burned area. Similarly, clear-cuts in Alaska eliminated red squirrels *(T. hudsonicus)* and shelterwood cutting treatments reduced the squirrels from 1.4/ha (0.6/acre) to 0.5/ha (0.2/acre) (Wolf 1975). Ground fires, however, such as in ponderosa pine *(Pinus ponderosa)*, probably had little influence on tree squirrels. The effect may have been favorable by perpetuating ponderosa pine, their primary food source. For example, Patton (1975) found that Abert squirrels *(Sciurus aberti aberti)* were most abundant in ponderosa pine trees that were 51 to 100 years of age [28- to 33-cm (11- to 13-in.) dbh range] and that nests generally were found in a tree that was one of a group of trees of similar size with interlocking branches.

Ground squirrels *(Spermophilus beecheyi, S. lateralis, Citellus tridecemlineatus)* increase in density after burning (Gashwiler 1970; Beck and Vogl 1972; Lowe 1975). Lowe found the golden-mantled ground squirrel *(S. lateralis)* increased 18-fold on a moderate burn and 8-fold on a severe burn when the areas were inventoried 2 years after a major fire. Beck and Vogl (1972) found that the 13-lined ground squirrel *(C. tridencemlineatus)* increased following fire in a northern pine-hardwood forest. Columbian ground squirrels *(Citellus columbianus)* were abundant in open meadows and burns in the Selway-Bitterroot Wilderness (Koehler and Hornocker 1977).

Chipmunks

Data on chipmunks *(Eutamias* spp.) are somewhat variable. In the Pacific Northwest Gashwiler (1970) found that Townsend's chipmunk *(Eutamias townsendii)* increased on burns. Similarly, Koehler and Hornocker (1977) found chipmunks *(E. amoenus)* on the most xeric sites. Lowe (1975), however, found that *E. cinereicollis* in the Southwest was always lower on burned than unburned areas. Thus, in mesic habitats, fire may be beneficial but not in xeric habitats.

Hares and Cottontails

Snowshoe hares *(Lepus americanus)* leave severely burned areas and occupy the moderately burned sites (Gashwiler 1970). They return to severely burned areas after the establishment of shrubs and saplings needed for food and cover (Keith and Surrendi 1971; Ream and Gruell 1979).

Cottontails *(Sylvilagus* spp.) generally exist in high density in open areas that contain large amounts of dead brush or other dense shrubby and herbaceous undergrowth (Ream and Gruell 1979). Populations increase after clear-cutting (Costa et al. 1976), but generally decrease after brush piles (escape and nesting areas) are removed (George 1977).They consume a large variety of vegetation that is enhanced by fire, but their refugia would be destroyed by severe fires. Thus moderate burns, which leave portions of the area unburned, may not be harmful to populations of snowshoe hares and cottontails.

BIRDS

Forest and shrub habitats in the middle successional or preclimax stages generally have the maximum diversity and numbers of bird species (Marshall 1963; Lawrence 1966; Evans 1978; Lowe et al. 1978). Wood and Niles (1978) characterized such habitats in the Southeast as "when the shrub and hardwood seedling component of the habitat is building and prior to crown closure of the pines." Such plant communities are complex and retain enough openness to maintain maximum diversity and numbers of birds (Szaro and Balda 1979a). Management units should be at least 34 ha (84 acres) to achieve maximum songbird species richness for a given vegetation type (Galli et al. 1976; Thomas et al. 1978).

Given the above optimum parameters, the immediate effect of fire on bird populations depends greatly upon the season and intensity of a burn. A relatively cool fire during the dormant season could greatly increase food sources and leave adequate nest sites for ground and brush-foraging birds (Lowe et al. 1978). An intermediate fire might have a similar effect for ground and brush-foraging birds, but would also create more openness for timber-drilling and flycatching birds (Lowe et al. 1978) and raptors (McAdoo and Klebenow 1978). A severe fire would seriously reduce the number and diversity of tree-foliage–searching and timber-gleaning birds (Bock and Lynch 1970).

The absence of fire or other destructive forces in forest communities will be accompanied by a decrease in bird niche diversity and carrying capacity (Marshall 1963; Wood and Niles 1978). However, in fire management, the most important concern is to have a wide variety of relatively small different-aged burns interspersed with some areas that have not been burned for several hundred years.

Nongame Birds

SONGBIRDS AND RAPTORS

Several studies in the Southwest by Marshall (1963), Franzreb (1977), Lowe et al. (1978), and Szaro and Balda (1979b) describe which species of birds in Arizona and northern Mexico are favored by open plant communities that were created either by burning or selective timber harvesting. Species favored by past fires or selective timber harvesting included the sparrow hawk *(Falco sparverius)*, roadrunner *(Geococcyx californianus)*, screech owl *(Otus asio)*, red-tailed hawk *(Buteo jamaicensis)*, common nighthawk *(Chordeiles minor)*, yellow-bellied sapsucker *(Sphyrapicus varius)*, hairy woodpecker *(Dendrocopos villosus)*, Cassin's kingbird *(Tyrannus vociferans)*, curve-billed thrasher *(Toxostoma curvirostre)*, brown towhee *(Pipilo fuscus)*, rufus-sided towhee *(P. erythrophthalmus)*, Harlequin quail *(Cyrtonyx montezumae)*, purple martin *(Progne subis)*, violet-green swallow *(Tachycineta thalassina)*, robin *(Turdus migratorius)*, western bluebird *(Sialia mexicana)*, eastern bluebird *(S. sialis)*, Mexican junco *(Junco phaeonotus)*, gray-headed junco *(J. caniceps)*, chipping sparrow *(Spizella passerina)*, Coues flycatcher *(Contopus pertinax)*, olive-sided flycatcher *(Nuttallornis borealis)*, western wood pewee *(Contopus sordidulus)*, rock wren *(Salpinctes obsoletus)*, and house wren *(Troglodytes aedon)*.

Untreated areas in the same study area were preferred by the dusty flycatcher *(Empidonax oberholseri)*, ash-throated flycatcher *(Myiarchus cinerascens)*, brown creeper *(Certhia familiaris)*, pygmy nuthatch *(Sitta pygmaea)*, white-breasted nuthatch *(S. carolinensis)*, blue-gray gnatcatcher *(Polioptila caerulea)*, black-throated gray warbler *(Dendroica nigrescens)*, yellow-rumped warbler *(D. coronata)*, Virginia's warbler *(Vermivora virginiae)*, red-faced warbler *(Cardellina rubrifrons)*, Scott's oriole *(Icterus parisorum)*, mountain chickadee *(Parus gambeli)*, solitary vireo *(Vireo solitarius)*, western tanager *(Piranga ludoviciana)*, golden-crowned kinglet *(Regulus satrapa)*, ruby-crowned kinglet *(R. calendula)*, and hermit thrush *(Hylocicha guttatus)*.

In sagebrush-grass communities, McAdoo and Klebenow (1978) found that birds such as yellowthroat *(Geothlypis trichas)*, yellow-breasted chat *(Icteria virens)*, Traill's flycatcher *(Empidonax trailli)*, and yellow-billed cuckoo *(Coccyzus americanus)* needed dense tangles of shrubs for ideal habitat. Vesper sparrows *(Pooecetes gramineus)* nest on the ground and need relatively little shrub cover. Brewer sparrows *(Spizella breweri)* need some big sagebrush *(Artemisia tridentata)* plants for nesting, but not thick areas. Common bushtits *(Psaltriparus minimus)*, blue-gray gnatcatchers, and pinyon jays *(Gymnorhinus cyanocephalus)* require trees and can be eliminated by burning. One could assume that a spotty burn that left enough shrubs and trees for nesting and cover would be most desirable for the greatest diversity of bird species in sagebrush-grass and pinyon *(Pinus* spp.)-juniper *(Juniperus* spp.) communities.

In grasslands, shrub cover is limited, and fires are often destructive to the nesting habitat for songbirds (Renwald 1978). For example, in the mixed prairie of Texas, the suppression of lotebush *(Ziziphus obtusifolia)* will drastically reduce songbird nesting areas for six to seven years. The primary species affected are cardinal *(Richmondena cardinalis)*, cactus wren *(Campylorhynchus brunneicapillus)*, mockingbird *(Mimus polyglottos)*, lark sparrow *(Chondestes grammacus)*, and brown towhee. Young honey mesquite *(Prosopis glandulosa* var. *glandulosa)* is of secondary importance to cardinals, cactus wrens, and mockingbirds for nesting, but is a key species used in the Rolling Plains for nesting by Bullock's oriole *(Icterus bullockii)*, ash-throated flycatcher, and scissor-tailed flycatcher *(Muscivora forficata)*.

Vogl (1973) observed that shoreline birds increased threefold between January and June, 1971 following a shoreline burn on a north Florida pond. Twenty-nine of the 35 species observed were more common on burned than unburned areas. The following species were significantly more common on burns: common egret *(Casmerodius albus)*, common crow *(Corvus brachyrhynchos)*, common gallinule *(Gallinula chloropus)*, great blue heron *(Ardea herodias)*, little blue heron *(Florida caerulea)*, common grackle *(Quiscalus quiscula)*, yellow-rumped warbler, eastern phoebe *(Sayornis phoebe)*, robin, and white-throated sparrow *(Zonotrichia albicollis)*. Stoddard (1963) reported similar results following burns for fewer species. In east Texas Michael and Thornburgh (1971) found that the robin increased dramatically after burning in a pine-hardwood unit.

In the Everglades National Park, three tree birds, eastern kingbird *(Tyrannus tyrannus)*, common grackle, and blue-gray gnatcatcher, increased following a December burn (Emlen 1970). Burned versus unburned ratios were 21:8, 12:3, and 8:2, respectively. The white-eyed vireo *(Vireo griseus)* and house wren were not favored by burns. They were twice as abundant on unburned areas as on burns in the shrub layer. A small sample indicated that the eastern meadowlark *(Sturnella magna)* a ground forager, may prefer unburned sites. In general, there was not a difference in the total number of birds between the burned and unburned sites. Lloyd (1938), however, reported that in Canada the house wren and eastern bluebird thrive on new burns. They began building nests as soon as the embers had cooled.

In the Sierra Nevada Mountains Bock and Lynch (1970) studied breeding bird populations on a burned and unburned conifer *(Pinus* and *Abies* spp.) forest. Species characteristic of low brush and relatively open ground in the severe burn included the house wren, mountain bluebird *(Sialia currucoides)*, Lazuli bunting *(Passerina amoena)*, green-tailed towhee *(Chlorura chlorura)*, chipping sparrow, and Brewer sparrow. Red-shafted flickers *(Colaptes cafer)*, robins, Cassin finches *(Carpodacus cassinii)*, and fox sparrows *(Passerella iliaca)* were more common in the burned area. Species which foraged among needles and twigs of conifers were most common on the unburned plot. These included mountain chickadee, golden-crowned kinglet, Nashville warbler *(Vermivora ruficapilla)*, and western tanager.

In a giant sequoia *(Sequoiadendron giganteum)* forest of the Sierra Nevada Mountains, Kilgore (1971) found that burning the understory material changed the species composition but not the total biomass of avifauna. Two species, the western wood pewee and the robin, showed the most definite increase in the understory. These are "edge" birds (Kendeigh 1944). On the other hand, the rufus-sided towhee and the hermit thrush disappeared from plots after burning. Rufus-sided towhees are generally restricted to chaparral or low brush cover. The hermit thrush likes dense thickets of young pine or white fir *(Abies concolor)*. Numbers of the Nashville warbler, Oregon junco *(Junco oreganus)*, western tanager, and black-headed grosbeak *(Pheucticus melanocephalus)* showed no change in response to burning.

KIRTLAND'S WARBLER

Kirtland's warbler *(Dendroica kirtlandii)* is an endangered songbird that requires fire for survival. It breeds only in the northern part of the lower peninsula of Michigan (Mayfield 1960) in a 135- by 160-km (85- by 100-mile) area. Nesting habitat is limited to homogeneous stands of jack pine *(Pinus banksiana)*, a scrubby pine that follows fire, over 130 ha (80 acres) in size. Pine-branch thickets near the ground of jack pine provide cover for Kirtland's warbler to enter and leave its nest unobserved. Only trees that occur in dense scattered patches that are 1.5 to 5 m (5 to 15 ft) tall (10 to 15 years old) provide this branch cover near the ground. Trees shorter than 1.5 m (5 ft) are too far apart and lack the thick lower branches. Trees taller than 5 m (16 ft) lose their lower branches by shading.

Other requirements for Kirtland's warbler seem to be dry, porous, sandy soil and ground cover (Mayfield 1960). Since the bird nests below ground level, a poorly drained soil could be fatal. Ground cover must be herbaceous material. Terrestrial lichens *(Cladonia* spp.) are not satisfactory (Mayfield 1960).

Modern-day logging and tree-planting practices do not promise to produce suitable nesting habitat for Kirtland's warbler. Resultant logged stands are unevenly aged and planted areas are usually smaller than that required by the warbler (Mayfield 1963). Also, because planted trees are not clustered and interspersed with small openings, there is some doubt as to whether the lower branches of planted jack pine stands will provide adequate cover for nests.

The Michigan Department of Conservation and the U.S. Forest Service have set aside forested areas that are managed to provide the appropriate nest habitat for Kirtland's warbler, as well as to produce commercial timber stands. On one area, the Huron National Forest, 1620 ha (4010 acres) were set aside in 1962. The area was divided into 12 130-ha (320 acre) blocks. One block will be burned every five years. The oldest block will be 60 years of age when harvested. Thus, at least two blocks, the 10- and 15-year blocks, will provide optimum nesting habitat for the warbler. The Michigan Department of Conservation set aside 3110 ha (7680 acres) of state forest land in three counties that will also be managed as habitat for Kirtland's warbler.

CALIFORNIA CONDOR

Cowells (1958) hypothesized that the California condor *(Gymnogyps californianus)*, one of the largest vultures in the western hemisphere, is also nearing extinction because of the lack of fire in its habitat. Due to man-induced disturbances (Verner 1978), it presently is restricted to a small area in the coastal mountains of California in the Los Padras National Forest. Vegetation is primarily chaparral.

Burning seems to be important to the survival of this large vulture for two reasons. The first is that after feeding, this bird needs a long runway to achieve flight in still air. Without burned openings in the chaparral, the bird is forced to feed in the lowlands. Observations by Miller et al. (1965) have shown that frequently after gorging themselves, condors cannot achieve flight and will spend the night in a tree before returning to their young. If the birds can take off in the mountains where there are strong thermal breezes they probably have no difficulty in achieving flight and returning to their nestlings.

The second point that seems to be important is the nutritive adequacy of their carrion food (Cowells 1967). Normally these birds will eat dead rabbits, squirrels, and other small mammals and deer in preference to large domesticated animals (cattle, sheep, and horses) (Miller et al. 1965). Their change in diet to a higher proportion of domestic carrion might reduce the dietary calcium requirement of the California condors (Cowell 1967). Theoretically, the condors can have a more than adequate supply of calcium from small mammals because they can ingest all of the small bones, but from large animals they probably cannot eat the large bones and thus a calcium deficiency could be developed. There is no observational or experimental proof for this statement (Cowells 1967; Verner 1978).

Nevertheless, the decline of the California condor remains unexplained and may be as low as 30 individuals (Verner 1978). Breeding efforts are below normal and have been attributed to lack of food sufficiently near nesting sites to permit efficient parental foraging for growing young (Sibley 1968; Wilbur 1972). Use of occasional fires in chaparral would create a mixture of edge and grasslands which could improve the habitat for rodents and deer severalfold (Chew et al. 1959; Cook 1959; Biswell 1963). Thus the problems relative to food supply, the potential need for dietary calcium, and the need for long takeoff areas could be answered with the use of fire in areas that are relatively free of human activity.

Condors do not begin breeding until they are 8 years old. They have an average life span of 15.5 years (Verner 1978). Thus with only 30 individuals and the capability of producing only one young every other year under ideal conditions (Wagner 1971) the problem of stabilizing this population is serious and may not be possible without the natural historic role of fire, if it is possible at all.

WOODPECKERS

Generally, woodpeckers thrive in open forests that have a 10 percent old-growth component scattered throughout their home range [80 to 400 ha (200 to 1000 acres)] (Jackson et al. 1979; McClelland et al. 1979). Recurring wildfires, thinning, and prescribed burning are factors that can create favorable habitats. In Montana old-growth western larch *(Larix occidentalis)* trees greater than 50 cm (20 in.) dbh are the preferred nest trees for the pileated woodpecker *(Dryocopus pileatus)*. Recurring wildfires not only create conditions favorable for western larch regeneration (McClelland et al. 1979), but western larch is also very resistant to fire and may live to be 700 years of age. Thus, fire can easily maintain a desirable habitat for the pileated woodpecker where western larch is a component of the forest.

Following fires in Engelmann spruce *(Picea engelmanni)*-subalpine fir *(Abies lasiocarpa)* forests, woodpeckers may increase 50-fold to seek insects in fire-killed trees (Koplin 1969). Increase in density was most pronounced in the northern three-toed woodpecker *(Picoides tridactylus)* and least in the downy woodpecker *(Dendrocopos pubescens)*. The hairy woodpecker was intermediate in abundance. All three woodpeckers remained in great abundance for at least two years. Hairy woodpeckers were the least competitive and left the burns within three years. Blackford (1955) also found three species, hairy woodpecker, northern three-toed woodpecker, and black-backed three-toed woodpecker *(Picoides articus)*, very active on a cutover and burned Douglas fir *(Pseudotsuga menziesii)* forest in the Kootenai National Forest. The three-toed woodpecker was normally rare in this area.

In the Southeast, Jackson et al. (1979) found that the average age of red-cockaded woodpecker *(Dendrocopos borealis)* cavity trees was 76 years for loblolly pine *(Pinus taeda)*, 85 years for pond pine *(P. serotina)*, and 95 years for slash pine *(P. elliottii)*. The age of the cavity trees ranged from 34 to 131 years. All of these trees are favored by fire, and, therefore, prescribed burning is an important management tool to maintain red-cockaded woodpecker habitat. Preferred foods in descending order include insect larvae, cockroaches, centipedes and/or millipedes, and spiders (Harlow and Lennartz 1977). Ligon (1970) tied the source of these foods to fungus *(Fomes pini)*. These foods would indicate that red-cockaded woodpeckers forage in pines of all ages and at times make extensive use of hardwoods (Jackson et al. 1979). As a consequence, a diversity of tree species and ages may be important in assuring a diverse and stable arthropod population as food sources for the birds (Jackson et al. 1979).

In the Central Coast Range Mountains of California, the acorn woodpecker *(Melanerpes formicivorus)* feeds heavily on acorn cotyledons (MacRoberts and MacRoberts 1976). Its diet averaged 23 percent animal and 77 percent vegetable matter in one study of which 53 percent of the total diet was acorn cotyledons. Thus they depend on several species of living oak *(Quercus* spp.) trees for food, but they also need dead trees for granaries to store acorns (MacRoberts and MacRoberts 1976). The granaries are usually large isolated trees and are primarily various species of oaks, California sycamore *(Patanus racemosa)*, and willows *(Salix* spp.). However, fence posts, pines *(Pinus* spp.), coastal redwood *(Sequoia sempervirens)*, and eucalyptus *(Globulus* spp.) have also been used as granaries. It appears that occasional severe fires in this savannah-type vegetation are desirable for the acorn woodpecker to provide some dead trees, yet leave most of the oak trees living as a source of food.

Game Birds

QUAIL

Bobwhite.   Bobwhite quail *(Colinus virginianus)* has been studied extensively in the Southeast and is a "fire bird" in this area (Stoddard 1963). Quail occupy the edges of burns before they stop smoking and the birds can fill their crops in a matter of minutes instead of hours (Stoddard 1963). Dead insects and seeds are abundant. Since feeding time is the danger time, the quicker that hunger can be satisfied, the better.

Bobwhites seldom feed more than 180 m (600 ft) from cover (Jackson 1969). Thus annual burns which are intermingled with 2- to 4-year-old roughs (about 10 to 20 percent of total area) create optimum habitat for quail in the Southeast. These roughs are generally 9.1 to 12.2 m (30 to 40 ft) in diameter and have been circled with a disk harrow. The balance between roughs and burns provides fruits and insects for summer foods as well as good nesting habitat (Harshbarger and Simpson 1970) and escape cover (Stoddard 1963).

Although quail numbers can be increased by feeding, cover plantings, food planting, and manipulation of farm patterns, control burning is the least expensive method based on cost-per-bird (Frye 1961). In the Southeast burning creates a habitat with bare ground, legumes, and isolated clumps of grass. The clumps of grass provide adequate passage ways for quail movement and screening for nests, which disappear after three years (Harshbarger and Simpson 1970). Dense roughs are not used for nesting. Almost all nests are located in bluestem *(Andropogon* spp.) clumps. Shrubs, however, can be a limiting factor. Nests are found in areas where shrub cover is less than 41 percent, the average being 14 percent (Harshbarger and Simpson 1970).

Legumes, prime food plants for quail, usually make up 20 percent of the cover on fresh burns, whereas only trace amounts are found on unburned areas (Reid and Goodrum 1959). Also, panic *(Panicum* spp.) and paspalum *(Paspalum* spp.) grasses are favored by burning. These seed-producers provide good winter food for quail. The diet of quail, however, is highly diversified and over 650 different seed foods have been recorded in crops of bobwhite quail on the Tall Timbers Research Station in Tallahassee, Florida (Landers and Johnson 1976).

In nonforested communities or communities that do not have dense stands of brush the results above do not apply to quail management. In many cases burning is detrimental to quail populations because they require some cover for escape, protection from winter weather, and loafing (Renwald et al. 1978). In prairies where shrubs are sparse, the quail population will decline following fire for five to six years and then eventually return to normal (George 1977; Renwald et al. 1978). Burns within prairies, which leave 15 percent preferred shrub cover, will create ideal feeding grounds if the shrub cover is distributed properly.

Gambel's, Scaled, and Masked Bobwhite.   Only general inferences can be made about the effect of fire on Gambel's quail *(Lophortyx gambelii)* and scaled quail

(Callipepla squamata) based on habitat requirements given by Goodwin and Hungerford (1977). Gambel's quail are generally found in dense overstories of velvet mesquite *(Prosopis glandulosa* var. *velutina)*, netleaf hackberry *(Celtis reticulata)*, wolfberry *(Lycium* sp.), and catclaw *(Acacia* greggii) with an understory of broom snakeweed *(Xanthocephalum sarothrae)* and burroweed *(Aaplopappus tenuisectus)*, whereas scaled quail prefer low grasses and shrubs with 10 to 50 percent ground cover. These species are mutually exclusive. Thus, we could assume that fire would be harmful to Gambel's quail and probably beneficial to scaled quail by increasing its area of suitable habitat.

Masked bobwhite quail *(Colinus virginianus ridgewayi)*, a re-introduced species in Arizona, use brush-piles, mesquite thickets, and dense grass-shrub "pockets" for cover (Goodwin and Hungerford 1977). They prefer edge habitat where mesquite-lined washes adjoin opened, grass-forb sites (Goodwin and Hungerford 1977). In Mexico, Tomlinson (1972) found that they preferred areas with a mixture of dense forb growth, dense grasses, and brush or trees. During the winter they moved to more woody thickets. These habitat preferences may reflect the need for a habitat with high humidity (Goodwin and Hungerford 1977), which would be drastically altered by fire, and thus be harmful to the masked bobwhite.

MOURNING DOVE

Mourning doves *(Zenaidura macroura)* are attracted to fresh burns immediately (Edwards and Ellis 1969). They are "pickers" instead of "scratchers," as are bob-white quail, and evidently find an abundant supply of dead insects and seeds on fresh burns. In the months following the fire they find an abundant supply of seed from such early succession plants as miner's lettuce *(Montia perfoliata)* (Lawrence 1966) and annual sunflower *(Helianthus annuus)*.

In addition to providing good feeding areas, fresh burns also provide excellent habitat for ground nests for at least two growing seasons (Soutiere and Bolen 1972). The number of young fledged per hectare was just as high on recently sprayed and burned treatments as on sprayed, unburned treatments. However, after two years the burns are not preferred nesting or feeding areas.

Live untreated trees produced only half as many young per hectare as standing sprayed trees, primarily because high winds in west Texas blow down the flimsy nests of mourning doves. Ground nests are not destroyed by high winds. Predators exerted equal pressure on both tree and ground nests (Soutiere and Bolen 1972).

CHUKAR PARTRIDGE

Chukar partridge *(Alectoris chukar)*, an introduced game species, has found its niche in the rugged Great Basin terrain from the valley floor to 3660 m (12,000 ft) (Harper et al. 1958; Christensen 1970). It is tied to deteriorated sites (often because of re-peated fires) of sagebrush-grass vegetation where cheatgrass *(Bromus tectorum)* is abundant. Cheatgrass, which can easily be increased through land misuse, fires, or drought (Wright et al. 1979), provides most of the chukar's food. In addition to cheatgrass, water is an essential part of the chukar's habitat during the summer (Christensen 1970). Thus rivers, streams, and water developments near rough terrain are key habitat areas. The chukar's range could easily be increased with additional water development in rough topography where cheatgrass is plentiful (Christensen 1970).

GROUSE

Greater prairie chickens *(Tympanuchus cupido)*, sharp-tailed grouse *(Pedioecetes phasianellus)*, ruffed grouse *(Bonasa umbellus)*, and other species of grouse have different habitat requirements. Prairie chickens thrive in grasslands where shrub and tree cover does not exceed 25 percent (Amman 1957). As the shrub cover becomes higher in northern United States and southern Canada, sharp-tailed grouse become more abundant, and do very well until the cover reaches 40 percent (Miller 1963). Ruffed grouse prefer localities where the deciduous cover exceeds 40 percent, although they avoid dense cover (Smith 1947). Blue grouse *(Dendrogapus obscurus* and *D. fulginosos)* use a variety of cover types in Douglas fir and subalpine vegetation types, including patches of dense conifers during winter (Bendell 1974).

Prairie Chickens and Sharp-tailed Grouse. In the case of greater prairie chickens and sharp-tailed grouse, fire not only maintains proper cover conditions in tallgrass prairie, but also promotes the growth of desirable subclimax foods (Miller 1963). Curly sedge *(Carex* sp.), chokecherry *(Prunus virginiana)*, pincherry *(P. pensylvanica)*, smartweed (*Polygonum* spp.) and blackberry *(Rubus* spp.), respond well to fire. For Attwater's prairie chicken *(Tympanuchus cupido attwateri)* in the coastal prairie of Texas, Chamrad and Dodd (1972) found that potential foods such as vegetation, seed, and insects all increased after fire.

Burns every three years are desirable to maintain ideal prairie chicken habitat in tallgrass prairie of Illinois (Westemeier 1972). This keeps the grass at a height where the chickens can see the surrounding area and also keeps the area open enough for the chickens to move about easily. In south Texas, Chamrad and Dodd (1972) found most Attwater's prairie chickens in vegetation that was about 46 cm (18 in.) tall and dense to moderately dense. Fire helps to maintain this height and facilitates movement and feeding activity of young broods.

Combining of seed crops in Illinois and Minnesota, such as redtop *(Agrostis alba)* and timothy *(Phleum pratense)*, provide a "substitute prairie" by keeping the stubble height at 25 to 46 cm (10 to 18 in.), instead of 71 cm (28 in.). Thus the plants stay erect for nesting and the habitat is open enough to provide good visibility but ample concealment (Westemeier 1972). Wheel tracks of combines provide travel lanes.

Litter buildups impede movement, promote cold and wet conditions, and reduce food availability for prairie chickens (Westemeier 1972). Insects, legumes, blackberries, and seeds of various kinds are easily accessible in Minnesota on burns or harvested stands of redtop and timothy.

Sharp-tailed grouse populations in the northern Great Plains should be managed so that the habitat contains adequate quantities of fleshy hawthorn *(Crataegus succulenta)* and western snowberry *(Symphoricarpos occidentalis)* for fall food. Silver buffaloberry *(Sheperdia argentea)* is by far the most important native winter food because the berries are high in metabolizable energy whether eaten fresh-frozen or air-dried (Evans and Dietz 1974). Fire will reduce the abundance and vigor of silver buffaloberry and harm the other species moderately (Wright and Bailey 1980). Thus burned areas may not be preferred habitat for sharp-tailed grouse for several years after a fire.

Ruffed Grouse. Ruffed grouse are adapted to the early successional stages in boreal forests (Grange 1949) and to Douglas fir-spruce *(Picea* sp.) communities (Marshall 1946). In Minnesota peak ruffed grouse populations occurred two to four and ten to twelve years after fires in aspen *(Populus* spp.) communities (Gullion 1970). The second and third years after fire are ideal brood habitat and after 10 to 12 years the habitat is suitable for year-round use by adult ruffed grouse. These areas generally have a rather broken tree canopy and a wide variety of shrubs growing in the many openings (Marshall 1946). After 20 to 25 years, burned areas are no longer suitable for year-round use, but are suitable for spring use by adult birds (Gullion 1970).

In Minnesota, Gullion (1970) found that staminate flower buds of the quaking and bigtooth aspens *(Populus tremuloides* and *P. grandidentata)*, which are 25 years old or older, are the primary food of breeding grouse in the spring. In Idaho, preferred ruffed grouse foods during the winter included leaves and buds of phacelia *(Phacelia* spp.), mountain ash *(Sorbus scopulina)*, serviceberry *(Amelanchier alnifolia)*, and Douglas maple *(Acer douglasii)* (Marshall 1946). Preferred foods in both of these areas must be regenerated by fire.

In Pennsylvania, ruffed grouse were found to feed on flower buds of blueberry plants *(Vaccinium* spp.), another fire genus (Sharp 1970). During summer, ruffed grouse feed on seeds of various forbs, particularly violets *(Viola* spp.), sedge *(Carex* sp.), and panic grasses, which are most abundant following fires. Extreme fires result in temporary loss of ruffed grouse habitat, but this level of burning intensity is essential in promoting regeneration of aspen and deciduous shrubs. Hot fires in forested habitats invariably leave a mosaic of burned and unburned areas. As succession advances, the burned areas become optimum habitat for ruffed grouse.

Fire has other benefits for ruffed grouse. First, it removes the litter which accelerates recycling of minerals (Ahlgren and Ahlgren 1960) and increases the nutrient quality of plant materials used by grouse. Second, fire removes accumulated branches, tree tops, and logs. Short vertical cover with litter-free ground provides ruffed grouse with the best opportunity to keep the landscape under effective surveillance for ground predators, whereas overhead cover is an effective barrier against surprise attack by raptors. Moreover, fire scorching may speed up the physiological aging of thrifty trees and thus provide greater quantities of aspen flower buds required by ruffed grouse in the spring (Gullion 1970).

Sage Grouse.   Sage grouse *(Centrocercus urophasianus)* in Idaho, which are associated with sagebrush-grass communities, avoid arid open areas (less than 10 percent
total shrub cover) and dense shrub cover (greater than 25 percent) for nesting sites
(Klebenow 1972; Braun et al. 1977). Average shrub cover of 87 nest sites in southeastern Idaho was 18.4 percent. About 15 to 20 percent seems optimum for nesting.
Following nesting, broods in these same areas are primarily dependent on fruits and
flowers of forbs for food during spring months (Klebenow and Gray 1968; Braun et
al. 1977) and, therefore, are attracted to burns (Klebenow and Beall 1977). Hens
with broods generally feed where sagebrush canopy cover averages 10.4 percent
compared with 30 percent on loafing sites (Braun et al. 1977). Adults use sites in
summer where sagebrush canopy averages 25 percent.

Burning can open up thick stands of big sagebrush and promote the growth of
forbs (Blaisdell 1953). Thus, the judicious use of prescribed burning with a mosaic
burning pattern will create cover and food (Klebenow and Beall 1977). Moreover,
spot burning of 1 ha (2.5 acres) in strategic locations has resulted in sage grouse moving into the burned area from a less satisfactory strutting ground nearby (Schlatterer
1960). Large, hot fires or repeated burns would be undesirable, because cover would
be limited (Klebenow 1972; Braun et al. 1977).

Blue Grouse.   Fire enhances spring, summer, and fall habitats by providing openings and promoting growth of herbs and deciduous shrubs. However, coniferous needles (primarily Douglas fir and subalpine fir) constitute 90 percent of the diet of blue
grouse in the winter (Beer 1943; Stewart 1944). Therefore fires are not particularly
detrimental to blue grouse as long as patches of unburned conifers remain, particularly on ridges where the primary winter feeding takes place (Bendell 1974).

Redfield et al. (1970) found that burns did not affect breeding densities and that
hens with older broods selected burned areas. During the summer and fall, blue
grouse eat fruits and seeds (17 to 44 percent of their diet) from a wide variety of fire-
tolerant forbs and shrubs (Beer 1943; Stewart 1944) indicating that occasional fires in
a portion of their range is beneficial. The juveniles depend on various kinds of insects
for 34 percent of their diet in early summer, but assume a diet similar to the adults
after they are 6 weeks old (Beer 1943). Again, burned areas that were a few years old
could increase such foods and would probably not be detrimental.

An open Douglas fir or spruce-fir community with a large component of forbs and
shrubs is ideal for blue grouse habitat. Such areas are usually in climax condition.
Small burns (a few hectares) in strategic locations might be beneficial, but holocausts
over large areas would be detrimental.

TURKEY

The eastern wild turkey *(Meleagris gallopavo)* needs fresh burns where it can find
green shoots and a good supply of insects for its young poults. Frequently turkeys
feed on fresh burns in the Rolling Plains of Texas. Zontek (1966) also noticed increased numbers of turkeys following burns on the Marks National Refuge in

Florida. Turkeys also need a portion of their habitat as parklike mature timber with an open understory (Miller 1963). During winter months they will consume acorns and other fruits in these understories as well as use the large trees for roosts. Fires should be used to create spring nesting and feeding grounds, but turkeys also need unburned climax hardwoods for winter mast and roosts.

MARSH SPECIES

Marsh burning is an important game management tool throughout North America. For example, clean burns are essential for goose *(Anser* sp.) management (Lynch 1941; Zontek 1966). They feed on roots and rhizomes of many sedges and grasses, as well as new green shoots (Lynch 1941). Burning makes these foods more available to geese.

Wetlands are a seral stage in the progression toward climax. As free water (ponds and shallow lakes) is filled with large accumulations of organic matter and rank, impenetrable growth by climax species such as common reed *(Phragmites communis),* bulrush *(Scirpus acutus* and *S. californicus),* sawgrass *(Cladium jamaicense),* wiregrass *(Spartina patens)* and common cattail *(Typha latifolia),* they become unsuitable for waterfowl and most other marsh species. With the large accumulations of organic debris and frequent droughts, fires must have been common in the past in these communities to maintain them as desirable marsh habitat (Lynch 1941). Today, fires frequently last for weeks (Lynch 1941) and months (Vogl 1964). In addition to new food plants becoming available after burns, seeds from dominant plants such as sawgrass are more accessible (Lynch 1941).

After fires occur in marsh communities in the Southeast, they are ideal habitats for muskrats *(Ondatra zibethica).* Brackish three-cornered grass *(Scirpus olneyi),* which follows fall burns, makes up 90 percent of the diets of muskrats in Louisiana (Perkins 1968). Likewise, clean rhizomes from these same plants are the predominant diet of blue geese *(Anser caerulescens caerulescens)* (Perkins 1968). Thus in Louisiana, management for muskrats and blue geese is essentially three-cornered grass management. For optimum muskrat production, three-cornered grass marshes should be burned annually between October 10 to January 1 with a 0- to 5-cm (0- to 2-in.) water level (O'Neal 1949). These cover burns not only increase food for wildlife, but allow easier access to food and facilitate trapping of muskrats (Lynch 1941).

If muskrat populations are high, there is a danger of muskrat "eat-outs" following burning. Under eat-out conditions there are so many runs and houses that they prevent the marsh from burning, and wiregrass will eventually dominate the area (O'Neal 1949). To avoid this, burning should be done in early October before the construction of houses begins. If eat-outs do occur, several years of burning treatments will be required to return the marsh to three-cornered grass (O'Neal 1949).

In the Delta Marsh in south-central Manitoba and in the Nebraska Sandhills, spring fires create more open water and more edge for nesting, reduce the density of common reed as much as 85 percent (Schlichtemeier 1967), which creates more feeding and loafing areas, increase desirable foods such as *Chenopodium rubrum,* and ease travel for all waterfowl, especially young broods. In the Nebraska Sandhills many desirable plants for waterfowl such as duckweed (*Lemna* sp.), pondweed *(Potamogeton* sp.), and wild rice *(Zizania* sp.) become more abundant following fire (Schlichtemeier 1967) because more open water is created. All fires should occur before April 20 to avoid destroying duck nests (Ward 1968). Mallards *(Anas platyrhynchos)* and pintails *(A. acuta)* start nesting at this time.

Along the east Texas Gulf Coast, Singleton (1951) observed that burning should be delayed until fall, after waterfowl plants such as wild millet *(Echinochloa* sp.) and smartweed have seeded. Burning in Gulf Coast marshes should not be done after mid-March, as waterfowl nesting begins at that time.

Root burns and peat burns during drought years can also have desirable effects for waterfowl (Lynch 1941). Deep clear pools are frequently carved in the humus (Ward 1968). These pools produce waterfowl foods and are ideal habitat for ducks (Lynch 1941). In Louisiana the shallow pools are preferred by puddle ducks such as mallards, pintails, and teals (*Anas* spp.). Emergent species such as spikerush (*Eleocharis* sp.), wild millet, and many other food plants occur in such areas (Lynch 1941). Deep burns are used by diving ducks. These deep pools produce submerged species such as watershield (*Brasenia* sp.), wigeongrass *(Ruppia* sp.), and other aquatic waterfowl foods (Lynch 1941). This type of burning can be dangerous unless the marsh can be flooded after burning has proceeded for a certain length of time. Small ponds and pools should not be larger than 0.2 to 0.8 ha (0.5 to 2 acres) for optimum waterfowl habitat (Yancey 1964).

Spotty cover burns are also desirable. The unburned portions of the marsh provide cover for ducks, muskrats, mink *(Mustela vison)*, raccoon *(Procyon lotor)*, and otter *(Lutra canadensis)* (Lynch 1941). While a fire is burning in the Southeast, most animals seek refuge in alligator holes.

LARGE MAMMALS

Deer

In dense brush and forest communities, except for south Texas (Steuter and Wright 1980), deer *(Odocoileus* spp.) numbers increase dramatically following fire, provided 40 percent or more escape cover remains after the burn (Clary 1972; Dealy 1975; Short et al. 1977; Wood 1977). Preferred plants are generally more productive and easily accessible (Bailey and Hines 1971; Ffolliott and Thill 1977; Fitzhugh et al. 1978; Lowe et al. 1978). In northwestern forest communities fire stimulates sprouts of quaking aspen, willow, and other hardwood species that are within reach of animals (Patton 1976; Blair and Feduccia 1977; Crouch 1979) and promotes seed germination of preferred species such as wedgeleaf ceanothus and redstem ceanothus *(Ceanothus sanguineus)* (Biswell 1969; Asherin 1973). The young shoots from sprouting species are more succulent and plentiful than old plant material and thus deer populations rise (Asherin 1973; Regelin and Wallmo 1978). Grasses and forbs also increase after burning but are generally only important during spring and early summer (Wood 1977). Browse is the most important food source for deer (Blair and Feduccia 1977; Crouch 1979).

Gibbens and Schultz (1963) found that old stands of mixed California chaparral produced from 15 to 119 kg/ha/year (13 to 106 lb/acre/year). After fall and spring burns, these same communities produced from 840 to 3090 kg/ha/year (750 to 2750 lb/acre/year) (Biswell 1969). Similarly, fruit growers in Lake County, California, were able to move deer out of their pear orchards by burning the chaparral adjacent to their property (Hendricks 1968).

Burning can be overdone. Fires in chaparral should not occur more often than 15 years or the nonsprouting shrubs will gradually disappear (Biswell 1969). Likewise, in southern forests deer depend on fruits (acorns) and other foods for half of their diet and care should be taken not to destroy all of the hardwoods (Short 1975). On winter ranges in the West, big sagebrush, curlleaf mahogany *(Cercocarpus ledifolius)*, and antelope bitterbrush *(Purshia tridentata)* can be killed by fire (Blaisdell 1953; Miller 1963). Antelope bitterbrush usually returns slowly after fire depending on genotype, season of burn, soil moisture, and burning intensity (Klebenow and Beall 1977; Wright et al. 1979).

Fire creates edge to provide more feeding areas as well as adequate cover. Black-tailed deer *(Odocoileus hemionus columbianus)* need patches of roughly 16 ha (40 acres) or more of dense brush in chaparral communities for adequate cover (Taber and Dasmann 1958) over at least 40 percent of the habitat (Wood 1977). Smaller areas will be used for cover in chaparral when feeding, but they will not be adequate if deer are harassed. Feeding areas or openings in dense shrublands should be no larger than 30 to 200 m (98 to 656 ft) in diameter for maximum use by deer (McCullock 1974; Klebenow and Beall 1977). A good balance between cover and feeding in chaparral communities in California is as follows (Taber and Dasmann 1958):

North-facing slopes	⅔ cover, ⅓ feeding area
South-facing slopes	⅓ cover, ⅔ feeding area.

Black-tailed deer feed on south-facing slopes during the winter and summer, whereas they feed on north-facing slopes only in the summer. On the other hand, they spend most of their time on south-facing slopes during the season of greatest harassment.

In western forests, openings should not be larger than 16 ha (40 acres) (Dealy 1975) to 30 ha (74 acres) (Lyon and Jenson 1980) over 20 percent of the area (Clary 1972). In western Montana, Lyon and Jenson (1980) found that the optimum size opening in timbered areas for mule deer *(Odocoileus hemionus hemionus)* was 24 ha (60 acres). Moreover, use of openings by mule deer increased as depth of slash decreased and as vegetation height increased over the range of 0.3 to 1.0 m (1 to 3 ft). Use of openings leveled off after vegetation height reached 1 m (3 ft). Thus cleaning up dead debris by burning would enhance use of openings by mule deer when vegetation height exceeded 0.3 m (1 ft). Cover quality in forests adjacent to openings is important for security and will also dictate the extent to which openings are used (Lyon and Jenson 1980).

Although increases in nutrient content of shrubs following burns have been discussed by several reseachers, for example, Taber and Dasmann (1958) in California, Lay (1957) in the Southeast, and Stransky and Hall (1978) in east Texas, these increases do not last for more than one to three years (Taber and Dasmann 1958; Asherin 1973) if they are present at all (Regelin et al. 1974). Deer in southeastern forests and in California chaparral appear to benefit most from nutrient increases after burning (Lay 1957; Taber and Dasmann 1958), but the overriding factor attracting deer to burns appears to be the succulence of the new shoots (Asherin 1973) and the availability of young sprouts from a variety of shrub species (Short et al. 1975; Regelin and Wallmo 1978). In Arizona use of preferred chaparral species was closely related to increases in moisture and crude protein (Reynolds 1967).

Protein content of grasses increases only slightly after burning and is short-lived. In the Southeast protein content of grasses following fire increased to 13 percent compared to 10 percent on the control (Hilmon and Hughes 1965b). Three months after the fire, protein content on the burn was 5 percent and had no advantage over unburned range (Hilmon and Hughes 1965a).

The increased abundance of succulent shoots of shrubs following burns will allow ovulation rates and weights of deer to increase. The ovulation rate on adult blacktailed does was 175 percent in opened brush and only 82 percent in untreated brush (Biswell 1963). The number of deer was 38 deer/km^2 (98 deer/mi^2) the first year after treatment of chaparral. This number rose to 50 (131) the second year, and dropped to 32 (84) by the sixth year. A nearby control area with untreated brush had a summer density of only 11 deer/km^2 (30 deer/mi^2). Other studies by Dasmann and Dasmann (1963) and Lowe et al. (1978) have shown that deer will increase 10- to 20-fold when forests are set back by fire or logging. Moreover, these areas remain as good deer range for at least 20 years (Regelin and Wallmo 1978).

The invasion of shrubs in grasslands of the Southwest has permitted mule deer and white-tailed deer (Odocioileus virginianus) to expand their range (Short 1977; Quinton et al. 1979). In these areas fire suppression has permitted deer to increase (Julander 1962; Rodgers et al. 1978). In south Texas, where native shrubs and predators are abundant, white-tailed deer thrive best where cover is at least 63 percent (Steuter and Wright 1980). Thus, fires in grasslands and south Texas shrublands may reduce white-tailed deer population about 50 percent (Steuter and Wright 1980).

White-tailed and mule deer have also spread northward in Canada following fires (Lloyd 1938). Aspen (Graham et al. 1963) and jack pine (Horton 1964) that follow fire are important browse species for deer, as are other associated plants in the early stages of succession (Dahlberg and Guettinger 1956; Miller 1963; Dealy 1975; Regelin and Wallmo 1978).

Some diseases of deer are often eliminated by fire. Following the Tillamook burn of 1933 in Oregon, deer were healthy and singularly free from liver fluke and lungworms that had plagued and debilitated the coast deer herds for years (Issac 1963).

The largest black-tail deer ever taken from the coast was killed on this burn. The fire killed the dry land snail which is an intermediate host for liver fluke and certain lungworms. After 30 years the deer population was again on the decline and probably both diseases and less browse were significant factors.

Elk

Rocky Mountain elk *(Cervus elaphus)* were scarce in climax forests in north-central Idaho when Lewis and Clark made their expedition in 1805–1806 (Hosmer 1917). Following the large 1910 fires in the upper Selway, elk herds started expanding (Young and Robinette 1939). Elk numbers continued to expand until the mid-1930s (McCullock 1955) and then started decreasing (Leege 1968). In the lower Selway, a similar pattern in elk numbers followed a 1934 fire (Beeman 1957), but recently they have been declining or holding their own (Leege 1968). Clear-cut logging operations have aided elk habitat; and where food is available, they will out-compete mule deer (Leege 1968).

Resprouts from Scouler willow *(Salix scouleriana)*, Sierra maple *(Acer glabrum)*, and serviceberry are the preferred foods of elk in northern Idaho (Leege 1968; Leege 1979). In the central and southern Rocky Mountains, aspen sprouts following fire are a key forage species (Patton and Avant 1970; Basile 1979; DeByle 1979) and can be maintained with a rotation burning program of 20 to 30 years (Patton and Jones 1977). Burning increases the availability of these browse foods by lowering the height of browse and increasing their succulence (Leege 1968; Asherin 1973). Redstem ceanothus and bittercherry *(Prunus emarginata)* seedlings were also abundant after fire, particularly redstem ceanothus which is very palatable (Leege 1968; Lyon 1969). Spring burns over the near-term were best for stimulating sprouting species, but fall burns were best for stimulating redstem ceanothus seeds to germinate (Leege 1968; Wright et al. 1979). Overall, fall burns appear to have the best long-term effect for shrub production; spring burns are often spotty and too cool to completely top-kill sprouting species.

In the Northwest elk are associated with a timbered habitat (Lyon 1975; Patton 1976), but need openings for feeding, preferably less than 40 ha (100 acres) and where the slash is less than 0.25 m (1 ft) deep (Lyon and Jenson 1980). However, elk do not use openings readily until the vegetation develops to a height of at least 1 m (3 ft) (Lyon 1977; Lyon and Jenson 1980). Maximum use occurs seven years after fire (Lowe et al. 1978). Elk prefer openings where they can move freely and are not required to move very far from forest cover (Lyon and Jenson 1980). About 20 years after a burn, elk use on burns is similar to that on the controls (Lowe et al. 1978).

Pronghorn Antelope

Relatively little is known about the effect of fire on pronghorn antelope *(Antilocapra americana)* although Klebenow and Beall (1977) have seen them concentrated on burns in areas where they had not been seen previously for many years. Historically, they reached their greatest numbers in short grassland habitats (Fautin 1946) and were codominant with bison *(Bison bison)* on the Great Plains and with elk on the Palouse and California Prairies (Evans and Probasco 1977). Presumably, they were compatible with these grazers because they prefer forbs (McAdoo and Klebenow 1978). In Alberta, Mitchell and Smoliak (1971) found that the year-long diets of pronghorns were 51 percent forbs, 35 percent browse, and 14 percent grasses and sedges. Most of the browse is grazed during late fall and winter. Thus, any treatment that would promote forbs, without destroying more than 30 to 50 percent of the shrubs, would enhance pronghorn antelope. In sagebrush-grass communities, fire will enhance forbs (Blaisdell 1963), and antelope were once the common ungulate in these communities (Fautin 1946).

Caribou and Moose

Barren-ground caribou *(Rangifer tarandus groenlandicus)* is a climax species of the boreal forest in northern Canada and is severely harmed by forest fires (Banfield 1951; Leopold and Darling 1953; Edwards 1954; Scotter 1964). Their principal high value winter food, terrestrial lichens, is easily destroyed by fire and takes from 70 to 100 years to recover (Scotter 1970), although Johnson and Rowe (1975) found that caribou eat all species of higher plants and that terrestrial lichens are not essential. Terrestrial lichens can be a high proportion of their diet, but selection of them may depend on abundance (Johnson and Rowe 1975). On the other hand, one cannot deny that caribou numbers have declined 60 percent and this appears to be related to man and fire (Banfield 1951).

When snow covers the ground with 1 m (3 ft) or more of snow in Canada, arboreal lichens *(Alectoria, Evernia,* and *Usnea)* are important winter forages for caribou. They are easily destroyed by fire. Moreover, they are slow to recover. In climax stands of black spruce *(Picea mariana)*, standing crops of terrestrial lichens and arboreal lichens may be as much as 785 kg/ha (700 lb/acre) and 680 kg/ha (605 lb/acre), respectively (Scotter 1970).

In British Columbia mountain caribou *(Rangifer tarandus caribou)* usually feed on arboreal lichens in mature lowland forests during winter months (Edwards 1954). Fires during the past 50 years have destroyed the winter food supply of arboreal lichens and thus the decline of mountain caribou during the 1930s and 1940s is no mystery (Banfield 1951; Edwards 1954).

Following fire in the Taiga, food supplies favor moose *(Alces alces)* over caribou (Chatlin 1951; Leopold and Darling 1953; Scotter 1970). The preferred moose forages—willow, paper birch *(Betula papyrifera)*, bog birch *(B. glandulosa)*, California hazel *(Corylus californica)*, and quaking aspen—abound after fire (Spencer and Hakala 1964). The plant species that thrive with fire are plentiful in three to four years (Buckley 1958) and remain in abundance for 25 to 50 years. Then they give way to lichens (Scotter 1970). Mature birch-spruce forests and spruce swamps have a winter population of 0.2 moose/km^2 (0.5 moose/mi^2), whereas burned winter ranges with willow and other browse have 2.0 to 22.0 moose/km^2 (5.2 to 57.2 moose/mi^2) (Chatlin 1951).

In the southern latitudes moose populations do not benefit significantly from vegetation following large wildfires for 60 years (Gruell 1980b). Increases appear to result from improved availability of winter forage, especially subalpine fir which increases with reduced fire periodicity (Cowan et al. 1950). Subalpine fir is a staple winter food and may be the only winter forage when snow depths reach 0.9 to 1.8 m (3 to 6 ft). When snow is this deep, most deciduous shrubs are unavailable. Thus shrub succession is slower in the southern latitudes than in the northern latitudes and moose enter burns in the United States many decades after a burn when climax species such as subalpine fir become a major component of the community. Nevertheless, rejuvenation of these winter ranges for moose can only be accomplished by fire or logging (Gruell 1980b).

Often the argument is raised in Canada as to whether the Taiga should be managed for caribou or moose. The potential meat yields are not necessarily the same (Scotter 1970). Barren-ground caribou spend approximately half of the year in the tundra, which might otherwise be unused. The other half of the year is spent in the Taiga, eating lichens which are a high-energy, low-protein forage that only caribou are adapted to using. Moose are more sedentary and generally remain in a localized area, but habitats such as birch-spruce and bog birch are ideally suited for them and should probably be maintained for moose with fire.

Other Mammals

GENERAL

Numbers of other large mammals change dramatically following fires in dense forests (Edwards 1954). Edwards found that cougars *(Felis concolor)* and coyotes *(Canis latrans)* flourished after fire in a red cedar *(Thuja plicata)*-western hemlock *(Tsuga heterophylla)* forest. Since they are predators of big game animals, which also increase after fire, their increase could be anticipated. Black bear *(Ursus americanus)* found huckleberries *(Vaccinium* spp.) and other foods plentiful (Edwards 1954; Hatler 1972). Mountain goats *(Oreamnos americanus)* ranged mostly above the fires and were affected little. Fire restricted the lowland wandering of wolverine *(Gulo luscus)* and grizzly bear *(Ursus horribilis)*. Bighorn sheep *(Ovis canadensis)* can maintain themselves successfully in either seral or climax communities, although they are associated wih fire communities in some parts of the Canadian Rocky Mountains (Cowan 1951; Stelfox 1971).

FISHER AND MARTEN

Marten eat voles *(Cleithrionomys* spp. and *Microtus* spp.) more than any other food item (Lensink et al. 1955; Quick 1955; Murie 1961; Weckwerth and Hawley 1962; Koehler and Hornocker 1977). Lensink et al. (1955) found that these small mammals comprised 74 percent of the diet in summer and 68 percent in winter. Because this food disappears the first year after a fire (Cook 1959), and does not reach maximum numbers until the seventh year after a burn (Buech et al. 1977), marten populations would be reduced for several years on large burns. Over the long-term, however, fire has been an important agent in establishing and maintaining a diversity of forest communities for marten (Koehler and Hornocker 1977). These carnivores winter in subalpine fir stands where the cover is greater than 30 percent, but feed extensively during the summer in openings created by fire. Fruits, insects, and ground squirrels provide food for marten during the summer (Koehler and Hornocker 1977).

Fisher *(Martes pennanti)* cover more territory than marten and appear to be better adapted to early successional stages of the forest than marten, although both species are essentially absent on recently logged and burned-over areas (de Vos 1952). The most abundant prey species in 32 stomachs and intestinal tracts of fishers in Ontario were snowshoe hare (31.5 percent), porcupine *(Erethizon dorsatum)* (20 percent), fish (4.5 percent), and mice (4.5 percent) (de Vos 1952). He did not find any red squirrels in fisher stomachs or intestines. The highest component of the fisher diet, snowshoe hares, is most abundant in the early stages of forest succession.

BEAVER

Modern forest fire control and intensive forest management practices are gradually reducing the areas of suitable beaver *(Castor canadensis)* habitat (Patric and Webb 1953). High beaver populations of many areas are the direct results of the extensive clear-cutting of timber and the widespread forest fires which were characteristic in the northern forests until recent years. Quaking aspen, willow, alder *(Alnus* sp.), and redosier dogwood *(Cornus stolonifera)* are prime beaver food trees. All resprout vigorously after fire and are of most value to beaver in these early growth stages (Patric and Webb 1953). As succession progresses, these trees become too large or are replaced by climax trees (Patric and Webb 1953).

ANTS X ACACIA

In Central America there is an interesting interaction between swollen-thorn acacia *(Acacia cornigera, A. sphaerocephala, A. collinsii, A. hindsii)* and obligate acacia-ant colonies *(Pseudomyrmex ferruginea, P. nigrocincta, P. spinicola)* (Janzen 1967). The ants live in enlarged stipular thorns of the acacia plants. Colonies feed on nectar from foliar nectaries of the plant and they eat the nutrient-rich Beltian bodies from the leaflet tips of the same leaves that bear the enlarged stipules (Janzen 1966). In turn, workers patrol the plant and attack nearly all insects and foreign matter that comes in contact with the acacia. They also clear herbaceous vegetation from around the base of the acacia, which protects it as well as themselves from the common, often annual, fires. If the plant and ant colony are killed by fire, the regenerating suckers are subject to reduced growth by phytophagous insects (Janzen 1967). Growth is also reduced by shading from dense stands of grass. Through reburns and competition, those plants without ant colonies usually meet their demise. Thus, the interaction between swollen-thorn acacia and ants is essential for both species to protect themselves from fire.

STREAM FAUNA

The effects of fire on stream fauna have to be evaluated as short-term and long-term effects. Short-term effects are often detrimental through loss of stream-side vegetation and increased sediment load to the stream(Burns 1970; Lyon et al. 1978). Removal of stream-side vegetation increases stream bank erosion, reduces available habitat and raises stream temperatures. Higher temperatures increase the faunal demand for oxygen, which decreases the dissolved oxygen content (Lyon et al. 1978). Moreover, increased incidence of fish disease is associated with increased stream temperatures (Fish and Rucker 1945). Increased sediment loads reduce the size of spawning gravels or deposit fine material that smothers eggs and prevents emergence of fry (Cordone and Kelly 1961; Phillips 1961; Cooper 1965).

Looking at fire's effect over time, however, it is evident that fire benefits stream fauna in many forests. The thinning and removal of conifers along streams by fire and stimulation of deciduous vegetation promotes cover, provides shading, and allows development of terrestrial insects important in the diet of fish. Of course, beaver are dependent upon deciduous vegetation, and their presence enhances fish habitat. Removal of conifers also reduces evapotranspiration. This can result in improved stream flows in drier environments during late summer when sustained flows are critical to survival of fish.

Concentrations of nutrients in a stream following a fire are seldom toxic, and the effects on productivity are usually beneficial (Fredriksen et al. 1975). Increased algae production at the bottom of the food chain appears to sustain a greater biomass and a more diverse population of insect larvae (Fredriksen 1978).

In general, short-term effects of fire are usually detrimental to stream fauna, but the long-term effects may be beneficial, especially in coniferous forests. With the judicious use of fire, removal of some conifers along streams will promote deciduous vegetation, whereas leaving areas unburned that contain only deciduous vegetation might be desirable. If possible, it would be best to minimize the proportion of steep slopes (above 30 percent) within a watershed that were burned to minimize sediment that may enter streams.

REFERENCES

Ahlgren, I. F., and C. E. Ahlgren. 1960. Ecological effects of forest fires. *Bot. Rev.* **26**:483–533.

Amman, G. A. 1957. The prairie grouse of Michigan. Mich. Dept. of Conserv. Tech. Bull.

Asherin, D. A. 1973. Prescribed burning effects on nutrition, production, and big game use of key northern Idaho browse species. M.S. Thesis. Univ. Idaho, Moscow.

Bailey, A. W., and W. W. Hines. 1971. A vegetation-soil survey of a wildlife-forestry research area and its application to management in northwestern Oregon. Game Rep. No. 2 Res. Div., Oregon State Game Comm., Corvallis.

Banfield, A. W. F. 1951. The Canadian Barren-Ground Caribou investigations. Can. Wildl. Serv. Rep.

Basile, J. V. 1979. Elk-aspen relationships on a prescribed burn. USDA For. Serv. Res. Note INT-271 Intermt. For. and Range Exp. Stn., Ogden, Utah.

Beck, A. M., and R. J. Vogl. 1972. The effects of spring burning on rodent populations in a brush prairie savannah. *J. Mammal.* **53**:336–346.

Beeman, R. D. 1957. Salt in the management of elk and other wildlife in the lower Selway River area. M. S. Thesis. Univ. Idaho, Moscow.

Beer J. 1943. Food habits of the blue grouse. *J. Wildl. Manage.* **7**:32–44.

Bendell, J. F. 1974. Effects of fire on birds and mammals. In T. T. Kozlowski and C. E. Ahlgren (eds.) *Fire and Ecosystems.* Academic Press, New York, pp. 73–138.

Biswell, H. H. 1963. Research in wildland fire ecology in California. *Proc. Tall Timbers Fire Ecol. Conf.* **2**:62–97.

Biswell, H. H. 1969. Prescribed burning for wildlife in California brushlands. *N. Amer. Wildl. and Natur. Res. Conf. Trans.* **34**:438–446.

Blackford, J. 1955. Woodpecker concentration in a burned forest. *Condor* **57**:28–30.

Blair, R. M., and D. P. Feduccia. 1977. Midstory hardwoods inhibit deer forages in loblolly pine plantations. *J. Wildl. Manage.* **41**:667–683.

Blaisdell, J. P. 1953. Ecological effects of planned burning of sagebrush-grass range on the Upper Snake River Plains. USDA Tech. Bull. No. 1075. Washington, D.C.

Bock, C. E., and J. F. Lynch. 1970. Breeding bird populations of burned and unburned conifer forest in the Sierra Nevada. *Condor* **72**:182–189.

Braun, C. E., T. Britt, and R. O. Wallestad. 1977. Guidelines for maintenace of sage grouse habitats. *Wildl. Soc. Bull.* **5**(3):99–106.

Buckley, J. L. 1958. Effects of fire on Alaskan wildlife. In Proc. Soc. Amer. For., Mills Bldg., Washington, D. C., pp. 123–126.

Buech, R. R., K. Siderits, R. E. Radtke, H. L. Sheldon, and D. Elsing. 1977. Small mammal populations after a wildfire in northeast Minnesota. USDA For. Serv. Res. Paper NC-151. North Central For. Exp. Stn., St. Paul, Minn.

Burns, J. E. 1970. Importance of streamside vegetation to trout and salmon in British Columbia. Dept. Rec. and Conserv. Vancouver Island Reg., Fish and Wildl. Br., Fish Tech. Circ. 1.

Chamrad, A. D., and J. D. Dodd. 1972. Prescribed burning and grazing for prairie chicken habitat manipulation in the Texas coastal prairie. *Proc. Tall Timbers Fire Ecol. Conf.* **12**:257–276.

Chatlin, E. F. 1951. Winter range problems of moose in the Susitna Valley. In Proc. Soc. Amer. For., Mills Bldg., Washington, D. C., pp. 343–347.

Chew, R. M., B. B. Butterworth, and R. Grechman. 1959. The effects of fire on the small mammal populations of chaparral. *J. Mammal.* **40**:253.

Christensen, G. C. 1970. The chukar partridge. Biol. Bull. No. 4.

Clary, W. P. 1972. A treatment prescription for improving big game habitat in ponderosa pine forests. *Proc. Ariz. Watershed Symp.* **16**:25–28. Phoenix.

Cook, S. F., Jr. 1959. The effects of fire on a population of small rodents. *Ecology* **40**:102–108.

Cooper, A. C. 1965. The effects of transported stream sediments on the survival of sockeye and pink salmon eggs and aelvins. Bull. Inter. Pac. Salmon Fish Comm. 18.

Cordone, A. J., and D. E. Kelley. 1961. The influence of inorganic sediment on the aquatic life of streams. *Calif. Fish and Game* **47** (2):189–228. Sacramento.

Costa, R., P. F. Ffolliott, and D. R. Patton. 1976. Cottontail responses to forest management in southwestern pine. USDA For. Serv. Res. Note RM-330. Rocky Mtn. For. and Range Exp. Stn., Fort Collins, Colo.

Cowan, I. M. 1951. Plant succession and wildlife management. In Proc. Soc. Amer. For., Mills Bldg., Washington, D. C., pp. 322–327.

Cowan, I. M., W. S. Hoar, and J. Hatter. 1950. The effect of forest succession upon the quantity and upon the nutritive values of woody plants used as food by moose. *Can. J. Res.* **28**:250–271.

Cowles, R. B. 1958. Starving the Condors? *Calif. Fish and Game* **44:**175–181. Sacramento.

Cowles, R. B. 1967. Fire suppression, faunal changes and Condor diets. *Proc. Tall Timbers Fire Ecol. Conf.* **7:**217–224.

Crouch, G. L. 1979. Food habits of black-tailed deer on forested habitats in the Pacific Northwest. In O. C. Wallmo and J. W. Schoen (eds.) Sitka black-tailed deer. Proc. of a Conf. in Juneau, Alaska. USDA For. Serv., Alaska Reg., in coop. with State of Alaska, Dept. Fish and Game.

Crowner, A. W., and G. W. Barrett. 1979. Effects of fire on the small mammal component of an experimental grassland community. *J. Mammal.* **60:**803–813.

Dahlberg, B. C., and R. C. Guettinger. 1956. The white-tailed deer in Wisconsin. Wisc. Dept. Conserv. Tech. Wildl. Bull. No. 14.

Dasmann, W. P., and R. F. Dasmann. 1963. Abundance and scarcity in California deer. *Calif. Game and Fish* **49:** 4–15. Sacramento.

Dealy, J. E. 1975. Management of lodgepole pine ecosystems for range and wildlife. In D. M. Baumgartner (ed.) Proc. Manage. of Lodgepole Pine Ecosystems Symp. Washington State Univ. Coop. Ext. Serv., Pullman, Wash., pp. 556–568.

DeByle, N. V. 1979. Potential effects of stable vs. fluctuating elk populations in the aspen ecosystem. In M. S. Boyce and L. D. Hayden-Wing (eds.) N. Amer. elk, ecology, behavior and manage. Univ. Wyoming, Laramie, pp. 13–19.

de Vos, A. 1952. The ecology and management of fisher and marten in Ontario. Ontario Dept. Lands For. Tech. Bull.

Edwards, R. Y. 1954. Fire and the decline of a mountain caribou herd. *J. Wildl. Manage.* **18:**521–526.

Edwards, W. R., and J. A. Ellis. 1969. Responses of three avian species to burning. *Wilson Bull.* **81:**338–339.

Emlen, J. T. 1970. Habitat selection by birds following a forest fire. *Ecology* **51:**343–345.

Evans, K. E. 1978. Oak-pine and oak-hickory forest bird communities and management options. In Proc. of the Workshop Management of Southern Forests for Nongame Birds, Atlanta, Ga. USDA For. Serv. Gen. Tech. Rep. SE-14. Southeastern For. Exp. Stn., Asheville, N. C., pp. 76–89.

Evans, K. E., and D. R. Dietz. 1974. Nutritional energetics of sharp-tailed grouse during winter. *J. Wildl. Manage.* **38:**622–629.

Evans, K. E., and G. E. Probasco. 1977. Wildlife of the prairies and plains. USDA For. Serv. Gen. Tech. Rep. NC-29. North Central For. Exp. Stn., St. Paul, Minn.

Fautin, R. W. 1946. Biotic communities of the northern desert shrub biome in western Utah. *Ecol. Monogr.* **16:**251–310.

Ffolliott, P. F., and R. E. Thill. 1977. Animal use of ponderosa pine openings. *J. Wildl. Manage.* **41:**782–784.

Fish, F. F., and R. Rucker. 1945. Columnaris as a disease of cold-water fishes. *Trans. Am. Fish Soc.* **73:**32–36.

Fitzhugh, E. L., J. E. Cornely, and J. T. Pealiew. 1978. The Rattle Burn: Disaster or blessing for wildlife? In Trans. New Mexico-Arizona Sec. Wildl. Soc., pp. 199–218.

Franzreb, K. E. 1977. Bird populations changes after timber harvesting of a mixed conifer forest in Arizona. USDA For. Serv. Res. Paper RM-184. Rocky Mtn. For and Range Exp. Stn., Fort Collins, Colo.

Fredriksen, R. L. 1978. Impacts of fire on stream habitat. Office Rep. Pac. Northwest For. and Range Exp. Stn., Corvallis, Oregon. (Mimeo.)

Fredriksen, R. L., C. G. Moore, and L. A. Norris. 1975. The impact of timber harvest, fertilization, and herbicide treatment on streamwater quality in western Oregon and Washington. In N. Amer. For. Soils Conf., Les Presses De L'Universite Laval, Quebec., pp. 283–313.

Frye, O. E., Jr. 1961. A review of bobwhite quail management in eastern North America. *N. Amer. Wildl. and Natur. Res. Conf. Trans.* **26:**273–281.

Galli, A. E., C. E. Leck, and R. T. Forman. 1976. Avian distribution patterns in forest islands of different sizes in central New Jersey. *Auk* **93**:356–364.

Gashwiler, J. S. 1970. Plant and mammal changes on a clearcut in west-central Oregon. *Ecology* **51**:1018–1026.

George, R. R. 1977. Use of controlled burning to maintain and restore bobwhite quail and cottontail rabbit habitat. Iowa State Conserv. Comm., Des Moines. Iowa-W-115R-3.

Gibbens, R. P., and A. M. Schultz. 1963. Brush manipulation on a deer winter range. *Calif. Fish and Game* **49**:95–118. Sacramento.

Goodwin, J. G., Jr., and C. R. Hungerford. 1977. Habitat use by native Gambel's and scaled quail and released masked bobwhite quail in southern Arizona. USDA For. Serv. Res. Paper RM-197. Rocky Mtn. For. and Range Exp. Stn., Fort Collins, Colo.

Graham, S. A., R. P. Harrison, Jr., and C. E. Westell, Jr. 1963. *Aspens: Phoenix Trees of the Great Lakes Region.* Univ. Mich. Press, Ann Arbor.

Grange, W. C. 1949. *The Way to Game Abundance.* Scribner's, New York.

Gruell, G. E. 1980a. Fire's influence on wildlife habitat on the Bridger-Teton National Forest, Wyoming. Vol. 1, Photographic record and analysis. USDA For. Serv. Res. Paper INT-235. Intermt. For. and Range Exp. Stn., Ogden, Utah.

Gruell, G. E. 1980b. Fire's influence on wildlife habitat on the Bridger-Teton National Forest, Wyoming. Vol. 2, Changes and causes, management implications. USDA For. Serv. Res. Paper INT-252. Intermt. For. and Range Exp. Stn., Ogden, Utah.

Gullion, G. W. 1970. Factors influencing ruffed grouse populations. *N. Amer. Wildl. and Natur. Res. Conf. Trans.* **35**:93–105.

Hamilton, W. J., Jr. 1941. The food of small forest mammals in eastern United States. *J. Mammal.* **22**:250–263.

Harlow, R. F., and M. R. Lennartz. 1977. Foods of nestling red-cockaded woodpeckers in coastal South Carolina. *Auk.* **94**:376, 377.

Harper, H. T., H. Beverly, H. Bailey, and W. D. Bailey. 1958. The chukar partridge in California. *Calif. Fish and Game* **44**(1):5–50. Sacramento.

Harshbarger, T. J., and R. C. Simpson. 1970. Late-summer nesting sites of quail in South Carolina. USDA For. Serv. Res. Note SE-131. Southeastern For. Exp. Stn., Asheville, N. C.

Hatler, D. F. 1972. Black bears in Interior Alaska. *The Can. Field-Natur.* **86**:17–31.

Hendricks, J. H. 1968. Control burning for deer management in chaparral in California. *Proc. Tall Timbers Fire Ecol. Conf.* **8**:218–233.

Hilmon, J. B., and R. H. Hughes. 1965a. Fire and forage in the wiregrass type. *J. Range Manage.* **18**:251–254.

Hilmon, J. B., and R. H. Hughes. 1965b. Forest Service research in the use of fire in livestock management in the south. *Proc. Tall Timbers Fire Ecol. Conf.* **4**:261–275.

Horton, K. W. 1964. Deer prefer jack pine. *J. For.* **62**:497–499.

Hosmer, J. K. 1917. *History of the Expedition of Captains Lewis and Clark.* Vols. 1 and 2. A. C. McClurg, Chicago.

Howard, W. E., R. L. Fenner, and H. E. Childs, Jr. 1959. Wildlife survival in brush burns. *J. Range Manage.* **12**:230–234.

Issac, L. A. 1963. Fire—A tool not a blanket rule in Douglas-fir ecology. *Proc. Tall Timbers Fire Ecol. Conf.* **2**:1–17.

Jackson, A. S. 1969. Quail management handbook. *Texas Parks and Wildl. Bull.* **48**.

Jackson, J. A., M. R. Lennartz, and R. G. Hooper. 1979. Tree age and cavity initiation by red-cockaded woodpeckers. *J. For.* **77**:102, 103.

James, E. W. Jr. 1952. Food of deer mice, *Peromyscus maniculatus* and *Peromycus boylei,* in the northern Sierra Nevada, California. *J. Mammal.* **33**:50–60.

Janzen, D. H. 1966. Coevolution of mutalism between ants and acacia in Central America. *Evolution* **20**:249–275.

Janzen, D. H. 1967. Fire, vegetation structure, and the ant X acacia interaction in Central America. *Ecology* **48**:26–35.

Johnson, E. A., and J. S. Rowe. 1975. Fire in the subarctic wintering ground of the Beverley Caribou Herd. *Amer. Midl. Natur.* **94**:1–14.

Julander, O. 1962. Range management in relation to mule deer habitat and herd productivity in Utah. *J. Range Manage.* **15**:278–281.

Keith, L. B., and D. C. Surrendi. 1971. Effects of fire on a snowshoe hare population. *J. Wildl. Manage.* **35**:16–26.

Kendeigh, S. C. 1944. Measurement of bird populations. *Ecol. Monogr.* **14**:67–106.

Kilgore, B. M. 1971. Response of breeding bird populations to habitat changes in a giant sequoia forest. *Amer. Midl. Natur.* **85**:135–152.

Klebenow, D. A. 1972. The habitat requirements of sage grouse and the role of fire in managment. *Proc. Tall Timbers Fire Ecol. Conf.* **12**:305–315.

Klebenow, D. A., and R. C. Beall. 1977. Fire impacts on birds and mammals on Great Basin Rangelands. In Proc. of the 1977 Rangeland Manage. and Fire Symp., Casper, Wyo., pp. 59–62.

Klebenow, D. A., and G. M. Gray. 1968. Food habits of juvenile sage grouse. *J. Range Manage.* **21**:80–83.

Koehler, G. M., and M. G. Hornocker. 1977. Fire effects on marten habitat in the Selway-Bitterroot Wilderness. *J. Wildl. Manage.* **41**:500–505.

Koplin, J. R. 1969. The numerical response of woodpeckers to insect prey in a subalpine forest in Colorado. *Condor* **71**:436–438.

Landers, J. L., and A. S. Johnson. 1976. Bobwhite food habits in the southeastern United States with a seed key to important foods. Misc. Pub. No. 4. Tall Timbers Res. Stn., Tallahassee, Fla.

Lawrence, G. E. 1966. Ecology of vertebrate animals in relation to chaparral fire in the Sierra Nevada foothills. *Ecology* **47**:278–291.

Lay, D. W. 1957. Browse quality and the effects of prescribed burning in southern pine forests. *J. For.* **55**:342–347.

Leege, T. A. 1968. Prescribed burning for elk in northern Idaho. *Proc. Tall Timbers Fire Ecol. Conf.* **8**:235–253.

Leege, T. A. 1979. Effects of repeated prescribed burns on northern Idaho elk browse. *Northwest Sci.* **53**:107–113.

Lensink, C. J., R. O. Skoog, and J. L. Buckley. 1955. Food habits of martens in Interior Alaska and their significance. *J. Wildl. Manage.* **19**:364–368.

Leopold, A. S. 1932. *Game Management.* Scribner's, New York.

Leopold, A. S., S. A. Cain, C. M. Cottam, I. N. Gabrielson, and T. L. Kimball. 1963. Wildlife management in the National Parks. *N. Amer. Wildl. Conf.* **28**:29–45.

Leopold, A. S., and F. E. Darling. 1953. Effects of land use on moose and caribou in Alaska. *Trans. N. Amer. Wildl. Conf.* **18**: 553–560.

Ligon, J. D. 1970. Behavior and breeding biology of the red-cockaded woodpecker. *Auk* **87**:255–278.

Lloyd, H. 1938. Forest fire and wildlife. *J. For.* **36**:1051–1054.

Lowe, P. O. 1975. Potential wildlife benefits of fire in ponderosa pine forests. M.S. Thesis. Univ. Ariz., Tucson.

Lowe, P. O., P. F. Ffolliott, J. H. Dieterick, and D. R. Patton. 1978. Determining potential wildlife benefits from wildfire in Arizona ponderosa pine forests. USDA For. Serv. Gen. Tech. Rep. RM-52. Rocky Mtn. For. and Range Exp. Stn., Fort Collins, Colo.

Lynch, J. J. 1941. The place of burning in management of the Gulf Coast refuges. *J. Wildl. Manage.* **5**:454–458.

Lyon, L. J. 1969. Wildlife habitat research and fire in the Northern Rockies. *Proc. Tall Timbers Fire Ecol. Conf.* **9**:213–227.

Lyon, L. J. 1975. Coordinating forestry and elk management in Montana: Initial recommendations. *Trans. N. Amer. Wildl. and Nat. Res. Conf.* **40**:193–201.

Lyon, L. J. 1977. Elk use as related to characteristics of clearcuts in Western Montana. In Proc. Elk-Logging-Roads Symp., Univ. Idaho, Moscow, pp. 69–72.

Lyon, L. J., H. S. Crawford, E. Czuhai, R. L. Fredriksen, R. F. Harlow, L. J. Metz, and H. A. Pearson. 1978. Effects of fire on fauna: A state-of-knowledge review. USDA For. Serv. Gen. Tech. Rep. WO-6. Washington, D. C.

Lyon, L. J., and C. E. Jenson. 1980. Management implications of elk and deer use of clearcuts in Montana. *J. Wildl. Manage.* **44**:352–362.

MacRoberts, M. H., and B. R. MacRoberts. 1976. Social organization and behavior of the acorn woodpecker in central coastal California. *Ornith. Monogr.* **21**:1–115.

Marshall, J. T., Jr. 1963. Fire and birds in the mountains of southern Arizona. *Proc. Tall Timbers Fire Ecol. Conf.* **2**:135–141.

Marshall, W. H. 1946. Cover preferences, seasonal movements, and food habits of Richardson's grouse and ruffed grouse in southern Idaho. *Wilson Bull.* **58**: 42–52.

Mayfield, H. F. 1960. The nesting ground. In The Kirtland's Warbler. Cransbrook Inst. of Sci., Bull. No. 40, pp. 9–33.

Mayfield, H. F. 1963. Establishment of preserves for the Kirtland's warbler in the state and national forest of Michigan. *Wilson Bull.* **75**: 216–220.

McAdoo, J. K., and D. A. Klebenow. 1978. Native faunal relationship in sagebrush ecosystems. In The Sagebrush Ecosystem: A symposium. Utah State Univ., Logan, pp. 50–61.

McClelland, B. R., S. S. Frissell, W.C. Fischer, and G.H. Halvarson. 1979. Habitat management for hole-nesting birds in forests of western larch and Douglas-fir. *J. For.* **77**: 480–483.

McCullock, C. Y., Jr. 1955. Utilization of winter browse on wilderness big game range. *J. Wildl. Manage.* **19**:206–215.

McCullock, C.Y., Jr. 1974. Control of pinyon-juniper as a deer management measure in Arizona. Completion Rep. Ariz. Red. Aid Proj. W-78-R18; WP 4, J 2 and 7. Ariz. Game and Fish Dept., Phoenix.

Michael, E. D., and P. I. Thornburgh. 1971. Immediate effect of hardwood removal and prescribed burning on bird populations. *Southwest Natur.* **15**:359–370.

Miller, A. H., I. I. McMillan, and E. McMillan. 1965. The current status and welfare of the California Condor. Res. Rep. No. 6, Nat. Audubon Soc.

Miller, H. A. 1963. Use of fire in wildlife management. *Proc. Tall Timbers Fire Ecol. Conf.* **2**: 19–30.

Mitchell, G. J., and S. Smoliak. 1971. Pronghorn antelope range characteristics and food habits in Alberta. *J. Wildl. Manage.* **35**:238–250.

Murie, A. 1961. Some food habits of the marten. *J. Mammal.* **42**: 516–521.

O'Neal, T. 1949. The muskrat in the Louisiana coastal marshes. Fed. Aid Sec., La. Dept. Wildl. and Fish., New Orleans.

Patric, E. F., and W. L. Webb. 1953. A preliminary report on intensive beaver management. *Proc. N. Amer. Wild. Conf.* **18**: 533–539.

Patton, D. R. 1975. Abert squirrel cover requirements in southwestern pine. USDA For. Serv. Res. Paper RM-145. Rocky Mtn. For. and Range Exp. Stn., Fort Collins, Colo.

Patton, D. R. 1976. Timber harvesting increases deer and elk use of a mixed conifer forest. USDA For. Serv. Res. Note RM-329. Rocky Mtn. For. and Range Exp. Stn., Fort Collins, Colo.

Patton, D. R., and H. D. Avant. 1970. Fire simulated aspen sprouting in a spruce-fir forest in New Mexico. USDA For. Serv. Res. Note RM-159. Rocky Mtn. For. and Range Exp. Stn., Fort Collins, Colo.

Patton, D. R., and J. R. Jones. 1977. Managing aspen for wildlife in the Southwest. USDA For. Serv. Gen. Tech. Rep. RM-37. Rocky Mtn. For. and Range Exp. Stn., Fort Collins, Colo.

Perkins, C. J. 1968. Controlled burning in the management of muskrats and waterfowl in Louisiana marshes. *Proc. Tall Timbers Fire Ecol. Conf.* **8:**269–280.

Phillips, R. W. 1961. The embryonic survival of coho salmon and steelhead trout as influenced by some environmental conditions in gravel beds. In 14th Annu. Rep., Pac. Mar. Fish Comm., pp. 60–73.

Quick, H. F. 1955. Food habits of marten *(Martes americana)* in northern British Columbia. *Can. Field-Nat.* **64:** 144–147.

Quinton, D. A., R. G. Horejsi, and J. T. Flinders. 1979. Influence of brush control on white-tailed deer diets in north-central Texas. *J. Range Manage.* **32:**93–97.

Ream, C. H., and G. E. Gruell. 1979. Influences of harvesting and residue treatments on small mammals and implications for forest management. In GTR Environ. Consequences of Timber Harvesting in the Rocky Mountain Coniferous Forest Symp. Proc. Sept. 11-13, Missoula, Mont. USDA For. Gen. Tech. Rep. INT-90. Intermt. For. and Range Exp. Stn., Ogden, Utah, pp. 455–467.

Redfield, J. A., F. C. Zwickel, and J. F. Bendell. 1970. Effects of fire on numbers of blue grouse. *Proc. Tall Timbers Fire Ecol. Conf.* **10:** 63–83.

Regelin, W. L., and O. C. Wallmo. 1978. Duration of deer forage benefits after clearcut logging of subalpine forest in Colorado. USDA For. Serv. Res. Note RM-356. Rocky Mtn. For. and Range Exp. Stn., Fort Collins, Colo.

Regelin, W. L., O. C. Wallmo, J. G. Nagy, and D. R. Dietz. 1974. Effects of logging on forage values for deer in Colorado. *J. For.* **72:** 4–7.

Reid, V. H., and P.D . Goodrum. 1959. Bobwhite quail on southern pine land. Final Prog. Rep. Denver Wildl. Res. Lab. USDI.

Renwald, J. D. 1978. The effect of fire on woody plant selection by nesting nongame birds. *J. Range Manage.* **31:** 467, 468.

Renwald, J. D., H. A. Wright, and J. T. Flinders. 1978. Effect of prescribed fire on bobwhite quail habitat in the Rolling Plains of Texas. *J. Range Manage.* **31:** 65–69.

Reynolds, H. G. 1967. Chemical constituents and deer use of some crown sprouts in Arizona chaparral. *J. For.* **65:** 905–908.

Rodgers, K. J., P. F. Ffolliott, and D. R. Patton. 1978. Home range and movement of five mule deer in a semidesert grass-shrub community. USDA For. Serv. Res. Note RM-355. Rocky Mtn. For. and Range Exp. Stn., Fort Collins, Colo.

Schlatterer, E. F. 1960. Productivity and movements of a population of sage grouse in southeastern Idaho. M.S. Thesis, Univ. Idaho, Moscow.

Schlichtemeier, G. 1967. Marsh burning for waterfowl. *Proc. Tall Timbers Fire Ecol. Conf.* **6:** 40–46.

Scotter, G. W. 1964. Effects of forest fires on the winter range of barren-ground caribou in northern Saskatchewan. *Can. Wildl. Serv. Wildl. Manage. Bull. Sec. 1,* **28:**1–111.

Scotter, G. W. 1970. Wildfires in relation to the habitat of barren ground caribou in the Taiga of northern Canada. *Proc. Tall Timbers Fire Ecol. Conf.* **10:**85–105.

Sharp, W. M. 1970. The role of fire in ruffed grouse habitat management. *Proc. Tall Timbers Fire Ecol. Conf.* **10:**47–61.

Short, H. L. 1975. Nutrition of southern deer in different seasons. *J. Wildl. Manage.* **39:** 321–329.

Short, H. L., W. Evans, and E. L. Boeker. 1977. The use of natural and modified pinyon pine-juniper woodlands by deer and elk. *J. Wildl. Manage.* **41:**543–559.

Silbey, F. C. 1968. The life history, ecology, and management of the California Condor *(Gymnogyps californianus)*. Annual Prog. Rep., Proj. No. 8-22. U.S. Bur. Sport Fish. and Wildl., Patuxent Wildl. Res. Cent.

Singleton, J. R. 1951. Production and utilization of waterfowl food plants on the east Texas Gulf Coast. *J. Wildl. Manage.* **15:** 46–56.

Smith, N. F. 1947. Controlled burning in Michigan's forest and game management programs. In Proc. Soc. Amer. For., Mills Bldg., Washington, D.C., pp. 200–205.

Soutiere, E. C., and E. G. Bolen. 1972. Role of fire in mourning dove nesting ecology. *Proc. Tall Timbers Fire Ecol. Conf.* **12:** 277–288.

Spencer, D. L., ad J. B. Hakala. 1964. Moose and fire on the Kenai. *Proc. Tall Timbers Fire Ecol. Conf.* **3:** 10–33.

Stelfox, J. G. 1971. Bighorn sheep in the Canadian Rockies: A history 1800–1970. *Can. Field-Natur.* **85:** 101–122.

Steuter, A. A., and H. A. Wright. 1980. White-tailed deer densities and brush cover on the Rio Grande Plain. *J. Range Manage.* **33:** 328–331.

Stewart, R. E. 1944. Food habits of blue grouse. *Condor* **46:**112–120.

Stoddard, H. L. 1963. Bird habitat and fire. *Proc. Tall Timbers Fire Ecol. Conf.* **2:**163–175.

Stransky, J. J., and L. K. Halls. 1978. Browse quality affected by pine site preparation in east Texas. *Proc. Ann. Conf. Southeast. Assoc. Fish Wild. Agencies* **30:** 507–512.

Szaro, R. C., and R. P. Balda. 1979a. Effects of harvesting ponderosa pine on nongame bird populations. USDA For. Serv. Res. Paper RM-212. Rocky Mtn. For. and Range Exp. Stn., Fort Collins, Colo.

Szaro, R. C., and R. P. Balda. 1979b. Bird community dynamics in a ponderosa pine forest. *Studies Avian Biol.* **3:** 1–66.

Taber, R. D., and R. F. Dasmann. 1958. The black-tailed deer of the chaparral. Calif. Dept. Fish and Game Bull. No. **8,**, Sacramento.

Tevis, L., Jr. 1956. Effects of a slash burn on forest mice. *J. Wildl. Manage.* **20:** 405–409.

Thomas, J. W., C. Moser, and J. E. Rodick. 1978. Edges—Their interspersion, resulting diversity and its measurement. In Proc. of the Workshop on Nongame Bird Habitat Management in the Coniferous Forests of the Western United States. USDA For. Serv. Gen. Tech. Rep. PNW-64. Pac. Northwest For. and Range Exp. Stn., Portland, Ore., pp. 91–100.

Tomlinson, R. E. 1972. Current status of the endangered masked bobwhite quail. *N. Amer. Wildl. and Nat. Resour. Conf.* **37:** 294–311.

Verner, J. 1978. California Condors: Status of the recovery effort. USDA For. Serv. Gen. Tech. Rep. PSW-28. Pac. Southwest For. and Range Exp. Stn., Berkeley, Calif.

Vogl, R. J. 1964. The effects of fire on a muskeg in northern Wisconsin. *J. Wildl. Manage.* **28:** 317–329.

Vogl, R. J. 1973. Effects of fire on the plants and animals of a Florida wetland. *Amer. Midl. Natur.* **89:** 334–339.

Vogl, R. J. 1977. Fire: A destructive menace or a natural process? In Cairns et al. (eds.) *Recovery and Restoration of Damaged Ecosystems.* Univ. Press of Virginia, Charlottesville, pp. 261–289.

Wagner, R. H. 1971. Fire and man. In *Environment and Man.* W.W. Norton, New York, pp. 79–92.

Ward, P. 1968. Fire in relation to waterfowl habitat of the delta marshes. *Proc. Tall Timbers Fire Ecol. Conf.* **8:** 254–267.

Weckwerth, R. P., and V. D. Hawley. 1962. Marten food habits and population fluctuations in Montana. *J. Wildl. Manage.* **26:**55–74.

Westemeier, R. L. 1972. Prescribed burning in grassland management for prairie chickens in Illinois. *Proc. Tall Timbers Fire Ecol. Conf.* **12:** 317–338.

Wilbur, S. R. 1972. Food resources of the California Condor. Admin. Rep., U.S. Bur. Sport Fish. and Wildl., Patuxent Wildl. Res. Cent.

Wolf, J. O. 1975. Red squirrel response to clearcut and shelterwood systems in Interior Alaska. USDA For. Serv. Res. Note PNW-255. Pac. Northwest For. and Range Exp. Stn., Portland, Ore.

Wood, G. W., and L. J. Niles. 1978. Effects of management practices on nongame bird habitat in longleaf-slash pine forests. In Proc. of the Workshop Management of Southern Forest for nongame birds, Atlanta, Ga. USDA For. Serv. Gen. Tech. Rep. SE-14. Southeastern For. and Range Exp. Stn., Asheville, N.C., pp. 40–49.

Wood, M. 1977. North Kings Deer Herd: A research and management program. Pac. Southwest For. and Range Exp. Stn., Berkeley, Calif.

Wright, H. A., and A. W. Bailey. 1980. Fire ecology and prescribed burning in the Great Plains—a research review. USDA For. Serv. Gen. Tech. Rep. INT-77. Intermt. For. and Range Exp. Stn., Ogden, Utah.

Wright, H. A., L. F. Neuenschwander, and C. M. Britton. 1979. The role and use of fire in sagebrush-grass and pinyon-juniper plant communities: A state-of-the-art review. USDA For. Serv. Gen. Tech. Rep. INT-58. Intermt. For. and Range Exp. Stn., Ogden, Utah.

Yancey, R. K. 1964. Matches and marshes. In J. P. Linduska (ed.) *Waterfowl Tomorrow*. USDI, U.S. Government Printing Office, Washington, D.C., pp. 619–626.

Young, V. A., and W. L. Robinette. 1939. A study of the range habits of elk on the Selway Game Preserve. Univ. Idaho Bull. No. 16. Moscow, Idaho.

Zontek, F. 1966. Prescribed burning on the St. Marks national wildlife refuge. *Proc. Tall Timbers Fire Ecol. Conf.* **5:**195–202.

CHAPTER

5

GRASSLANDS

The historical prevalence of fire in grasslands cannot be denied (Fidler 1793; Moss 1932; Sauer 1944; Stewart 1951; 1953; Dix 1960; Humphrey 1962; Jackson 1965; Nelson and England 1971; Kirsch and Kruse 1972; Seevers et al. 1973). For example, the diary of Fidler (1793), who observed fires in the fescue grassland of southwestern Alberta, contains these statements:

> These large plains either in one place or another is constantly on fire and when the grass happens to be long and the wind high, the sight is grand and awful, and it drives along with amazing swiftness. The lightning in the spring and fall frequently lights the grass, and in winter it is done by the Indians . . . These fires among the long grass is very dangerous.

Although lightning caused many fires (Haley 1929; Komarek 1966) and the Indians set some (Nelson and England 1971; Heady 1972), most documented conflagrations of the shortgrass prairie in the late 1880s were the result of carelessness by trail outfits, cowboys, and cooks (Haley 1929). Today, dry lightning storms and people are the major causes of fire in grasslands. Regardless of origin, fires have always been common and widespread on prairies during drought years.

In the semiarid areas, big prairie fires in the past usually occurred during drought years that followed one to three years of above average precipitation, because of the abundant and continuous fuel. Consequently, wildfires traveled for many kilometers when the winds and air temperatures were high and relative humidity was low. An example is an account of a fire (Haley 1929) that started in the fall of 1885 in the

Arkansas River country of western Kansas. It jumped the Cimarron River, burned across the North Plains of Texas, and did not stop until it reached the rugged Canadian River Breaks, a distance of 282 km (175 miles). About 0.4 million ha (1.0 million acres) of the XIT Ranch alone burned in Texas. Haley (1929) gave several other accounts of large fires [32 by 97 km (20 by 60 miles)] on the High Plains of Texas.

Many grassland fires still start during dry weather but they do not travel as far as in the past. Cultivated land breaks up the continuous grass cover of the prairie. One of the largest fires observed in the Texas-New Mexico area started from a broken power line in Lea County, New Mexico in April, 1974. Wind was 90 km (55 miles)/hr, relative humidity was 5 percent, and air temperature was 38°C (100°F). Herbaceous growth from the previous year was abundant. The fire traveled 42 km (26 miles), burning 13,470 ha (52 sections), and crossed three major highways. It was stopped by a plowed field. This typifies the prairie fires of today and the conditions under which they will travel long distances.

FIRE HISTORY OF THE GRASSLANDS

There are no reliable historical records of fire frequencies in the Great Plains grassland because there are no trees to carry fire scars from which to estimate fire frequency. However, we know that fire frequency was high because explorers and settlers were concerned about the danger of prairie fires. We can extrapolate fire frequency data for grasslands from forests having grassland understories, such as ponderosa pine *(Pinus ponderosa)* in the West and longleaf pine *(Pinus palustris)* in the Southeast. A variety of sources (Chapman 1926; 1944; Weaver 1951; Wagener 1961; Hall 1976; Arno 1976) indicate that fire frequency in pine forests varied from 2 to 25 years. Because prairie grassland is typically of level to rolling topography, a natural fire frequency of 5 to 10 years seems reasonable. In topography dissected with breaks and rivers, such as the Rolling Plains and Edwards Plateau of Texas, fire frequency may have been 20 to 30 years. The latter assumption is based on historical accounts by Marcy (1849) of large honey mesquite *(Prosopis glandulosa* var. *glandulosa)* trees in the Rolling Plains and the frequency of fire that we know is necessary to keep large Ashe juniper trees *(Juniperus ashei)* out of prairies in the Edwards Plateau (Wink and Wright 1973).

How important have fires been in maintaining grasslands? Stewart (1951; 1953) and Sauer (1944) proposed that treeless grasslands are a product of repeated fires set by aborigines. Wedel (1957) and Hastings and Turner (1966) make a strong case for climate being the primary influence on American grasslands. Winter rainfall decreases rapidly from the Southeast to the center of the United States. Snowfall decreased rapidly from the eastern edge of the Rockies to the Southwest and across the northern margin of the Great Plains, making this area unsuitable for tree growth (Wedel 1957). Wells (1970) presented evidence that the origin and maintenance of grasslands is directly related to topography. He stated that the "rougher and more dissected the topography, the greater the former extent and the current spread of woody vegetation at the expense of grasslands.''

Although we feel that climate is the dominant factor controlling North American grasslands, wide fluctuations in woody vegetation would occur if it was the only factor (Albertson and Weaver 1945; Bragg and Hulbert 1976). The impact of drought on the maintenance of grasslands has been illustrated by Albertson and Weaver. They surveyed the mortality of native trees, timber belts, old shelterbelts, and hedgerows following the drought of the 1930s. Their studies ranged from Oklahoma to Nebraska, and they recorded mortality rates ranging from 30 to 93 percent among the native deciduous trees [elm *(Ulmus* sp.), ash *(Fraxinus* sp.), hackberry *(Celtis occidentalis)*] and from 35 to 80 percent or higher among juniper.

On the eastern edge of the Great Plains ''the balance between forest and grassland is so delicate that a little higher water content of soil, a slightly greater humidity, or protection from drying winds throws this balance in favor of tree growth, while the reverse conditions exclude it'' (Albertson and Weaver 1945). Thus, there is good reason to believe that climate is the major factor in maintaining grasslands.

On level to undulating topography in the southern mixed prairie, drought (not necessarily severe), fire, insects, rodents, lagomorphs, and competition from grass apparently interact to maintain grasslands (Wright 1974a). For example, in 1969 we burned mesquite trees that had been top-killed by herbicide in 1966. Of the 1200 trees marked, 26 percent of them were killed over a 5-year period (Wright et al. 1976a). Part of the mortality may be attributed to fire, but more than half of it seems to have resulted from the interaction of fire, a mild drought in 1970 and a severe one in 1971, insects, wood rats, and competition from grass. A natural fire every 15 to 30 years in the southern mixed prairie could significantly reduce shrubs.

Shrubs and trees have always existed as scattered individuals or mottes on grasslands and along drainageways. In the Great Plains they are most abundant in the southern mixed prairie, mesic edge of the northern mixed prairie, eastern edge of the tallgrass prairie, and throughout the fescue prairie. Shrubs and trees are also present on rocky breaks or heavily grazed areas where fires are least frequent. Droughts can control shrub abundance where grass is healthy, but shifts from grasslands to shrubs and trees could occur on a 100-year cycle if vegetation was controlled solely by climate. Fire seems to have restricted shrub and tree growth in the past (Malin 1953), not so much as a single influence but in concert with other factors.

The restriction of shrub and tree growth by droughts, fire, and biotic factors did not eliminate trees in the southern mixed prairie nor on the eastern edge of the tallgrass prairie (Malin 1953). Open groves of large honey mesquite trees existed in the Rolling Plains of central Texas in 1849 (Marcy 1849). Historical evidence documents that these areas were subjected to fire (Michler 1850). Based on our experience, large honey mesquite and interior live oak *(Quercus virginiana)* trees are very tolerant of hot fires (the crowns of the trees remain alive), whereas trees less than 3 to 4 m (10 to 12 ft) tall can easily be top-killed. Thus one can surmise that before the arrival of the Europeans, it was feasible for a prairie fire to leave a mosaic of large trees, which left the appearance of a savannah, in the southern mixed prairie and the eastern tallgrass prairie. Trees that were top-killed probably had a broad genetic base, as we will show in a later section, which permitted reburns to kill some trees and only suppress the resprouts of others.

Nonsprouting species such as Ashe juniper were more susceptible to fire. Wink and Wright (1973) have shown that hot fires can easily kill large juniper trees. Subsequent fires within 10 to 15 years will kill new juniper trees before they are old enough to produce seed, eliminating the potential local seed source of such a nonsprouting species. Thus with occasional fires, such areas could easily have been converted to grasslands, with occasional patches of juniper on rocky sites. Protection from fire has favored the reestablishment of dense stands of juniper.

ECOLOGICAL CHARACTERISTICS AND EFFECTS OF FIRE

The vast North American Grassland lies between the Rocky Mountains and the eastern forest and extends from south-central Texas to the aspen-parkland in central Alberta and Saskatchewan (Fig. 5.1). The grassland may be divided from west to

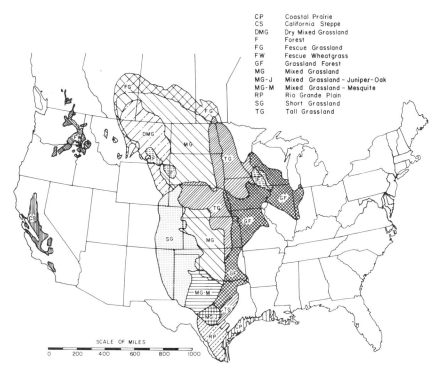

CP	Coastal Prairie
CS	California Steppe
DMG	Dry Mixed Grassland
F	Forest
FG	Fescue Grassland
FW	Fescue Wheatgrass
GF	Grassland Forest
MG	Mixed Grassland
MG-J	Mixed Grassland – Juniper-Oak
MG-M	Mixed Grassland – Mesquite
RP	Rio Grande Plain
SG	Short Grassland
TG	Tall Grassland

Fig. 5.1. Map of natural grasslands. (Modified from Küchler 1964 and Rowe 1972. Reproduced with permission of the American Geographical Society and the Minister of Supply and Services Canada.)

east into the shortgrass, mixed, and tallgrass prairies and grassland-forest combinations (Launchbaugh 1972). Because there is such a wide variation in species combinations across the Great Plains, for purposes of discussion we will separate the area into the Southern Great Plains, Central Great Plains, and Northern Great Plains. The northern Great Plains will include aspen-parkland, a transitional zone of vegetation between the Great Plains grasslands and the boreal forest. Other grasslands to be discussed include the Rio Grande Plains and Coastal Prairie, California Prairie, Palouse Prairie, and Mountain Grasslands.

Southern Great Plains

DISTRIBUTION, CLIMATE, SOILS, AND VEGETATION

The southern Great Plains includes the eastern third of New Mexico, the northern two-thirds of Texas, and most of Oklahoma. Within the region, the shortgrass prairie (High Plains) (Fig. 5.2) lies west of the 100° meridian. Annual precipitation in the shortgrass prairie varies from 38 to 51 cm (15 to 20 in.). Except for the sandy soils in southeastern New Mexico and the Canadian River country in northern Texas and western Oklahoma, soils are primarily clay loams, silt loams, and sandy loams. A caliche layer is frequently present at 50 to 90 cm (20 to 36 in.) in the fine-textured soils. Most of the area is tableland that is 1200 to 1830 m (4000 to 6000 ft) in elevation (south to north) on the western edge and slopes eastward to 915 m (3000 ft) on the edge of the Caprock in Texas. Dominant grasses are buffalograss *(Buchloe dactyloides)*, and blue grama *(Bouteloua gracilis)*, with varying amounts of threeawns *(Aristida* spp.), lovegrass *(Eragrostis* spp.), tridens *(Tridens* spp.), sand dropseed

Fig. 5.2.   Shortgrass prairie in Panhandle and northern Texas. Dominant species are blue grama and buffalograss.

(Sporobolus cryptandrus), sideoats grama *(Bouteloua curtipendula)*, tobosagrass *(Hilaria mutica)*, galleta *(H. jamesii)*, vine-mesquite *(Panicum obtusum)*, bush muhly *(Muhlenbergia porteri)*, and Arizona cottontop *(Digitaria californica)*. Forbs can be abundant during wet years, but they are seldom a major component of the shortgrass prairie.

Major forbs include annual broomweed *(Xanthocephalum dracunculoides)*, false mesquite *(Hoffmanseggia densiflora)*, western ragweed *(Ambrosia psilostachya)*, horsetail conyza *(Conyza canadensis)*, warty euphorbia *(Euphorbia spathulata)*, silverleaf nightshade *(Solanum elaeagnifolium)*, manystem evax *(Evax multicaulis)*, woolly plantago *(Plantago purshii)*, dozedaisy *(Aphanostephus* spp.), goosefoot *(Chenopodium* spp.), croton *(Croton* spp.), summercypress *(Kochia scoparia)*, and globemallow *(Sphaeralcea* spp.).

Dominant woody plants are honey mesquite, sand shinnery oak *(Quercus havardii)*, sand sagebrush *(Artemisia filifolia)*, broom snakewood *(Xanthocephalum sarothrae)*, yucca *(Yucca* spp.), and fourwing saltbush *(Atriplex canescens)*. Cactus *(Opuntia* spp.) is also abundant. The most prevalent species include plains pricklypear *(Opuntia polyacantha)*, brownspine pricklypear *(O. phaecantha)*, walkingstick cholla *(O. imbricata)*, and tasajillo *(O. leptocaulis)*.

East of the shortgrass plains is the mixed prairie (Rolling Plains and Edwards Plateau). It includes most of west-central Texas and western Oklahoma. Elevation drops from 915 m (3000 ft) along the western edge to about 275 m (900 ft) along the eastern edge in central Texas and Oklahoma. Topography is undulating, with occasional breaks and rivers. The zone is about 240 km (150 miles) wide and precipitation varies from 51 cm (20 in.) on the western edge to 71 cm (28 in.) on the eastern edge. Soil textures are primarily clay loams, silt loams, and sandy loams.

Honey mesquite (Fig. 5.3) and Ashe juniper (Fig. 5.4) dominate the overstory in Texas, but these species are not prevalent in Oklahoma. In the Rolling Plains honey mesquite dominates the overstory. Lotebush *(Zizyphus obtusifolia)* is an important subdominant shrub that provides cover for bobwhite quail *(Colinus virginianus)* and nesting for many songbirds (Renwald 1977; Renwald et al. 1978). Other shrubs include fourwing saltbush, elbowbush *(Foresteria pubescens)*, ephedra *(Ephedra* spp.), skunkbush sumac *(Rhus trilobata)*, dalea *(Dalea* sp.), and acacia *(Acacia* sp.). Breaks throughout the Rolling Plains contain large amounts of redberry juniper *(Juniperus pinchoti)*, which resprouts from crown buds after fire. Cactus species similar to those in the shortgrass prairie are also present throughout the Rolling Plains (Fig. 5.5 and 5.6).

Dominant grasses include sideoats grama, tobosagrass, buffalograss, little bluestem *(Schizachyrium scoparium)*, and the cool season grass, Texas wintergrass *(Stipa leucotricha)*. Other grasses include vine-mesquite, Arizona cottontop, sand dropseed, white tridens *(Tridens albescens)*, threeawn species, plains bristlegrass *(Setaria leucopila)*, and green sprangletop *(Leptochola dubia)*. Many annual forbs and some annual grasses are abundant during wet winters. Annual broomweed, bitterweed *(Hymenoxys odorata)*, Carolina canary grass *(Phalaris carolinensis)* and little barley *(Hordeum pusillum)* are the most prevalent annual species over a wide

Fig. 5.3. A mesquite-tobosa community (dormant season), with 4570 kg/ha (4070 lb/acre) of fine fuel, in the mixed prairie near Colorado City, Texas.

Fig. 5.4. A heavily grazed stand of Ashe juniper in the mixed prairie of central Texas that will need a major reclamation program to revert it back to a natural grassland. Communities such as this need fire every 20 years or so and good management to keep Ashe juniper out.

Fig. 5.5.   Pricklypear (*Opuntia phaecantha*) can easily be killed with fire. Generally 70 to 80 percent of the plants are dead two to three years after a fire because of a fire-insect interaction.

Fig. 5.6.   Cholla are easily killed if less than 0.6 m (2 ft) tall. About 50 percent of the plants die after being burned.

area. Perennial forbs include *Englemannia, Gaillardia, Oenothera, Aphanostephus, Chenopodium, Gaura, Helianthus, Plantago, Solanum, and Sphaeralcea.*

In the Edwards Plateau, southeast of the Rolling Plains, Ashe juniper dominates the overstory. Other major species include interior live oak, Texas oak *(Quercus shumardii* var. *texana)*, post oak *(Q. stellata)*, blackjack oak *(Q. marilandica)*, smoothleaf sumac *(Rhus glabra)*, Mexican redbud *(Cercis canadensis* var. *mexicana)*, and shin oak *(Quercus* spp.). Dominant grasses include little bluestem, sideoats grama, Texas wintergrass, tall grama, *(Bouteloua pectinata)*, vine-mesquite, buffalograss, and meadow dropseed *(Sporobolus asper* var. *hookeri)*. Forbs are similar to those in the Rolling Plains.

The tallgrass prairie is mixed with various amounts of the "Cross Timbers" from central Texas and Oklahoma to their eastern boundaries. The Cross Timbers are dominated by post oak and blackjack oak and occur on sandy soil. Precipitation varies from 69 to 114 cm (27 to 45 in.). Deep sandy loam and silt loam soils are common in the pure grasslands. Elevation varies from 150 to 300 m (500 to 1000 ft), sloping generally to the east. Dominant grasses are little bluestem, big bluestem *(Andropogon gerardi)*, Indiangrass *(Sorghastrum nutans)*, and switchgrass *(Panicum virgatum)*. Shrubs vary in abundance, with the important ones being smoothleaf sumac, leadplant *(Amorpha canescens)*, and wild plum *(Prunus* sp.). Forbs are similar to those mentioned for tallgrass prairie of the central Great Plains.

FIRE EFFECTS–SHORTGRASS PRAIRIE

Grasses.   During dry years, most species of the shortgrass prairie are harmed by fire. Following a spring wildfire, when soil was dry, Launchbaugh (1964) found that the recovery time for a buffalograss-blue grama community took three growing seasons. Recovery was 35, 62, and 97 percent following the first, second, and third growing seasons, respectively. Hopkins et al. (1948) reported similar results in west-central Kansas. Western wheatgrass *(Agropyron smithii)* recovered more slowly—18, 27, and 77 percent for the three growing seasons (Launchbaugh 1964). Following a wildfire in New Mexico when the moisture balance was more favorable, Dwyer and Pieper (1967) found that the production of blue grama was reduced only 30 percent the first year. With above average precipitation the second year after burning, blue grama had essentially recovered. Results from prescribed burns in Texas during years with above normal winter and spring precipitation show that buffalograss and blue grama can tolerate fire with no loss in herbage yield at the end of the first growing season (Trlica and Schuster 1969; Heirman and Wright 1973; Wright 1974b) (Table 5.1).

Tolerance of most grass species to fire in the shortgrass prairie, under different moisture regimes, appears to be similar to that for buffalograss and blue grama. Red threeawn *(Aristida longiseta)* and sand dropseed are usually harmed by fire (Hopkins et al. 1948; Dwyer and Pieper 1967; Trlica and Schuster 1969). By contrast, Wright (1974b) found that sand dropseed tolerated fire when winter and spring precipitation were 40 percent above normal. Other species that Dwyer and Pieper (1967) found to

Table 5.1.   Yields [kg/ha (lb/acre)] of Buffalograss and Blue Grama After Burning During a Year With Above Normal Precipitation (Wet Year) and a Year With Below Normal Precipitation (Dry Year)

Year After Burn	Burned		Unburned	
	Current Growth	Litter	Current Growth	Litter
Wet years (Texas data)[a]				
Buffalograss				
First (1968)	1894 (1686)	—	1679 (1494)	818 (728)
Second (1969)	2718 (2063)	344 (306)	2166 (1928)	515 (458)
Third (1970)	1571 (1398)	1766 (1572)	1494 (1330)	1018 (906)
Blue grama				
First (1970)	1888 (1680)	—	1606 (1429)	2780 (2474)
Second (1971)	1538 (1369)	785 (699)	1401 (1247)	2904 (2584)
Third (1972)	2407 (2142)	1966 (1750)	1971 (1754)	2171 (1932)
Dry years (Kansas data)[b]				
Buffalograss-Blue grama				
First (1959)	1236 (1100)	—	3539 (3150)	562 (500)
Second (1960)	2067 (1840)	281 (250)	3371 (3000)	562 (500)
Third (1961)	2921 (2600)	371 (330)	2978 (2650)	618 (550)

[a]Data from Wright 1974b.
[b]Data from Launchbaugh 1964. (Reproduced by permission of Journal of Range Management.)

be harmed by fire included slim-stemmed muhly *(Muhlenbergia filiculmis)*, ring muhly *(M. torreyi)*, wolftail *(Lycurus phleoides)*, and galleta. These species were harmed by a wildfire during a year of below normal precipitation. Tumble windmillgrass *(Schedonnardus paniculatus)* was not harmed by fire (Trlica and Schuster 1969). Weeping lovegrass *(Eragrostis curvula)*, an introduced species, is increased 14 percent by burning. But the greatest benefit from burning is a 53 percent increase in utilization (Klett et al. 1971).

In the southern shortgrass plains, sandy lands are common among the heavy clay soils that are dominated by buffalograss and blue grama. The sandy soils are dominated by sand bluestem *(Andropogon hallii)*, little bluestem, switchgrass, and sand shinnery oak. Burning generally increases production of sand bluestem and switchgrass about 337 kg/ha (300 lb/acre) and similarly decreases production of little bluestem with a net increase in total forage of 20 percent (McIlvain and Armstrong 1968).

Forbs.   Grasses provide the major portion of prairie vegetation but many species of forbs occur during years with above normal precipitation. Heirman and Wright (1973) found that spring burning was temporarily detrimental to many forbs: annual broomweed, silverleaf nightshade, western ragweed, and horsetail conyza. Warty euphorbia, manystem evax, and woolly plantago were not affected. False mesquite was favored by the burn.

Total forb yields are usually reduced more by spring burns than fall burns (Hopkins et al. 1948). In all cases, however, forb composition will be altered the least by burning when plants are dormant. Young, active, growing forbs will be severely harmed by fire (Wright 1974a).

Shrubs.   The shortgrass prairie does not have any species of shrubs, but shrubby mesquite is abundant on native ranges in the southern mixed prairie. Mesquite has not always been a noticeably prevalent shrub on the High Plains (shortgrass prairie) of Texas. The following observations were made by Captain R.B. Marcy (1849) as he traveled with his command over the northern part of the Llano Estacado near present day Amarillo, Texas.

> When we were on the high tableland, a view presented itself as boundless as the ocean. Not a tree, shrub, or any other object, either animate or inanimate, relieved the dreary monotony of the prospect; it was a vast illimitable expanse of desert prairie — the dreaded "Llano Estacado" of New Mexico; or, in otherwords, the great Zahara of North America. It is a region almost as vast and trackless as the ocean — a land where no man, either savage or civilized, permanently abides; it spreads forth into a treeless, desolate waster of uninhabited solitude, which always has been and must continue, uninhabited forever; even the savages dare not venture to cross it except at two or three places, where they know water can be found. The only herbage upon these barren plains is a very short buffalograss, and, on account of a scarcity of water, all animals appear to shun it.

Today, honey mesquite is not only prevalent on the High Plains, but it is almost impossible to kill with fire after it is 0.3m (1 ft) tall (Wright et al. 1976a). Even the seedlings are very tolerant of fire (Fisher 1947). Honey mesquite on the High Plains has an exceptional ability to resprout, compared to mesquite in the Rolling Plains (mixed prairie). Based on fire tolerance and the very few recorded observations of honey mesquite by early explorers on the High Plains (Malin 1953; Box 1967), we believe that before the arrival of Europeans, honey mesquite maintained a low-growth form and high frequency. On the High Plains honey mesquite seems to be genetically adapted to fire because of the historical necessity to survive frequent fires. Thus it is possible that the combination of fire, drought, competition from grasses, and damage from small mammals, particularly jackrabbits (*Lepus* spp.) and wood rats (*Neotoma* spp.), combined to maintain a low-growth form of mesquite.

On sandy loams in eastern New Mexico, northern Texas, and western Oklahoma, sand shinnery oak is abundant, whereas in southeastern New Mexico the sandy land is dominated by shrubby honey mesquite. Sand shinnery oak is fire-tolerant. Density of its stems increases 15 percent after burning (McIlvain and Armstrong 1966). However, acorns are not formed during the year of the burn, which could reduce feed available for lesser prairie chickens *(Tympauchus pallidicinctus)* and wild turkeys *(Meleagris gallopavo)*.

Algerita *(Berberis trifoliata)*, fourwing saltbush, winterfat *(Ceratoides lanata)*, and skunkbush sumac resprout vigorously after fire in New Mexico (Dwyer and Pieper 1967). Chickasaw plum *(Prunus angustifolia)* and aromatic sumac *(Rhus aromatica)* sprout vigorously after burning in the southern Great Plains (Jackson 1965). In the northern panhandle of Texas, sand sagebrush is a sprouter and seedlings appear soon after fire (Jackson 1965).

Cacti are readily killed by fire, but two years may be required for mortality. Mortality of tasajillo following burning may exceed 80 percent (Bunting et al. 1980). Walkingstick cholla and brownspine pricklypear are also easily killed by fire if they are less than 0.3 m (1ft) high. Dwyer and Pieper (1967) found that chollas less than 0.3 m(1ft) high were reduced by 50 percent after burning in New Mexico, but cacti over 0.3 m (1 ft) high were hardly damaged. Heirman and Wright (1973) reported similar data for west Texas. They attributed the high mortality of the shorter plants [0.46 to 0.61 m (1.5 to 2 ft)] to flame heights that easily engulf the plants. Taller plants are not burned at the higher levels and survive fire in the shortgrass prairie. Mortality of tall walkingstick cholla plants is greatly increased if they are chained before burning (Heirman and Wright 1973).

FIRE EFFECTS—MIXED PRAIRIE

Grasses.   Most grasses of the mixed prairie tolerate fire during years with normal to above normal precipitation, but sideoats grama and Texas wintergrass can be severely damaged. The rhizomatous form of sideoats grama is almost always reduced 40 to 50 percent by fire during dry years and may require three years for full recovery (Hopkins et al. 1948; Wright 1974b). It tolerates fire reasonably well during exceptionally wet years with no significant reduction in yield (Wink and Wright 1973). Texas wintergrass is severely harmed by sweeping hot fires (Dahl and Goen 1973) but increases following creeping, cool fires (Bean et al. 1975). Fire will cause little bluestem to decrease as much as 58 percent during dry years (Hopkins et al. 1948) or increase as much as 81 percent during wet years (Wink and Wright 1973).

Tobosagrass (Fig. 5.3), a southern desert species, is prevalent on clay sites in the southern mixed prairie. It is a highly productive species (Paulsen and Ares 1962; Dwyer 1972) until it accumulates large amounts of litter (Wright 1969) which decays slowly (Weaver and Albertson 1956). Young tobosagrass leaves are palatable but as plants mature and acccumulate litter they become coarse and unpalatable (Herbel and Nelson 1966; Wright 1972a; Heirman and Wright 1973).

Burning can greatly increase the production and palatability of tobosagrass during normal-to-wet years (Heirman and Wright 1973). During wet years after burning tobosagrass will increase over 2250 kg/ha (2000 lb/acre). Over a series of dry, normal, and wet years (5 years) tobosagrass production increased an average of 1160 kg/ha (1030 lb/acre) the first growing season following a burn (Wright 1972a). Total production on burned areas increased over control areas for a three-year period and then reached equilibrium during the fourth year after burning (Wright 1972a; Neuenschwander et al. 1978). However in southern New Mexico, where annual precipitation is only 23 cm (9 in.), an increase in tobosagrass yields cannot be expected after burning (Dwyer 1972).

Since tobosagrass is such a coarse grass, it should be grazed with in a few weeks after burning. If tobosagrass is rested for 3 or 4 months, as we generally recommend for most grasses, it will be too coarse for animals to eat. Cattle normally like to eat this grass during the spring and fall when it is growing rapidly. Heirman and Wright (1973) found that tobosagrass utilization could be increased many-fold following a burn. Normally, cattle only eat about 10 percent of the tobosagrass, but following a burn they will eat as much as 60 percent of the herbage (Heirman and Wright 1973).

Since tobosagrass and buffalograss often grow in combination and cattle will eat tobosagrass in preference to buffalograss during the spring after a fire (Heirman and Wright 1973), burning can be a means to increase the vigor of desirable grasses and improve the condition of the range. However, tobosagrass cannot take heavy utilization for an extended number of years (Canfield 1939). Maximum utilization of tobosagrass during any year probably should not exceed 50 percent.

As long as soil moisture is adequate, vine-mesquite, Arizona cottontop, plains bristlegrass, Texas cupgrass *(Eriochloa sericea)*, the bunchgrass form of sideoats grama, and meadow dropseed thrive after fire (Box et al. 1967; Wink and Wright 1973; Wright 1974b). Tall grama, a potentially susceptible bunchgrass after burning, yields as much forage as unburned controls during years with normal to above normal precipitation (Wright 1974b).

Cool season annual grasses are severely harmed by spring burning. In the southern mixed prairie, spring fires severely reduce yields of the principal cool season grasses: little barley and Carolina canary grass. Care should be taken to not burn an entire field in early spring where these species occur. Otherwise there will be very little green feed for animals during the subsequent winter months, although little barley is not a very palatable species.

Forbs.   Forbs that begin growth before the burning season are usually harmed by fire, whereas those that initiate growth after the burning season are usually not harmed by fire. In the southern mixed prairie, species usually harmed during the first growing season after burning include annual broomweed, horsetail conyza, plains dozedaisy *(Aphanostephus ramossissimus)*, scarlet globemallow *(Sphaeralcea coccinea)*, and bitterweed. Species common on burns include slimleaf lambsquarters *(Chenopodium leptophyllum)*, silverleaf nightshade, Carolina horsenettle *(Solanum carolinense)*, and annual sunflower *(Helianthus annuus)*. During the second growing season after burning, plains dozedaisy and redseed plantain *(Plantago rhodosperma)* reach their maximum importance value (Neuenschwander 1976).

In west-central Kansas, Hopkins et al. (1948) found that spring burning severely harmed wild onion *(Allium nuttallii)* and broom snakewood but left western ragweed and ashy goldenrod *(Solidago mollis)* unharmed.

Trees and Shrubs.   The presence of honey mesquite and other shrubs in the southern mixed prairie before the arrival of Europeans has been well documented in the journals of Marcy (1849) and Michler (1850). Mesquite was present throughout the southern mixed prairie on uplands and bottomlands in the Rolling Plains (mixed prairie) but usually not on stream banks (Michler 1850). A map of Marcy's expedition shows vegetation marked as ''mesquite timber'' from Big Spring to the junction of the Clear Fork of the main Brazos River. This area was approximately 190 to 240 km (120 to 150 miles) long and 80 km (50 miles) wide. Throughout the rest of the Rolling Plains they recorded the continuous presence of mesquite, frequently as a low-growing shrub at the northern and southern extremities.

Fire was a part of the ''mesquite timber'' country. Michler (1850) gave the following description after leaving the Double Mountain Fork and the Clear Fork of the Brazos on his way to Big Spring, Texas.

There was but little timber upon these streams upon first leaving the main fork, but the further we advanced the more we found, elm being the principal growth. The whole country was well timbered with mesquite, but most of it had been killed by prairie fires.

Evidently, this must have been a recent fire because Marcy and his command had traveled through the same country the previous year and did not mention the fire.

We have done considerable research on honey mesquite near Colorado City, Texas. It is in the ''mesquite timber'' country so designated by Marcy. Mesquite is moderately affected by fire, depending upon its age, number of dead basal stems with insect borer activity, weather at time of burning, and the amount of fine fuel for burning (Wright 1972a; 1972b; Wright et al. 1976a). Unless very young, green mesquite trees are difficult to kill with one fire. Mesquite trees 1.5 years of age or younger were easily killed by fire when soil surface temperatures were above 93°C (200°F). At 2.5 years of age, they were severely harmed, but trees older than 3.5 years of age are fire-resistant (Table 5.2).

Large mesquite trees that had previously been top-killed with 2,4,5-T (2,4,5-Trichlorophenoxy acetic acid) were killed more easily than small trees with resprouts (Britton and Wright 1971). Insect borers had infested the larger trees after the drought of the 1950's. Fire more readily consumed the wood perforated by wood borer holes. Percentage of mortality of 1200 trees that had resprouted after spraying in 1966 and were burned in 1969 was

Year	1969	1970	1971	1972	1973
Mortality (%)	10.8	17.7	22.6	25.7	26.4

Table 5.2. Percentage of Mortality of Young Mesquite Trees After Burning, in Relation to Age and Maximum Soil Surface Temperature

Age (Years)	Maximum Soil Surface Temperature				
	93°C (200°F)	260°C (500°F)	427°C (800°F)	593°C (1100°F)	Unburned
0.5	43	91	100	100	14
1.5	60	100	100	100	0
2.5	20	40	64	72	0
3.5	8	8	8	8	4
10 (approx.)	0	0	4	8	0

First year mortality resulted from the ignition of dead mesquite stems, which served as a fuel source to burn down into living root crowns. To achieve this effect, winds must be in excess of 13 km (8 mph)/hr and the relative humidity must be below 40 percent at the time of burning. After the initial fire-related kill, more trees died from the weakened condition caused by interactions among fire, drought, and biotic suppression.

Junipers are quite common thoughout the prairie on rocky slopes such as escarpments, ridges, or rimrocks (Wells 1970) and on areas that have been protected from fire (Penfound 1964). In Oklahoma, Arend (1950) found that fire was the worst natural enemy of eastern redcedar *(Juniperus virginiana)*, a nonsprouting species. Similarly Dalrymple (1969) and Wink and Wright (1973) have found Ashe juniper, another nonsprouting species, to be highly susceptible to fire in southern Oklahoma and central Texas. These species cannot maintain themselves in areas that burn frequently (Fig. 5.7) because the leaves are very flammable, especially during fall months, and the bark is so thin that heat from one surface fire usually kills all trees.

With 560 to 1120 kg/ha (500 to 1000 lb/acre) of herbaceous fuel, Dalrymple (1969) obtained a 100 percent mortality of trees less than 0.6 m (2 ft) tall, 77 percent mortality of trees 0.6 to 1.8 m (2 to 6 ft) tall, and 27 percent mortality of trees over 1.8 m (6 ft) tall, for an average mortality of 68 percent. Where fine fuel was at least 1240 kg/ha (1100 lb/acre), Wink and Wright (1973) found that 99 percent of the Ashe juniper trees less than 1.8 m (6 ft) tall were killed by fire under the following weather conditions: air temperatures 24 to 29°C (75 to 85°F), relative humidity 25 to 35 percent and wind 16 to 24 km/hr (10 to 15 mph). If fine fuel was above 2890 kg/ha (2500 lb/acre) all juniper trees were killed by fire. With 840 kg/ha (750 lb/acre) of fine fuel, Dwyer and Pieper (1967) reported that 70 percent of the juniper trees exposed to high temperatures of a summer wildfire had died by the following year.

Fig. 5.7.   Following dozing and prescribed burning, dense stands of Ashe juniper can be reverted to grasslands (compare with Fig. 5.4). The oak trees were left for aesthetic purposes.

Redberry juniper, a sprouting species on rough breaks in the Rolling Plains, is very difficult to kill by fire unless the trees are under 12 years of age (Smith et al. 1975). However fires reduce the sphere of influence of the trees. Very little forage grows under juniper trees. When the trees are burned, they shade less area, and grasses and forbs encroach.

Several shrub species are present in the mixed prairie of the southern Great Plains, but they are less abundant than trees. Fourwing saltbush, a palatable shrub, thrives after fire. It is a vigorous sprouter and appears to fully recover within three years after a fire. Lotebush also sprouts after a fire, but requires about six years to recover 75 percent of its original canopy cover (Neuenschwander 1976). Littleleaf sumac *(Rhus microphylla)* and algerita sprout following fires, but we have little research data on these species in west Texas. Smoothleaf sumac and all species of oak are vigorous sprouters in Ashe juniper communities of the Edwards Plateau. Cacti are abundant in the mixed prairie and equally susceptible to fire as mentioned for the shortgrass prairie.

FIRE EFFECTS—MIXED TALLGRASS—FOREST

The Cross Timbers region occupies a sandy belt of land in east-central Texas and eastern Oklahoma. It contains post oak and blackjack oak as well as many tallgrass species. Both oak species are easily top-killed with fire but resprout vigorously. Many ranchers feel that it is no longer economical to use goats or chemicals to keep oak sprouts suppressed in bluestem pastures. Presently there is interest in determining whether a four-year burning rotation will be effective. However fire research has not been done in the area and the natural role of fire is unclear.

Central Great Plains

DISTRIBUTION, CLIMATE, SOILS, AND VEGETATION

The central Great Plains extend from the foothills of the Rockies in eastern Colorado and southeastern Wyoming eastward through Kansas and Nebraska to grassland-forest communities in northwest Missouri, southern Iowa, and Illinois. Shortgrass prairie (Fig. 5.8) lies primarily in eastern Colorado, but it also occurs in western Kansas, southeastern Wyoming, and the extreme portion of western Nebraska. Annual precipitation varies from 28 to 46 cm (11 to 18 in.). Surface soil textures are largely sand, sandy loam, loamy sand, loam, and silt loam. Elevation drops from 1520 or 1830 m (5000 or 6000 ft) along the foothills in Colorado to 915 m (3000 ft) on the eastern edge of the shortgrass prairie in Kansas. In Nebraska the eastern elevations vary from 1220 to 1520 m (4000 to 5000 ft).

Buffalograss, blue grama, western wheatgrass, and scarlet globemallow are the dominants on loams and silt loams. Ranges in good to excellent condition will also support green needlegrass *(Stipa viridula)*. Sandy-textured soils are dominated by blue grama, prairie sandreed *(Calamovilfa longifolia)*, and needle-and-thread *(Stipa comata)*. Other grasses and grasslike species include red threeawn, sand dropseed, and sun sedge. Well managed ranges in eastern Colorado could also support sand bluestem, switchgrass, and Indiangrass. Dominant forbs and shrubs are western rag-

Fig. 5.8. Shortgrass prairie in the central Great Plains of eastern Colorado. Dominant grasses are blue grama, buffalograss, and western wheatgrass.

weed, bush morninglory *(Ipomoea leptophila)*, herbaceous sage *(Artemisia ludoviciana)*, and fourwing saltbush.

East of the shortgrass prairie lies mixed prairie in western Nebraska and Kansas. There is tallgrass prairie, however, in the Sandhills of northwestern Nebraska. Elevation along the western edge of the mixed prairie in Kansas is 915 m (3000 ft) but rises to as much as 1520 m (5000 ft) along the western edge of the Sandhills of Nebraska. Elevation starts at 400 m (1300 ft) along the eastern edge in southern Kansas and rises to as much as 610 m (2000 ft) in north-central Nebraska. Precipitation varies from 46 to 71 cm (18 to 28 in.) in Kansas and 46 to 64 cm (18 to 25 in.) in Nebraska. Topography varies from undulating to rolling ridge-tops, gently sloping, and hilly with steeply sloping valley sides. Soil textures are sand, silt, loam, silt loam, silty clay loam, and clay uplands.

Dominant grasses in the mixed prairie of the central Great Plains are blue grama, little bluestem, sand dropseed, tall dropseed *(Sporobolus asper)*, western wheatgrass, buffalograss, sideoats grama, purple threeawn *(Aristida purpurea)*, needle-and-thread, junegrass *(Koeleria cristata)*, and occasional sand bluestem, prairie sandreed, and switchgrass plants. Western wheatgrass and needle-and-thread become more prevalent northward from Kansas into Nebraska. Common forbs include scarlet globemallow, western ragweed, resindot skullcap *(Scutellaria resinosa)*, prairie coneflower *(Ratibida columnaris)*, heath aster *(Aster ericoides)*, black sampson *(Echinacea angustifolia)*, prairie phlox *(Phlox pilosa)*, prairie clover *(Petalostemum purpureum)*, dotted gayfeather *(Liatris punctata)*, slim-flowered scurfpea *(Psoralea tenuiflora)*, Missouri goldenrod *(Solidago missouriensis)*, and many others. Western ragweed and annual sunflowers are abundant on heavily grazed sites.

Tallgrass prairie in the eastern third of Nebraska, northern Iowa, and east-central Kansas varies in elevation from 305 to 610 m (1000 to 2000 ft). Annual precipitation varies from as low as 58 cm (23 in.) in eastern Nebraska to as much as 89 cm (35 in.) along the eastern edge of the tallgrass prairie. The Sandhills in western Nebraska, a westward extension of the tallgrass prairie, has precipitation as low as 46 cm (18 in.) per year. With the exception of the Sandhills of Nebraska and the Flint Hills of Kansas, most of the soils are medium-textured. Soils in the Flint Hills are primarily Lithosols and topography is gently rolling.

Grasses of the tallgrass prairie (Fig. 5.9) are primarily little bluestem, big bluestem, switchgrass, Indiangrass, and prairie dropseed *(Sporobolus heterolepis)*. Other grasses include Canada wildrye *(Elymus canadensis)*, porcupine grass *(Spartina pectinata)*, and eastern gamagrass *(Tripsacum dactyloides)*. Additional species in the Sandhills of Nebraska are prairie sandreed and sand bluestem. Important shrubs include western snowberry *(Symphoricarpos occidentalis)*, inland ceanothus *(Ceanothus ovatus)*, lead plant, willow *(Salix* spp.), gooseberry *(Ribes* spp.), and prairie rose *(Rosa arkansana)*. A wide variety of forbs occur in tallgrass prairie (Weaver and Clements 1938; Weaver and Albertson 1956). Typical genera include *Aster, Solidago, Silphium, Helianthus, Astragalus, Baptisia, Callinhoe, Phlox, Sisyrinchium, Lithospermum, Viola, Anemone, Tradescantia, Psoralea, Erigeron,*

Fig. 5.9. Tallgrass prairie in the Flint Hills of Kansas (top). Surface soils are very rocky, which prevents cultivation. Tallgrass prairie in central Missouri (bottom). Dominant grasses in both areas are little and big bluestem.

Petalostemum, Glycyrrhiza, Echinacea, Liatris, Vernonia, Coreopsis, Bidens, Kuhnia, and *Carduus.* Trees of the tallgrass prairie include American elm *(Ulmus americana),* hackberry *(Celtis occidentalis),* eastern redcedar, bur oak *(Quercus macrocarpa),* chinquapin oak *(Q. muhlenbergii),* eastern redbud *(Cercis canadensis),* bitternut hickory *(Carya cordiformis),* and roughlead dogwood *(Cornus drummondii)* (Smith and Owensby 1972; Bragg and Hulbert 1976).

Fig. 5.10.   Mixed grassland-forest combination in Missouri. This range is in poor condition, but would support tallgrasses if managed properly.

Tallgrass prairie and forest combinations (Fig. 5.10) extend eastward into eastern Kansas, northwestern Missouri, southern Iowa, and Illinois. The tallgrass species are those of the tallgrass prairie and the forest is Oak-Hickory *(Quercus-Carya)*. Precipitation increases to as much as 102 cm (40 in.) per year and elevation drops to 150 m (500 ft).

FIRE EFFECTS—SHORTGRASS AND MIXED GRASS PRAIRIE

Prescribed fire research has not been conducted in plant communities of the central Great Plains. Hopkins et al. (1948) and Launchbaugh (1964) studied the effects of wildfire. Following dry years, Hopkins et al. found that the cover and yield of big bluestem, little bluestem, hairy grama *(Bouteloua hirsuta)*, sideoats grama, buffalograss, hairy sporobolus *(Sporobolus pilosus)*, and blue grama were reduced by fire; undesirable broadleaved plants, principally western ragweed, increased. Similarly, Launchbaugh found that buffalograss, blue grama, and western wheatgrass did not fully recover after fire until the third growing season.

Based on data recorded during wet years in the southern Great Plains, there is no benefit to burning the shortgrass prairie unless there is a need to improve grazing distribution or there are unusually heavy accumulations of litter that need to be removed. Therefore it is difficult to justify prescribed burning research in the essentially shrubless shortgrass prairie of the central Great Plains. Most of the central mixed prairie region is in wheat.

FIRE EFFECTS—TALLGRASS PRAIRIE

Where the eastern plains have a permanently moist subsoil, Shantz and Zon (1924) suggested that the grassland had been induced by fire and drought. The environment is suited to trees when fire is absent. Work by Kucera (1960), Blan (1970), and Bragg and Hulbert (1976) support this theory. Aerial photography by Bragg and Hulbert in the Flint Hills bluestem prairie showed that on unburned pastures the combined tree and shrub cover increased 34 percent from 1937 to 1969. Tree cover alone increased 24 percent from 1856 to 1969. Tree cover alone increased 24 percent from 1856 to 1969. Invasion by trees was greatest on the deep, permeable, lowland soils. Woody plants increased only slightly on the droughty upland soils. Based on this data, Bragg and Hulbert concluded that burning had been effective in restricting woody plants to natural, presettlement levels. For Minnesota and Nebraska, however, Weaver (1954) concluded that "over most of the territory it seems probable that shrubs and woodland could not extend their areas greatly even if unhandicapped by mowing and prairie fires."

The incentive for burning in the Flint Hills of Kansas was stimulated initially by lease arrangements in the 1880s for transient steer grazing. Lessees required that the lands be burned (Kollmorgen and Simonett 1965) because the forage had higher nutritional value after burning. Livestock gains were 11 kg (25 lb)/steer higher on late spring burns than on adjacent unburned pastures (Smith and Owensby 1972) and growth began 7 to 10 days earlier on burned plots (Kucera and Ehrenreich 1962). Penfound and Kelting (1950) demonstrated that cattle will eat more little bluestem on burned rangeland.

Grasses.   The effect of fire on grasses depends on the site, the amount of soil moisture, and the frequency of burning. However, there is reasonably good agreement among many authors about the effect of fire on grasses in the tallgrass prairie. Big bluestem almost always increases after burning (Robocker and Miller 1955; Kucera and Ehrenreich 1962; McMurphy and Anderson 1965; Hulbert 1969; Anderson et al. 1970). Hadley and Kieckhefer (1963) noted a 275-percent increase in big bluestem one year after burning in Illinois, which is indicative of a decadent plant community before it is burned. Likewise, Indiangrass increases after burning (Dix and Butler 1954; Robocker and Miller 1955; Kucera and Ehrenreich 1962; Hadley and Kieckhefer 1963; Anderson et al. 1970).

Switchgrass has not been studied as intensively as the previous two species. However, Robocker and Miller (1955) found that it increased after burning. In a study by Anderson et al. (1970) there was relatively little switchgrass in the plots and no change was detected. In another study where mulching was applied to plots, the yields of switchgrass decreased with increased mulching (Weaver and Rowland 1952).

Little bluestem also increases after single burns in the true prairie (Hensel 1923; Aldous 1934; Penfound and Kelting 1950; Dix and Butler 1954; Robocker and Miller 1955; Kucera and Ehrenreich 1962). Following eight years of consecutive annual burning, however, Anderson et al. (1970) did not find and change in production of little bluestem, provided the burns were conducted in late spring (May 1). Early spring (March 20) burns reduced yields as much as 25 percent (McMurphy and Anderson 1965; Owensby and Anderson 1967). Soil moisture has to be considerably below normal in this rainfall zone for fire to harm little bluestem (Box and White 1969).

Sideoats grama generally does not change in yield after burning (Hensel 1923; Robocker and Miller 1955; Anderson et al. 1970; Smith and Owensby 1972). Other grasses that increase following early spring or winter burning include prairie junegrass (McMurphy and Anderson 1965), sand dropseed (Hensel 1923), blue grama, and hairy grama (Anderson et al. 1970). Buffalograss was unchanged after 17 consecutive annual burns (Anderson et al. 1970).

Cool-season grasses, particularly the introduced species, are severely harmed by spring burning. Many authors (Hensel 1923; Ehrenreich 1959; Hadley and Kieckhefer 1963; Old 1969) have reported that Kentucky bluegrass *(Poa pratensis)* decreased 80 percent or more following a spring burn. Curtis and Partch (1948) found that Canada bluegrass *(Poa compressa)* and Kentucky bluegrass were severely damaged by spring burning. Similarly, Canada wildrye and Virginia wildrye *(Elymus virginicus)* (Robocker and Miller 1955), Japanese brome *(Bromus japonicus)* (McMurphy and Anderson 1965), and smooth brome *(B. inermis)* (Old 1969) are all damaged by fire. Smooth brome and similar species which begin growth about mid-May are only inhibited by burning, whereas early growing species such as Kentucky bluegrass are almost eliminated by burning (Old 1969). Fall witchgrass *(Leptoloma cognatum)* is favored by spring fires (Penfound 1964).

Kucera (1970) proposed a three-year burning interval to maintain tallgrass dominance as well as to retain the species diversity typical of the native prairie community.

Forbs.   Late spring burning reduces all forbs (McMurphy and Anderson 1965), although the composition of forbs is changed relatively little (Anderson 1965). Major forbs that are harmed by fire include *Petalostemum* species (Hadley 1970), heath aster, and *Solidago* species (Kucera and Koelling 1964). Plains wildindigo *(Baptisia leucophaea)* is favored by fire (Anderson 1965).

Wolfe (1972) studied the effects of a spring wildfire on a prairie sandreed-bluestem association in the Nebraska Sandhills. Herbage growth was reduced as much as 45 percent. Most of the decreaser forbs increased after fire while the increaser forbs declined. Forbs increasing after burning included prairie sunflower *(Helianthus petiolaris)*, dotted gayfeather, Missouri goldenrod, false boneset *(Kuhnia eupatorioides)*, and silky prairieclover *(Petalostemum villosum)*. Those that decreased were pepperweed *(Lepedium* sp.), Virginia dayflower *(Commelina virginica)*, woolly plantain, goosefoot, prairie coneflower, pigweed *(Amaranthus* spp.), and puccoon *(Lithospermum ruderale)*. Western ragweed and Missouri spurge *(Euphorbia missurica)* remained unchanged.

Shrubs and Trees.   There are relatively few species of shrubs in the tallgrass prairie. The shrubs favored by fire include smoothleaf sumac, lead plant (Anderson et al. 1970), and western snowberry (Pelton 1953). Coralberry *(Symphoricarpos*

Fig. 5.11.   Redcedar will invade tallgrass prairie in the absence of fire.

orbiculatus) is slightly harmed by annual spring burning, but will increase dramatically if protected from fire (McMurphy and Anderson 1965).

Eastern redcedar (Fig. 5.11) and western snowberry will invade a protected prairie (Penfound 1964). American elm seedlings establish early after a burn (McMurphy and Anderson 1965), but a later fire will remove the seedlings.

Seed Yields.   Several articles (Burton 1944; Curtis and Partch 1950; Kucera and Ehrenreich 1962; Old 1969) have shown that herbage removal or burning increases flower stalk production. Burton reported an increase in seed yield in burned, grazed, or mowed prairie compared with ungrazed, unburned areas. Curtis and Partch (1950), working in Wisconsin, reported a sixfold increase in flowering on clear-cut sites when compared to unburned areas. The increase was equivalent to that recorded on burned areas. Kucera and Ehrenreich (1962) also found that flower stalks were more numerous on burned areas. The main species affected were big bluestem, little bluestem, and Indiangrass. Quadrant counts in 1960 showed percentage increases attributable to burning of 270, 1200, and 400, respectively.

A detailed study of the effect of litter and burning on flower stalk production was conducted by Old (1969). Cutting and raking increased the number of flower stalks for two years. Burning caused a greater increase in flowering, however, than did cutting (Table 5.3). Seed yields increased because of litter removal, the removal of competing cool-season plants, and increased nitrification. The increased soil temperatures after fire stimulated nitrification. The addition of ash had no effect, but the addition of 225 kg/ha (200 lb/acre) of nitrogen fertilizer increased seed yield more than burning.

Table 5.3.   The Effect of Various Herbage Removal Treatments on Flowering in the Tallgrass Prairie

	Number of Flower Stalks	
Treatment	First Year	Second Year
Burning	102	74
Cut and Raked	66	63
Cut and Left	29	52
Undisturbed	11	34

Data from Old 1969. Copyright 1969, the Ecological Society of America.

Litter.   Prairie closed to both grazing and fire soon begins to deteriorate (Anderson et al. 1970). Accumulation of mulch depresses herbage yield and reduces the number of plant species (Weaver and Tomanek 1951; Ehrenreich 1959). Most of the decreases are associated with lower soil temperatures (Peet et al. 1975).

Reduced herbage yield associated with increased litter is due in part to the amount of ammonium nitrogen. Rice and Pancholy (1973) found that the amount of ammonium nitrogen was lowest in the first successional stage, intermediate in the intermediate successional stage, and highest in the climax. The amount of nitrate nitrogen was highest in the first successional stage, intermediate in the intermediate successional stage, and lowest in the climax. The data indicate that the nitrifiers are inhibited in the climax stage so that ammonium nitrogen is not oxidized to nitrate as readily in the climax as in earlier successional stages.

Northern Great Plains

DISTRIBUTION, CLIMATE, SOILS, AND VEGETATION

The northern Great Plains include the eastern two-thirds of Montana, eastern third of Wyoming, North Dakota, South Dakota, and the western edge of Minnesota. A large part of Minnesota, however, is a mixture of grassland and forest communities. This region extends 400 km (250 miles) into southeastern Alberta, southern Saskatchewan, and the southwestern tip of Manitoba (Moss 1955; Rowe 1972). The plant associations include mixed prairie, tallgrass prairie, fescue prairie, and aspen parkland. The shortgrass association does not occur in the northern Great Plains. The mixed prairie includes eastern Montana, eastern Wyoming, and all but the eastern edges of North Dakota and South Dakota, as well as the southeastern portion of Alberta and southern Saskatchewan. Precipitation varies from 25 to 61 cm (10 to 24 in.), with the driest areas on the western edge of the mixed prairie. Soil textures are primarily sand, sandy loam, silt loam, silty clay loam, and loam. Most soils have developed from glacial till north of the 48th parallel. The last continental ice sheet left the region about 12,000 years ago. Elevation varies from 400 to 1130 m (1300 to 4000 ft). The highest elevations occur in eastern Wyoming and southeastern Montana. From this high, rolling topography, elevation drops to 915 m (3000 ft) in northern Montana and southern Alberta and then to 400 m (1300 ft) in the eastern parts of the northern Great Plains. Topography in northern Montana and the central Dakotas is level to gently rolling with abrupt breaks in many areas.

Annual precipitation averages 25 to 30 cm (10 to 12 in.) in most of the arid parts of eastern Wyoming, and in Montana, southeastern Alberta, and southwestern Saskatchewan (Fig. 5.12). The grasses and sedges are mainly blue grama, needle-and-thread, green needlegrass, western wheatgrass, thickspike wheatgrass *(Agropyron dasystachyum)*, threadleaf sedge *(Carex filifolia)*, Sandberg bluegrass *(Poa sandbergii)*, plains muhly *(Muhlenbergia cuspidata)*, little bluestem, and junegrass. Forbs are not very abundant, but pussytoes *(Antennaria* spp.), moss phlox *(Pholx hoodii)*, little club moss *(Selaginella densa)*, scarlet globemallow, black sampson, silverleaf scurfpea *(Psoralea argophylla)*, prairie crocus *(Anemone patens* var. *wolfgangiana)*, Missouri goldenrod, and others are representative species. Shrubs are predominantly fringed sagebrush *(Artemisia frigida)*, silver sagebrush *(A. cana)*, Nuttall saltbush *(Atriplex nuttallii)*, winterfat, and plains pricklypear.

The more mesic portions of the mixed prairie (Fig. 5.13) average 33 to 46 cm (13 to 18 in.) annual precipitation. This section occupies an arc to the north and east of the more arid areas, occurring in Alberta, Saskatchewan, North Dakota, and South Dakota. Many grass species of the arid section occur in mesic mixed prairie along

Fig. 5.12.   Semiarid mixed prairie (dry mixed prairie) in eastern Wyoming. Dominant species are blue grama, western wheatgrass, and needle-and-thread.

Fig. 5.13. Mesic mixed prairie in North Dakota. Dominant grasses are thickspike wheatgrass, western wheatgrass, junegrass, green needlegrass, western wheatgrass, junegrass, green needlegrass, western porcupine grass, Canada wildrye, sedges, and little bluestem. Fire can be used in this vegetation to control introduced cool-season grasses such as Kentucky bluegrass and smooth bromegrass.

with other species. Western porcupine grass *(Stipa spartea* var. *curtiseta)* and thickspike wheatgrass are usually codominants in the Canadian sections (Coupland 1950). In North Dakota these same species are present with increasing amounts of needle-and-thread, little bluestem, junegrass, blue grama, green needlegrass, and prairie sandreed. In South Dakota the dominant grasses are little bluestem, big bluestem, needle-and-thread, gum needlegrass, junegrass, blue grama, and western wheatgrass, as well as numerous sedges. Kentucky bluegrass and smooth brome, cool-season exotics, have invaded much of the northern mixed prairie in the absence of fire (Kirsch and Kruse 1972).

A variety of forbs (about 25% of total yield) normally occur in the mixed prairie, but major species include locoweed *(Astragalus* and *Oxytropis* spp.), heath aster, aromatic aster *(Aster oblongifolius)*, wild onion, American vetch *(Vicia americana)*, Missouri goldenrod, woolly plantain, penstemon *(Penstemon* sp.), slim-flowered scurfpea, hairy golden-aster *(Chrysopsis villosa)*, moss phlox, little club moss, wild lettuce *(Lactuca pulchella)*, western yarrow *(Achillea millefolium)*, plains erysimum *(Eyrsimum capitatum)*, scarlet gaura *(Gaura coccinea)*, white milkwort *(Polygala alba)*, annual sunflower, herbaceous sage, and dotted gayfeather. Dominant invaders include summercypress, yellow sweetclover *(Melilotus officinalis)*, gumweed *(Grindelia squarrosa)*, and foxtail barley *(Hordeum jubatum)*. A number of shrubs are present including fringed sagebrush, western snowberry, russet buffaloberry *(Shepherdia canadensis)*, silverberry *(Elaeagnus commutata)*, rose, plains pricklypear,

Fig. 5.14. Tallgrass prairie in eastern North Dakota with marsh area.

broom snakewood, Nuttall saltbush, fourwing saltbush, lead plant, willow, and wild plum.

The tallgrass prairie region, which is mostly under cultivation (Fig. 5.14), occupies the eastern edge of North Dakota and South Dakota, the western edge of Minnesota, and the southwestern corner of Manitoba. Surface soil textures are primarily loam, silt loam, and silty clay loam. Elevation varies from 245 to 550 m (800 to 1800 ft), and topography varies from a level to gently rolling glacial plain. Precipitation varies from 46 cm (18 in.) in southwestern Manitoba to 76 cm (30 in.) in south-central Minnesota. Most tallgrasses on native lands are similar to those in the tallgrass prairie of the central Great Plains. However, cool season grasses such as porcupine grass *(Stipa spartea)*, bearded wheatgrass *(Agropyron subsecundum)*, quackgrass *(A. repens)*, slender wheatgrass *(A. trachycaulum)*, smooth brome, and Kentucky bluegrass are most abundant as codominants with the bluestems, switchgrass, prairie cordgrass, and Indiangrass. The shrubs, silverberry, and fringed sagebrush occur in addition to western snowberry, prairie rose, and smoothleaf sumac. Forbs are very similar to those mentioned for tallgrass prairie in the central Great Plains.

Fescue prairie occupies the eastern foothills of the Rocky Mountains in northwestern Montana and southwestern Alberta. It also occurs in the aspen parkland of central Alberta and Saskatchewan (Fig. 5.15), extending eastward to southwestern Manitoba. Throughout the mixed prairie in Canada patches of fescue grassland occur on northerly facing slopes and at higher elevations where the precipitation is most effective (i.e., there is the least evaporation). Annual precipitation ranges from 36 to 46 cm (14 to 18 in.) in the aspen parkland to 38 to 61 cm (15 to 24 in.) in the foothills. Soil textures are primarily sandy loam, silt loam, and loam. Most of the area has been

Fig. 5.15. Fescue prairie in the aspen parkland near Edmonton, Alberta. Dominant grass is rough fescue with some western porcupine grass. A colony of western snowberry is in the foreground.

glaciated. Fescue grasslands occur at elevations as high as 2290 m (7500 ft) in the Rocky Mountains in Canada. Most foothill fescue grasslands are at 1070 to 1830 m (3500 to 6000 ft). Fescue grasslands in the parklands are at 610 m (2000 ft) in central Alberta, descending gradually to 365 m (1200 ft) in southeastern Manitoba. The greatest topographic relief is in the Rocky Mountain foothills. Plains topography ranges from level to sharply rolling.

The fescue prairie can be divided into two sections: a foothills and mountain flora; and a plains or aspen parkland flora. Mountain glaciers left a number of areas uncovered, providing a refugia for the flora. The foothills fescue prairie is a species-rich flora. Continental glaciers apparently eliminated some species from the plains region. About 50 species present in the fescue prairie of the foothills are absent from the fescue prairie in central Alberta and Saskatchewan (Moss and Campbell 1947).

Grasses in the foothills region include the dominant, rough fescue *(Festuca scabrella)*, Parry oatgrass *(Danthonia parryi)*, Idaho fescue *(Festuca idahoensis)*, bluebunch wheatgrass *(Agropyron spicatum)*, intermediate oatgrass *(Danthonia intermedia)*, slender wheatgrass, and hooker's oatgrass *(Helictotrichon hookeri)*. Timothy *(Phleum pratense)*, Kentucky bluegrass, and smooth brome are now common exotic species in valley bottoms and on lower slopes. In foothills, rough fescue is a 0.9-m (3-ft) tall, 25- to 51-cm (10- to 20-in.) diameter bunchgrass (Moss and Campbell 1947), but in parklands is a 46-cm (18-in.) tall, 2.5- to 5.1-cm (1- to 2-in.) diameter bunchgrass having short rhizomes. Looman and Best (1979) have differentiated *Festuca scabrella* Torr. into *Festuca doreana* Loom., the foothills rough fescue, and *Festuca hallii* (Vassey) Piper, the plains rough fescue.

Western porcupine grass is frequently codominant with rough fescue in central Alberta and Saskatchewan. Parry oatgrass, Idaho fescue, and bluebunch wheatgrass are absent. Several grasses from the mixed prairie frequently occur. They include thickspike wheatgrass, blue grama, needle-and-thread, and prairie sandreed. Cool-season grasses from the boreal forest that occur throughout the fescue prairie on more mesic sites include narrow reedgrass *(Calamagrostis neglecta)*, northern reedgrass *(C. inexpansa)*, tufted hairgrass (Deschampsia caespitosa), hairy brome *(Bromus ciliatus)*, sweetgrass *(Hierochloe odorata)*, and purple oatgrass *(Schizachne purpurascens)*. The common sedges are blunt sedge *(Carex obtusata)*, sun sedge, Pennsylvania sedge, and low sedge.

Foothill grasslands are rich in forbs, including yarrow, pussytoes, herbaceous sage, purple aster *(Aster laevis)*, milk vetch, arrowleaf balsamroot *(Balsamorhiza sagittata)*, sticky geranium *(Geranium viscosissimum)*, hedysarum *(Hedysarum americanum)*, dotted gayfeather, puccoon, lupine, peavine *(Lathyrus* spp.), and death camas *(Zygadenus gramineus)*. Forbs are less important in fescue grasslands of the parklands but include yarrow, pussytoes, herbaceous sage, purple aster, milk vetch, Missouri goldenrod, buffalo bean, prairie crocus, moss phlox, American vetch, northern bedstraw *(Galium boreale)* and wild strawberry *(Fragaria virginiana)*.

Shrubby cinquefoil *(Potentilla fruticosa)* increases with grazing intensity in the fescue grasslands of the foothills but does not extend into the plains northeast of Calgary. Other shrubs common throughout the fescue grasslands include silverberry, western snowberry, willow, roses, fringed sagebrush, and serviceberry. The trees, aspen *(Populus tremuloides)* and balsam poplar *(P. balsamifera)*, are invading the more mesic sites throughout the fescue grasslands. Willow and conifers [white spruce *(Picea glauca)* and Douglas fir *(Pseudotsuga menziesii)*] are also invading many grasslands in the Rocky Mountain foothills because of the cessation of wildfires. If trees remain, the soils are expected to change from the productive black chernozems to less productive, less fertile, grey luvisols (Dormaar and Lutwick 1966).

FIRE EFFECTS—SEMIARID MIXED PRAIRIE

Prescribed burning has been studied in the arid portion [less than 33 to 38 cm (13 to 15 in.) precipitation] of the mixed prairie by Clarke et al. (1943). Their study evaluated effects of spring and fall burning of *Stipa-Bouteloua* and *Agropyron smithii-A. dasytachyum* ranges in southeastern Alberta. Coupland (1973) and DeJong and MacDonald (1975) have studied the effect of an August wildfire on equivalent *Agropyron* sites in southwestern Saskatchewan. All fires reduced herbage yield. Coupland (1973) found that one year after the burn production of western thickspike wheatgrass was reduced 19 percent, junegrass was reduced 63 percent, low sedge increased 36 percent, and green needlegrass increased 45 percent. Total grass production was reduced 13 percent. At the end of three growing seasons, current growth was 69 to 73 percent of unburned areas.

DeJong and MacDonald (1975) also studied the effects that an August wildfire had on the soil moisture regime. Major effects of burning were a reduction in soil moisture recharge the first winter after the fire and lower water use the growing season following burning. Less efficient moisture storage on the burned site was probably due to snow blowing off the area. By the second winter soil moisture recharge was the same on burned and unburned grasslands. However soil moisture continued to be lower under burned grassland throughout the first and second growing season after fire. Apparently the wildfire reduced the soil moisture infiltration rate on the clay soils for at least the first two years.

Increased water stress on plants of the burned site was measured by Redmann (1978). Both water potential and osmotic potential of leaves of thickspike wheatgrass and junegrass were lower on the burned areas. The primary effect of burning appeared to be an alteration of microclimate, resulting in the development of an unfavorable plant and soil water status. The duration of this detrimental effect is not known. However, annual herbage production was still 28 percent less on burned than on unburned plots by the third growing season after the wildfire.

Clarke et al. (1943) found that prescribed burning in the spring reduced yield 50 percent the first year and 15 percent the second year, with full recovery the third year. Fall burning decreased yield 30 percent the first year after burning with no significant reduction thereafter. The wheatgrass type (western wheatgrass-thickspike wheatgrass) was more detrimentally affected than the communities dominated by blue grama and needle-and-thread. Vegetation did not recover as rapidly on grazed pastures as on the ungrazed experimental plots. It took three to five years for a burned and grazed pasture to regain normal productivity. This vegetation response is similar to that reported by Hopkins et al. (1948) and Launchbaugh (1964) in the shortgrass prairie of the central Great Plains.

Research results indicate no apparent benefits from burning herbaceous species in the arid mixed prairie where wheatgrasses predominate. Research on needle-and-thread and green needlegrass is inadequate to determine response to prescribed fire. However, data from Coupland (1973) and comments by Clarke et al. (1943) lead us to believe that the needlegrasses are not as detrimentally affected by fire as are the wheatgrasses. On the other hand, fire will kill silver sagebrush (Rowe 1969), fringed sagebrush and little club moss (Dix 1960), and will reduce the vigor of western snowberry. More than three years must be allowed for full recovery of a burned range under grazing (Clarke et al. 1943). Such a long rest would be a high price to pay for shrub control and may not be necessary. Precipitation on the northern Great Plains follows dry and wet cycles. Prescribed burning may be a useful management tool for the control of shrubs and little club moss following winters with above normal precipitation. Only meticulous research will provide the answers.

FIRE EFFECTS—MESIC MIXED PRAIRIE

Dix (1960) studied the effect of three wildfires (May, August, September—all subject to trespass grazing) in western North Dakota where annual precipitation averaged 41 cm (16 in.). He found that a hot, late-May wildfire reduced the frequency of bearded wheatgrass, blue grama, prairie sandreed, needle-and-thread, and green needlegrass. The frequency remained the same or higher for western wheatgrass, low sedge, threadleaf sedge, Pennsylvania sedge, junegrass, and Plains muhly. However, herbage yield might have been reduced for several years. Frequencies of fringed sagebrush, Arkansas rose, and silverleaf scurfpea were significantly reduced, whereas most forbs remained unchanged. Following a fall burn, the long-term evaluations (4 years) showed relatively few changes in botanical composition except for a possible reduction in threadleaf sedge, fringed sagebrush, leafy spurge *(Euphorbia esula)*, and little club moss. Rough hedeoma *(Hedeoma hispida)*, stickseed *(Lappula redowski)*, and herbaceous sage increased. Pennsylvania sedge, prairie sandreed, hairy golden-aster, and wild lettuce were harmed by the late summer fire, but no explanation was given.

On a similar mixed prairie site in Wind Cave National Park in South Dakota, Schripsema (1977) found that an early spring burn (April 21, 1976) increased western wheatgrass, needle-and-thread, buffalograss, and blue grama. Green needlegrass, bluegrasses, and forbs decreased, whereas sedges showed no change. Yields, following the same burn in a drier year (1977), showed similar trends except for decreases in needle-and-thread and sedges. Forb composition was similar to the controls. Thus green needlegrass and the exotic bluegrasses are quite susceptible to early spring burns, but most other grasses are tolerant of such burns if there is moisture in the soil at the time of the burn.

Bluestem sites of the forest-grass ecotone (Fig. 5.16) in the Black Hills area are more tolerant of prescribed fires (Gartner and Thompson 1972; Schripsema 1977). Annual precipitation is 38 to 43 cm (15 to 17 in.). A late spring burn (May 27, 1976) increased little bluestem 31 percent, big bluestem 20 percent, forbs 108 to 405 per-

Fig. 5.16. Mixed prairie that borders the Black Hills of South Dakota. Little bluestem and big bluestem are the dominant grasses. Fire can be used in this vegetation to keep young pines from spreading into the grasslands.

cent, and decreased bluegrasses 65 percent (Schripsema 1977). Silverleaf scurfpea and slim-flowered scurfpea were two forbs that increased noticeably. Minor amounts of western wheatgrass, buffalograss, blue grama, and needlegrasses were present. Western wheatgrass decreased and the other species increased. On the other hand, a winter burn (March 1, 1977) had almost the reverse effect. Big and little bluestem were severely harmed. Western wheatgrass and the needlegrasses increased, and bluegrasses were reduced only 28 percent. Thus late spring burns during years with normal to above average precipitation are the most preferable to increase herbaceous yields of desirable grasses and achieve desired compositional changes.

Kirsch and Kruse (1972) studied the effects of a prescribed spring fire on vegetation at the extreme eastern edge of the mixed prairie in east-central North Dakota, near the tallgrass prairie. Annual precipitation averages 44 cm (17.5 in.). Their purpose in burning was to improve wildlife habitat. The warm-season grasses, big and little bluestem, prairie sandreed, blue grama, and Leiberg panicum *(Panicum leibergii)*, increased in cover after burning. So did the three needlegrasses: needle-and-thread, western porcupine grass, and green needlegrass. Bearded and western wheatgrass maintained themselves with burning but declined in unburned areas (L. Kirsch, pers. comm.). Kentucky bluegrass had high foliar cover before burning in 1969 but was nearly eliminated by three consecutive fires. Quack grass and smooth brome continued to expand on unburned plots but were not present on burned plots.

Most forbs either increased in cover or did not change. Major forb increasers after burning included western ragweed, meadow anemone *(Anemone canadensis)*, candle anemone *(A. cylindrical)*, prairie crocus, heath aster, prairie chickweed *(Cerastium arvense)*, Maximilian sunflower *(Helianthus maximiliani)*, purple prairie clover, and silverleaf scurfpea. Most other forbs, including western yarrow, herbaceous sage, bedstraw, and goldenrod, did not change appreciably. The shrubs, fringed sagebrush, prairie rose, and western snowberry, also did not change appreciably. Canada thistle decreased greatly after burning.

FIRE EFFECTS—TALLGRASS PRAIRIE

A classic account of the effect of fire on tallgrass prairie concerns rejuvenation of the Curtis Prairie in Wisconsin (Curtis and Partch 1948; R. Anderson 1972). This particular area was abandoned from cultivation in 1932. By 1936 the principal perennials were quack grass, Kentucky bluegrass, and Canada bluegrass (Curtis and Partch 1948). Various annuals were also present. A number of prairie forbs and grasses were transplanted into the area in 1936 and 1937 at a density of 2.8 plants per 10 m^2 (2.6 plants per 100 ft^2). It was soon evident that the native species were not maintaining themselves against the bluegrass sod. Fire was introduced experimentally to test the desirability of burning, the best season, and frequency of burning. Since 1950, half to two-thirds of the 8.1-ha (20-acre) prairie has been burned every year.

Annual burning between 1941 and 1946 reduced bluegrass sod by 80 percent and permitted big bluestem, rattlesnake master *(Eryngium yuccifolium)*, goldenrod, tall gayfeather *(Liatris aspera)*, common ragweed *(Ambrosia artemisiifolia)*, heath as-

Table 5.4. Frequency of Plant Species in Curtis Prairie, Stand A, 1961

Species	Frequency (%) 1951	Frequency (%) 1961
Prairie species		
Achillea millefolium	44	55
Ambrosia artemisiifolia	79	53
Andropogon gerardi	6	44
Andropogon scoparius	—	47
Aslepias verticillata	59	61
Eryngium yuccifolium	1	69
Helianthus grosseseratus	—	5
Lactuca canadensis	48	84
Liatris aspera	2	37
Monarda fistulosa	53	73
Ratibida pinnata	6	32
Rudbeckia hirta	—	3
Silphium terebithinacium	2	21
Solidago gigantea	—	3
Solidago nemoralis	—	81
Solidago rigida	—	8
Sorghastrum nutans	—	68
Other species		
Agrostis alba	—	8
Aster pilosus	71	31
Pastinaca sativa	32	11
Poa compressa	79	97
Solidago altissima	32	48
Weeds		
Agropyron repens	29	11
Oxalis stricta	54	44
Poa pratensis	60	13
Trifolium repens	74	43

Data from Cottam and Wilson 1966. Copyright 1966, the Ecological Society of America.

ter, and annual fleabane *(Erigeron annuus)* to increase. Atlantic wildindigo *(Baptisia leucantha)* was not affected by fire and purple coneflower *(Echinacia purpurea)* was the only species reduced by burning. A more recent botanical composition evaulation of the prairie from Cottam and Wilson (1966) illustrates the increase in big bluestem, little bluestem, Indiangrass, *Solidago* sp., *Eryngium* sp., *Lactuca* sp., tall gayfeather, *Ratibida* spp., and *Silphium* sp. (Table 5.4). Quack grass and Kentucky bluegrass were still declining as of 1961. Wild parsnip *(Pastinaca sativa)*, a trouble-some weedy species, is also declining (R. Anderson 1972).

Hadley (1970) reported an increase in big bluestem and little bluestem and a de-crease in Kentucky bluegrass following one spring burn in eastern North Dakota. Other species that increased were prairie dropseed, prairie cordgrass, saltgrass *(Distichlis stricta)*, western snowberry, and Arkansas rose. Wheatgrasses increased on upland sites but declined on lowland sites. Those species that declined were por-cupine grass, needle-and-thread, *Poa* spp., *Muhlenbergia* spp., foxtail barley, prai-rie clover, and heath aster. All of this data was based on yields taken in 1966, a drier than normal year (Hadley 1970). Total herbage yield was 3350 kg/ha (2980 lb/acre) on the unburned plot and 4310 kg/ha (3830 lb/acre) on the burned plot.

Total production of herbage on the Curtis Prairie in Wisconsin one year after burn-ing was 9530 kg/ha (8480 lb/acre) compared to 4700 kg/ha (4180 lb/acre) on the un-burned site (R. Anderson 1972). Similar differences for the same prairie have been shown by Peet et al. (1975). Removal of litter, increased soil temperatures, and day-light in early spring were the main factors contributing to the increased plant growth.

Shrub oak *(Quercus ellipoidalis)* is very abundant in Wisconsin prairies that have been protected from fire for 25 to 80 years (Vogl 1967).

FIRE EFFECTS–FESCUE PRAIRIE

The complexity of fire effects on fescue prairie does not allow a simple assessment of relative detriment and benefits in the Rocky Mountain foothills and the plains area of the aspen parkland (Bailey 1978). Prescribed burning research is very recent and much is still unpublished. Wildfires reduced forage production, plant vigor, basal area, leaf length, and number of flowering culms in the rough fescue-Parry oatgrass community in the foothills of southwestern Alberta (A. Johnston, pers. comm.). A mid-April, 1976 prescribed burn, prior to growth of rough fescue, maintained forage production, reduced basal area, leaf length, and number of flowering culms in the first year when growing season precipitation was normal (S. G. Klumph, pers. comm.). The second year, when precipitation was below normal, forage production was significantly higher on burned areas than on unburned areas.

A July wildfire near Missoula, Montana decreased cover and frequency of bluebunch wheatgrass and Idaho fescue, but cover and frequency of rough fescue was maintained (Mitchell 1958).

The foothill ecotypes of rough fescue are 0.6 to 0.9 m (2 to 3 ft) tall and have 20- to 51-cm (8- to 22-in.) diameter bunches (Moss and Campbell 1947). Response of the big bunches to burning depends upon severity of the fire. Mitchell (1958) observed a tall stubble on the burned rough fescue clumps. This stubble probably insulated the crown from an excessive rise in temperature during burning. He attributed the cause of the maintenance of rough fescue and decline in Idaho fescue and bluebunch wheatgrass under burning to differences in crown characteristics. The fire burned closer to the crown of Idaho fescue and bluebunch wheatgrass. Moreover Sinton (1980) found a greater reduction in herbage yield when weather conditions favored burning plains rough fescue stubble close to the soil surface.

Not all fires burn the foothill ecotypes of rough fescue in the same manner as reported by Mitchell (1958). Wright (1971) found that a bunchgrass with high density of fine fuels maintained the heat longer within the clump than did one with coarse culms and a lower density of fine fuels. Under certain conditions, severe damage and high mortality can be inflicted upon a big bunchgrass like rough fescue. The fire may burn down into the crowns of rough fescue clumps under very dry conditions. A backfire, a slow-moving headfire, or continued burning of clumps after the fire front passes may generate hot fires within plants, resulting in high mortality. Apparently, Idaho fescue's growth form is more susceptible to continuous burning after the passage of a flame front than is foothills rough fescue or bluebunch wheatgrass (Wright and Britton 1976). The latter two grasses develop larger bunches, coarser stems, and less fine fuel close to the growing point than does Idaho fescue. According to Conrad and Poulton (1966), a hot July wildfire in northeastern Oregon killed many more Idaho fescue plants than bluebunch wheatgrass plants.

A complex relationship exists between prescribed fire and plains rough fescue-western porcupine grass communities in the aspen parkland of central Alberta (Bailey and Anderson 1978). The first growing season after early spring or late fall burning total herbage yields were the same as the unburned control. However, there were significant changes in plant composition. Spring burning reduced foliar cover 22 percent and seed production of rough fescue 97 percent; but fall burning was detrimental to western porcupine grass, particularly number of seed heads (reduced 97%). Cool-season plants such as sedge and Hooker's oatgrass were reduced by a spring fire. Fire usually increased cover and frequency of most forbs. The half-shrub fringed sagebrush was reduced by both fires.

Another year, when spring growth was initiated earlier and burning was conducted when rough fescue leaves were about 10 cm (4 in.) tall, rough fescue, bearded wheatgrass, and slender wheatgrass cover was reduced 60 percent the first growing season. It took three years for the species to recover. Perennial forbs more than doubled, particularly milkvetch *(Astragalus striatus, A. flexusus)*, three-flowered avens *(Geum triflorum)*, and western yarrow.

Sinton (1980) conducted burning and mowing trials on plains rough fescue near the study area of M. Anderson (1972). She confirmed Bailey and Anderson's (1978) observations that mid-spring burning after rough fescue had initiated active growth was more detrimental than early spring burning soon after snowmelt. Burning had essentially the same effect on rough fescue as mowing between early April and mid-October. Either treatment at any season reduced herbage production during the first growing season, primarily because of the reduction in leaf length. Burning was least detrimental in early spring or late fall but burning or mowing in the middle of the

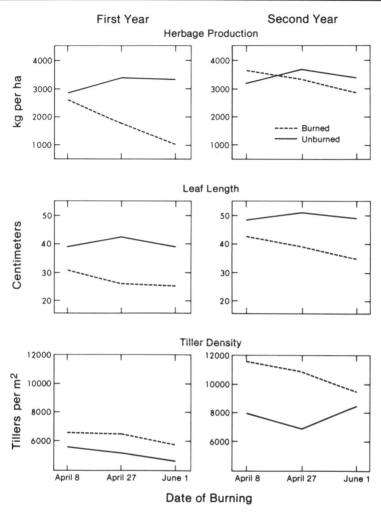

Fig. 5.17. Response of plains rough fescue in the first and second growing season following early-, mid-, and late-spring burning. (Data from Sinton 1980.)

growing season reduced herbage production for at least two years. Rough fescue responded to burning or mowing by increasing the numbers of tillers. For two years after treatment the sward was shorter but more dense. Herbage production surpassed the control in the second growing season in an early spring (April 8) burn because of the increased tiller density and partial recovery of leaf length (Fig. 5.17). A nearly linear relationship was found between herbage production, leaf length, and tiller density following burning one week (April 8), four weeks (April 27), or eight weeks (June 1) after snowmelt. The further into the growing season that burning occurred, the greater was the detrimental effect on this cool-season grass.

The plains rough fescue-western procupine grass community appears to have developed as a fire-climax grassland. It is well adapted to fire. The cessation of burning because of strict fire control enforcement after settlement in the early 1900s has resulted in tree and shrub invasion into fescue prairie (Maini 1960; Johnston and Smoliak 1968; Bailey and Wroe 1974; Scheffler 1976). Annual early spring burning for 25 to 30 years in central Alberta eliminated few plant species, increased the diversity of herbaceous species, and maintained forest cover at about presettlement levels (Anderson and Bailey 1980) (Fig. 5.18). In unburned areas aspen forest cover in-

Fig. 5.18. Top—unburned aspen parkland with dense cover of *Elaeagnus, Symphoricarpos,* and *Festuca* between populus groves. Bottom—aspen parkland burned annually in the spring for 30 years. Grasses predominate with a high density of *Elaeagnus* and a few groves of aspen.

creased from 5 percent in 1940 to 68 percent in 1975. Surface soil organic matter content was 14.7 percent in burned areas versus 10.4 percent in unburned areas. Annual burning, normally conducted two or three weeks after snowmelt, reduced herbage production about the same amount as a single burn at that season of the year. Herbage production of annually burned areas was 66 percent of that of unburned grasslands. Average herbage production on all lands under annual burning was much higher than in adjacent burned areas where 90 percent of the area was in aspen forest and shrubland that produced little herbage.

Annual early spring burning caused an increase in the frequency of many grasses and forbs (Table 5.5). Although canopy cover of the codominant rough fescue and western procupine grass declined under annual burning, frequency increased. Cover of the two grasses was lower primarily because leaf blade length was 40 to 45 percent

Table 5.5. Frequency and Canopy Cover of Common Species Following Early Burning of Grassland Shrublands in Canada

Species[a]	Frequency (%)		Canopy Cover (%)	
	Unburned	Burned	Unburned	Burned
Grasses				
Decreasers				
Festuca scabrella	65	82	36a[b]	18b
Stipa spartea var. *curtiseta*	45	68	15c	11d
Agropyron subsecundum	30	20	3a	1.2b
Stipa viridula	6	0	1.6a	0b
Bromus inermis	5	2	1.2c	0.2d
Bromus ciliatus	4	1	0.3c	+d[c]
Increasers				
Calamovilfa longifolia	3	49	0.7b	41a
Helictotrichon hookeri	2	48	0.1b	2a
Oryzopsis hymenoides	3	16	0.1b	0.6a
Muhlenbergia richardsonis	2	4	0.9d	1.3c
Bouteloua gracilis	0	13	0b	1.3a
Agropyron smithii	0	10	0b	.0.5a
Sedges				
Decreasers				
Carex siccata	14	3	4a	0.5b
Carex deflexa	6	0	1.1a	0b
Carex douglasii	4	0	0.7c	0d
Increasers				
Carex obtusata	10	91	1.1b	18a
Carex heliophila	+	26	+b	14a

(continued)

Table 5.5. Continued

Species[a]	Frequency (%)		Canopy Cover (%)	
	Unburned	Burned	Unburned	Burned
Forbs				
Decreasers				
Aster laevis var. *geyeri*	9	6	1.2e	0.5f
Equisetum laevigatum	15	8	0.5e	0.3f
Agoseris glauca	5	2	0.6a	0.1b
Lathyrus ochroleucus	3	+	0.2a	+b
Anemone patens	2	5	0.1a	+b
Atragalus striatus	5	2	0.2	0.1f
Increasers				
Solidago missouriensis	18	50	1.7b	27a
Thermopsis rhombifolia	14	47	0.7b	3a
Achillea millefolium	23	36	1.1b	3a
Galium boreale	22	20	1.0b	4a
Lathyrus venosus	2	15	0.1b	1.2a
Commandra pallida	9	37	0.3b	2a
Monarda fistulosa	5	9	0.5f	1.3e
Artemisia ludoviciana	24	31	2b	3a
Campanula rotundifolia	17	24	0.5f	0.9e
Vicia americana	4	10	0.2b	0.5a
Vicia sparsifolia	4	8	0.1f	0.3e

From Anderson and Bailey (1980).
[a]The terms "decreasers" and "increasers" refer only to species that had significant changes in canopy cover.
[b]Means in the same row followed by the ssame letter are not significantly different.
[c]Plus (+) refers to less than 0.1 percent frequency or cover.

of that of unburned plants. Litter was reduced from 4500 kg/ha (4000 lb/acre) to 260 kg/ha (230 lb/acre) on the annually burned area (Anderson and Bailey 1980). Annual burning probably created a more arid microenvironment than normally found in the fescue prairie, favoring the mixed prairie grasses: prairie sandreed, Indian ricegrass, blue grama, western wheatgrass, several sedges, Missouri goldenrod, and buffalo bean.

Shrubs of the aspen parkland are well adapted to fire. Annual spring burning reduced the canopy cover of most species, but the frequency and stem density increased for silverberry, serviceberry, prairie rose, and aspen suckers (Table 5.6). Only western snowberry, prickly rose *(Rosa acicularis)*, and wild raspberry declined in frequency and stem density. No shrubs were eliminated by annual burning.

Table 5.6. Frequency, Canopy Cover, and Stem Density of Shrubs Under Early Spring Annual Burning

Shrub	Frequency (%)		Canopy Cover (%)		Stem Density (no./m^2)	
	Unburned	Burned	Unburned	Burned	Unburned	Burned
Symphoricarpos occidentalis	56	52	31a[a]	2b	25y	18z
Elaeagnus commutata	20	34	4a	2b	1.2z	6.4y
Amelanchier alnifolia	8	16	4a	1.4b	1.2z	6.2y
Rosa arkansana	11	24	1.3a	0.6b	1.0y	1.2y
Rosa acicularis	13	1	3a	0.4b	1.4y	0.8z
Rubus strigosus	5	2	1.2a	0.1b	0.4y	0.2z
Populus tremuloides	3	18	0.2b	0.9a	0.2z	2.6y
Artemisia frigida	3	7	0.2b	0.3a	[b]	[b]
Prunus virginiana	6	15	1.9a	1.5a	[b]	[b]
Prunus pensylvanica	2	2	0.4a	0.1a	0.2y	0.2y

Data from Anderson and Bailey (1980).
[a]Means in the same row followed by the same letter are not significantly different.
[b]No data.

Fig. 5.19. Effect of grazing by cattle on burned and seeded aspen forest: (A) ungrazed, 15 months after treatment; (B) grazed, 15 months later; (C) ungrazed, two years later; (D) grazed, two years later (2-m stake marked in decimeters).

Shrubs of the fescue prairie respond differently to a single fire than to annual burn-ing. Anderson and Bailey (1979) burned dense colonies of western snowberry that grow in patches in the grassland. When burned by a single fire the snowberry recov-ered to preburn canopy coverage levels three months after the burn. Stem density of western snowberry and wild raspberry remained at two to five times that of unburned stands for at least three years after burning. One year after spring burning, cover of perennial forbs increased, including American vetch, cream peavine *(Lathyrus ochroleucus)*, vetchling *(L. venosus)*, woundwort *(Stachys palustris* var. *pilosa)*, northern bedstraw, and herbaceous sage. An annual forb, Russian pigweed *(Axyris amaranthoides)*, grew only in burned areas the first year after burning and decreased during the second and third years.

Based on present knowledge, fire has the potential to remove litter buildup on ungrazed fescue grassland and to control shrubs with only moderate damage to the grasses. Forage losses can be minimized by using fire only during those springs that are preceded by above normal precipitation. The desired interval for use of fire to control shrubs is probably five to ten years. More judicious use of these ranges by cat-tle, sheep, and game in combination with burning may enhance the healthiness of these grasslands.

FIRE EFFECTS—ASPEN FOREST

Aspen forest that has invaded the fescue prairie of the aspen parkland and Rocky Mountain foothills can be controlled with fire only (Anderson and Bailey 1980) or fire in combination with grazing and herbicides. Aspen, western snowberry, wil-lows, roses, wild raspberry, wild gooseberry, serviceberry, and cherries *(Prunus* spp.) sucker profusely after burning. New suckers of all species except for western snowberry are palatable to cattle. Forage seed can be broadcast soon after the burn before the ash crusts (Bailey 1977). Heavy grazing will cause severe damage to the woody suckers, provide much forage, and reduce woody competition releasing seeded forages (Fig. 5.19). Beginning with 6280 kg/ha (5600 lb/acre) of available forage 15 months after a burn, cattle consumed 4170 kg/ha (3720 lb/acre), leaving 2110 kg/ha (1880 lb/acre) of woody stems.

As an initial treatment, herbicides can be applied to top-kill the aspen forest. If a prescribed fire follows in two to five years, it should consume trees, logs, dead shrubs, and part of the litter layer, kill woody stems, destroy any residual herbicides above ground, and create an ash seedbed for seeding forages.

Rio Grande Plains and Coastal Prairie

DISTRIBUTION, CLIMATE, SOILS, AND VEGETATION

The Rio Grande plains covers 8 million hectares (20 million acres) in south Texas (Fig. 5.20) and joins the coastal prairie (Fig. 5.21) region [4 million hectares (10 mil-lion acres)] along the gulf coast of Texas and a small portion of Louisiana (Gould

Fig. 5.20.   Mixed brush in the Rio Grande plains of south Texas.

Fig. 5.21.   Coastal prairie near Sinton, Texas. Dominant grass is gulf cordgrass. (Photo courtesy of Rob and Bessie Welder Wildlife Foundation, Sinton, Texas.)

1975). Except for the salt marsh areas along the coast, most of the region is a mixture of brush and prairie. Many areas of the Rio Grande plains are completely dominated by brush.

Precipitation varies from 41 cm (16 in.) along the western edge of the plains to 127 cm (50 in.) along the coast (Gould 1975). Rains occur during all months of the year with peaks in the spring and fall. There are very few days of subfreezing weather. Elevation varies from sea level along the coast to 275 m (900 ft) at the southern end of the Edwards Plateau. Soils in the Rio Grande plains vary from sandy loams to clays, but along the coast clays or clay loams are most prevalent with some sands and sandy loams (Box 1961; Gould 1975).

Vegetation has been described by Gould (1975), Gould and Box (1965), and Box and Chamrad (1966). Near the coast is a salt marsh type that is dominated by gulf cordgrass *(Spartina spartinae)*. Saline sandy soils support seashore dropseed *(Sporobolus virginicus)*, filly panic *(Panicum filipes)*, and sea oats *(Uniola paniculata)*. A few miles inland on less saline soils the dominants become seacoast bluestem *(Schizachyrium scoparium* var. *littoralis)*, big bluestem, Indiangrass, eastern gamagrass, gulf muhly *(Muhlenbergia capillaris* var. *filipe)*, Texas wintergrass, silver bluestem *(Andropogon saccharoides)*, and various species of *Panicum* and *Paspalum*. Further inland on the Rio Grande Plains dominant genera include species of *Setaria, Paspalum, Tridens, Chloris, Trichloris, Cenchrus,* seacoast bluestem, longspike silver bluestem *(Andropogon saccharoides* var. *longipaniculata)*, tanglehead *(Heteropogon contortus)*, Arizona cottontop, buffalograss, curly mesquite *(Hilaria belangeri)*, and species of *Pappophorum* and *Bouteloua*. A wide variety of forbs are also present, especially during wet years.

Shrubs increase dramatically from the coast across the Rio Grande plains. Some of the most prevalent shrubs include huisache *(Acacia farnesiana)*, blackbrush *(Acacia rigidula)*, algerita, granjeno *(Celtis pallida)*, lotebush, brasil *(Condalia obovata)*, honey mesquite, oaks *(Quercus* spp.), Mexican persimmon *(Diospyros texana)*, lycium *(Lycium berlandieri)*, creeping mesquite *(Prosopis reptans)*, prickly ash *(Zanthoxylum fagara)*, and cactus *(Opuntia leptocaulis* and *O. lindheimeri)* (Box et al. 1967).

FIRE EFFECTS

Brush density has increased dramatically in south Texas over the past 50 years (Lehmann 1965; White 1969) and the reduced fire frequency in recent years has probably been the major contributing factor (Lehmann 1965; Box 1967; Box et al. 1967). Lehmann (1965) gives numerous accounts of fire in south Texas and the coastal prairie region during the 1800s and early 1900s.

Research by Box et al. (1967), Box and White (1969), White (1969), and Drawe (1980) shows that fire can be an effective agent to reduce brush density in the Rio Grande plains without harming herbaceous yields. Burning reduced brush cover 24 percent on areas that had not been pretreated by mechanical control. Burning that followed mechanical treatments had reduced brush cover 51 to 83 percent one year after burning. Mortality was low, generally 4 to 14 percent for the 16 most prevalent shrubs, regardless of treatment combinations.

Some treatment combinations with fire, however, can cause high mortality for a few species. Pricklypear species *(Opuntia leptocaulis* and *O. lindheimeri)* were reduced 50 percent by burning (Box et al. 1967) and if burned after being treated with 0.56 kg/ha (0.5 lb/acre) of Tordon 225, mortality was essentially 100 percent (Steuter 1978). The same treatment of Tordon 225 killed 35 to 43 percent of the honey mesquite and fire increased the mortality to 48 percent. For blackbrush and lotebush mortality following spraying and burning was 10 and 4 percent, respectively (Steuter 1978). Bontrager (1977) reported that the application of picloram pellets (0.128 g/cm^2 basal area) to huisache killed only 47 and 16 percent of the small- and medium-sized plants, respectively. However if this treatment was followed by burning 66 percent of the plants died. Burning alone killed only 3 percent of the plants.

Thus the use of fire to control shrubs in the Rio Grande plains seems feasible but initial burning treatments should follow mechanical or chemical treatments (Scifries 1980). Thereafter fires may be conducted every three to ten years, depending on precipitation and rate of regrowth, to keep the shrubs suppressed. Most of the shrubs are vigorous sprouters so high frequencies of shrubs will remain, but burning can improve access to forage.

Grass and forb yields increase 10 to 17 percent after fall or winter burning on the Rio Grande plains (Box and White 1969). Filly panic, Roemer threeawn *(Aristida roemeriana)*, curly mesquite, vine-mesquite, plains bristlegrass, white tridens, and meadow dropseed were more abundant on burns than unburned areas (Box and White 1969). Tumble windmillgrass *(Chloris verticillata)*, hairyseed paspalum *(Paspalum pubiflorum)*, Texas wintergrass, and pink tridens *(Tridens congestus)* were less abundant on burns. Grasses remaining unchanged included little bluestem and buffalograss.

Western ragweed and Texas broomweed *(Xanthocephalum texanum)* are usually harmed by fall burns (Box et al. 1967), whereas saltmarsh aster *(Aster subulatus)*, sawtooth fogfruit *(Phyla incisa)*, crownbeard *(Verbesina microptera)*, and falsemallow *(Malvastrum aurantiacum)* usually increase (Box et al. 1967; Box and White 1969). Other genera that either remained unchanged or decreased included *Cienfuegosia, Commelina, Croton, Desmanthus, Lythrum, Portulaca, Ratibida, Ruellia,* and *Solanum* (Box et al. 1967, Box and White 1969).

Gulf cordgrass, a highly productive grass in the coastal prairie, recovers within four months after early spring burning (April 11) (McAtee et al. 1979). Growth continues throughout most of the year and production doubled 11 months after burning. Then yield begins to decline, but current growth was still 29 percent above the control 27 months after burning. However, McAtee et al. (1979) found that June and July burns were less desirable than early spring burns. Following a burn on June 6, 1976 recovery was 42 percent of control two months after burning and 86 percent of control after 12 months.

Macartney rose *(Rosa bracteata)*, a native of China and Formosa, is a severe management problem in coastal prairies and marshes (Gordon and Scifres 1977). Gordon and Scifres found that Macartney rose could be effectively burned in February with the least damage to grasses and forbs. Herbage yields recovered fully by July 1, increased 16 percent above controls by August 17, and remained 29 percent above controls 15 months after burning. Not only did winter burning result in the highest total herbage yields, but it also provided the best conditions for forb establishment (Gordon and Scifres 1977). Texas croton *(Croton texensis)* and snow-on-the-prairie *(Euphorbia bicolor)* are especially favored after a burn (Scifres 1975).

Cover of Macartney rose was replaced at the rate of 10 to 15 percent per month. However after one year the rose canopies tended to assume a low-spreading growth habit instead of the upright growth-form of unburned plants. Thus burning improved access of forage to livestock for more than one year.

California Prairie

DISTRIBUTION, CLIMATE, SOILS, AND VEGETATION

Annual grasslands of California (Fig. 5.22) cover about 7.1 million hectares (17.5 million acres) in foothills surrounding the Sacramento-San Joaquin Valleys and the coastal mountains of California (Biswell 1956). Elevation varies from near sea level in the interior valleys to 1220 m (4000 ft) in foothills surrounding the valleys and in coastal mountains of California (Heady 1972). Climate is Mediterranean with cool-wet winters and hot-dry summers. Precipitation falls between October and May with two-thirds coming from December to March (Talbot et al. 1939). Average rainfall decreases from west to east and north to south with a maximum of 102 cm/year (40 in./year) occurring in the northern coastal mountain grasslands and a minimum of 7.5 cm/year (5 in./year) along the southwestern side of the San Joaquin Valley (Sampson et al. 1951). Soil textures vary from sandy loam to clays (Biswell 1956).

Original vegetation was dominated principally by purple needlegrass *(Stipa pulchra)* and associated species such as California oatgrass *(Danthonia californica)*, tufted hairgrass, and two perennial *Festucas* (Burcham 1957). Following the introduction of annuals in the nineteenth century, the perennials all but disappeared in the

Fig. 5.22. California prairie grassland in the foothills of northern California. (Photo courtesy of University of California, Hopland Field Station.)

Central Valley and surrounding foothills. Today the climax dominant grasses are California wild oat *(Avena fatua)*, slender wild oat *(Avena barbata)*, and soft chess *(Bromus mollis)*. Medusahead *(Taeniatherum asperum)* will invade these "climax species," but is considered a very undesirable grass. Next in succession sequence includes ripgut brome *(Bromus rigidus)*, Italian rye grass *(Lolium multiflorum)*, red brome *(Bromus rubens)*, Mediterranean barley *(Hordeum hystrix)*, mouse barley *(Hordeum leporinum)*, redstem filaree *(Erodium cicutarium)*, broadleaf filaree *(Erodium botrys)*, and bur clover *(Medicago hispida)*. At the bottom of the successional ladder are various annual fescues *(Festuca* spp.), rabbits foot *(Polypogon monspeliensis)*, silver hairgrass *(Aira caryophyllea)*, and turkey mullein *(Eremocarpus setigerus)*. Yields on annual grasslands vary from 2250 to 5060 kg/ha (2000 to 4500 lb/acre) in a 46- to 102-cm (18- to 40-in.) rainfall belt.

FIRE EFFECTS

Very little is known about the effects of fire on the original composition of California prairie. We suspect that purple needlegrass was fire-tolerant because pure stands of it can be found along railroad tracks in the Central Valley. However, we do not know anything about the ability of the other original climax species to tolerate fire.

Hervey (1949) found that fire reduced yields of annual grasses about 25 percent during the first growing season after a burn. The combined yield of grasses such as California wild oat, slender wild oat, meadow barley *(Hordeum brachyantherum)*, soft chess, and Italian ryegrass was reduced 26 percent, whereas the combined yield of bur clover, redstem filaree, and broadleaf filaree increased sevenfold, or from 4 to 29 percent of the vegetation. The shift in composition to filarees is not desirable since they have a short growing season and then dry up and disintegrate. Bur clover, however, is palatable, high in protein (29% when green and 17% when dry, leached) and stays green later in the growing season than most annual forage plants (Hart et al. 1932). It produces seeds that average 21 percent protein. Pure stands of bur clover will produce from 720 to 2540 kg/ha (640 to 2260 lb/acre) of seeds. Since 3.6 kg (8 lb) of burs will produce 0.45 kg (1 lb) of lamb, a shift in composition to bur clover can be desirable. Most of the annual grasses only contain 11 percent protein when green and 3 to 6 percent when dry (Hart et al. 1932). Soft chess cures with the highest percentage of protein because it retains seed in the plants when dry. It is also the most seriously harmed by fire.

In the 1950s and early 1960s burning was considered a desirable range improvement practice to control medusahead in California (Furbush 1953, Murphy and Turner 1959, Major et al. 1960). A recent study by Young et al. (1972) indicates that fire does not reduce the density of medusahead, even though most of the caryopses produced during the year of the burn can be killed by fire while in the soft dough stage under weather conditions described by McKell et al. (1962). Adequate viable seed for restocking germinate from the previous years' crops that remain in the soil and litter after a burn (Young et al. 1972).

Since the late 1960s burning of medusahead in California has been abandoned (Heady 1972). Medusahead requires after-ripening temperatures below 10°C (50°F) before germination will take place. Thus a lack of germination during the after-ripening period may have misled earlier researchers into believing they were accomplishing more through burning than actually was the case (Young et al. 1972).

The best methods to control stands of medusahead that have invaded stands of climax annuals is by tillage and seeding (Young et al. 1969) or by herbicide application and seeding (Evans et al. 1969). Good stands of perennial grasses remain relatively stable in areas that were formally dominated by medusahead, but will deteriorate to medusahead if the density of perennial grasses becomes depleted (Young et al. 1972).

Palouse Prairie

DISTRIBUTION, CLIMATE, SOILS, AND VEGETATION

The true Palouse prairie (Fig. 5.23) is primarily in eastern Washington and covers about 1.6 million hectares (4 million acres). Most of this area is presently cultivated for commerical wheat. Reminants of this grassland are dominated by bluebunch wheatgrass and Idaho fescue. They are a continuous carpet of plants in which many species are present and the individual plants are difficult to separate. A few big sagebrush *(Artemisia tridentata)* plants occasionally are present. Species composition is very much like that of the sagebrush-grass vegetation with a mixture of grasses and

Fig. 5.23.   Palouse prairie in eastern Washington. (Photo courtesy of University of Idaho.)

forbs except that the latter grows in a more xeric environment with distinct individual bunchgrasses and predominance of sagebrush.

Elevation of the major plateau is 305 to 610 m (1000 to 2000 ft), but drops to 150 m (500 ft) near rivers. Precipitation varies from 30 to 64 cm (12 to 25 in.) annually, most of which falls during the late fall, winter, and early spring months (Daubenmire 1942). Soil is composed mostly of windblown loess.

FIRE EFFECTS

Relatively little fire research has been done in this grassland, although fire has probably played a significant role in eliminating sagebrush (Daubenmire 1942). Daubenmire (1970) has reported on some of the effects of cool fires and hot summer fires. Cool fires generally do not affect the cover of the dominants: bluebunch wheatgrass or Idaho fescue. Since these species grow in a sodlike form in the Palouse prairie, this is a most likely response under desirable prescribed burning conditions. Also cool fires favor rock star *(Lithophragma bulbifera)*, woolly plantain *(Plantago patagonica)*, and Japanese bromegrass. Species reduced by the fire included tansymustard *(Descurania pinnata)*, tumble mustard *(Sisymbrium altissimum)*, and Plains pricklypear.

Following hot summer fires considerable damage is done to perennial grasses (Daubenmire 1970). In one fire, Daubenmire reported that half of the bluebunch wheatgrass plants were killed and only a small part of the basal area of the rest of the plants survived. A similar response could be anticipated for Idaho fescue. Annuals such as *Amsinckia, Helianthus, Lactuca, Leptilon,* and *Sisymbrium* invade abundantly after a fire. Normally, however, all perennials and shrubs sprout readily after a fire in the Palouse prairie, indicating that this community is adapted to fire. Rabbitbrush *(Chrysothamnus* spp.) resprouts vigorously after fire (Daubenmire 1942).

Mountain Grasslands

Mountain grasslands are intermixed with timber stands between the upper limits of the ponderosa pine zone and the alpine ecosystem, usually at elevations from 2440 to 3500 m (8000 to 11,500 ft) (Paulsen 1975). They do not comprise more than 10 percent of the total area, but are very important grazing areas. They produce 1120 to 2250 kg/ha (1000 to 2000 lb/acre) of forage (Paulsen 1975) and contain either a nearly pure stand of grasses or a rich and colorful mixture of grasses and forbs. Some of the forb genera are *Potentilla, Geranium, Osmoriza, Castellejia, Ligusticum, Delphinum, Wyethia, Erigeron, Achillea, Aster,* and *Senecio*. Grasses are generally slender wheatgrass, mountain brome, alpine timothy *(Phleum alpinum)*, Columbia needlegrass *(Stipa columbiana)*, and spike trisetum *(Trisetum spicatum)*. Carexes are also present.

Nimir and Payne (1978) found that after considerable growth had begun in the spring, fire damaged sedges, Idaho fescue, alpine bluegrass *(Poa alpina)*, three-flowered avens *(Geum triflorum)*, tufted phlox *(Phlox caespitosa)* big sagebrush, and shrubby cinquefoil. Species favored included slender wheatgrass, common yarrow, Crag aster *(Aster scopulorum)*, and northwest cinquefoil *(Potentilla gracilis)*. Most other species only showed minor changes in cover or different responses, depending on intensity of burn. Grass yield had recovered about 12 weeks after the burns. Yield of common avens had decreased 89 percent, but was the only forb showing a significant decrease in yield.

For fall burns the effect of fire is essentially unknown. Presumably fall burns would not harm most species for more than a year or two. The forbs would be dormant and most grasses are small in diameter, which would minimize heat damage. The major effect would be similar to top removal in relation to season.

MANAGEMENT IMPLICATIONS

Beneficial Effects

Where fire is an appropriate management tool, the major benefits of prescribed burning in grasslands are to control undesirable shrubs and trees, burn dead debris, increase herbage yields, increase utilization of coarse grasses, increase availability of forage, improve wildlife habitat (more food with unburned patches for cover), and control cool-season species where warm-season grasses are dominant (Wright 1974a). Several objectives can be achieved simultaneously with one burn, the major advantage of using fire as a management tool.

When large amounts of litter cause stagnation in the tallgrass and southern mixed prairie grasslands, fire is an effective tool for increasing plant growth (Weaver and Rowland 1952; Sharrow and Wright 1977a). Removal of the litter permits soil temperatures to rise 6° to 17°C (10° to 30°F) in early spring (Peet et al. 1975) which stimulates nitrification by bacteria (Sharrow and Wright 1977a). The high population of soil bacteria after fire (Neuenschwander 1976) decomposes organic matter to produce additional nitrates. This sequence of events, plus optimum growing temperatures created by the bare soil, allows warm-season plants to grow at an optimum rate if moisture is adequate (Sharrow and Wright 1977a; 1977b). Thus most of the fertilizing effect after a fire comes from nitrates released when bacteria consume organic matter, not from nutrients in ash (Old 1969; Sharrow and Wright 1977a).

The young, tender growth after fire is naturally more palatable and easily accessible to livestock and wildlife. It enables us to use fire effectively to attract livestock to grasses that are normally too coarse and contain too much litter to be palatable. Undesirable cool-season grasses and forbs can be reduced with spring burns. Forbs, which provide an important food source for many upland game birds, are frequently more readily available on burned areas (Kirsch and Kruse 1972). Resprouts of shrubs are not only more accessible but may be more nutritious up to three years after the burn.

Controlling shrubs in the mixed prairie, tallgrass prairie, coastal prairie, and fescue prairie is a major objective for using fire. Fire can be used to top-kill mesquite, completely kill 25 percent of the mesquite on upland sites, kill 50 to 80 percent of all cactus species, and kill Ashe juniper (Britton and Wright 1971; Wink and Wright 1973; Wright et al. 1976a; Bunting et al. 1980). Bragg and Hulbert (1976) have shown conclusive evidence that without fire the tallgrass prairie is easily invaded by shrubs and trees.

Prescribed burning is of importance in the fescue prairie to prevent the further encroachment of aspen groves and to control the density of fire-adapted shrubs. Prescribed fire is not yet a common practice in this grassland primarily because of the vigorous suppression policies that have been in place for decades. Burning of aspen groves followed by broadcast seeding of forages will provide an abundance of low-cost forage and heavy grazing will reduce woody competition and promote establishment of grass and legume seedlings (FitzGerald and Bailey 1981).

In the shortgrass prairie, semiarid northern mixed prairie, California prairie, Palouse prairie, and mountain grasslands fire does not appear to produce major beneficial effects, except in very special situations where it is desirable to control shrubs, improve livestock distribution, or remove litter that has stagnated plant growth.

Potential Impacts

Detrimental effects of fire are generally associated with its misapplication. For example, if burning is conducted during dry years, desirable herbaceous species are seriously harmed and have less than normal herbage yields for at least two years. This is especially true in the shortgrass and arid portions of the mixed prairie. In the northern Great Plains, more cool-season species may be harmed for two to five years, depending upon season of the burn and whether the plants are under moisture stress.

Nevertheless, dry years are often best for long-term effectiveness in killing some shrubs when there are minimum amounts of fine fuel. Eliminating juniper is a good example. Fire may be needed every 20 to 30 years to control nonsprouting juniper, and burning during droughts can be tolerated if the ranges are rested from grazing until herbaceous species have fully recovered.

Erosion following fires is not likely to be a serious problem unless the slopes are steeper than 20 percent (Wright et al. 1976b) or the soils are sandy. Sandy soils are subject to wind erosion when protective plant cover is removed. If burning is considered desirable on such soils, low intensity fires should be used and burning planned so that a mosaic of unburned areas remain.

Smoke can be objectionable near highways and populated areas. Burning should be done when there is a steep adiabatic lapse rate (daytime) so that the smoke will rise and disperse at high altitudes. Burning should be done when winds are blowing away from nearby towns. Nighttime burning is particularly undesirable because smoke will drift and settle in low-lying valleys and stay there until the middle of the next day. These situations can lead to complaints to pollution boards and cause further restrictions on burning. Such incidents can be avoided by burning within the prescribed burning regulations and by burning as rapidly as possible. Most people will tolerate smoke for a few hours, but not for several days.

Fires should not be used more frequently than every five to ten years in the western and central portions of the Great Plains. Long-term declines in production of grasses can result from frequent burns (Neuenschwander et al. 1978). As one goes eastward and the precipitation increases, fire frequency may be every one to three years without harming the grasses (Anderson et al. 1970).

REFERENCES

Albertson, F. W., and J. E. Weaver. 1945. Injury and death or recovery of trees in prairie climate. *Ecol. Monogr.* **15**:393–433.

Aldous, A. E. 1934. Effect of burning on Kansas bluestem pastures. Kansas Agric. Exp. Stn. Tech. Bull. 38. Manhattan.

Anderson, H. G., and A. W. Bailey. 1980. Effects of annual burning on vegetation in the aspen parkland of east central Alberta. *Can. J. Bot.* **58**:985–996.

Anderson, K. L. 1965. Fire ecology—some Kansas prairie forbs. *Proc. Tall Timbers Fire Ecol. Conf.* **4**:153–160.

Anderson, K. L., E. F. Smith, and C. E. Owensby. 1970. Burning bluestem range. *J. Range Manage.* **23**:81–92.

Anderson, M. L. 1972. Effect of fire on grasslands in the Alberta aspen parkland. M.Sc. Thesis. Univ. of Alberta, Edmonton.

Anderson, M. L., and A. W. Bailey. 1979. Effect of fire on a *Symphoricarpos occidentalis* Hook. shrub community in central Alberta. *Can. J. Bot.* **57**:2819–2823.

Anderson, R. C. 1972. The use of fire as a management tool on the Curtis Prairie. *Proc. Tall Timbers Fire Ecol. Conf.* **12**:23–35.

Arend, J. L. 1950. Influence of fire and soil on distribution of eastern red cedar in the Ozarks. *J. For.* **48**:129, 130.

Arno, S. F. 1976. The historical role of fire on the Bitterroot National Forest. USDA For. Serv. Res. Paper INT-187. Intermt. For. and Range Exp. Stn., Ogden, Utah.

Bailey, A. W. 1977. Prescribed burning as an important tool for Canadian rangelands. In S. B. R. Peters and A. W. Bailey (eds.) Range improvement in Alberta: A literature review. Univ. of Alberta Bookstore, Edmonton, pp. 361–381.

Bailey, A. W. 1978. Use of fire to manage grasslands of the Great Plains: Northern Great Plains and adjacent forests. In Proc. 1st Int. Rangeland Cong., Denver, Colo., pp. 691–693.

Bailey, A. W., and M. L. Anderson. 1978. Prescribed burning of a *Festuca-Stipa* grassland. *J. Range Manage.* **31**:446–449.

Bailey, A. W., and R. A. Wroe. 1974. Aspen invasion in a portion of the Alberta parklands. *J. Range Manage.* **27**:263–266.

Bean, W. E., J. P. Goen, and B. E. Dahl. 1975. Livestock response to mechanical brush control and seeding on tobosa ranges. *Research Highlights: Noxious Brush and Weed Control. Range Wildlife Manage.* **6**:22. Texas Tech Univ., Lubbock.

Biswell, H. H. 1956. Ecology of California grasslands. *J. Range Manage.* **9**:19–24.

Blan, K. R. 1970. Evaluation of eastern redcedar *(Juniperus virginiana)* infestations in the northern Kansas Flint Hills. M.S. Thesis. Kansas State Univ., Manhattan.

Bontrager, O. E. 1977. Manipulation of huisache communities with herbicides, burning, and mechanical methods. M.S. Thesis. Texas A & M Univ., College Station.

Box, T. W. 1961. Relationships between plants and soils of four plant communities in south Texas. *Ecology* **42**:794–810.

Box, T. W. 1967. Brush, fire, and west Texas rangeland. *Proc. Tall Timbers Fire Ecol. Conf.* **6**:7–19.

Box, T. W., and A. D. Chamrad. 1966. Plant communities of the Welder Wildlife Refuge. Contr. 5, Series B., Welder Wildlife Foundation, Sinton, Texas.

Box, T. W., J. Powell, and D. Lynn Drawe. 1967. Influence of fire on south Texas chaparral communities. *Ecology* **48**:955–961.

Box, T. W., and R. S. White. 1969. Fall and winter burning of south Texas brush ranges. *J. Range Manage.* **22**:373–376.

Bragg, T. B., and L. C. Hulbert. 1976. Woody plant invasion of unburned Kansas bluestem prairie. *J. Range Manage.* **29**:19–24.

Britton, C. M., and H. A. Wright. 1971. Correlation of weather and fuel variables to mesquite damage by fire. *J. Range Manage.* **24**:136–141.

Bunting, S. C., H. A. Wright, and L. F. Neuenschwander. 1980. The long-term effects of fire on cactus in the southern mixed prairie of Texas. *J. Range Manage.* **33**:85–88.

Burcham, L. T. 1957. California rangeland. Calif. Div. For., Sacramento.

Burton, G. W. 1944. Seed production of several southern grasses as influenced by burning and fertilization. *Amer. Soc. Agron. J.* **36**:523–529.

Canfield, R. H. 1939. The effect of intensity and frequency of clipping on density and yield of black grama and tobosagrass. USDA Tech. Bull. 681. Washington, D.C.

Chapman, H. H. 1926. Factors determining natural reproduction of longleaf pine on cut-over lands in La Salle Parish, La. Yale Univ. School of Forest., New Haven, Conn. Bull. 16.

Chapman, H. H. 1944. Fires and pines. *Amer. For.* **40**:62–64; 91–93.

Clarke, S. E., E. W. Tisdale, and N. A. Skoglund. 1943. The effects of climate and grazing on shortgrass prairie vegetation. *Can. Dominion Dept. Agric. Tech. Bull.* **46**:53.

Conrad, E. C., and C. E. Poulton. 1966. Effect of wildfire on Idaho fescue and bluebunch wheatgrass. *J. Range Manage.* **19**:138–141.

Cottam, G., and H. C. Wilson. 1966. Community dynamics on an artificial prairie. *Ecology* **47**:88–96.

Coupland, R. T. 1950. Ecology of mixed prairie in Canada. *Ecol. Monogr.* **20**:271–315.

Coupland, R. T. 1973. Producers: I. Dynamics of above-ground standing crop. Matador Project. Canadian IBP Prog. Tech. Rep. No. 27. Saskatoon, Sask.

Curtis, J. T., and M. L. Partch. 1948. Effect of fire on competition between bluegrass and certain prairie plants. *Amer. Midl. Natur.* **39**:437–443.

Curtis, J. T., and M. L. Partch. 1950. Some factors affecting flower stalk production in *Andropogon gerardi*. *Ecology* **31**:488, 489.

Dahl, B. E., and J. P. Goen. 1973. 2,4,5-T plus fire for management of tobosa grassland. *Research Highlights: Noxious Brush and Weed Control. Range Wildlife Manage.* **4**:20. Texas Tech Univ., Lubbock.

Dalrymple, R. L. 1969. Prescribed grass burning for Ashe juniper control. Prog. Rep. from Noble Foundation, Inc., Ardmore, Okla.

Daubenmire, R. 1942. An ecological study of the vegetation of south-eastern Washington and adjacent Idaho. *Ecol. Monogr.* **12**:53–80.

Daubenmire, R. 1970. Steppe vegetation of Washington. Washington Agric. Exp. Stn., Tech. Bull. 62. Pullman, Wash.

DeJong, E., and K. B. MacDonald. 1975. The soil moisture regime under native grassland. *Geoderma* **14**:207–221.

Dix, R. L. 1960. The effects of burning on the mulch structure and species composition of grasslands in western North Dakota. *Ecology* **41**:49–56.

Dix, R. L., and J. E. Butler. 1954. The effects of fire on a dry, thin-soil prairie in Wisconsin. *J. Range Manage.* **7**:265–268.

Dormaar, J. F., and L. E. Lutwick. 1966. A biosequence of soils of the rough fescue prairie-poplar transition in southwestern Alberta. *Can. J. Earth Sci.* **3**:457–471.

Drawe, L. D. 1980. The role of fire in the coastal prairie. In C. W. Hanselka (ed.) Prescribed range burning in the Coastal Prairie and eastern Rio Grande Plains of Texas. Texas Agric. Exp. Serv. Bull. College Station, pp. 101–113.

Dwyer, D. D. 1972. Burning and nitrogen fertilization of tobosagrass. New Mexico State Univ. Agric. Exp. Stn. Bull. 595. Las Cruces.

Dwyer, D. D., and R. D. Pieper. 1967. Fire effects of blue grama-pinyon-juniper rangeland in New Mexico. *J. Range Manage.* **20**:359–362.

Ehrenreich, J. H. 1959. Effect of burning and clipping on growth of native prairie in Iowa. *J. Range Manage.* **12**:133–137.

Evans, R. A., J. A. Young, and R. E. Eckert, Jr. 1969. Herbaceous weed control and revegetation of semi-arid rangelands of western United States. *Outlook on Agric.* **6**(2):60–66.

Fidler, P. 1793. Diary of Peter Fidler for the period 1792–1793. *Glenbow-Alberta Inst.*, Calgary.

Fisher, C. E. 1947. Present information on the mesquite problem. Texas Agric. Exp. Stn. Rep. 1056. College Station.

FitzGerald, R. D., and A. W. Bailey. 1981. Influence of grazing with cattle in establishment of forage in aspen brushland following burning. XIV Int. Grassland Congress. (In press.)

Furbush, P. 1953. Control of medusahead on California ranges. *J. For.* **51**:118–121.

Gartner, F. G., and W. W. Thompson. 1972. Fire in the Black Hills forest-grass ecotone. *Proc. Tall Timbers Fire Ecol. Conf.* **12**:37–68.

Gordon, R. A., and C. J. Scifres. 1977. Burning for improvement of Macartney rose-infested coastal prairie. Texas A. & M. Exp. Stn. Pub. No. B-1183. College Station.

Gould, F. W. 1975. Texas plants: A checklist and ecological summary. Texas Agric. Exp. Stn. MP-585. College Station.

Gould, F. W., and T. W. Box. 1965. *Grasses of the Texas Coastal Bend.* Texas A & M Univ. Press, College Station.

Hadley, E. B. 1970. Net productivity and burning responses of native eastern North Dakota prairie communities. *Amer. Midl. Natur.* **84**:121–135.

Hadley, E. B., and B. J. Kieckhefer. 1963. Productivity of two prairie grasses in relation to fire frequency. *Ecology* **44**:389–395.

Haley, J. E. 1929. Grass fires of the southern Great Plains. *West Texas Hist. Year Book* **5**:23–42.

Hall, F. C. 1976. Fire and vegetation on the Blue Mountains—implications for land managers. *Proc. Tall Timbers Fire Ecol. Conf.* **15**:155–170.

Hart, G. H., H. R. Guilbert, and H. Goss. 1932. Seasonal changes in the chemical composition of range forage and their relation to nutrition of animals. Calif. Agric. Exp. Stn. Bull. 543. Berkeley.

Hastings, J. R., and R. M. Turner. 1966. *The changing mile.* Univ. of Ariz. Press, Tucson.

Heady, H. F. 1972. Burning and the grasslands in California. *Proc. Tall Timbers Fire Ecol. Conf.* **12**:97–107.

Heirman, A. L., and H. A. Wright. 1973. Fire in the medium fuels of West Texas. *J. Range Manage.* **26**:331–335.

Hensel, R. L. 1923. Recent studies on the effect of burning on grassland vegetation. *Ecology* **4**:183–188.

Herbel, C. H., and A. B. Nelson. 1966. Species preference of Hereford and Santa Gertrudis cattle on a southern New Mexico range. *J. Range Manage.* **19**:177–181.

Hervey, D. F. 1949. Reaction of a California annual-plant community to fire. *J. Range Manage.* **2**:116–121.

Hopkins, H., F. W. Albertson, and A. Riegel. 1948. Some effects of burning upon a prairie in west-central Kansas. *Kansas Acad. of Sci. Trans.* **51**:131–141.

Hulbert, L. C. 1969. Fire and litter effects in undisturbed bluestem prairie in Kansas. *Ecology* **50**:874–877.

Humphrey, R. R. 1962. *Range Ecology.* Ronald Press., New York.

Jackson, A. S. 1965. Wildfires in the Great Plains grasslands. *Proc. Tall Timbers Fire Ecol. Conf.* **4**:241–259.

Johnston, A., and S. Smoliak. 1968. Reclaiming brushland in southwestern Alberta. *J. Range Manage.* **21**:404–406.

Kirsch, L. M., and A. D. Kruse. 1972. Prairie fires and wildlife. *Proc. Tall Timbers Fire Ecol. Conf.* **12**:289–303.

Klett, W. E., D. Hollingsworth, and J. L. Schuster. 1971. Increasing utilization of weeping lovegrass by burning. *J. Range Manage.* **24**:22–24.

Kollmorgen, W. M., and D. S. Simonett. 1965. Grazing operations in the Flint Hills bluestem pastures of Chase County, Kansas. *Annals. Assoc. Amer. Geogr.* **55**:260–290.

Komarek, E. V., Sr. 1966. The meteorological basis for fire ecology. *Proc. Tall Timbers Fire Ecol. Conf.* **5**:85–125.

Kucera, C. L. 1960. Forest encroachment in native prairie. *Iowa State J. Sci.* **34**:635–639.

Kucera, C. L. 1970. Ecological effects of fire on tallgrass prairie. Proc. Symp. of Prairie and Prairie Restoration. Knox College, Galesburg, Ill.

Kucera, C. L., and J. H. Ehrenreich. 1962. Some aspects of annual burning on central Missouri prairie. *Ecology* **43**:334–336.

Kucera, C. L., and M. Koelling. 1964. The influence of fire on composition of central Missouri prairie. *Amer. Midl. Natur.* **72**:142–147.

Küchler, A. W. 1964. Potential natural vegetation of the conterminous United States. Manual to accompany the map. Amer. Geogr. Soc. Spec. Pub. 36. (With map, rev. ed., 1965, 1966.)

Launchbaugh, J. L. 1964. Effects of early spring burning on yields of native vegetation. *J. Range Manage.* **17**:5–6.

Launchbaugh, J. L. 1972. Effect of fire on shortgrass and mixed prairie species. *Proc. Tall Timbers Fire Ecol. Conf.* **12**:129–151.

Lehmann, R. W. 1965. Fire in the range of Attwater's prairie chicken. *Proc. Tall Timbers Fire Ecol. Conf.* **4**:127–143.

Looman, J., and K. F. Best. 1979. Budd's flora of the Canadian prairie provinces. Can. Agric. Pub. 1662. Ottawa, Ont.

Maini, J. S. 1960. Invasion of grassland by *Populus tremuloides* in the northern Great Plains. Ph.D. Diss., Univ. Saskatchewan, Saskatoon.

Major, J., C. M. McKell, and L. J. Berry. 1960. Improvement of medusahead infested rangeland. Calif. Agric. Exp. Stn., Ext. Serv. Leaf. 123. Davis.

Malin, J. C. 1953. Soil, animal, and plant relations of the grasslands, historically reconsidered. *Sci. Monthly* **76**:207–220.

Marcy, R. B. 1849. Report of Captain R. B. Marcy. House Exec. Doc. 45, 31st Congr., 1st Session, Pub. Doc. 577. Washington, D.C.

McAtee, J. W., C. J. Scifres, and D. L. Drawe. 1979. Improvement of gulf cordgrass range with burning or shredding. *J. Range Manage.* **32**:372–375.

McIlvain, E. H., and C. G. Armstrong. 1966. A summary of fire and forage research on shinnery oak rangelands. *Proc. Tall Timbers Fire Ecol. Conf.* **5**:127–129.

McIlvain, E. H., and C. G. Armstrong. 1968. Progress in range research. Woodward Brief 542. Woodward, Okla. (mimeo.)

McKell, C. M., A. M. Wilson, and B. L. Kay. 1962. Effective burning of rangelands infested with medusahead. *Weeds* **10:**125–131.

McMurphy, W. E., and K. L. Anderson. 1965. Burning Flint Hills range. *J. Range Manage.* **18:**265–269.

Michler, N., Jr. 1850. Routes from the western boundary of Arkansas to Santa Fe and the valley of the Rio Grande. House Exec. Doc. 67, 31st Congr., 1st Session, Public Doc. 577. Washington, D.C.

Mitchell, W. W. 1958. An ecological study of the grasslands in the region of Missoula, Montana. M.A. Thesis. Univ. Montana, Missoula.

Moss, E. H. 1932. The vegetation of Alberta. IV. The poplar association and related vegetation of central Alberta. *J. Ecol.* **20:**380–415.

Moss, E. H. 1955. Vegetation of Alberta. *Bot. Rev.* **21:**493–567.

Moss, E. H., and J. A. Campbell. 1947. The fescue grassland of Alberta. *Can. J. Res.* **25**(C):209–227.

Murphy, A. H., and D. Turner. 1959. A study on the germination of medusahead seed. *Calif. Dept. Agric. Bull.* **48**(1):6–10. Davis.

Nelson, J. C., and R. E. England. 1971. Some comments on the causes and effects of fire in the northern grasslands area of Canada and the nearby United States, Ca. 1750–1900. *Can. Geogr.* **15:**295–306.

Neuenschwander, L. F. 1976. The effect of fire on a tobosagrass-mesquite community on Stamford clay soils. Ph.D. Diss. Texas Tech Univ., Lubbock.

Neuenschwander, L. F., H. A. Wright, and S. C. Bunting. 1978. The effect of fire on a tobosagrass-mesquite community in the Rolling Plains of Texas. *Southwestern Natur.* **23:**315–338.

Nimir, M. B., and G. F. Payne. 1978. Effects of spring burning on a mountain range. *J. Range Manage.* **31:**259–263.

Old, S. M. 1969. Microclimate, fire, and plant production on an Illinois prairie. *Ecol. Monogr.* **39:**355–384.

Owensby, C. E., and K. L. Anderson. 1967. Yield response to time of burning in the Kansas Flint Hills. *J. Range Manage.* **20:**12–16.

Paulsen, H. A., Jr. 1975. Range Management in the central and southern Rocky Mountains: A summary of the status of our knowledge by range ecosystems. USDA For. Serv. Res. Pap. RM-154. Rocky Mountain For. and Range Exp. Stn., Fort Collins, Colo.

Paulsen, H. A., and F. N. Ares. 1962. Grazing values and management of black grama and tobosa grasslands and associated shrub ranges of the southwest. USDA Tech. Bull. 1270. Washington, D.C.

Peet, M., R. Anderson, and M. S. Adams. 1975. Effect of fire on big bluestem production. *Amer. Midl. Natur.* **94:**15–26.

Pelton, J. 1953. Studies on the life-history of *Symphoricarpos occidentalis* in Minnesota. *Ecol. Monogr.* **23:**17–39.

Penfound, W. T. 1964. The relation of grazing to plant succession in the tallgrass prairie. *J. Range Manage.* **17:**256–260.

Penfound, W. T., and R. W. Kelting. 1950. Some effects of winter burning on a moderately grazed pasture. *Ecology* **31:**554–560.

Redmann, R. E. 1978. Plant and soil water potentials following fire in a northern mixed grassland. *J. Range Manage.* **31:**443–445.

Renwald, J. D. 1977. Effect of fire on lark sparrow nesting densities. *J. Range Manage.* **30:**283–285.

Renwald, J. D., H. A. Wright, and J. T. Flinders. 1978. Effect of prescribed fire on bobwhite quail habitat in the Rolling Plains of Texas. *J. Range Manage.* **31:**65–69.

Rice, E. L., and S. K. Pancholy. 1973. Inhibition of nitrification by climax ecosystems. II. Additional evidence and possible role of tannins. *Amer. J. Bot.* **60:**691–702.

Robocker, C. W., and B. J. Miller. 1955. Effects of clipping, burning, and competition on establishment and survival of some native grasses in Wisconsin. *J. Range Manage.* **8**:117–121.

Rowe, J. S. 1969. Lightning fires in Saskatchewan grassland. *Can. Field Natur.* **83**:317–342.

Rowe, J. S. 1972. Forest regions of Canada. Dept. of the Environment, Can. For. Serv., Pub. No. 1300.

Sampson, A. W., A. C. Chase, and D. W. Hedrick. 1951. California grasslands and range forage grasses. Calif. Agric. Exp. Stn. Bull. 724. Berkeley.

Sauer, C. O. 1944. A geographic sketch of early man in America. *Geogr. Rev.* **34**:529–573.

Scheffler, E. J. 1976. Aspen forest vegetation in a portion of the east central Alberta parklands. M.Sc. Thesis, Univ. of Alberta, Edmonton.

Schripsema, J. R. 1977. Ecological changes on pine-grassland burned in spring, late spring, and winter. M.A. Thesis. Biology-Botany Dept., South Dakota State Univ., Brookings.

Scifres, C. J. 1975. Systems for improving Macartney rose infested coastal prairie rangeland. Texas A. & M. Exp. Stn. Pub. No. MP-1225. College Station.

Scifres, C. J. 1980. Integration of prescribed burning with other brush control methods: The systems concept of brush management. In L. D. White (ed.) Prescribed burning in the Rio Grande Plains of Texas. Texas Agric. Ext. Serv. Bull. College Station, pp. 44–50.

Seevers, P. M., P. N. Jensen, and J. V. Drew. 1973. Satellite imagery for assessing range fire damage in the Nebraska Sandhills. *J. Range Manage.* **26**:462, 463.

Shantz, H. L., and R. Zon. 1924. Atlas of American agriculture. Part I(E): The natural vegetation of the United States. USDA. Washington, D.C.

Sharrow, S. H., and H. A. Wright. 1977a. Effects of fire, ash, and litter on soil nitrate, temperature, moisture, and tobosagrass production in the Rolling Plains. *J. Range Manage.* **20**:266–270.

Sharrow, S. H., and H. A. Wright. 1977b. Proper burning intervals for tobosagrass in west Texas based on nitrogen dynamics. *J. Range Manage.* **30**:343–346.

Sinton, H. M. 1980. Effect of burning and mowing on *Festuca hallii* (Vassey) Piper *(Festuca scabrella* Torr.). M.Sc. Thesis. Univ. of Allberta, Edmonton.

Smith, E. F., and C. E. Owensby. 1972. Effects of fire on true prairie grasslands. *Proc. Tall Timbers Fire Ecol. Conf.* **12**:9–22.

Smith, M. A., H. A. Wright, and J. L. Schuster. 1975. Reproductive characteristics of redberry juniper. *J. Range Manage.* **28**:126–128.

Steuter, A. A. 1978. Response of wildlife to brush control in the Rio Grande Plains. M.S. Thesis. Texas Tech Univ., Lubbock.

Stewart, O. C. 1951. Burning and natural vegetation in the United States. *Geogr. Rev.* **41**:317–320.

Stewart, O. C. 1953. Why the Great Plains are treeless. *Colo. Quart.* **1**:40–50. Univ. Colorado, Boulder.

Talbot, M. W., H. H. Biswell, and A. L. Hormay. 1939. Fluctuations in the annual vegetation of California. *Ecology* **20**:394–402.

Trlica, M. J., and J. L. Schuster. 1969. Effects of fire on grasses of the Texas High Plains. *J. Range Manage.* **22**:329–333.

Vogl, R. J. 1967. Controlled burning for wildlife in Wisconsin. *Proc. Tall Timbers Fire Ecol. Conf.* **6**:47–96.

Wagener, W. 1961. Past fire incidence in the Sierra Nevada forests. *J. For.* **59**:739–747.

Weaver, H. 1951. Fire as an ecological factor in the southwestern ponderosa pine forests. *J. For.* **49**:93–98.

Weaver, J. E. 1954. *North America Prairie.* Johnsen Publ. Co., Lincoln, Nebr.

Weaver, J. E., and F. W. Albertson. 1956. Grasslands of the Great Plains: Their nature and use. Johnsen Publ. Co., Lincoln, Nebr.

Weaver, J. E., and F. E. Clements. 1938. *Plant Ecology,* 2nd ed. McGraw-Hill, New York.

Weaver, J. E., and N. W. Rowland. 1952. Effects of excessive natural mulch on development, yield and structure of native grassland. *Bot. Gaz.* **114:**1–19.

Weaver, J. E., and G. W. Tomanek. 1951. Ecological studies in a mid-western range: The vegetation and effects of cattle on its composition and distribution. Nebr. Cons. Bull. 31.

Wedel, W. R. 1957. The central North American grassland: man-made or natural? In Social Sci. Monogr., Ill., Anthropol. Soc. Wash., pp. 39–69.

Wells, P. V. 1970. Postglacial vegetational history of the Great Plains. *Science* **167:**1574–1582.

White, R. S. 1969. Fire temperatures and the effect of burning on south Texas brush communities. M. S. Thesis. Texas Tech Univ., Lubbock.

Wink, R. L., and H. A. Wright. 1973. Effects of fire on an Ashe juniper community. *J. Range Manage.* **26:**326–329.

Wolfe, C. W. 1972. Effects of fire on a Sandhills grassland environment. *Proc. Tall Timbers Fire Ecol. Conf.* **12:**241–255.

Wright, H. A. 1969. Effect of spring burning on tobosagrass. *J. Range Manage.* **22:**425–427.

Wright, H. A. 1971. Why squirreltail is more tolerant to burning than needle-and-thread. *J. Range Manage.* **24:**277–284.

Wright, H. A. 1972a. Fire as a tool to manage tobosa grasslands. *Proc. Tall Timbers Fire Ecol. Conf.* **12:**153–167.

Wright, H. A. 1972b. Shrub response to fire. In Wildland and Shrubs—Their Biology and Utilization. An Int. Symp., Utah State Univ., Logan. USDA For. Serv. Gen. Tech. Rep. INT-1. Intermt. For. and Range Exp. Stn., Ogden, Utah, pp. 204–217.

Wright, H. A. 1974a. Range burning. *J. Range Manage.* **27:**5–11.

Wright, H. A. 1974b. Effect of fire on southern mixed prairie grasses. *J. Range Manage.* **27:**417–419.

Wright, H. A., and C. M. Britton. 1976. Fire effects on vegetation in western rangeland communities. In Use of Prescribed Burning in Western Woodland and Range Ecosystems: A Symposium. Utah State Univ., Logan, pp. 35–41.

Wright, H. A., S. C. Bunting, and L. F. Neuenschwander. 1976a. Effect of fire on honey mesquite. *J. Range Manage.* **29:**467–471.

Wright, H. A., F. M. Churchill, and W. Clark Stevens. 1976b. Effect of prescribed burning on sediment, water yield, and water quality from dozed juniper lands in central Texas. *J. Range Manage.* **29:**294–298.

Young, J. A., R. A. Evans, and R. E. Eckert, Jr. 1969. Population dynamics of downy brome. *Weed Sci.* **17:**20–26.

Young, J. A., R. A. Evans, and J. Robison. 1972. Influence of repeated annual burning on medusahead community. *J. Range Manage.* **25:**372–375.

CHAPTER

6

SEMIDESERT GRASS–SHRUB

G rasslands in the semidesert grass-shrub type have gradually given way to higher
and higher densities of shrubs during the past 75 years (Fig. 6.1), but the mechanisms by which the invasion has taken place are not well understood (Buffington and
Herbel 1965; Martin 1975). The historical role of fire is especially perplexing because fires that kill shrubs usually kill grasses too (Martin 1975). Buffington and
Herbel (1965) were also skeptical about the role of fire in desert grasslands and did
not consider fire as a factor in the maintenance of brush-free range in southern New
Mexico. However, Thornber (1907, 1910), Griffiths (1910), Wooton (1916), Leopold (1924), and Humphrey (1958) were convinced that fire controlled shrubs in
those portions of the semidesert grasss-shrub type that had sufficient fine fuel. ''That
such fires burning over the mesas and foothills have not been uncommon in times
past may be judged by the fact that in many places abundant remains of charred
stumps of at least 10 years duration are frequently met with'' (Thornber 1910).
Wooten (1916) commented on fires severe enough to kill plants 3.0 to 3.7 m (10 to 12
ft) high.
   Despite the skepticism about fire in controlling shrubs in desert grasslands, climate has not changed enough to account for the rapid increase of shrubs (Gardner
1951; Paulsen 1956; Humphrey 1958; Buffington and Herbel 1965). Moreover, once
velvet mesquite *(Prosopis glandulosa* var. *velutina)* or honey mesquite *(P. glandulosa* var. *glandulosa)* seed-trees were present, mesquite seedlings increased in
Arizona and New Mexico whether pastures were protected or grazed (Griffiths 1910;

Fig. 6.1.   (a) Open grass-covered area on the Santa Rita Experimental Range in 1903. There are very few widely scattered velvet mesquite trees. (b) Same area as 6.1a in 1941 with a well developed stand of velvet mesquite. (Photos courtesy of USDA Forest Service.)

Wooten 1916; Leopold 1924; Glendening 1952; Buffington and Herbel 1965). This leaves the distinct possibility that occasional fires, in combination with drought, competition, rodents, and lagomorphs, played a significant role in controlling shrubs in the semidesert grass-shrub type (Griffiths 1910; Wooton 1916; Leopold 1924; Branscomb 1956; Humphrey 1958; Bock et al. 1976), except on black grama *(Bouteloua eriopoda)* uplands (Buffington and Herbel 1965).

DISTRIBUTION, CLIMATE, SOILS, AND VEGETATION

Distribution

Semidesert grass-shrub vegetation occurs in broad basins, slightly sloping drainages, and lower slopes of the southern Rocky Mountains in southeastern Arizona, southern New Mexico, and southwestern Texas (Fig. 6.2) (Humphrey 1958; Martin 1975;

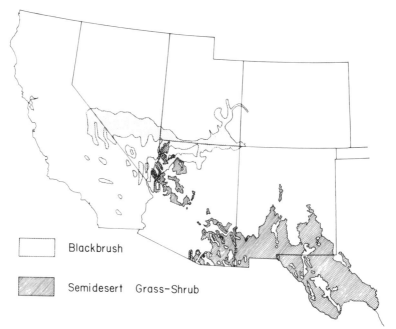

Fig. 6.2. Distribution of semidesert-grass shrub vegetation (from Küchler 1964, reproduced with permission of the American Geographical Society) and blackbrush vegetation (from Bowns and West 1976, reproduced with permission of the authors and the Utah Agricultural Experiment Station).

Bunting 1978). Approximately 17,813,765 ha (44,000,000 acres) occur in the United States but the center of the semidesert grass-shrub type lies in Mexico (Clements 1920). Elevation of this vegetation type usually ranges from 915 to 1400 m (3000 to 4500 ft) in Arizona and New Mexico (Martin and Cable 1974), but occurs as high as 1740 m (5710 ft) in southwestern Texas (Bunting 1978). Below the semidesert shrub-grassland lies the Chihuahuan desert and above it the vegetation gives way to chaparral, pinyon (*Pinus* spp.)-juniper (*Juniperus* spp.), oak (*Quercus* spp.) woodland, or occasionally grassland (Martin 1975).

Climate

Average annual precipitation ranges from 20 cm (8 in.) at the western edge of Tucson to 50 cm (20 in.) on the lower slopes of mountain ranges in southeastern Arizona and southwestern Texas (Martin 1975). In south-central New Mexico, average annual rainfall on the Jornada Experimental Range is 23 cm (9.1 in.). Throughout the semidesert grass-shrub region over 50 percent of the annual rainfall occurs between July 1 and September 30 (Hinckley 1944; Buffington and Herbel 1965; Cable 1972).

Soils

Soils vary widely from sandy or gravelly loams to clays in both the surface and sub-soil (Buffington and Herbel 1965; Cable 1972). They have developed primarily in al-luvium from the adjacent igneous and limestone mountains and are characteristically immature, light-colored, and low in organic matter (Bunting 1978). Light precipitation and high evaporation often result in an accumulation of salts in and below the soil in basins. Thus concentrations of $CaCO_4.2H_2O$ (gypsum) and $CaCO_3$ (lime) occur in many soils throughout the region (Carter and Cory 1930).

Mesa and upland soils of the black grama community are compacted sands, shallow, and usually 30 cm (12 in.) or less in depth. Frequently they are underlain by caliche (Nelson 1934; Martin 1975). Mixed grama *(Bouteloua* spp.) occurs on a wide variety of soils (Martin 1975) and mesquite grassland occurs mainly on sandy soils (Buffington and Herbel 1965). Creosotebush *(Larrea tridentata)* grows best on limestone-derived alluvial fans (Fosberg 1940) and is absent on soils having gypsum in the profile (Waterfall 1946; Buffington and Herbel 1965). Tarbush *(Flourensia cernua)* also grows on limestone-derived soils but predominates on deep, well-drained soils (Buffington and Herbel 1965). Clay soils and fine silts, generally in swales and basins, are deep to moderately deep, poorly drained, calcareous, and contain appreciable quantities of readily soluble salts. They usually support tobosagrass *(Hilaria mutica)* and/or one of the sacaton species *(Sporobolus airoides, S. wrightii)* (Buffington and Herbel 1965; Bock and Bock 1978).

Vegetation

Major vegetational communities include the black grama uplands (Fig. 6.3.) of New Mexico and West Texas, the floodplains of tobosagrass (Fig. 6.4), sacaton *(Sporobolus wrightii)* (Fig. 6.5), or alkali sacaton *(S. airoides)* that occur along water courses, mixed grama *(Bouteloua gracilis, B. hirsuta, B. chondrosioides)* grasslands, mesquite-infested grasslands, the widespread creosotebush stands, and the tarbush, whitethorn *(Acacia constricta)*, and creosotebush areas of adjoining and included portions of the Chihuahuan desert (Martin 1975).

Fig. 6.3.   Black grama upland in southern New Mexico. Associated plants are mesa dropseed, broom snakeweed, and soapweed. (Photo courtesy of Science and Education Administration.)

Fig. 6.4.   Tobosagrass floodplain that is typical of southern Arizona, New Mexico, and southwestern Texas. Soils are usually nonsaline. (Photo courtesy of Science and Education Administration.)

Fig. 6.5.   Alkali sacaton grassland in southeastern New Mexico. It is typical of saline floodplains.

Black grama and tobosagrass are the most common grasses in the semidesert shrub-grassland (Paulsen and Ares 1962). A rothrock grama *(Bouteloua rothrockii)* community is abundant in Arizona and a curly mesquite *(Hilaria belangeri)* community is scattered throughout the desert grassland on well-drained clay soils (Shantz and Zon 1924; Humphrey 1953). Other species in Arizona include annual needle grama *(B. aristidoides)*, annual six-weeks threeawn *(Aristida adscenionis)*, tall threeawns *(A. hamulosa* and *A. ternipes)*, Santa Rita threeawn *(A. glabrata)*, bush muhly *(Muhlenbergia porteri)*, Arizona cottontop *(Digitaria californica)*, tanglehead *(Heteropogon contortus)*, hairy grama *(B. hirsuta)*, sideoats grama *(B. curtipendula)*, and plains lovegrass *(Eragrostis intermedia)*. Principal plant species in New Mexico on well-drained sites include black grama, mesa dropseed *(Sporobolus flexuosus)*, threeawns *(Aristida* spp.), and fluffgrass *(Erioneuron pulchellum)*. Low-lying areas contain tobosagrass and burrograss *(Scleropogon brevifolius)* (Buffington and Herbel 1965). In Texas, blue grama *(B. gracilis)*, sideoats grama, vine-mesquite *(Panicum obtusum)*, Warnock grama *(B. warnockii)*, bush muhly, dropseed *(Sporobolus* sp.), hairy tridens *(Erioneuron pilosum)*, and Wooton three-awn *(Aristida pansa)* may be added to the predominant stands of black grama, alkali sacaton, and tobosagrass (Bunting 1978). Lehmann lovegrass *(Eragrostis lehmanniana)* and Boer lovegrass *(Eragrostis chloromelas)* are the most commonly seeded grasses in the semidesert type (Martin 1975). Forbs are not very prevalent (Bunting 1978).

Common shrubs include mesquite, creosotebush, tarbush, fourwing saltbush *(Atriplex canescens)*, soapweed *(Yucca elata)*, and broom snakeweed *(Xanthocephalum sarothrae)* (Buffington and Herbel 1965; Martin 1975). Burroweed *(Aplopappus tenuisectus)* is prevalent in southern Arizona and southwestern New Mexico and some acacias *(Acacia* sp.) may be found throughout the entire region (Martin 1975). Walkingstick cholla *(Opuntia imbricata)*, jumping cholla *(O. fulgida)*, and Engelmann's pricklypear *(O. engelmannii)* are common cactus species (Cable 1972).

Herbage Yields

Herbage yields are quite variable. Martin (1966) found that herbage yields ranged from 6 to 280 kg/ha (5 to 250 lb/acre) on a mesquite-infested range that received 25 to 30 cm (10 to 12 in.) of precipitation annually. In a higher annual rainfall area [41 cm (16 in.)] yields ranged from 340 to 1350 kg/ha (300 to 1200 lb/acre) under a scattered stand of velvet mesquite. In another case on the Santa Rita Experimental Range, where precipitation averaged from 19 cm (7.4 in.) in the driest pasture to 27 cm (11 in.) in the wettest pasture, yields ranged from 19 to 560 kg/ha (17 to 500 lb/acre) (Martin 1975). On the Jornada Experimental Range, where precipitation averages 23 cm (9.1 in.) annually, Herbel (cited by Martin 1975) reported yields as high as 900 kg/ha (800 lb/acre) on black grama upland sites and 3930 kg/ha (3500 lb/acre) on tobosagrass-alkali sacaton flood plains.

POSSIBLE ROLE OF FIRE IN SEMIDESERT GRASS-SHRUB TYPE

The environmental and biological factors that may have limited the invasion of shrubs into desert grasslands before the arrival of Europeans in North America have to be looked at simultaneously to evaluate the possible role of fire. Vigorous perennial grasses compete strongly with mesquite seedlings (Martin 1975; Wright et al. 1976). Experiments on the Santa Rita Experimental Range showed that 16 times as many mesquite seedlings were established on bare areas as in vigorous stands of perennial grasses (Glendening and Paulsen 1955). Wright et al. (1976) found similar results in west Texas with no survivors in tobosagrass. Moreover, once established, growth of young mesquite plants is severely restricted in good stands of grass; for example, the senior author observed a mesquite plant 30 cm (12 in.) tall in black grama grass planted 18 years before on the Santa Rita Experimental Station. Thus competition is a key factor in keeping shrubs suppressed. Moreover frequent droughts in the semidesert grass-shrub type (Nelson 1934) were just as hard on young mesquite plants as they were on the grasses (Bogusch 1952).

Droughts can have devastating effects on black grama, the most prevalent grass species in the semidesert grassland (Cottle 1931; Nelson 1934). However black grama can recover quickly following protection and a couple of years of less than average to above average precipitation (Cottle 1931; Nelson 1934; Cable 1975). In two years yield increased from 13 to 131 g/m^2 in southwestern Texas (Cottle 1931) and area of sets increased from 3.0 to 78.5 cm/m^2 in southeastern Arizona (Nelson 1934). Regrowth is usually slow the first year of rest but accelerates the second year (Cottle 1931; Nelson 1934). Therefore when livestock was not a factor, a very susceptible plant such as black grama could have quickly reestablished itself with new stoloniferous plants and have been competitive with shrubs, if it had good vigor at the time of the catastrophy and was followed by average or better than average precipitation (Cottle 1931; Nelson 1934; Cable 1975). During the severe drought of 1951 to 1956, however, nearly all of the black grama on deep sandy and low hummocky sites was lost in an ungrazed exclosure and will not recover in our lifetime even under complete protection (Herbel et al. 1972). On shallow soils that were underlain by caliche, black grama was much more resistant to the severe drought.

Mesquite seedlings are most prevalent following warm summers and good fall rains (Wright et al. 1976). Because grassland fires usually occurred during dry seasons that followed one or two years of average to above average precipitation (Wright and Bailey 1980), a high percentage of young mesquite plants could easily have been killed by fire (Glendening and Paulsen 1955; Wright et al. 1976). The few surviving black grama plants (Reynolds and Bohning 1956) on black grama ranges might have recovered quickly if ungrazed, because they received good summer rains the year before the fire (Cottle 1931; Cable 1975). However, findings by Reynolds and Bohning (1956), where moderate grazing was confounded with fire effects, leaves one to doubt whether grazing can be permitted after a burn until black grama is completely recovered. Intervening droughts can lengthen the recovery period for several years (Nelson 1934; Reynolds and Bohning 1956).

For those areas that escaped fire, competition from healthy grass would reduce the number of mesquite plants 94 percent (Martin 1975). Those that survived would be fed on by jackrabbits (*Lepus alleni* and *L. californicus*) (Vorhies and Taylor 1933) and wood rats (*Neotoma* spp.) (Wright and Bailey 1979), especially during dry seasons, to meet metabolic moisture needs. In southeastern Arizona velvet mesquite constituted 36 to 56 percent of all food consumed by jackrabbits (Vorhies and Taylor 1933). In well-preserved black grama grasslands, there are relatively small numbers of rodents and lagomorphs (Buffington and Herbel 1965). Thus competition from grass and feeding by jackrabbits and wood rats appear to have, historically, been major factors that controlled the density and vigor of mesquite in southern desert grasslands.

Frequent droughts, insects, and diseases would also have taken their toll on young mesquite trees (Bogusch 1952; Glendening and Paulsen 1955; Wright and Bailey 1980). Velvet mesquite seedlings rarely survived the first spring drought on well-grassed sites (Glendening and Paulsen 1955). Those areas that escaped fire for 10 to 20 years could easily have kept young mesquite suppressed (via biotic and nonbiotic factors) to less than 1.3 cm (0.5 in.) in diameter. A fire at this time would kill 52 percent of such trees (Glendening and Paulsen 1955) and probably have kept most of them in a non-seed–producing state (Humphrey 1958; Martin 1975). Thus several factors interacting together with the help of fire and no grazing by domestic livestock could have kept shrubs, particularly mesquite, out of the semidesert grasslands. Even black grama could have theoretically tolerated occasional fires, when grazing was not a factor.

Overgrazing in Arizona, as practiced by forest administrators in the early 1900s to reduce fire hazard and promote the growth of trees (Leopold 1924), helped to prevent fires and let brush encroach upon the grassland (Griffiths 1910). Overgrazing on open range in desert grasslands, particularly during droughts, had a similar effect (Chew and Chew 1965). Griffiths (1910) and Leopold (1924) concluded that before 1880 the southern desert grasslands produced more grass and that fires occurred at approximately 10-year intervals. Initially fire harmed the grasses, but 10 years was plenty of time for a lusty growth of grass to come back and accumulate the fuel for another fire (Leopold 1924). The poor seed source, slow establishment, and slow growth rate of shrubs and mesquite trees would have permitted a fire every 10 years to control the shrubs and mesquite trees (Griffiths 1910). The key seems to be to burn at frequent enough intervals to prevent the production of seed by the shrubs (Humphrey 1958; Chew and Chew 1965; Martin 1975). No seeds are borne by creosotebush younger than 13 years and significant numbers of fruits appear only after 18 to 20 years of growth (Chew and Chew 1965). With competition from biotic and nonbiotic factors, mesquite could also take this long to have seed (Humphrey 1958; Martin 1975). This reasoning, however, does not seem to apply for southern New Mexico because there is no historical evidence of fire (Buffington and Herbel 1965).

Today grazing by domestic livestock is the biggest hinderance to the potential use of fire in semidesert grass-shrub vegetation, especially black grama ranges. Grazing has reduced fine fuel for fires and allowed shrubs to invade (Chew and Chew 1965; Martin 1975). Mesquite trees have become well established on former black grama ranges and have further reduced the chance for a site to produce enough fine fuel to carry a fire (Martin 1975). Fire might be used to prevent reinvasion (Martin 1975). In most cases, however, a major reclamation program involving brush control and improved grazing systems would be required to reclaim semidesert grass-shrub ranges to grass before fire could be introduced into a management program (Martin 1975; Wright et al. 1976).

EFFECTS OF FIRE ON VEGETATION

Grasses

Following a 15-year burning study on the Santa Rita Experimental Range, Cable (1967) concluded that fire had no lasting effects, beneficial or detrimental, on perennial grass cover. Generally the detrimental effects of fire on most of the perennial grasses only lasted one to two years. Annual grasses (predominantly needle grama and six-weeks threeawn), due to the reduction in burroweed, doubled their yields during wet years (average or above average precipitation) but remained the same as the control during dry years (below average precipitation). Humphrey (1949) reported similar results about the response of annual grasses to fire.

Rothrock grama tolerates fire well unless burned during dry years. Reynolds and Bohning (1956) and Cable (1967) found that burning during a dry year caused a 30 percent reduction in rothrock grama. However it had fully recovered by the end of the second growing season. In an earlier study, Humphrey (1949) found that numbers of rothrock grama plants were more abundant on two different burns near the Santa Rita Experimental Range than on controls two years after the burns.

Black grama is harmed most seriously of all the southern desert grasses and is very slow to recover. Following a hot June fire, Cable (1965) found that 90 percent of the black grama plants died. Even during a wet year, black grama only recovered 23 percent of its preburn basal area after the first growing season (Reynolds and Bohning 1956). Following two subsequent drought years and moderate grazing, basal cover of black grama had increased to 33 percent. However basal area of black grama dropped to 22 percent of the preburn basal area at the end of the fourth year when precipitation was above average. These data indicate that droughts following fire will lengthen the recovery period for black grama (Nelson 1934; Reynolds and Bohning 1956) and if droughts following fire are compounded with moderate grazing, black grama will never recover to its preburn basal area (Canfield 1939).

Canfield (1939) found that moderate grazing (simulated by clipping plants to a stubble height of 5 cm) in combination with droughts, regardless of frequency or season of harvesting, reduced the yield of black grama to zero. Moderate grazing entirely outweighed the beneficial effects of above average rainfall. The result was deterioration of black grama sites through excessive wind and water erosion. Thus if black grama ranges are burned, they should be completely rested until after two consecutive years of above average summer precipitation (Nelson 1934; Cable 1975). Then, if grazing is resumed, it should be light.

Rooting of stolons of black grama is the primary method of reproduction. Once these stolons are destroyed, forage is lost and the sand mulch is swept away by winter and spring winds (Herbel et al. 1972) unless other grasses can be established on the site. In many cases, other forage does not become established.

Further north near Flagstaff, Arizona, where the precipitation is higher than southern New Mexico, Jameson (1962) only noticed a 25 percent reduction in black grama by prescribed fires. Even hot summer fires did not show excessive damage. The senior author has observed similar effects in a 38-cm (15-in.) rainfall area in southeastern New Mexico.

Santa Rita threeawn is favored by fire during wet or dry years, but tall threeawns are generally reduced 30 to 50 percent of their original basal cover (Humphrey 1949; Reynolds and Bohning 1956; Cable 1967). During the first growing season after burning, a dry year, Reynolds and Bohning found that the density of Santa Rita threeawn had increased 34 percent over the control. Following another dry year at the end of the second growing season the density had doubled. And after the third growing season, a wet year, the density of Santa Rita threeawn had increased 350 percent. Increases, although not as dramatic, were also reported by Cable (1967). The reason that Santa Rita threeawn is more tolerant of fire than other threeawns has been explained by Cable. Santa Rita threeawn generally grows in open areas between burroweed plants, whereas the other threeawns generally grow within the burroweed crowns. As a consequence, the tall threeawns are subjected to more heat and easily harmed by fire, whereas many Santa Rita threeawn plants are not burned. Generally, threeawn species are easily harmed by fire because their root-shoot region is close to the soil surface.

Arizona cottontop and tanglehead are mildly harmed by fire during dry years but recover quickly during wet years (Reynolds and Bohning 1956; Cable 1967). Neither of these species show a gain in production following burning. However, in the southern mixed prairie, where the average rainfall is twice as high as in the southern desert grasslands, Arizona cottontop responds favorably to fire during wet years (Wright 1974). Based on limited data, bush muhly appears to be seriously harmed by fire (Humphrey 1949).

Tobosagrass, a dominant of the southern desert grasslands in southern New Mexico and southeastern Arizona, is severely harmed by burning during dry years (it produces only 30 to 60 percent as much as the control), but will recover fully by the end of the third growing season if normal precipitation follows the dry year (Dwyer 1972). In the southern mixed prairie, yield of this species increases twofold to threefold if the soil is moist at the time of burning (Wright 1969; 1972). These differences in response to fire between the two vegetation types can be attributed to differences in precipitation.

Alkali sacaton and sacaton communities, which are similar in density, coarseness, and structure to tobosagrass, were probably burned more frequently in their natural state than tobosagrass communities (Humphrey 1962). Records of fire occurrence are extremely rare. Mesquite and acacia have taken over many sacaton communities. Contributing factors have been overgrazing and channel cutting, which eliminates periodic flooding and lowers the water table. Sacaton grasslands require two years to fully regain plant cover and 54 percent of its original height (Bock and Bock 1978). In this community vine-mesquite was significantly more abundant on a winter burned area than on an unburned area (Bock and Bock 1978).

On the northern boundary of the semidesert grass-shrub type, galleta *(Hilaria jamesii)* is slightly harmed by fire (Jameson 1962; Dwyer and Pieper 1967). Following winter burns with adequate soil moisture, galleta yielded 75 percent as much forage the first growing season after burning as the unburned control.

Wolf plants of Lehmann lovegrass, an introduced species, are severely reduced by hot wildfires in June, but seedlings quickly reestablish on the burned areas (Cable 1965). A burn that was followed by unusually favorable moisture for plant growth reduced a stand of Lehmann lovegrass about one-third (Humphrey and Everson 1951). Seedlings were abundant, however, and the reduction in forage yield was only temporary. Following a February burn, Pase (1971) found that lovegrass in a chaparral community was essentially unaffected by burning. Because this species of lovegrass is a bunchgrass, damage by fire would be related to intensity of the fire, amount of dead fuel in plant crowns, soil moisture, and precipitation that followed the burn.

Forbs

Forbs are generally not mentioned in the literature for semidesert grass-shrub communities, but Bock and Bock (1978) reported that forbs that were more common in sacaton communities after winter or summer burning included *Amaranthus, Ipomoea, Bidens, Convolvulus, Solidago, Portulaca, Chenopodium,* and *Ambrosia.*

Cacti

Cactus species are relatively susceptible to fire. Using data from three studies—Humphrey (1949), Reynolds and Bohning (1956), and Cable (1967)—average kills two growing seasons after a burn were jumping cholla, 50 percent; walkingstick cholla, 45 percent; pricklypear, 30 percent; and barrel cactus *(Echinocactus wislizenii),* 65 percent. These mortalities usually included the interactive effects between fire, insects, drought, and grazing by rodents, lagomorphs, and domestic animals. For example, when the spines of barrel cactus are burned off, cattle eat them readily (Reynolds and Bohning 1956).

Walkingstick cholla and pricklypear do not recover from the effect of initial burns for at least 13 years after a burn (Cable 1967). However jumping cholla increased 17 percent 13 years after a burn. Reburns that were three years apart did not increase the mortality of cactus species (Cable 1967). The reason given was that the first burn removed the accumulated weeds and litter from the base of cactus plants.

A study by Glendening (1952) also supports the need for fire to control cactus species. His 17-year study showed that walkingstick cholla, pricklypear, and barrel cactus increased under protection and under moderate grazing. Jumping cholla was the only species that decreased under protection. However it appears that jumping cholla has a life span of about 40 years and may go through rapid die-off cycles due to population buildups of bacillus *(Edwinea carnegieana)* (Tschirley and Wagle 1964; Martin and Turner 1977). Die-offs of 25 to 35 percent also occur in the other cactus species from time to time (Humphrey and Everson 1951). To date there is no evidence to support the common belief that reduced grass competition, resulting from grazing, has caused cactus species to flourish. More likely, lack of fire has caused them to flourish, as the evidence seems to suggest.

Shrubs

Griffiths (1910) and Wooton (1916) believed that fires almost entirely prevented the establishment of undesirable shrubs in the southern desert. Griffiths stated that because of the slow growth of shrubs, he believed they could be controlled by fires that occurred only once in 10 years. Wooton (1916), working in the same area (Santa Rita Experimental Range, Arizona), saw occasional fires that were hot enough to kill velvet mesquite trees 3.0 to 3.7 m (10 to 12 ft) high. In his opinion, fire had been the only restricting influence on the spread of trees and shrubs. Although grasses, with the possible exception of black grama, recovered quickly from such burning, shrubs were usually just reappearing by the time another fire occurred. Regrowth from small mesquites that are merely top-killed can be rapid, however.

One of the most prevalent shrub-tree species in the Southwest is velvet mesquite. This species is moderately affected by fire, depending on the size of mesquite and amount of fine fuel available for burning (Cable 1961; 1965; 1967; White 1969). Following a June 28, 1963 wildfire, Cable (1965) reported a 21-percent kill of velvet mesquite less than 5 cm (2 in.) in diameter and a 10-percent kill of trees larger than 5 cm (2 in.) in diameter. Using artificial fuels for controlled fires, Glendening and Paulsen (1955) obtained a 52-percent kill on young mesquites having basal stem diameters of 1.3 cm (0.5 in.) or less. Only 8 to 18 percent of the larger trees were killed by fire. Reynolds and Bohning (1956) killed 9 percent of the mesquite trees by using a prescribed burn on June 30, 1952. In a wildfire near Sasabe, Arizona, White (1969) reported a 20-percent kill of mesquite trees in moderate and severe burns.

Occasionally fire may be more damaging to velvet mesquite than is normally expected. Humphrey (1949) has reported mesquite kills of 50 percent on the Beach Ranch Study and 75 percent on the Sierrita Mountain Study. After 15 years, Humphrey revisited these same areas and still found mesquite drastically reduced. High kills, such as these reported by Humphrey, are rare. Part of this variation in mesquite kills, however, may be due to the amount of fuel available. On areas having 5055 kg/ha (4500 lb/acre) of fine fuel, fire killed 25 percent of the mesquites, but on areas having 2470 kg/ha (2200 lb/acre) of fine fuel, only 8 percent of the mesquites were killed (Cable 1965). Another source of variation is the general vigor of the plants. Mesquite plants with low vigor, growing on dense rocky subsoils, do not have the recovery potential of more vigorous trees. Another factor may be the degree to which erosion has removed soil around the base of the tree, thereby exposing the bud zone to heat. Lastly, summer burns are more damaging to mesquite than winter burns (Glendening and Paulsen 1955; Blydenstein 1957).

In addition to the mortality of plants, burning in some way seems to inhibit the establishment of mesquite seedlings. Mesquite numbers on an unburned area increased from 40 to 128/ha (from 16 to 52/acre) within a 13-year period while they only increased from 59 to 62/ha (from 24 to 25/acre) on a burned area (Cable 1967).

Reduced yield of mesquite seed on trees that were partially top-killed may be one reason for such an effect. The reduction in numbers of Merriam kangaroo rats *(Dipodomys merriami)*, resulting from the loss of cactus and other shrubs that formerly sheltered the wood rats, would also reduce the amount of seed cached on a burned area. Lastly, jackrabbits eat young mesquite plants (Vorhies and Taylor 1933) and the mortality of mesquite seedlings is higher on areas grazed by cattle and jackrabbits than in cattle-jackrabbit exclosures (Glendening 1952).

Generally, velvet mesquite in southeastern Arizona is more susceptible to fire than honey mesquite, which grows in New Mexico and Texas. Two successive fires are necessary to kill 27 percent of the large honey mesquite trees on upland sites in the Rolling Plains of Texas, but trees are rarely killed with successive fires on bottomland sites (Wright et al. 1976). Seedlings of honey mesquite are easy to kill with moderate fires until they reach 1.5 years of age, can be severely harmed at 2.5 years of age, and are very tolerant of intense fires after 3.5 years of age (Wright et al. 1976). Based on data by Cable (1961), it appears that at these young ages velvet mesquite might be slightly more tolerant of fire, but his seedlings had been transplanted when three or four weeks old. On the High Plains (shortgrass prairie) of Texas, where mesquite was not reported by early scouts, honey mesquite is very tolerant to fire. We have observed no mortality, indicating that the plants have evolved in a fire environment and were kept suppressed by fire, droughts, competition, rodents, and lagomorphs (Wright et al. 1976).

Other shrubs only moderately affected by fire are false-mesquite *(Calliandra eriophylla)* and velvet-pod mimosa *(Mimosa dysocarpa)*. Very few of these plants (2–10%) died on severe burns and no plants died on light and moderate burns (White 1969). Reynolds and Bohning (1956) found that false-mesquite recovered within two years after burning and had almost doubled its crown density, compared to unburned areas, by the third year after burning.

Soapweed can be adversely affected by fire. Humphrey (1949) reported a 25-percent kill following a wildfire on Sierrita Mountain. In general, however, most *Yucca* species are tolerant of fires and appear to hold their position in various plant communities despite fire.

Ocotillo *(Fouquieria splendens)* and Wheeler sotol *(Dasylirion wheeleri)* are severely reduced by fire (White 1969). In a June, 1963 wildfire, many plants of ocotillo died—67 percent in severe burns, 40 percent in moderate burns, and 50 percent in light burns. Only 3 percent of the Wheeler sotol plants survived severe burns, but all survived in moderate and light burns.

Larchleaf goldenweed *(Aplopappus laricifolius)* is also easily killed by fire (White 1969). Severely damaged plants were completely killed and did not sprout by the end of the second growing season. Only 10 percent of the moderately damaged plants survived following fire. About 90 percent of the lightly damaged plants survived in the first growing season, but the number of survivors declined to 80 percent in the second growing season.

Paloverde *(Cercidium floridum)*, broom snakeweed, and burroweed are three more species that can be severely damaged by fire. Humphrey (1949) reported a

90-percent mortality of paloverde on the Sierrita Mountain Study. Wooton (1916) observed that broom snakeweed was easily killed by fire. A later study showed that mortalities of broom snakeweed and burroweed following a July control burn were 95 percent or higher (Humphrey and Everson 1951). Cable (1967) and Reynolds and Bohning (1956) have also reported 95- to 100-percent kills for burroweed when burned in June. One study on burroweed in which burning was done at all seasons of the year showed burning to be reasonably effective from mid-April to mid-September, but most effective about June 1 (Tschirley and Martin 1961).

After six years, Cable (1967) found that burroweed was only 25 to 30 percent of preburn density on a June burn (Fig. 6.6), although it fluctuated upward during wet years and downward during dry years. After four years, burroweed exceeded preburn densities (Fig. 6.6). However in another study, Humphrey (1949) found that burroweed failed to reinvade after 15 years following a wildfire. The most striking effect of June burns in southern desert grasslands is the elimination of burroweed, at least temporarily, and the increase in annual grasses during wet years (Cable 1967).

Thornber (1907) noted that fire was effective in killing catclaw *(Acacia greggii)*, creosotebush, longleaf ephedra *(Ephedra trifurca)*, and graythorn *(Condalia lycioides)*. Except for creosotebush, however, no research studies document the extent to which these species are affected by fire. Creosotebush (Fig. 6.7) can resprout after burning, but intense fires, particularly during June, will cause 100-percent mortality (White 1968; White and Ehrenreich 1968).

Algerita *(Berberis trifoliata)*, fourwing saltbush, winterfat *(Ceratoides lanata)*, and skunkbush sumac *(Rhus trilobata)* resprout vigorously after fire (Dwyer and Pieper 1967). Wright baccharis *(Baccharis wrightii)*, a highly palatable shrub, also resprouts vigorously and appears to be unaffected by fire (Humphrey 1949).

Desert blackbrush *(Coleogyne ramosissima)* (Fig. 6.8), a nonsprouter in northern Arizona, southern Nevada, and Utah (a transitional zone of vegetation between the salt desert shrub and the semidesert grass-shrub types), is very susceptible to fire and is slow to reinvade after fires in southern Nevada and Utah (Jenson et al. 1960; Beatley 1966). Plant communities that succeed desert blackbrush are highly variable (Bowns and West 1978). Bowns and West found that some plant communities will return to a mixture of shrubs such as turpentine bush *(Thamnosma montana)*, desert bitterbrush *(Purshia glandulosa)*, desert almond *(Prunus fasciculata)*, and big sagebrush *(Artemisia tridentata)*. Other areas return to pure stands of threadleaf snakeweed *(Xanthocephalum microcephalum)* or big sagebrush. Even though desert blackbrush is not a preferred shrub, widescale burning is not recommended as a desirable management policy for this type (Bowns and West 1978). The vegetation that may follow is too unpredictable.

MANAGEMENT IMPLICATIONS

The southern desert grass-shrub type is a delicate ecosystem with wide swings in herbage yields because severe droughts are common. Moreover, droughts frequently

Fig. 6.6. (a) A stand of 4443 burroweeds per acre on the Santa Rita Experimental Range in June, 1952. (b) Same as 6.6a, July, 1952, eight days after burning. Fire killed 94 percent of burroweeds. (c) Same as 6.6a, September, 1958. The area is dominated by native perennial grasses. (d) Same as 6.6a, September, 1966, 14 years after burning. The site has been reinvaded by burroweed, 10,433 plants per acre. (Photos courtesy of USDA Forest Service.)

last two or three years. When moderate to heavy grazing is imposed on black grama ranges, grass competition and vigor of grasses are drastically reduced (Canfield 1939). These factors favor high mortality of herbs during drought years and the eventual establishment of shrubs following wet years when other climatic factors, such as soil temperature, are favorable. It appears that use of fire would compound the existing problems on black grama ranges and may not have a place for shrub control on good ranges.

Fig. 6.6   *(continued)*

Our problem is to reclaim poor rangelands (predominately brush) and to properly manage our good rangelands. Poor rangelands cannot be managed with fire. These rangelands must first be restored using other reclamation techniques. However, once the rangelands are in good condition, fire can be used as an effective management tool in special situations during wet weather cycles to control burroweed, broom snakeweed, creosotebush, and young mesquite trees. Fires can also be used to suppress cactus species. Most burning should be done in June, but only following two previous years of better than average plant growth. This is especially important for grasses to recover quickly after burning.

Desirable shrubs that are either favored or not harmed by fire include false-mesquite, velvet-pod mimosa, Wright baccharis, and fourwing saltbush. Wheeler sotol and barrel cactus are easily harmed by fire and should be protected, if possible.

Today fire should be used only on a selective basis, or in combination with other methods, to achieve specific management objectives in the semidesert grass-shrub type. Fire probably has the greatest value to manage tobosagrass, sacaton, alkali sac-

Fig. 6.7.   Creosotebush is a widespread plant community in the semidesert zone and has very little herbaceous vegetation. However it is susceptible to fire in June, indicating that wildfires could have easily kept it from invading grasslands in the past.

Fig. 6.8.   Typical view of a pure stand of blackbrush. Note the absence of herbaceous vegetation. (Photo courtesy of Utah State University.)

aton, and mixed grama ranges. Good black grama grasslands appear to be too delicate to manage with fire. If fire is used, three to four years rest might be required after a burn.

REFERENCES

Beatley, J. C. 1966. Ecological status of introduced brome grasses *(Bromus* spp.) in desert vegetation of southern Nevada. *Ecology* **47**:548–554.

Blydenstein, J. 1957. The survival of velvet mesquite *(Prosopis juliflora* var. *velutina)* after fire. *J. Range Manage.* **10**:221–223.

Bock, C. E., and J. H. Bock. 1978. Response of birds, small mammals, and vegetation to burning sacaton grasslands in southeastern Arizona. *J. Range Manage.* **31**:296–300.

Bock, J. H., C. E. Bock, and J. R. McKnight. 1976. A study of the effects of grassland fires at the research ranch in southeastern Arizona. *Ariz. Acad. Sci.* **11**(3):49–57.

Bogusch, E. R. 1952. Brush invasion on the Rio Grande Plain of Texas. *Texas J. Sci.* **1**:85–90.

Bowns, J. E., and N. E. West. 1976. Blackbrush *(Coleogyne ramosissima* Torr.) on southwestern Utah rangelands. Utah Agric. Exp. Stn., Res. Rep. 27. Logan.

Branscomb, B. L. 1956. Shrub invasion of a New Mexico desert grassland range. M.S. Thesis. Univ. of Ariz., Tucson.

Buffington, L. C., and C. H. Herbel. 1965. Vegetational changes on a semidesert grassland range from 1858 to 1963. *Ecol. Monogr.* **35**:139–164.

Bunting, S. C. 1978. The vegetation of the Guadalupe Mountains. Ph.D. Diss. Texas Tech Univ., Lubbock.

Cable, D. R. 1961. Small velvet mesquite seedlings survive burning. *J. Range Manage.* **14**:160–161.

Cable, D. R. 1965. Damage to mesquite, Lehmann lovegrass, and black grama by a hot June fire. *J. Range Manage.* **18**:326–329.

Cable, D. R. 1967. Fire effects on semidesert grasses and shrubs. *J. Range Manage.* **20**:170–176.

Cable, D. R. 1972. Fire effects on southwestern semidesert grass-shrub communities. *Proc. Tall Timbers Fire Ecol. Conf.* **12**:109–127.

Cable, D. R. 1975. Influence of precipitation on perennial grass production in the semidesert southwest. *Ecology* **56**:981–986.

Canfield, R. H. 1939. The effect of intensity and frequency of clipping on density and yield of black grama and tobosa grass. USDA Tech. Bull. 681. Washington, D.C.

Carter, W. T., and V. L. Cory. 1930. Soils of the Trans-Pecos, Texas and some of their vegetative relations. *Trans. Texas Acad. Sci.* **15**:19–37.

Chew, R. M., and A. E. Chew. 1965. The primary productivity of a desert-shrub *(Larrea tridentata)* community. *Ecol. Monogr.* **35**:355–375.

Clements, F. E. 1920. *Plant Indicators: The Relation of Plant Communities to Process and Practice.* Carnegie Inst. Wash., Washington, D.C.

Cottle, H. C. 1931. Studies in the vegetation of southern Texas. *Ecology* **12**:105–155.

Dwyer, D. D. 1972. Burning and nitrogen fertilization of tobosagrass. New Mexico State Univ., Agric. Exp. Stn. Bull. 595. Las Cruces.

Dwyer, D. D., and R. D. Pieper. 1967. Fire effects of blue grama-pinyon-juniper rangeland in New Mexico. *J. Range Manage.* **20**:359–362.

Fosberg, F. R. 1940. The aestival flora of the Mesilla Valley region, New Mexico. *Am. Midl. Natur.* **23**:573–593.

Gardner, J. L. 1951. Vegetation of the creosote area of the Rio Grande Valley in New Mexico. *Ecol. Monogr.* **21**:379–402.

Glendening, G. E. 1952. Some quantitative data on the increase of mesquite and cactus on a desert grassland range in southern Arizona. *Ecology* **33**:319–328.

Glendening, G. E., and H. A. Paulsen, Jr. 1955. Reproduction and establishment of velvet mesquite as related to invasion of semidesert grasslands. USDA Tech. Bull. 1127. Washington, D.C.

Griffiths, D. A. 1910. A protected stock range in Arizona. USDA Bur. Plant Indus. Bull. 177. Washington, D.C.

Herbel, C. H., F. N. Ares, and R. A. Wright. 1972. Drought effects on a semidesert grassland range. *Ecology* **53**:1084–1093.

Hinckley, L. C. 1944. The vegetation of the Mount Livermore area in Texas. *Am. Midl. Natur.* **32**:236–250.

Humphrey, R. R. 1949. Fire as a means of controlling velvet mesquite, burroweed, and cholla on southern Arizona ranges. *J. Range Manage.* **2**:175–182.

Humphrey, R. R. 1953. The desert grassland, past and present. *J. Range Manage.* **6**:159–164.

Humphrey, R. R. 1958. The desert grassland. *Bot. Rev.* **24**:193–253.

Humphrey, R. R. 1962. Fire as a factor. In *Range Ecology.* Ronald Press, New York, pp. 148–189.

Humphrey, R. R., and A. C. Everson. 1951. Effect of fire on a mixed grass-shrub range in southern Arizona. *J. Range Manage.* **3**:264–266.

Jameson, D. A. 1962. Effects of fire on a galleta-black grama range invaded by juniper. *Ecology* **43**:760–763.

Jenson, D. E., M. W. Butan, and D. E. Dimock. 1960. Blackbrush burns. Report on field examinations. Las Vegas Grazing Dist., Nevada.

Küchler, A. W. 1964. Potential natural vegetation of the conterminous United States. Manual to accompany the map. Amer. Geogr. Soc. Spec. Pub. 36. (With map, rev. ed., 1965, 1966.)

Leopold, A. 1924. Grass, brush, timber and fire in southern Arizona. *J. For.* **22**:1–10.

Martin, S. C. 1966. The Santa Rita Experimental Range: A center for research on improvement and management of semidesert rangelands. USDA For. Serv. Res. Paper RM-22. Rocky Mtn. For. and Range Exp. Stn., Fort Collins, Colo.

Martin, S. C. 1975. Ecology and management of southwestern semidesert grass-shrub ranges: The status of our knowledge. USDA For. Serv. Res. Paper RM-156. Rocky Mtn. For. and Range Exp. Stn., Fort Collins, Colo.

Martin, S. C., and D. R. Cable. 1974. Managing semidesert grass-shrub ranges: vegetation responses to precipitation, grazing, soil texture, and mesquite control. USDA Tech. Bull. 1480. Washington D.C.

Martin, S. C., and R. M. Turner. 1977. Vegetation changes in the Sonoran Desert region, Arizona and Sonora. *J. Ariz. Acad. Sci.* **12**(2):59–69.

Nelson, E. W. 1934. The influence of precipitation and grazing upon black grama grass range. USDA Tech. Bull. 409. Washington, D.C.

Pase, C. P. 1971. Effect of a February burn on Lehmann lovegrass. *J. Range Manage.* **24**:454–456.

Paulsen, H. A., Jr. 1956. The effect of climate and grazing on black grama. In Range Day pub. N. M. Agric. Exp. Stn. and USDA Sci. Educ. Admin. and USDA For. Serv., pp. 17–24.

Paulsen, H. A., Jr., and F. N. Ares. 1962. Grazing values and management of black grama and tobosa grasslands and associated shrub ranges of the southwest. USDA Tech. Bull 1270. Washington, D.C.

Reynolds, H. G., and J. W. Bohning. 1956. Effects of burning on a desert grass-shrub range in southern Arizona. *Ecology* **37**:769–777.

Shantz, H. L., and R. Zon. 1924. Atlas of American agriculture Part I, Section E. Natural vegetation. USDA Bur. Agric. Econ. Washington, D.C.

Thornber, J. J. 1907. 18th Annual Report. Ariz. Exp. Stn., Tucson.

Thornber, J. J. 1910. Grazing ranges of Arizona. In Univ. of Ariz. Agric. Exp. Stn. Bull. 65. Tucson, pp. 245–360.

Tschirley, F. H., and S. C. Martin. 1961. Burroweed on southern Arizona rangelands. Ariz. Agric. Exp. Stn. Tech. Bull. 146. Tucson.

Tschirley, F. H., and R. F. Wagle. 1964. Growth rate and population dynamics of jumping cholla (*Opuntia fulgida* Engelm.). *J. Ariz. Acad. Sci.* **3**:67–71.

Vorhies, C. T., and W. P. Taylor. 1933. Life history and ecology of jackrabbits, *Lepus alleni* and *Lepus californicus*, spp., in relation to grazing in Arizona. In Ariz. Agric. Exp. Stn. Tech. Bull. 49. Tucson, pp. 470–587.

Waterfall, V. T. 1946. Observations on the desert gypsum flora of southwestern Texas and adjacent New Mexico. *Am. Midl. Natur.* **36**:456–466.

White, L. D. 1968. Factors affecting susceptibility of creosotebush [*Larrea tridentata* (D.C.) cov.] to burning. Ph.D. Diss. Univ. of Ariz., Tucson.

White, L. D. 1969. Effects of a wildfire on several grassland shrub species. *J. Range Mange.* **22**:284,285.

White, L. D., and J. H. Ehrenreich. 1968. Factors affecting susceptibility of creosotebush to burning. Abstr. of Papers, Amer. Soc. Range Manage., Albuquerque, N. Mex., pp. 51, 52.

Wooton, E. O. 1916. Carrying capacity of grazing ranges in southern Arizona. USDA Bull. 367. Washington, D.C.

Wright, H. A. 1969. Effect of spring burning on tobosagrass. *J. Range Manage.* **22**:425–427.

Wright, H. A. 1972. Fire as a tool to manage tobosa grasslands. *Proc. Tall Timbers Fire Ecol. Conf.* **12**:153–167.

Wright, H. A. 1974. Effect of fire on southern mixed prairie grasses. *J. Range Manage.* **27**:417–419.

Wright, H. A., and A. W. Bailey. 1980. The role and use of fire in the Great Plains: A state-of-the-art review. USDA For. Serv. Gen. Tech. Rep. INT-77. Intermt. For. and Range Exp. Stn., Ogden, Utah.

Wright, H. A., S. C. Bunting, and L. F. Neuenschwander. 1976. Effect of fire on honey mesquite. *J. Range Manage.* **29**:467–471.

CHAPTER

7

SAGEBRUSH–GRASS

Sagebrush (*Artemisia* spp.)–grass vegetation covers at least 39.1 million hectares (96.5 million acres) in the western United States (USDA For. Serv. 1936), but probably considerably less than the 109 million hectares (270 million acres) estimated by Beetle in 1960 (Tisdale et al. 1969). The largest contiguous area lies in eastern Oregon, southern Idaho, southwestern Wyoming, northern Utah, and northern Nevada (Vale 1975). Most of this area (Fig. 7.1) occurs below the pinyon-juniper (*Pinus-Juniperus* spp.) zone, but in the absence of a pinyon-juniper zone, sagebrush-grass vegetation will border curlleaf mahogany *(Cercocarpus ledifolius)*, Gambel oak *(Quercus gambelii)*, ponderosa pine *(Pinus ponderosa)* or Douglas fir *(Pseudotsuga menziesii)*. Sagebrush-grass communities also occur above the pinyon-juniper zone in the Great Basin (Billings 1951) and throughout most mountain plant communities in the Rocky Mountain and Intermountain Regions (Beetle 1960).

FIRE HISTORY

Before the influence of humans, fire covered contiguous units of sagebrush-grass communities in northern Yellowstone National Park at an average frequency of 32 to 70 years (Houston 1973). Within a large portion of this locale, however, fire swept smaller areas at least every 17 to 41 years.

Dating all fires that occurred within a locale, Houston theorized that the frequency of fire in sagebrush communities within Yellowstone Park was 20 to 25 years. This estimated frequency was based on the assumption that the record of fire scars for any one tree underestimated the frequency of fire because not all trees were scarred by ev-

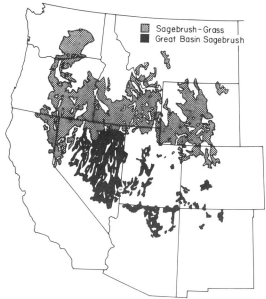

Fig. 7.1. Distribution of sagebrush and sagebrush-grass communities in the western United States. (From Küchler 1964. Reproduced with permission of the American Geographical Society.)

ery fire. Houston's findings that many but not all trees had fire scars with similar dates suggests that once burned, an area was unlikely to have sufficient fuel to reburn for several years.

Based on the vigorous response of horsebrush *(Tetradymia canescens)* to fire (Fig. 7.2) and the 30-plus years that are needed for it to decline to a low level after a fire in eastern Idaho (Harniss and Murray 1973), probable frequency of fire for that area

Fig. 7.2. Horsebrush resprouts vigorously after burning in southeastern Idaho. (Photo courtesy USDA Forest Service.)

would be about 50 years. If fires occurred every 20 to 25 years, as Houston implies, many sagebrush-grass communities in eastern Idaho could be dominated by horsebrush and rabbitbrush (*Chrysothamnus* spp.). In the driest sagebrush communities [e.g., Wyoming big sagebrush *(Artemisia tridentata* ssp. *wyomingensis)*], we suspect that the fire frequency could have been as low as 100 years.)

DISTRIBUTION, CLIMATE, SOILS, AND VEGETATION

Most of the sagebrush-grass zone is found at elevations from 610 to 2130 m (2000 to 7000 ft). The sagebrush-grass zone also occurs below 305 m (1000 ft) in southcentral Washington and British Columbia and mixes with all vegetation zones to varying degrees up to 3050 m (10,000 ft) (Beetle 1960), including the subalpine herbland. Where sagebrush-grass prevails below 2130 m (7000 ft), annual precipitation varies between 20 to 50 cm (8 to 20 in.) (Tisdale et al. 1969). Soil texture varies from loamy sand to clay (Tisdale et al. 1969). Most soils are derived from basalt, although extensive areas have soils dervied from rhyolite (southeastern Oregon and Nevada), loess, lacustrine, alluvium, and limestone. Interactions of soils, precipitation, and elevation results in many distinct combinations of sagebrush-grass dominated ecosystems.

Three subspecies of big sagebrush—basin big sagebrush *(Artemisia tridentata* ssp. *tridentata)*, Wyoming big sagebrush, and mountain big sagebrush *(Artemisia tridentata* ssp. *vaseyana)*—dominate the sagebrush-grass zone. Basin big sagebrush [1 to 5 m (3.3 to 16.4 ft) tall] and Wyoming big sagebrush [0.4 to 0.8 m (1.5 to 2.4 ft) tall] are the dominants from 610 to 2130 m (2000 to 7000 ft), with the latter being the most drought-tolerant (McArthur et al. 1974). Basin big sagebrush occupies a 25- to 40-cm (10- to 16-in.) precipitation zone on deep, well-drained alluvial soils, whereas Wyoming big sagebrush occupies a 20- to 30-cm (8- to 12-in.) precipitation zone on shallow soils (Tisdale et al. 1969). Mountain big sagebrush [0.8 to 1.2 m (2.4 to 4.0 ft) tall] is the most mesic subspecies (Fig. 7.3) and can occur at elevations from 1524 to 3050 m (5000 to 10,000 ft) (McArthur et al. 1974) where precipitation varies from 35 to 50 cm (14 to 20 in.) per year.

Other species of sagebrush, in decreasing order of economic importance, are low sagebrush *(Artemisia arbuscula)*, three-tip sagebrush *(A. tripartita)*, black sagebrush *(A. nova)*, silver sagebrush *(A. cana* ssp. *viscidula* and *A. cana* ssp. *bolanderii)*, alkali sagebrush *(A. longiloba)*, Bigelow sagebrush *(A. bigelovii)*, and scabland sagebrush *(A. rigida)* (Tisdale et al. 1969; McArthur et al. 1974). The first three species generally grow below 1830 m (6000 ft) elevation, although they can occur at higher elevations. Low sagebrush occurs on shallow soils or those that possess a restrictive B horizon, largely in southern Idaho, Nevada, southeastern Oregon, and northeastern California (Fosberg and Hironaka 1964). Three-tip sagebrush occurs east of this region on mesic or dry soils in a precipitation zone of 25 to 40 cm (10 to 16 in.). Black sagebrush is usually associated with calcareous soils on dry sites, but can occur on mesic sites as high as the Douglas fir zone in eastern Idaho. Silver sagebrush oc-

Fig. 7.3.   Mountain big sagebrush in Idaho with an understory of Idaho fescue. (Photo courtesy University of Idaho.)

curs primarily in spring-flooded bottomlands (Fig. 7.4), and at high elevations where snow drifts. All species except three-tip sagebrush and silver sagebrush are non-sprouters. Three-tip sagebrush is a weak sprouter and silver sagebrush is a vigorous sprouter. /

Major shrubs associated with big sagebrush include antelope bitterbrush *(Purshia tridentata)*, horsebrushes *(Tetradymia* spp.), rabbitbrush and broom snakeweed *(Xanthocephalum sarothrae)*. Greasewood *(Sarcobatus vermiculatus)* grows around the perimeter of saline playas throughout the sagebrush-grass region. Spiny hopsage

Fig. 7.4.   Silver sagebrush, a sprouter, occurs primarily in spring-flooded bottomlands. (Photo courtesy University of Oregon.)

(Grayia spinosa) and Mormon tea *(Ephedra nevadensis)* are sporadically present in the lower rainfall areas near the Salt Desert.

Dominant grasses include bluebunch wheatgrass *(Agropyron spicatum)*, Idaho fescue *(Festuca idahoensis)*, needle-and-thread *(Stipa comata)*, Thurber needlegrass *(Stipa thurberiana)*, and, to a lesser extent, Indian ricegrass *(Oryzopsis hymenoides)*. All of these species or only one may be present in a particular understory. Needle-and-thread and Indian ricegrass dominate sandy soils throughout the sagebrush-grass zone. On other soils bluebunch wheatgrass dominates areas with moderate annual precipitation [22 to 35 cm (9 to 14 in.)] and Idaho fescue dominates the most mesic sites, generally with more than 35 cm (14 in.) of annual precipitation. Temperature interacts with precipitation, so the moisture threshold that separates bluebunch wheatgrass from Idaho fescue can vary from 30 to 40 cm (12 to 16 in.) of precipitation annually. Thurber needlegrass occurs on medium-textured soils in a 20- to 30-cm (8- to 12-in.) precipitation zone. Sandberg bluegrass *(Poa sandbergii)* and bottlebrush squirreltail *(Sitanion hystrix)* are the most common subdominant bunchgrasses. Junegrass *(Koeleria cristata)* and other *Poa* species are often present if annual precipitation is above 28 cm (11 in.). Rhizomatous grasses that occupy local-ized areas include thickspike wheatgrass *(Agropyron dasystachyum)*, plains reedgrass *(Calamagrostis montanensis)*, and riparian wheatgrass *(Agropyron riparium)*. Cheatgrass *(Bromus tectorum)*, an introduced annual, occupies millions of hectares on disturbed ranges (Klemmedson and Smith 1964). Medusahead *(Taeniatherum asperum)*, another introduced annual, occupies disturbed clay sites which have well developed profiles (Dahl 1966).

Forbs are present in great variety and abundance in climax communities (Fig. 7.5) where the precipitation is in excess of 28 to 30 cm (11 to 12 in.) per year. They may

Fig. 7.5. Arrowleaf balsamroot and lupine make up a large component of the herbaceous yield in this three-tip sagebrush community. Fall burning would help to maintain the forb component. (Photo courtesy USDA Forest Service.)

account for as much as 50 percent of the herbaceous production in eastern Idaho. For this reason, herbicides, at least in our opinion, are undesirable to manage sagebrush-grass communities in eastern Idaho where arrowleaf balsamroot *(Balsamorhiza sagittata)* and lupine *(Lupinus* spp.) are typically the most abundant forbs (Hurd 1955; Blaisdell and Mueggler 1956a). Forbs account for only 5 to 15 percent of the herbaceous vegetation in eastern Oregon. Lambstongue groundsel *(Senecio integerrimus)*, tapertip hawksbeard *(Crepis acuminata)*, western yarrow *(Achillea millefolium)*, and locoweed *(Astragalus* spp.) are the most common forbs and re-cover within three to four years after the use of herbicides (Wright et al. 1979).

EFFECTS OF FIRE

Grasses

The effect of fire on grasses depends largely on their growth form and season of burn-ing. Bunchgrasses with densely clustered culms, such as Idaho fescue and needle-and-thread, can be severely harmed by fire (Blaisdell 1953; Countryman and Cornelius 1957; Wright and Klemmedson 1965; Daubenmire 1970; Wright 1971), especially if burned during June or July (Fig. 7.6) (Wright 1971). Their dense culms will burn for 2 to 3 hr after a fire passes. Temperatures as high as 538°C (1000°F) will be reached 45 min after a fire ignites the plants (Wright 1971). Thus many leafy bunchgrasses often die following a fire or have only a few culms that survive, regard-less of the intensity of the passing fire. Late summer and fall burns are the least harm-ful (Wright and Klemmedson 1965).

Threadleaf sedge *(Carex filifolia)* also has a compact growth form and is severely harmed by fire (Blaisdell 1953; Conrad and Poulton 1966; Vallentine 1971). Idaho fescue is very sensitive to summer and fall fires where the precipitation is marginal for its existence (Blaisdell 1953; Conrad and Poulton 1966). Preliminary research in eastern Oregon indicates that Idaho fescue will recover in two to three years follow-ing each early spring burning if the soil is moist (Wright et al. 1979).

Bluebunch wheatgrass, bottlebrush squirreltail, and the crested wheatgrasses *(Agropyron cristatum, A. desertorum,* and *A. sibericum)* are less susceptible to fire injury than Idaho fescue or *Stipa* species (Blaisdell 1953; Conrad and Poulton 1966; Vallentine 1971; Wright 1971) because they are composed primarily of coarse stems with a minimum of leafy material (Fig. 7.6). They burn quickly with little heat transferred below the soil surface (Wright 1971). Moreover, the small size of Sandberg bluegrass and bottlebrush squirreltail in climax communities helps them survive fires (Wright and Klemmedson 1965); therefore they usually increase in abundance after a fire. All rhizomatous grasses such as thickspike wheatgrass and plains reedgrass increase immediately after a fire (Fig. 7.7) (Blaisdell 1953). Produc-tion from rhizomatous grasses on burned plots will be above that on controls for about 30 years (Harniss and Murray 1973).

Bluebunch wheatgrass (Fig. 7.8) will return to preburn production in one to three years after a fire (Blaisdell 1953; Moomaw 1957; Daubenmire 1963; Conrad and

Fig. 7.6. Comparative effects of summer fires on fine-stemmed (A & B) and coarse-stemmed (C & D) bunchgrass. A. Dense plant material at the base of a needle-and-thread plant. B. Severe charring down to the growing points of needle-and-thread during summer months is common. C. Plant material is coarse and not very dense at the base of squirreltail plants. D. After burning, charring is less severe and less damaging to the growing point.

Fig. 7.6 *(continued)*

Poulton 1966; Uresk et al. 1976); needle-and-thread in three to eight years, depending on site (Blaisdell 1953; Dix 1960; Wright et al. 1979); and Idaho fescue in two to twelve or more years depending on soil moisture, season, and intensity of the fire (Blaisdell 1953; Conrad and Poulton 1966; Harniss and Murray 1973). The response of prairie junegrass to fire is similar to that of needle-and-thread (Vallentine 1971). Cusick bluegrass *(Poa cusickii)* is reduced to 50 percent the first growing sea-

Fig. 7.7.   Dense stand of thickspike wheatgrass and plains reedgrass three years after burning a mountain big sagebrush community near Dubois, Idaho. Patches of sagebrush in the background were not burned by the fire. (Photo courtesy USDA Forest Service.)

Fig. 7.8.   Excellent stand of bluebunch wheatgrass with a moderate amount of Basin big sagebrush in Idaho. Sagebrush communities such as this do not impede animal movements and do not need to be burned. (Photo courtesy University of Idaho.)

son after burning (Uresk et al. 1976). Indian ricegrass is only slightly damaged by fire (Vallentine 1971).

Repeated burning every few years or burning in early summer will deplete a stand of perennial grasses and allow annual grasses, primarily cheatgrass (Fig. 7.9), to increase sharply (Pickford 1932; Wright and Klemmedson 1965). Once a sagebrush-grass community is depleted of perennial plant cover, secondary succession goes

Fig. 7.9. Repeated summer fires will deplete a stand of perennial grasses and allow cheatgrass, as shown in this photo, to dominate. (Photo courtesy University of Idaho.)

from Russian thistle *(Salsola kali)* to mustard *(Sisymbrium* and *Descurainia* spp.) to cheatgrass within five years (Piemeisel 1951). Pechanec and Hull (1945) found that burning near Boise, Idaho reduced cheatgrass plants in varying numbers, depending on the month of the burn, as follows

Burn	Cheatgrass plants	
	m^2	ft^2
June	151	14
July	118	11
August	441	41
October	484	45
November	1334	124

Early summer burns are only a temporary setback for cheatgrass at a time of the year when climax perennials are easily killed by fire (Wright and Klemmedson 1965). Hence the density of cheatgrass increases over time while fewer perennials survive after each fire.

Such areas can only be reclaimed by chemical fallow techniques (Eckert and Evans 1967) or plowing and then seeding. Most seeding has been done with wheatgrasses (Hull 1971). Fairway wheatgrass *(Agropyron cristatum)*, crested wheatgrass *(A. desertorum)*, and Siberian wheatgrass *(A. sibericum)* are well adapted to this zone. Crested wheatgrass is best adapted to the driest sites in the 20- to 30-cm (8- to 12-in.) precipitation zone.

**Table 7.1.   Susceptibility of Forbs to Fire by Three
Classifications at Dubois, Idaho**

Severely Damaged	Slightly Damaged	Undamaged
Antennaria dimorpha	*Astragalus* spp.	*Achillea millefolium*
Antennaria microphylla	*Castilleia angustifolia*	*Allium* spp.
Arenaria uintahensis	*Crepis acuminata*	*Arnica fulgens*
Erigeron engelmannii	*Geranium viscosissimum*	*Balsamorhiza sagittata*
Eriogonum caespitosum	*Lupinus caudatus*	*Comandra umbellata*
Eriogonum heracleoides	*Penstemon radicosus*	*Erigeron corymbosus*
Phlox canescens	*Sphaeralcea munroana*	*Lupinus leucophyllus*
		Phlox longifolia
		Senecio integerrimus
		Sisymbrium linifolium
		Zygadenus paniculatus

From Pechanec et al. (1954). Reproduced by permission of USDA Forest Service.

Forbs

Fall burning does not harm most forbs because many of them are dry and often disintegrated by this time. However, some forbs remain green and are very susceptible to fire. Pechanec et al. (1954) classified forbs according to their susceptibility to fire (Table 7.1). Some variations from this data are reported in the literature. Daubenmire (1970) found that *Erigeron corymbosus* and *Madia exiguas* decreased after burning and that *Arenaria congesta* and *Fransera albicaulis* decreased. Unpublished data in Utah shows that late summer or fall burning can kill Indian paintbrush *(Castilleja angustifolia)* (Wright et al. 1979).

After 12 years Blaisdell (1953) found that only the heavy sagebrush-grass burn (all sagebrush plants are consumed by fire) supported more forbs than the control. By the end of 30 years after burning, forbs had returned to preburn levels (Harniss and Murray 1973), although both burned and unburned plots contained at least five times as many forbs as before the burn. This may indicate that the year of the burn in 1936 was a dry one.

Shrubs

Fires can have a devastating and long-lasting effect on shrubs in sagebrush-grass communities. Big sagebrush, a nonsprouter, is highly susceptible to fire injury (Pickford 1932; Blaisdell 1953). Blaisdell found that the production of this species on burned areas in Idaho was only 10 percent of that on the control 12 years after the burn, but was near preburn levels 30 years after the burn (Harniss and Murray 1973).

Some areas, however, recover much more quickly. Differences in recovery rates may be related to season of burn as it affects seed production (Johnson and Payne 1968), summer precipitation, and completeness of burn. Three-tip sagebrush is also damaged by fire, but some plants resprout (Blaisdell 1953).

Antelope bitterbrush (Fig. 7.10) is severely damaged by burning (Blaisdell 1953; Pechanec et al. 1954; Countryman and Cornelius 1957; Nord 1965; Daubenmire 1970). In Idaho, 12 to 15 years after an experimental burn, antelope bitterbrush was only producing 50 to 60 percent as much as the control (Blaisdell 1953). If soil is wet at the time of burning or shortly after the burn, antelope bitterbrush regularly resprouts (Blaisdell 1953; Blaisdell and Mueggler 1956b; Nord 1965). If the plants resprout, they will regain original growth in 9 to 10 years (Blaisdell 1953). Where fires are not followed by rain, antelope bitterbrush seldom sprouts (Nord 1965; Daubenmire 1970). Seeds of antelope bitterbrush germinate best following fall burns (Wright et al. 1979). Thus burning after a rain in the fall should minimize damage to antelope bitterbrush and enhance the establishment of seedlings. In southern California, desert bitterbrush *(Purshia glandulosa)* resprouts vigorously and abundantly (Nord 1965), even without postfire rains.

Other species severely damaged by fire include Mexican cliffrose *(Cowania mexicana* var. *mexicana)*, curlleaf mahogany, granite gilia *(Gilia pungens)*, and broom snakeweed (Pechanec et al. 1954; Vallentine 1971; Klebenow et al. 1976). Curlleaf mahogany is a weak sprouter (Wright et al. 1979). Burning damages mountain snowberry *(Symphoricarpos oreophilus)*, but yield remains unchanged 15 years after burning (Pechanec et al. 1954; Blaisdell 1953). Creeping barberry *(Berberis repens)* is favored by burning, especially following intense fires (Blaisdell 1953). Greasewood is easily top-killed, but resprouts vigorously after a fire (Daubenmire 1970).

Fig. 7.10. When bitterbrush (dark plants) is a major component of a plant community, caution should be used before burning. It is very susceptible to fire and should either be protected from fire or burned when the soil is wet. Photo taken near Boise, Idaho. (Photo courtesy USDA Forest Service.)

Rabbitbrush, a common genus in the sagebrush-grass zone, is usually enhanced by fire (Cottam and Stewart 1940; Blaisdell 1953; McKell and Chilcote 1957; Countryman and Cornelius 1957; Chadwick and Dalke 1965; Young and Evans 1974). An exception to this general response is apparent in rubber rabbitbrush *(Chrysothamnus nauseosus)*. Robertson and Cords (1957) reported no recovery of this species two years after a burn on September 3, 1942 near Mono Lake, California and after a burn on November 7, 1943 near McGill, Nevada. However burning in the latter area was repeated on an unburned area the following year on the same date and 95 percent of the plants resprouted. Evidently the intensity of the fire is important, because most of the sprouting after fire is epicormic (stem sprouting), not basal or root-sprouting (Wright et al. 1979).

Chrysothamnus viscidiflorus (all of its varieties and subspecies) resprouts vigorously. Generally production is reduced for one to three years after burning, then it increases dramatically. On the U.S. Sheep Station near Dubois, Idaho burning reduced production 59 percent the first year after burning (Blaisdell 1953). Three years after burning, production had doubled and was tripled at the end of 12 years (Blaisdell 1953). Similarly Chadwick and Dalke (1965) found that the cover of *C. viscidiflorus* had increased four to nine times on 8- to 18-year-old burns on sandy soils in northeast Idaho. Production of *C. bloomeri (C. viscidiflorus)* doubled five years after a burn in northern California (Countryman and Cornelius 1957). In western Nevada Young and Evans (1974) found that green rabbitbrush *(C. viscidiflorus* var. *viscidiflorus)* continued to dominate burns and reestablish itself periodically for 15 years. Communities 40 to 50 years old were dominated by big sagebrush and contained reduced populations of green rabbitbrush.

Production of horsebrush was reduced 50 percent the first year after burning, but doubled at the end of three years (Blaisdell 1953). At the end of 12 years it had increased fivefold. After 30 years many of these plants were dying out, but production was still 60 percent above that in the unburned control (Harniss and Murray 1973). Fire greatly enhances the dominance of this species.

Desirable shrubs such as serviceberry *(Amelanchier alnifolia)*, snowbrush *(Ceanothus velutinus)*, and true mountain mahogany *(Cercocarpus montanus)* are only temporarily set back by fire (Vallentine 1971).

MANAGEMENT IMPLICATIONS

Prescribed fire can be a useful tool in many big sagebrush communities if the fires are carefully planned and livestock do not graze the burn for two growing seasons. Removal of tall, thick sagebrush by fire will greatly enhance movement of livestock and will release grasses and forbs from competition, resulting in increased yields. However depending on the vegetation, fires should not be too frequent and should be planned in early spring or after late summer. Caution should be exercised where antelope bitterbrush is dominant and where horsebrush or rabbitbrush is abundant. Where forbs are abundant (Fig. 7.5) or sagebrush is tall and thick (Fig. 7.11), fall burning may be preferred over chemical treatments as a management tool.

Fig. 7.11. Dense stands of Basin big sagebrush suppress the yield of herbaceous species and impede livestock movements. Fire would create variety in the habitat and enhance forbs and grasses. Photo taken in southern Idaho. (Photo courtesy University of Idaho.)

In the sagebrush-grass region, much burning can be done without firelines, especially where patches of big sagebrush grow in swales or small ravines surrounded by low sagebrush. Fires will not carry in low sagebrush even when winds are 32 km/hr (20 mi/hr). This condition is highly desirable because only big sagebrush growing on the most productive sites is burned and such burning creates an ideal mosaic for wildlife. Similarly early spring burning can be done at higher elevations in stands of big sagebrush and bunchgrass where snow patches can be used as firelines. Reduction of big sagebrush, but not low sagebrush, is desirable for game management. Livestock should be prevented from concentrating on the burned areas.

Burning followed by seeding will suppress medusahead, but the practice is not very successful where cheatgrass is dominant. Most cheatgrass areas must be treated with chemicals or plowed and then seeded, if perennials are desired. Fire will not convert pure stands of cheatgrass to native perennials. Likewise pure stands of cheatgrass will not revert back to native perennials with rest. Such areas have a high fire frequency and will decline in site potential (Young et al. 1976).

Areas that are primarily sagebrush and cheatgrass can be successfully seeded after fire if the burns are hot enough to consume the sagebrush plants (Young et al. 1976) and at the same time to destroy cheatgrass seed. If much live cheatgrass seed remains, the burn should be chemically fallowed (Eckert and Evans 1967) or the germinating cheatgrass plowed before drilling with perennial wheatgrasses.

REFERENCES

Beetle, A. A. 1960. Study of sagebrush—the section tridentatae of Artemisia. Univ. Wyo. Agric. Exp. Stn. Bull. 368. Laramie.

Billings, W. D. 1951. Vegetational zonations in the Great Basin. In *Les Bases Ecologiques de la Regeneration de la Vegetation des Zones Arides*. U.I.S.B. Paris, pp. 101,102.

Blaisdell, J. P. 1953. Ecological effects of planned burning of sagebrush-grass range on the upper Snake River Plains. USDA Tech. Bull. 1075. Washington, D.C.

Blaisdell, J. P., and W. F. Mueggler. 1956a. Effect of 2,4-D on forbs and shrubs associated with big sagebrush. *J. Range Manage.* **9**:38–70.

Blaisdell, J. P., and W. F. Mueggler. 1956b. Sprouting of bitterbrush *(Purshia tridentata)* following burning or top removal. *Ecology* **37**:365–370.

Chadwick, H. W., and P. D. Dalke. 1965. Plant succession on dune sands in Fremont County, Idaho. *Ecology* **46**:765–780.

Conrad, E. C., and C. E. Poulton. 1966. Effect of a wildfire in Idaho fescue and bluebunch wheatgrass. *J. Range Manage.* **19**:138–141.

Cottam, W. O., and G. Stewart. 1940. Plant succession as a result of grazing and of meadow desiccation by erosion since settlement in 1862. *J. For.* **38**:613–626.

Countryman, C. M., and D. R. Cornelius. 1957. Some effects of fire on a perennial range type. *J. Range Manage.* **10**:39–41.

Dahl, B. E. 1966. Environmental factors related to medusahead distribution. Ph.D. Diss. Univ. Idaho, Moscow.

Daubenmire, R. 1963. Ecology and improvement of brush-infested range. W-25 Annual Prog. Rep. Single page summary.

Daubenmire, R. 1970. Steppe vegetation of Washington. Wash. Agric. Exp. Stn. Tech. Bull. 62. Pullman.

Dix, R. L. 1960. Effects of burning on the mulch structure and species composition of grasslands in western North Dakota. *Ecology* **41**:49–56.

Eckert, R. E., Jr., and R. A. Evans. 1967. A chemical-fallow technique for control of downy brome and establishment of perennial grasses on rangeland. *J. Range Manage.* **20**:35–41.

Fosberg, M. A., and M. Hironaka. 1964. Soil properties affecting the distribution of big and low sagebrush communities in southern Idaho. *Amer. Soc. Agron. Spec. Pub.* **5**:230–236.

Harniss, R. O., and R. B. Murray. 1973. 30 years of vegetal change following burning of sagebrush-grass range. *J. Range Manage.* **26**:322–325.

Houston, D. B. 1973. Wildfires in northern Yellowstone National Park. *Ecology* **54**:1111–1117.

Hull, A. C. 1971. Grass mixtures for seeding rangelands. *J. Range Manage.* **24**:150–152.

Hurd, R. M. 1955. Effect of 2,4-D on some herbaceous range plants. *J. Range Manage.* **8**:126–128.

Johnson, J. R., and G. F. Payne. 1968. Sagebrush re-invasion as affected by some environmental influences. *J. Range Manage.* **21**:209–212.

Klebenow, D. A., R. Beall, A. Bruner, R. Mason, B. Roundy, W. Stager, and K. Ward. 1976. Controlled fire as a management tool in the pinyon-juniper woodland, Nevada. Annual Prog. Rep., Univ. Nev., Reno.

Klemmedson, J. O., and J. G. Smith. 1964. Cheatgrass *(Bromus tectorum* L.). *Bot. Rev.* **30**:226–262.

Küchler, A. W. 1964. Potential natural vegetation of the conterminous United States. Manual to accompany the map. Am. Geogr. Soc. Spec. Pub. 36. (With map, rev. ed., 1965, 1966.)

McArthur, E. D., B. C. Guinta, and A. P. Plummer. 1974. Shrubs for restoration of depleted ranges and disturbed areas. *Utah Sci.* **35**:28–33.

McKell, C. M., and W. W. Chilcote. 1957. Response of rabbitbrush following removal of competing vegetation. *J. Range Manage.* **10**:228–230.

Moomaw, J. C. 1957. Some effects of grazing and fire on vegetation in the Columbia Basin region, Washington. *Diss. Abstr.* **17**(4):733.

Nord, E. C. 1965. Autecology of bitterbrush in California. *Ecol. Monogr.* **35**:307–334.

Pechanec, J. F., and A. C. Hull, Jr. 1945. Spring forage lost through cheatgrass fires. *Nat. Wool Grower* **35**(4):13.

Pechanec, J. F., G. Stewart, and J. P. Blaisdell. 1954. Sagebrush burning—good and bad. USDA Farmer's Bull. 1948. Washington, D.C.

Pickford, G. D. 1932. The influence of continued heavy grazing and of promiscuous burning on spring-fall ranges in Utah. *Ecology* **13:**159–171.

Piemeisel, R. L. 1951. Causes affecting changes and rate of change in a vegetation of annuals in Idaho. *Ecology* **32:**53–72.

Robertson, J. H., and H. P. Cords. 1957. Survival of rabbitbrush, *Chrysothamnus* sp., following chemical, burning, and mechanical treatments. *J. Range Manage.* **10:**83–89.

Tisdale, E. W., M. Hironaka, and M. A. Fosberg. 1969. The sagebrush region in Idaho: A problem in range resource management. Idaho Agric. Coll. Ext. Bull. 512. Moscow.

Uresk, D. W., J. F. Cline, and W. H. Rickard. 1976. Impact of wildlife on three perennial grasses of south-central Washington. *J. Range Manage.* **29:**309, 310.

USDA. Forest Service. 1936. The western range. U.S. Congress 74th, 2nd Session, Senate Doc. 199.

Vale, T. R. 1975. Presettlement vegetation in the sagebrush-grass area of the Intermountain West. *J. Range Manage.* **28**(1):32–36.

Vallentine, J. F. 1971. *Range Development and Improvements.* Brigham Young Univ. Press, Provo, Utah.

Wright, H. A. 1971. Why squirreltail is more tolerant to burning than needle-and-thread. *J. Range Manage.* **24:**277–284.

Wright, H. A., and J. O. Klemmedson. 1965. Effects of fire on bunchgrasses of the sagebrush-grass region in southern Idaho. *Ecology* **46:**680–688.

Wright, H. A., L. F. Neuenschwander, and C. M. Britton. 1979. The role and use of fire in sagebrush-grass and pinyon-juniper plant communities: A state-of-the-art review. USDA For. Serv. Gen. Tech. Rep. INT-58. Intermt. For. and Range Exp. Stn., Ogden, Utah.

Young, J. A., and R. A. Evans. 1974. Population dynamics of green rabbitbrush in disturbed big sage-brush communities. *J. Range Manage.* **27**(2):127–132.

Young, J. A., R. A. Evans, and R. A. Weaver. 1976. Estimating potential downy brome competition after wildfires. *J. Range Manage.* **29:**322–325.

CHAPTER

8

CHAPARRAL AND OAKBRUSH

Throughout the world plant geographers regard chaparral as a fire-induced vegetation type (Fig. 8.1) (Shantz 1947). Typical of Mediterranean climates, it is known in the Mediterranean region and South Africa as *macchia* or *fymbos*, in southwest France as *heath*, in south Australia as *brigalow-scrub*, in Spain as *tomillares*, and in the Balkans as *phrygana*. Chaparral communities also exist in North Africa, Asia Minor, Mexico, and central Chile. All these communities are similar to California chaparral and the species depend and behave similarly as regards fire (Shantz 1947). Arizona chaparral (Fig. 8.2) has a higher proportion of sprouting shrubs than California chaparral (Carmichael et al. 1978), but both types have a common origin in the Madro-Teritiary geoflora of the Cenozoic Era on the North American Continent (Axelrod 1958).

Shantz (1947) states "that this type was ever free of fire seems unlikely." Shantz further states that every plant species in chaparral communities survives because it can survive fire. The numerous brush seedlings after fire in California chaparral stands (Sampson 1944; Sweeny 1956) as well as the vigorous sprouting ability of many species are unmistakable signs of their adaptability to fire (Phillips 1965; Hanes 1971; Keeley and Zedler 1978). Axelrod (1958) states that these communities burn readily and have done so for over 2,000,000 years. Their relatively large amount of loosely arranged small material (Fig. 8.3)—much of it becoming dead as the plants mature—and high volatile oil content (Philpot 1969) make them very flam-

Fig. 8.1. Chaparral brushfield in southern California. Dominant species is chamise. Burned before 1940. Photo taken in 1974. Elevation is approximately 610 m (2000 ft). (Photo courtesy USDA Forest Service.)

Fig. 8.2. Arizona chaparral—a shrub live oak-mixed shrub community. (Photo courtesy USDA Forest Service.)

Fig. 8.3. Chamise presents an ideal fuel arrangement for high fire intensities. (Photo courtesy USDA Forest Service.)

mable, particularly in late fall when fire weather is hazardous (Countryman and Philpot 1970). Thus fires are natural and inevitable (Hanes 1971).

Oak-brush communities in the Rocky Mountains will be mentioned briefly in this chapter, but fires are less frequent and relatively little is known about the role of fire in these communities.

FIRE HISTORY

Rogers (1961) correlated the size of burns in California chaparral with age of brush. His data showed that until brush is more than 20 years old, few fires exceed 400 ha (1000 acres). After a wildfire, reburns can occur in the herbaceous and dead fuels left for about five years. Then the woody plants close in and for about 15 years the chaparral is fire-resistant. At 20 years of age the proportion of dead fuels becomes great enough to support big fires under adverse conditions. As a consequence, the recurrence interval of large fires [those of more than 2000 ha (5000 acres)] is 20 to 40 years (Sweeny 1956; Muller et al. 1968; Byrne 1978) but most fires larger than 12,000 ha (30,000 acres) occur in brushfields that are 30 years old or older (L. R. Green, pers. comm.). At high elevations [1200 m (4000 ft)] on northern aspects, fire frequency in Eastwood manzanita *(Arctostaphylos glandulosa)* chaparral may be as low as 50 to 100 years (Vogl and Schorr 1972). Less is known about the interval of recurrent fires in the woodland-grass chaparral that borders true chaparral in California.

Arizona chaparral burns periodically, but has a lower fire frequency than California chaparral. Burned chaparral areas that are left to recover naturally seldom support a reburn for at least 20 years, and many chaparral stands, particularly those containing shrub live oak *(Quercus turbinella)*, are 80 to 100 years old (Cable 1975; Carmichael et al. 1978).

The oak-brush zone that is just above the pinyon-juniper *(Pinus-Juniperus* spp.) zone of the central and southern Rocky Mountains probably has a fire frequency of 50 to 100 years, although this is speculative. Most fires in Gambel oak *(Quercus gambelii)* occur after a buildup of litter and mulch under the shrub mottes and generally occur during dry periods (J. F. Vallentine, pers. comm.). Most fires are spotty and irregular.

CALIFORNIA CHAPARRAL

Distribution, Climate, Soils, and Vegetation

Chaparral is a major plant association in California and in a small part of south-central Oregon (Fig. 8.1), composed almost entirely of shrubs 0.6 to 3.0 m (2 to 10 ft) tall (Fig. 8.4). It covers 3.45 million ha (8.5 million acres) of foothill and low-elevation mountain terrain near the coast and valleys and on inland mountains of southern California west of the deserts (Wieslander and Gleason 1954; Bentley 1967). Some chaparral mixes with woodland on the foothills of the Sierra Nevada but occurs in smaller blocks. Another million hectares (2.5 million acres) of coastal sage

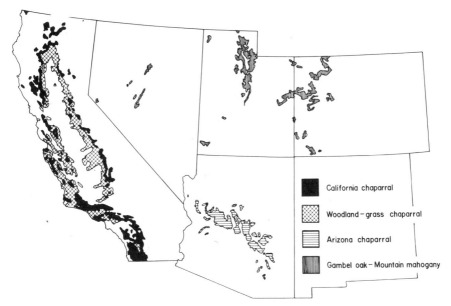

California chaparral

Woodland–grass chaparral

Arizona chaparral

Gambel oak–Mountain mahogany

Fig. 8.4. Distribution of chaparral and oakbrush vegetation. (Adapted from Küchler 1964 and Carmichael et al. 1978. Reproduced by permission of the American Geographical Society and the USDA Forest Service.)

Fig. 8.5. Coastal sage vegetation containing black sage, chamise, and California sagebrush in southern California. (Photo courtesy of USDA Forest Service.)

(often referred to as chaparral) (Fig. 8.5) and minor conifer with chaparral occur along southern coastal areas of California (Wieslander and Gleason 1954).

The climate is distinctly Mediterranean. Winters are mild and rain occurs in late fall, winter, and early spring; the summers are hot and dry. Limits of chaparral distribution are associated with 25 to 130 cm (10 to 50 in.) of precipitation; average temperatures are above freezing but below 38°C (100°F); and elevation ranges from 150 to 1530 m (500 to 5000 ft) (Sampson 1944). Extended periods of cold and snow in the winter or several consecutive days of temperature higher than 38°C (100°F) recurring throughout the summer appear to limit the chaparral association. Soils are usually rocky, coarse-textured, and shallow (Hanes 1971).

Chaparral communities generally are bounded above by forests and below by grasslands. Though fire is part of the chaparral environment, but probably not necessary to maintain its identity (Hanes 1971), fire carries this type far beyond its natural boundaries into forest communities (Burcham 1959). Thus in forest communities, chaparral is induced by fire, but in true chaparral communities it is most likely induced by climate. There is no evidence to support Horton and Kraebel's (1955) claim that, in the absence of fire, chamise-chaparral will be succeeded by a live oak woodland below 1220 m (4000 ft) (Hanes 1971). Above this elevation, however, there is evidence to indicate that mesic sites would gradually succeed to oak woodland in the absence of fire (L. R. Greene, pers. comm.).

Chamise *(Adenostema fasciculatum)* is the most abundant and widespread of all chaparral shrubs in California (Biswell 1974), and the genus manzanita *(Arctostaphylos* spp.), comprised of many sprouting and nonsprouting species, is the second

Fig. 8.6. Mixed chaparral at 1070 m (3500 ft) elevation containing chamise, manzanita, ceanothus, and other species. (Photo courtesy of USDA Forest Service.)

most important group of shrubs. Chamise, a sprouting species and prolific seeder, is present throughout California. It usually occupies the drier, south-facing slopes, whereas manzanita usually occupies the more moist, north-facing exposures. Chamise decreases in abundance with elevation and gives way to manzanita at higher altitudes (Fig. 8.6) on south exposures (Wilson and Vogl 1965). Other common shrubs include Christmas berry *(Heteromeles arbutifolia)*, wedgeleaf ceanothus *(Ceanothus cuneatus)*, desert ceanothus *(C. greggii)*, scrub oak *(Quercus dumosa)*, and western mountain-mahogany *(Cercocarpus betuloides)*. Of these latter shrubs, only the ceanothus species are nonsprouters. On south-facing slopes of coastal exposures, common shrubs include California sagebrush *(Artemisia californica)*, black sage *(Salvia mellifera)*, white sage *(S. apiana)*, California buckwheat *(Eriogonum fasciculatum)*, and deerweed *(Lotus scoparius)* (Fig. 8.5) (Biswell 1974).

Fire Effects

California chaparral is well adapted to fire and a fire every 20 or 30 years keeps it healthy. Many shrub species sprout, several will sprout and reproduce from seed, and many others, as well as forbs, only reproduce from seed that can lie dormant in the soil for many decades (Biswell 1974; Keeley and Zedler 1978). Without fire, a high proportion of the nonsprouting shrubs eventually die (Keeley and Zedler 1978) and the communities become unproductive. For example, Gibbens and Schultz (1963) found that a 60-year-old stand of decadent mixed chaparral brush produced 15 to 119 kg/ha (13 to 106 lb/acre) of browse annually. After fall and early spring burns, similar communities produced from 1840 to 3090 kg/ha (750 to 2750 lb/acre) annually for the next five years (Sampson 1944; Biswell 1969).

During the sixth year after burning, the rate of recovery of both chamise and chaparral slows, and the annual growth increment declines appreciably by the eighth year, when the stand has recovered 50 percent of its preburn biomass (Sampson 1944). During the next 10 to 15 years, chamise continues growing, but begins to level off at 20 to 25 years of age (Hedrick 1951; Horton and Kraebel 1955). At 37 years of age, chamise stops growing and senescence begins (Rundel and Parsons 1979). Thereafter it becomes a declining climax that shows no sign of being replaced by another plant community (Hanes 1971).

Succession after fire is basically autogenic, although forbs, grasses, and low-growing shrubs pass through distinct seral stages as the climax shrubs continue to increase in size (Hanes 1971). Numerous shrubs resprout and establish seedlings after fire. Nonsprouting species usually have the highest number of seedlings (Keeley and Zedler 1978), but most seedlings, regardless of species or habitat, have a high mortality that begins during the first summer after germination, and continues though at a diminished rate as the stand matures (Hanes 1971). Hanes states that "succession in chaparral is more a gradual elimination of individuals present from the outset than a replacement of initial shrubs by new species." Numerous annual forbs and grasses are present on chaparral burns for the first few years, but they rapidly give way to the shrubs after five years (Sampson 1944; Sweeny 1956). Then the understory of the shrubs has very few scattered annuals, if any (Hanes 1971).

SHRUBS

Chamise is a vigorous sprouter, although 20 to 40 percent of the plants usually die following a fire (Keeley and Zedler 1978). Those that die are the oldest plants, which might be nature's way of keeping the chamise population vigorous and healthy (Biswell 1974). To compensate for this loss, numerous seedlings germinate (Table 8.1). Many of them eventually die, but an adequate number survive to replace the old plants killed by the fire (Biswell 1974). Since chamise is best adapted to the lower elevations, a higher proportion of seedlings germinate at 305 to 610 m (1000 to 2000 ft) than at 610 to 1220 m (2000 to 4000 ft) (Horton and Kraebel 1955).

The reproductive potential of manzanita varies by species. Eastwood manzanita is the primary sprouting species in southern California (Hanes 1971). Only 1 percent of the plants die after a fire and it produces relatively few seedlings, which have one of the highest mortalities among chaparral species (Keeley and Zedler 1978). Woollyleaf manzanita *(Arctostaphylos tomentosa)* and greenleaf manzanita *(A. patula)* are also sprouting species. The former is common throughout the chaparral zone and the latter is usually mixed with forest species (Biswell 1974). Greenleaf manzanita also has the unique ability to reproduce by layering (Biswell 1974).

Among the nonsprouting species of manzanita, bigberry manzanita *(A. glauca)*, common in southern California, is one species that has been studied intensively. Fol-

Table 8.1.   Average Numbers of Seedlings on Three Plots
[9.3 m² (100 ft²)] for Four Years in North-Coastal
Region of California Following a Burn in August, 1949

| Species | 1950 | | 1951 | 1952 | 1953 |
	February 23	June 30	June	June	June
Adenostema fasciculatum	1340	49	22	17	19
Ceanothus cuneatus	108	17	11	9	10
Ceanothus parryi	76	9	4	3	3
Cercocarpus betuloides	29	11	11	10	8

From Sweeny (1956). Reprinted by permission of the University of California Press.

lowing a fire, 6400 to 10,600 seedlings per hectare (2590 to 4290 seedlings per acre) may occur and 33 to 44 percent will surive after the first year (Keeley and Zedler 1978). This demonstrates the tenacity with which nonsprouting species maintain their presence in a mixed chaparral community. In northern California, common manzanita *(A. manzanita)* is the most common species of its genus but there are many other species throughout the state.

Scrub oak is a vigorous sprouter that usually has no adult mortality and no seedling establishment after a fire (Keeley and Zedler 1978). Even so, it occupies more area as a stand is burned more times (Hanes 1971). New scrub oak plants appear to be rare, but probably become established in a mesic microenvironment (Keeley and Zedler 1978). Leather oak *(Quercus durata)* is also a vigorous sprouter and responds similarly to fire.

Yerbasanta *(Eriodictyon angustifolium)* is common after fire and comes from seed and an extensive rhizomatous root system (Sampson 1944; Hanes 1971). Other vigorous sprouting species common after fire, depending on the site, include deerbrush *(Ceanothus integerrimus)*, Christmas berry, Interior live oak *(Quercus wislizenii)*, western mountain-mahogany, and silk-tassel *(Garrya veatchii)*. Deerbrush also reproduces prolifically from seed. Species of lesser abundance that sprout vigorously after fire are laurel sumac *(Rhus laurina)*, sugar sumac *(R. ovata)*, whitethorn *(Ceanothus leucodermis)*, evergreen cherry *(Prunus ilicifolia)*, redberry *(Rhamnus crocea)*, coffeeberry *(R. californica)*, poison oak *(Rhus diversiloba)*, canyon oak *(Q. chrysolepis)*, chaparral pea *(Pickeringia montana)*, California laurel *(Umbellaria californica)*, blueblossom *(Ceanothus thyrsiflorus)*, California huckleberry *(Vaccinium ovatum)*, black sage, white sage, and chilocote *(Marah macrocarpus)*.

Coyote bush *(Baccharis pilularis)*, a weak sprouter, has become increasingly abundant in the San Francisco Bay area due to the exclusion of fire (McBride and Heady 1968). In a simulated burning study, 8 out of 11 plants burned at the base died; small shrubs were killed more easily than large shrubs. These data were similar to those commonly observed on wildfires by McBride and Heady (1968) and give ample evidence that coyote bush is reduced, but not eliminated, by wildfires.

Nonsprouting species thrive on new burns as seedlings, depending on aspect, elevation, and proximity to the coast. Manzanita comprises a large portion of the nonsprouting species but many others are present. Some of the other most prevalent species include chamise, deerweed, wedgeleaf ceanothus, desert ceanothus, hoaryleaf ceanothus *(Ceanothus crassifolius)*, wavyleaf ceanothus *(C. foliousus)*, California buckwheat, California ceanothus *(C. lemmoni)*, goldenbush *(Aaplopappus squarrosus)*, and monkey flower *(Mimulus spp.)*.

Seeds of many chaparral species, such as deerweed (Went et al. 1952), require a heat treatment by fire before they will germinate, even though heat destroys many. This species is particularly abundant in sunny locations on burns near the coast during the first decade after fire, but then diminishes. Only isolated individuals remain by the end of the second decade (Hanes 1971).

Seeds of other chaparral species, such as hoaryleaf ceanothus, California sagebrush, black sage, white sage, chaparral yucca *(Yucca whipplei)*, and California buckwheat, do not require scarification for germination. However they can endure fire and often germinate better after a fire (Sweeny 1956). Better germination of those seeds not requiring a heat treatment may be related to the removal of competition, litter, and particularly phytotoxic substances in the soil that are produced by certain shrubs (Muller et al. 1968; Hanes 1971; Christensen and Muller 1975). Soil fertility does not appear to increase germination, although the plants are considerably more luxuriant in fertilized and ash-treated plots (Christensen and Muller 1975). Hellmers et al. (1955) reported nitrogen and phosphorus deficiencies sufficient to retard native plant growth in southern California brushland soils. Burning will generally increase ammonium nitrate, phosphate, sulfate, and potassium (Christensen and Muller 1975).

FORBS AND GRASSES

Annual forbs and grasses are sparse (less than 1% of cover) in mature stands of chaparral (Sampson 1944). After fire, however, endemic forbs become very abundant during the first year (Fig. 8.7). During the second year, annual grasses and forbs are prevalent (Fig. 8.8) (Sampson recorded 106 species) and then decline gradually until the fifth year. By this time the herbaceous species are so sparse that they resemble the density in mature stands (Hanes 1971). Chaparral seedlings and sprouts have largely recaptured all burned areas at the end of five years.

Forbs are numerous after fires in chaparral and occasional fires are essential for their survival (Sweeny 1956). As the shrubs increase in stature and begin producing toxins again, seed germination of herbs is again inhibited (Muller et al. 1968). Thus annual species are gradually eliminated, but the seeds of most species remain viable until the next fire.

Several perennial forbs which are common after burns include purplehead brodea

Fig. 8.7. Forbs are usually the most prevalent herbaceous species after a fire. Plants in this photo include *Eschscholtzia* sp., *Phacelia* sp., *Lotus* sp., *Chaenactus* sp., *Artermisiaefolia* sp., and *Salvia mellifera*. (Photo courtesy of USDA Forest Service.)

Fig. 8.8. During the second and third years after fire in California chaparral, grasses invade and become prevalent with the forbs. (Photo courtesy of USDA Forest Service.)

(Brodia pulchella), Mariposa and white globe lilies *(Calochortus albus* and *C. weedii)*, soap plant *(Chlorgalum pomeridianum)*, and death camas *(Zygadenus fremontii)* (Biswell 1974; Christensen and Muller 1975). Most plants germinate from seed, but some occasionally germinate from old plants (Christensen and Muller 1975).

Common annual forbs include whispering bells *(Emmantha penduliflora)*, phacelia *(Phacelia brachyloba, P. divaricata, P. suavcolens,* and *P. grandiflora)*, western poppy *(Papaver californicum)*, snapdragon *(Antirrhinum cornutum)*, monkey flower tuberous skullcap *(Scutellaria tuberosa)*, Spanish clover *(Lotus americanus)*, and navelseed *(Navarretia mellita)* (Sampson 1944; Sweeny 1956; Biswell 1974; Christensen and Muller 1975). Weedy species from nearby roadsides may also invade the burned areas. These species may include woolly-yarrow *(Eriophyllum lanatum)*, tarweed *(Hemizonia* sp.), Napa star thistle *(Centaurea melitensis)*, western thistle *(Cirsium occidentata)*, spreading fleabane *(Erigeron divergens)*, and common groundsel *(Senecio vulgaris)*.

Annual grass seed (exotic species) does not remain dormant for long periods and must invade from areas adjacent to the burn. Thus few annual grasses usually appear on burned areas the first year after a fire (Fig. 8.7) (Biswell 1974). They increase rapidly during the second and third years (Fig. 8.8), however, and usually include silver hairgrass *(Aira caryophyllea)*, nitgrass *(Gastridium ventricosum)*, foxtail fescue *(Festuca megalura)*, red brome *(Bromus rubens)*, and ripgut brome *(B. rigidus)* (Sampson 1944).

Conversion to Grasslands

Conversion of chaparral to grasslands is often tried in California to increase grass and water yields. This usually involves seeding of perennial grasses after burning, followed by three consecutive years of spraying with 2,4-D and 2,4,5-T. This practice increases water yields and forage for livestock and game if the slopes are under 30 percent (Bentley 1967). Perennial grasses that are most commonly seeded on dry sites in northern California are hardinggrass *(Phalaris tuberosa* var. *stenoptera)* and smilograss *(Oryzopsis miliacea)*. At higher elevations wheatgrasses *(Agropyron* spp.) are seeded. Annual clovers *(Trifolium subterraneum, T. hirtum, T. incarnatum)* and bur clover *(Medicago hispida)* are often seeded with the grasses. Seedings of perennial grasses are generally done in the fall, but have limited success in southern California below 1400 m (3500 ft).

In northern California runoff generally increases water yield [1.5 surface-cm (0.6 surface-in.)] on most chaparral sites to 15 surface-cm (6.1 surface-in.) on deep soils after conversion from brush to grass (Bentley 1967). However, such conversion should be limited to gentle and moderate slopes to minimize soil losses and generally cannot be justified for water yield alone (Bentley 1967). Southern California receives less precipitation and would not receive significant increases in water yield following brush-to-grass conversion, except possibly at the higher elevations.

ARIZONA CHAPARRAL

Distribution, Climate, Soils, and Vegetation

Arizona chaparral (Fig. 8.2, 8.4) covers between 1.3 and 1.6 million ha (3.2 to 4.9 million acres) of the middle elevation lands in central Arizona (Cable 1975; Carmichael et al. 1978). Most of the vegetation is on rough, broken terrain south of the Mogollon rim at elevations that range from 915 to 1830 m (3000 to 6000 ft) (Hibbert and Ingebo 1971; Carmichael et al. 1978). The upper limit borders ponderosa pine *(Pinus ponderosa)* or pinyon-juniper, whereas the lower limit borders desert grassland or southern desert shrubs (Carmichael et al. 1978).

Average annual precipitation ranges from 40 cm (16 in.) on the driest sites to 65 cm (26 in.) on the wettest sites (Carmichael et al. 1978). About 55 percent of the precipitation occurs between November and April, as rain and some snow. Convectional storms are common during July, August, and September, but May, June, and October are dry months (Carmichael et al. 1978). Thus Arizona chaparral grows primarily during the summer, whereas California chaparral grows primarily during the winter and spring (Clements 1920).

Topography varies from low hills to deep canyons with long, steep slopes. Occasional benches and mesas occur at the higher elevations. The lower areas have highly erodible coarse soils that are poorly developed, shallow, and derived from granitic parent material (Pase and Pond 1964). Over half the soils of Arizona chaparral are derived from granite (Carmichael et al. 1978). At higher elevations the soils are finer textured, contain more organic matter, and are derived from shales, sandstones, and limestones (Pase and Pond 1964).

Most shrubs in the Arizona chaparral zone are prolific crown- or root-sprouters and produce few seedlings (Pase 1969), although the few nonsprouters produce abundant seed (Carmichael et al. 1978). Shrub live oak, a vigorous sprouter, is by far the most abundant shrub and is present throughout most of the chaparral zone (Cable 1975).

At the lower elevations and drier exposures, shrub live oak and pointleaf manzanita *(Arctostaphylos pungens)* dominate, with small amounts of skunkbush sumac *(Rhus trilobata)*, sugar sumac, redberry, desert ceanothus, hairy mountain-mahogany *(Cercocarpus breviflorus)*, and other species of desert shrubs (Pond and Cable 1962; Cable 1975). On northern exposures and upper elevations codominants with shrub live oak in ascending order include western mountain-mahogany, yerbasanta, desert ceanothus, Wright silktassel *(Garrya wrightii)*, deerbrush, pringle manzanita *(Arctostaphylos pringlei)*, yellowleaf silktassel *(G. flavescens)*, Emory oak *(Quercus emoryi)*, Arizona oak *(Q. arizonica)*, and Arizona cypress *(Cupressus arizonica)* (Cable 1975; Carmichael et al. 1978).

Canopy cover of shrubs may vary from less than 40 percent on dry sites to more than 80 percent on the wettest sites (Carmichael et al. 1978). Few understory forbs and grasses are present after canopy cover reaches 60 percent, but they may be fairly common in open stands of chaparral, particularly annuals (Pase and Johnson 1968). Common grasses at lower elevations include black grama *(Bouteloua eriopoda)*, blue grama *(B. gracilis)*, sideoats grama *(B. curtipendula)*, threeawns *(Aristida* spp.), wolftail *(Lycurus phleoides)*, muhlys *(Muhlenbergia* spp.), red brome, and various species that have been seeded such as weeping lovegrass *(Eragrostis curvula)*, Lehmann lovegrass *(E. lehmanniana)*, and Boer lovegrass *(E. chloromelas)* (Cable 1975; Carmichael et al. 1978). Carmichael et al. listed common forbs and half-shrubs as penstemon *(Penstemon* sp.), redstar morning glory *(Ipomoea coccinea)*, dark spurge *(Euphorbia melanadenia)*, mustards *(Descurania* sp. and *Sysimbrium* sp.), buckwheats *(Eriogonum* spp.), asters *(Aster* spp.), fleabanes *(Erigeron* spp.), bluedicks *(Dichelostemma pulchellum)*, broom snakeweed *(Xanthocephalum sarothrae)*, rough menodora *(Menodora scabra)*, and broom menodora *(M. scoparia)*.

Fire Effects

SHRUBS

Shrub live oak, sugar sumac, skunkbush sumac, Wright silktassel, redberry, catclaw mimosa *(Mimosa biuncifera)*, Emory oak, and yerbasanta resprout vigorously after fire (Cable 1975; Carmichael et al. 1978). Other sprouters include true mountain mahogany, western mountain mahogany, and hairy mountain mahogany (Cable 1975), which also recover rapidly following fire (Carmichael et al. 1978). Desert ceanothus and Mexican cliffrose (*Cowania mexicana* var. *mexicana*) seldom sprout following fire. Deerbrush sprouts occasionally, but is primarily dependent upon seed for regeneration (Carmichael et al. 1978). These two species, as well as pointleaf manzanita, pringle manzanita, and Arizona cypress, depend upon seed for reestablishment. Yerbasanta also reproduces from seed, even though it is a vigorous sprouter. Broom snakeweed is a weak sprouter. Most broom snakeweed plants are killed by fire (Cable 1975), but this species reestablishes itself easily from seed.

Seeds of several chaparral species are stimulated to germinate by heat. They include the two species of manzanita, deerbrush, desert ceanothus, and yerbasanta. All of these species are considered fire-induced types (Carmichael et al. 1978). Yellowleaf silktassel also appears to be a fire-induced species (Carmichael et al. 1978). Light burns favor germination and survival of yellowleaf silktassel seedlings (Cable 1975). Shrub live oak appears to require 38 cm (15 in.) of fall and winter precipitation and 25 cm (10 in.) of summer precipitation before oak seedlings can successfully germinate and become established (Cable 1975). Fire does not favor its establishment.

Although shrub live oak is a vigorous sprouter, most (68%) shrub live oak-mixed shrub stands have not been burned for 80 years (Carmichael et al. 1978). Similarly Carmichael et al. (1978) found that stands dominated by Emory oak and Arizona oak do not have a recent fire history. Ultimately they dominate other fire-induced types at elevations above 1645 m (5400 ft). Arizona cypress grows at the highest elevations of the chaparral zone and is the least tolerant to fire of all trees and shrubs in the chaparral zone (Carmichael et al. 1978).

Shrub cover generally recovers rapidly after a fire. About 76 percent of the preburn crown cover is reached in six years (Pase and Pond 1964), but it may take more than 11 years for the shrub cover to reach preburn levels (Hibbert et al. 1974). Despite the rapidity with which unseeded chaparral stands recover, they will not support a repeat burn for at least 20 years (Cable 1975). Thus all evidence indicates that the natural fire frequency in Arizona chaparral is considerably less than for the low- to mid-elevational ranges of California chaparral.

The fire-type communities designated by Carmichael et al. (1978) for Arizona chaparral contain a high proportion of nonsprouting species. Generally these fire-type communities appear after burns in communities that were dominated by sprouting species and had a previous fire-free period in excess of 20 years. This sequence of plant succession supports the theory of Keeley and Zedler (1978) for California chaparral in which they state that long fire-free periods (e.g., 100 years) favor those shrubs that are dependent on seed survival. Long fire-free intervals tend to decrease the density of the dominant sprouting species thus creating larger openings among the sprouting shrubs after fire and providing more space for species with an obligate seeding strategy (Keeley and Zedler 1978).

FORBS AND GRASSES

Grass yields averaged 110 kg/ha (100 lb/acre) on burned oak-chaparral sites of Mingus Mountain with maximum yields of 110 to 225 kg/ha (100 to 200 lb/acre) coming after the third year of burning (Pase and Pond 1964). This production came primarily from weeping lovegrass and crested wheatgrass *(Agropyron desertorum)* that were seeded after the fire, and sideoats grama. Bottlebrush squirreltail *(Sitanion hystrix)* and longtongue muttongrass *(Poa longiligula)* were also present. In other areas Lehmann lovegrass is the dominant seeded species. Cool fires in late winter do not lower yields of Lehmann lovegrass (Pase 1971) but hot June fires will kill 90 percent of the plants (Cable 1965).

Forbs are not nearly as common in Arizona chaparral as in California chaparral. In a study by Pase and Pond (1964) the most common forbs were Palmer penstemon *(Penstemon palmeri)*, few-flowered goldenrod *(Solidago sparsiflora)*, Wright verbena *(Verbena wrightii)*, and hoarhound *(Marrubium vulgare)*. However, forb composition varies widely. Unlike grasses, which maintain their yields for five to seven years after a burn, particularly in open shrub communities, the forbs reach their peak in two to three years and then decline rapidly (Pase and Pond 1964).

OAKBRUSH

Distribution, Climate, Soils, and Vegetation

Oakbrush (Fig. 8.9), which is dominated by Gambel oak, is widely distributed in the central and southern Rocky Mountains primarily in Utah, Colorado, New Mexico, and Arizona. It is usually bounded at the lower elevation of 1520 m (5000 ft) by pinyon-juniper and at the upper elevation of 2805 m (9000 ft) by ponderosa pine or aspen *(Populus tremuloides)* (McKell 1950; Brown 1958). Frequently it is a major component of the ponderosa pine understory in Arizona and New Mexico (Reynolds et al. 1970). Elevation extremes for Gambel oak vary from 990 m (3250 ft) to 3050 m (10,000 ft) (Christensen 1949).

In the central part of its range (much of Utah), oakbrush dominates elevations where ponderosa pine would normally be expected (Baker and Koristan 1931), apparently because the late spring and summer months of this region are normally dry (Baker and Koristan 1931; Markham 1939). To the north, May precipitation is ample for reproduction of ponderosa pine and, to the south, July and August precipitation is ample for reproduction of ponderosa pine (Baker and Koristan 1931). Gambel oak requires mean annual temperatures that are about 8°C (15°F) cooler than its southern relative, shrub live oak, in Arizona chaparral. Upper elevational limits for Gambel oak appear to be determined by long periods of persistent subfreezing temperatures (Grover et al. 1970) and lower elevational limits are determined by high temperatures (Christensen 1949).

Fig. 8.9. Gambel oak brush community in eastern Colorado. (Photo courtesy of Bureau of Land Management.)

σ / Annual precipitation varies from 38 to 58 cm (15 to 23 in.), most of which falls during fall, winter, and early spring (Horton 1975). Soils are derived from limestones, sandstones, coarse conglomerates, and shales with surface textures varying from fine sandy loam to clay (Markham 1939; Brown 1958). The A horizon is generally 33 to 53 cm (13 to 21 in.) deep, depending on the slope (Brown 1958). However, the total soil mantle is in excess of 2.4 m (8 ft) deep (Tew 1969).[

Shrubs that are frequently associated with Gambel oak include serviceberry *(Amelanchier utahensis)*, snowberry *(Symphoricarpos vaccinoides; S. oreophilus)*, antelope bitterbrush *(Purshia tridentata)*, Wood rose *(Rosa woodsii)*, and creeping barberry *(Berberis repens)* (McKell 1950). True mountain mahogany often dominates the upper elevations of oakbrush in the southern two-thirds of its range (Cronquist et al. 1972). Northward, the upper elevational limits may be dominated by curlleaf mahogany *(Cercocarpus ledifolius)* (Hayward 1948; McMillian 1948) or mountain lover *(Pachystima myrsinites)* and ninebark *(Physocarpus malvaceous)* (Küchler 1964; Wright 1972).

Common grasses include western wheatgrass *(Agropyron smithii)*, Kentucky bluegrass *(Poa pratensis)*, elk sedge *(Carex geyeri)*, Letterman needlegrass *(Stipa lettermanii)*, and muttongrass *(Poa fendleriana)*. Forbs are common and include many species.

Fire Effects

SHRUBS

Gambel oak is a very fire-tolerant species that has experienced occasional burning for hundreds of years (McKell 1950). It appears to be a true climatic climax with fire playing no significant role in sustaining the community or extending its boundaries (Brown 1958). Oak stems have a natural mortality of about 80 years. The most abundant age class is one to nine years, so there is a continual replenishment of old stems with new stems, regardless of the frequency of fire. It is our guess that fires would occur in this community once or twice every 100 years during extended dry periods, but are not necessary for any particular purpose.

The most common effect of fire on Gambel oak in west-central Colorado is to stimulate suckering with a resultant thickening of open stands and merging of scattered stands into continuous thickets (Brown 1958). This is largely because Gambel oak reproduces primarily by suckers from surface roots and root crown sprouts. However, as oak is protected from fire, it tends to thin out and retreat (McKell 1950; Brown 1958). In Utah McKell (1950) found that Gambel oak grew rapidly the first two growing seasons (50% recovery) after burning, but it had only recovered 75 percent of its original cover 18 years after a burn. The number of shoots increased fourfold the first year after burning and then declined until they were equal on both burned and unburned areas after 18 years.

In addition to Gambel oak, McKell (1950) followed the rate of recovery of several other shrub species associated with Gambel oak on 1-, 2-, 9-, and 18-year-old burns. Cover of chokecherry *(Prunus virginiana)* and Wood rose exceeded that of adjacent unburned areas the second year after burning. After 18 years the cover of both of these species was twice as high on the burned area as the control. Snowberry was slightly harmed by fire for the first few years after burning. After nine years both cover and plant numbers of snowberry plants were nearly equal to those of the unburned area.

Recovery of ninebark from fire was similar to that of snowberry. Serviceberry recovers rapidly, but then gradually declines to 60 percent of the cover found on unburned ranges. Part of this decline may be due to an interaction between the effect of fire and use by big game. Creeping barberry and mountain lover remained drastically reduced on burns after nine years. But after 18 years creeping barberry had increased threefold over unburned plots, whereas mountain lover was still only 50 percent recovered.

FORBS AND GRASSES

Perennial grasses within the mottes, particularly bunchgrasses, are virtually eliminated by intense heat from wildfires (J. F. Vallentine, pers. comm.). Annuals, however, increase dramatically the first year after burning and then decline rapidly (McKell 1950). In the early as well as late stages of succession the most common grass and grasslike plants are elk sedge, Kentucky bluegrass, and cheatgrass *(Bromus tectorum)*. Elk sedge is most common under oak whereas the other two species dominate openings. Other grasses present mainly in the openings are western wheatgrass, Letterman needlegrass, Columbia needlegrass *(Stipa columbiana)*, Canada wildrye *(Elymus canadensis)*, and Thurber fescue *(Festuca thurberi)*. In the lower Gambel oak zone thickspike wheatgrass *(Agropyron dasystachyum)* and western wheatgrass are frequently present and tolerate fire well, even under mottes. Intermediate wheatgrass *(A. intermedium)* and pubsecent wheatgrass *(A. pubescens)*, introduced seeded species, also survive fires well (J. F. Vallentine, pers. comm.).

Dominant forbs after a fire include western yarrow *(Achillea millefolium)*, cutleaf balsamroot *(Balsamorhiza macrophylla)*, Canada thistle *(Cirsium arvense)*, chlorocrambe *(Chlorocrambe hastata)*, goldenrod *(Solidago canadensis)*, American vetch *(Vicia americana)*, Fremont geranium *(Geranium fremontii)*, showy goldeneye *(Viguiera multiflora)*, meadow rue *(Thalictrum fendleri)*, lambsquarters *(Chenopodium album)*, collomia *(Collomia linearis)*, gayophytum *(Gayophytum ramosissimum)*, knotweed *(Polygonum douglasii)*, and bedstraw *(Galium bifolium)*. In later years all of these plants thin out and the dominant forbs in mature stands are fleabanes, western yarrow, American vetch, knotweed, lupine *(Lupinus* spp.), cutleaf balsmroot, peavine *(Lathyrus* sp.), aster *(Aster canescens)*, goldenrod, showy goldeneye, and lambstongue groundsel *(Senecio integerrimus)*.

Average production of herbage on mature oakbrush sites in west-central Colorado was 390 kg/ha (345 lb/acre) (Brown 1958). Moreover productivity was about the same under oakbrush plants as in the openings. Fire most likely increases this production for two or three years, but then it drops quickly. On a long-term basis, natural fires that are not followed with a seeding treatment do not seem to improve herbage yield or species composition in Gambel oak communities. Burns in thickets that are followed by seeding, however, may increase grass yields fivefold. Plummer et al. (1970) demonstrated this kind of seeding success following mechanical shredding treatments, using intermediate wheatgrass, smooth brome *(Bromus inermis)*, and fairway wheatgrass *(Agropyron cristatum)*. Intermediate wheatgrass, pubescent wheatgrass, and smooth brome seem to tolerate the shade of Gambel oak very well.

VALUE AND CONTROL OF GAMBEL OAK

Most land managers see Gambel oak as an important shrub species to stabilize soil, retard snowmelt, and provide browse for deer and elk. It is still too early to tell whether fire and seeding of perennial grasses has the potential to improve these values significantly. Use of Tordon 225 one or two years after a fire shows promise as an effective control of Gambel oak on deep soils (J. F. Vallentine, pers. comm.).

MANAGEMENT IMPLICATIONS

California chaparral has a history of burning every 20 to 40 years. Thus management plans should take this into consideration with planned prescribed burns to avoid costly catastrophic fires. Fire exlusion will not prevent fire; it will only forestall fire (Hanes 1971). Dead fuel, which is high in ether extractives, continues to build up with time (Philpot 1969), and therefore when a fire starts in dry, windy weather, it is explosive and ends in an uncontrollable holocaust that can destroy many homes. Also, protecting chaparral communities from fire does not keep them at their optimum growth rate (Gibbens and Schultz 1963) for productivity of deer (Hendricks 1968; Biswell 1969; Longhurst 1978).

Based on our present knowledge of fire frequencies in California chaparral, we recommend concentrating prescribed burns in stands that exceed 30 years of age. These are the least productive areas and are the most dangerous when wildfires occur in them. Prescribed burning under proper weather conditions would be the most sensible approach to managing old stands of chaparral. The greatest benefits would be reduction in costs of fire control, loss of fewer homes, loss of fewer human lives, and improvement of wildlife habitat.

If fires are more frequent than every 10 to 15 years, sprouting shrubs will be favored over those that depend on seed for reestablishment. Many nonsprouting chaparral species need about 10 years to produce adequate supplies of seed that will survive the next fire. Once seed is produced, it seems to have the capacity to lie dormant for hundreds of years. Seeds of many species require heat to break dormancy or improve percentage germination.

Long fire-free intervals favor nonsprouting shrubs in California and Arizona. Usually they live for shorter periods of time (though there are exceptions) than the sprouting species. As the sprouting species grow larger and compete with each other, some plants die. The areas with dead plants create more space for the obligate seed species. Thus recently burned plant communities often have a preponderance of nonsprouting species.

Fire frequency in Arizona chaparral averages between 20 and 100 years. Since burned communities are fireproof for at least 20 years, it seems that rotational systems of burning could be developed easily. However many Arizona chaparral communities are open and accumulate little dead material. Therefore they do not need to be burned very often, if at all. Fire can be an effective tool to thin thick chaparral and to stimulate germination of palatable nonsprouting species, such as desert ceanothus.

Gambel oak is very tolerant to fire, and fire does not appear to improve herbage yield or species composition, unless the areas are seeded to introduced grasses. For the most part, these areas should be left untreated to stabilize soil, retard snowmelt, and provide browse for deer. Accidentally burned areas have been seeded with intermediate wheatgrass, smooth brome, and fairway wheatgrass which increased grass yields and reduced oak growth (Plummer et al. 1970).

REFERENCES

Axelrod, D. I. 1958. Evolution of the Madro-Tertiary geoflora. *Bot. Rev.* **24**:433–509.

Baker, F. S., and C. Koristan. 1931. Suitability of brushlands in the Intermountain Region for the growth of natural or planted western yellow pine forests. USDA Tech. Bull. 256. Washington, D.C.

Bentley, J. R. 1967. Conversion of chaparral areas to grassland: Techniques used in California. USDA Handb. 328. Washington, D. C.

Biswell, H. H. 1969. Prescribed burning for wildlife in California brushlands. In Trans. Thirty-Fourth N. Amer. Wildlife and Natur. Resour. Conf. **34**:438–446.

Biswell, H. H. 1974. Effects of fire on chaparral. In T. T. Kozlowski and C. E. Ahlgren (eds.) *Fire and Ecosystems*. Academic Press, New York, pp. 321–364.

Brown, H. E. 1958. Gambel oak in west-central Colorado. *Ecology* **39**:317–327.

Burcham, L. T. 1959. Planned burning as a management practice for California wildlands. Calif. Dept. Natur. Resour., Div. For. Sacramento.

Byrne, R. 1978. Fossil record discloses wildfire history. *Calif. Agric.* **10**:13, 14. Sacramento.

Cable, D. R. 1965. Damage to mesquite, Lehmann lovegrass, and black grama by a hot June fire. *J. Range Manage.* **18**:326–329.

Cable, D. R. 1975. Range management in the chaparral type and its ecological basis: The status of our knowledge. USDA For. Serv. Res. Paper RM-155. Rocky Mtn. For. and Range Exp. Stn., Fort Collins, Colo.

Carmichael, R. S., O. D. Knipe, C. P. Pase, and W. W. Brady. 1978. Arizona chaparral: Plant associations and ecology. USDA For. Serv. Res. Paper RM-202. Rocky Mtn. For. and Range Exp. Stn., Fort Collins, Colo.

Christensen, E. M. 1949. Distributional observations of oakbrush *(Quercus gambelii* Nutt.) in Utah. *Proc. Utah Acad. Sci. Arts and Lett.* **27**:22–25.

Christensen, H. L., and C. H. Muller. 1975. Effects of fire on factors controlling plant growth in *Adenostoma* chaparral. *Ecol. Monogr.* **45**:29–55.

Clements, F. E. 1920. Plant indicators—the relation of plant communities to process and practice. Carnegie Inst. of Wash. Pub. No. 290.

Countryman, C. M., and C. W. Philpot. 1970. Physical characteristics of chamise as a wildland fuel. USDA For. Serv. Res. Paper PSW-66. Pac. Southwest For. and Range Exp. Stn., Berkeley, Calif.

Cronquist, A., A. H. Holmgren, N. A. Holmgren, and J. L. Reveal. 1972. *Intermountain Flora—Vascular Plants of the Intermountain West, USA* (Vol. 1). Hafner, New York.

Gibbens, R. P., and A. M. Schultz. 1963. Brush manipulation on a deer winter range. *Calif. Fish and Game* 49:95–118. Sacramento.

Grover, B. L., E. A. Richardson, and A. R. Southard. 1970. *Quercus gambelii* as an indicator of climatic means. *Proc. Utah Acad. Sci. Arts Sci. Arts & Lett.* 47(1):187–191.

Hanes, T. L. 1971. Succession after fire in the chaparral of southern California. *Ecol. Monogr.* 41:27–52.

Hayward, C. L. 1948. Biotic communities of the Wasatch Chaparral, Utah. *Ecol. Monogr.* 18:473–506.

Hedrick, D. W. 1951. Studies on the succession and manipulation of chamise brushlands in California. Ph.D. Diss. Texas A & M Univ., College Station.

Hellmers, H., J. F. Bonner, and J. K. Kelleher. 1955. Soil fertility: A watershed management problem in the San Gabriel Mountains of southern California. *Soil Sci.* **80:** 189–197.

Hendricks, J. H. 1968. Control burning for deer management in California. *Proc. Tall Timbers Fire Ecol. Conf.* **8:**219–233.

Hibbert, A. R., E. A. Davis, and D. G. Scholl. 1974. Chaparral conversion potential in Arizona, Part I: Water yield response and effects on other resources. USDA For. Serv. Res. Paper RM-126. Rocky Mtn. For. and Range Exp. Stn., Fort Collins, Colo.

Hibbert, A. R., and P. A. Ingebo. 1971. Chaparral treatment effects on stream flow. *Proc. Ariz. Watershed Symp., Phoenix* **15:**25–34.

Horton, J. S., and C. J. Kraebel. 1955. Development of vegetation after fire in the chamise chaparral of southern California. *Ecology* 36:244–262.

Horton, L. E. 1975. An abstract bibliography of Gambel oak *(Quercus gambelii* Nutt.). USDA For. Serv., Intermt. Region. Ogden, Utah.

Keeley, J. E., and P. H. Zedler. 1978. Reproduction of chaparral shrubs after fire: A comparison of sprouting and seeding strategies. *Amer. Midl. Natur.* 99(1):142–161.

Küchler, A. W. 1964. Potential natural vegetation of the conterminous United States. Manual to accompany the map. Amer. Geogr. Soc. Spec. Pub. 36. (With map, rev. ed., 1965, 1966.)

Longhurst, W. M. 1978. Responses of bird and mammal populations to fire in chaparral. *Calif. Agric.* 10(10):9–12. Sacramento.

Markham, B. S. 1939. A preliminary study of the vegetative cover in Spanish Fork Canyon, Utah. M. S. Thesis. Brigham Young Univ., Provo, Utah.

McBride, J., and H. F. Heady. 1968. Invasion of grassland by *Baccharis pilularis* DC. *J. Range Manage.* 21:106–108.

McKell, C. M. 1950. A study of plant succession in the oak brush *(Quercus gambelii)* zone after fire. M. S. Thesis. Univ. Utah, Salt Lake City.

McMillian, C. 1948. A taxonomic and ecological study of the flora of the Deep Creek Mountains of central western Utah. M. S. Thesis. Univ. Utah, Salt Lake City.

Muller, C. H., R. B. Hanawalt, and J. K. McPherson. 1968. Allelopathic control of herb growth in the fire cycle of California chaparral. *Bull. Torr. Bot. Club* 95:225–231.

Pase, C. P. 1969. Survival of *Quercus turbinella* and *Q. emoryii* seedlings in an Arizona chaparral community. *Southwest Natur.* 14(2):149–155.

Pase, C. P. 1971. Effect of a February burn on Lehmann lovegrass. *J. Range Manage.* 24:454–456.

Pase, C. P., and R. R. Johnson. 1968. Flora and vegetation of the Sierra Ancha Experimental Forest, Arizona. USDA For. Serv. Res. Paper RM-41. Rocky Mtn. For. and Range Exp. Stn., Fort Collins, Colo.

Pase, C. P., and F. W. Pond. 1964. Vegetational changes following the Mingus Mountain burn. USDA For. Serv. Res. Note 18. Rocky Mtn. For. and Range Exp. Stn., Fort Collins, Colo.

Phillips, J. 1965. Fire—as master and servant: Its influence in the bioclimatic regions of Trans-Saharan Africa. *Proc. Tall Timbers Fire Ecol. Conf.* **4**:7–109.

Philpot, C. W. 1969. Seasonal changes in heat content and ether extrative content of chamise. USDA For. Serv. Res. Paper INT-61. Intermt. For. and Range Exp. Stn., Ogden, Utah.

Plummer, A. P., D. A. Christensen, and K. R. Jorgensen. 1970. Highlights, results, and accomplishments of game range restoration studies. In Utah State Dept. of Fish and Game, Pub. No. 70-3, pp. 26–30.

Pond, F. W., and D. R. Cable. 1962. Recovery of vegetation following a wildfire on a chaparral area in Arizona. USDA For. Serv. Res. Note 72. Rocky Mtn. For. and Range Exp. Stn., Fort Collins, Colo.

Reynolds, H. G., W. P. Clary, and P. F. Ffolliott. 1970. Gambel oak for southwestern wildlife. *J. For.* **68**:454–547.

Rogers, D. H. 1961. Measuring the efficiency of fire control in California chaparral. *J. For.* **49**:697–703.

Rundel, P. W., and D. J. Parsons. 1979. Structural changes in chamise *(Adenostoma fasciculatum)* along a fire-induced age gradient. *J. Range Manage.* **32**:462–466.

Sampson, A. W. 1944. Plant succession on burned chaparral lands in northern California. Calif. Agric. Exp. Stn. Bull. 685. Berkeley.

Shantz, H. L. 1947. The use of fire as a tool in the management of brush ranges of California. Calif. Div. For. Bull. Sacramento.

Sweeny, J. R. 1956. Responses of vegetation to fire. A study of herbaceous vegetation following chaparral fires. *Univ. Calif. Pub. Bot.* **28**:143–216.

Tew, R. K. 1969. Converting Gambel oak sites to grass reduces soil-moisture depletion. USDA For. Serv. Res. Note 104. Intermt. For. and Range Exp. Stn., Ogden, Utah.

Vogl, R. J., and P. K. Schorr. 1972. Fire and manzanita chaparral in the San Jacinto Mountains, California. *Ecology* **53**:1179–1188.

Went, F. W., G. Juhren, and M. C. Juhren. 1952. Fire and biotic factors affecting germination. *Ecology* **33**:351–364.

Wieslander, A. W., and C. H. Gleason. 1954. Major brushland areas of the Coast Ranges and Sierra-Cascade Foothills in California. Calif. For. and Range. Exp. Stn. Misc. Paper No. 15. Berkeley.

Wilson, R. C., and R. J. Vogl. 1965. Manzanita chaparral in the Santa Ana Mountains, California. *Madrona* **18**:47–62.

Wright, H. A. 1972. Shrub response to fire. In Wildland and Shrubs—Their Biology and Utilization. An Int. Symp., Utah State Univ., Logan. USDA For. Serv. Gen. Tech. Rep. INT-1. Intermt. For. and Range Exp. Stn., Ogden, Utah, pp. 204–217.

9

PINYON–JUNIPER

The pinyon-juniper (*Pinus-Juniperus* spp.) association covers from 17.4 to 30.8 million hectares (43 to 76 million acres) in western North America, depending whether you use Küchler's (1964) map (Fig. 9.1) of potential pinyon-juniper woodland (which gives the lower figure) or the earlier Senate Document 199 (Clapp 1936) where all western lands then having juniper, with or without pinyon pines, were included. West et al. (1975) have taken a liberal definition of pinyon-juniper woodlands and included both Küchler's "juniper-pinyon woodlands" and "juniper steppe"-types, plus the areas within the Great Basin big sagebrush (*Artemisia tridentata* ssp. *tridentata*), sagebrush-grass steppe, and the Trans-Pecos shrub savannah where junipers were abundant. Leaving out the juniper woodlands in Texas, the figure by West et al. (1975) for the western United States is 30.4 million hectares (75 million acres). Except for the central Great Basin where pinyon and juniper appear to be superimposed upon a large *Artemisia*/cool-season grass zone (Billings 1952), pinyon-juniper woodlands are located below Gambel oak *(Quercus gambelii)* and ponderosa pine *(Pinus ponderosa)* zones, but above the sagebrush or grassland areas into which juniper, the least mesic of the two dominants (pinyon and juniper), is continually spreading.

FIRE HISTORY

Historically fire has been the dominant force controlling the distribution of pinyon-juniper, particularly juniper, but fire cannot be separated from the effects of drought and competition (Leopold 1924; Johnsen 1962; Burkhardt and Tisdale 1976). All

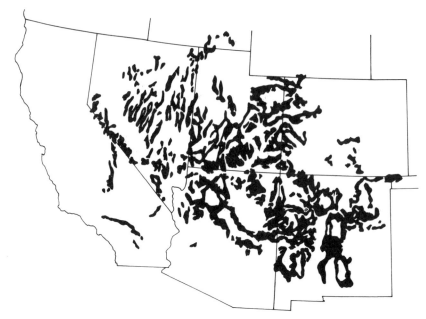

Fig. 9.1. Distribution of pinyon-juniper woodlands in the western United States. (From Küchler 1964. Reproduced with permission of the American Geographical Society.)

three forces seem to have played a complementary role in limiting the distribution of juniper before grazing by domestic livestock was a factor. However, droughts and competition from grass probably only served to slow the invasion and growth of junipers in adjacent grasslands, since the trees are easily established during wet years (Johnsen 1962; Smith et al. 1975), especially where shade is present (Meagher 1943). Then fire, occurring about every 10 to 30 years (Leopold 1924), kept the junipers restricted to shallow, rocky soils and rough topography (Arend 1950; Burkhardt and Tisdale 1969; O'Rourke and Ogden 1969). For the last 70 years, however, heavy livestock grazing has reduced grass competition as well as fuel for fires. Reduced competition from grasses has permitted pinyon and juniper to invade adjacent communities rapidly (Nabi 1978) and the reduced number of fires, each of a lower intensity than the fires before heavy grazing, has left the juniper invasion unchecked (Fig. 9.2).

DISTRIBUTION, CLIMATE, SOILS, AND VEGETATION

Pinyon-juniper woodlands extend from the east slope of the Sierra Nevada Mountains, eastward throughout the mountains of the Great Basin in Nevada and Utah, and on both flanks of the Rocky Mountains in Colorado (as well as on the mesas of Colorado Plateau and interior valleys), then southward into Arizona, New Mexico, and

Fig. 9.2.   Pinyon and juniper invading a sagebrush community in Nevada. (Photo courtesy of University of Nevada.)

northern Mexico (West et al. 1975). Dense stands of juniper alone that join the pinyon-juniper woodlands extend further north into eastern Oregon, southern Idaho, and Wyoming.

Pinyon-juniper woodlands are generally found at elevations from 1370 to 2285 m (4500 to 7500 ft) (Springfield 1976), but they are best developed between 1525 to 2135 m (5000 to 7000 ft) (Woodbury 1947). Nevertheless, pinyon-juniper may be found growing from 760 to 915 m (2500 to 3000 ft) in the upper parts of deserts (Johnsen 1962; Franklin and Dyrness 1969) to 2745 m (9000 ft) in ponderosa pine forests (Lanner 1975). The highest elevation for pinyon-juniper woodland is nearly 3050 m (10,000 ft) in the White Mountains of California where the Sierra Nevada Mountains create an extreme rain-shadow effect (St Andre et al. 1965). The upper altitudinal limit of the pinyon-juniper zone is determined by low temperature and the lower range by a deficiency in moisture (Pearson 1920; Daubenmire 1947).

Annual precipitation varies from 25 to 38 cm (10 to 15 in.) over most of the pinyon-juniper zone where stands are open (Fig. 9.3) (Woodbury 1947), but dense stands may receive 40 to 55 cm (16 to 22 in.) of precipitation (Springfield 1976). On xeric sites at high elevations, annual precipitation of pinyon-juniper communities may be as high as 65 cm (26 in.) (Pearson 1931). Pinyon *(Pinus edulis* and *P. monophylla)* is usually most abundant at the higher elevations of the pinyon-juniper zone with a mixing of pinyon and juniper at middle elevations (Reveal 1944; West et al. 1975; Springfield 1976). Because pinyon is very susceptible to fire damage (Leopold 1924), as well as being the more moisture-demanding of the two genera, fire, in addition to lower moisture, probably keeps pinyon out of the lower elevations.

Fig. 9.3. An open stand of pinyon-juniper with blue grama understory in eastern New Mexico.

Distribution of the pinyon-juniper association is not limited by parent material (Emerson 1932; Springfield 1976) or soil texture (Pearson 1931; Springfield 1976). The trees grow on residual and transported soils derived from sandstones, limestones, basalt, granite, and mixed alluvium (Springfield 1976). Soil textures vary from stony, cobbly, and gravelly loams to clay loam and clay, and in depth from shallow to deep (Springfield 1976), but the best stands of pinyon are found on coarse gravel, gravelly loam, or coarse sands (Phillips 1909). The soils are generally calcareous and alkaline, but juniper can grow easily on acid soils down to a pH of 4.7 (Arend 1950). Except for those soils derived from basalt, most are low in fertility (Howell 1941; Springfield 1976) and often shallow and rocky.

Dominant tree species of the pinyon-juniper woodlands are Utah juniper *(Juniperus osteosperma)*, one-seed juniper *(J. monosperma)*, Rocky Mountain juniper *(J. scopulorum)*, alligator juniper *(J. deppeana)*, doubleleaf pinyon *(Pinus edulis)*, and singleleaf pinyon *(P. monophylla)*. Doubleleaf pinyon is associated with the southern Rocky Mountains and extends westward to the eastern edge of the Great Basin. Throughout the Great Basin, eastern slope of the Sierra Nevada Mountains, and western portions of Arizona, singleleaf pinyon is associated primarily with Utah juniper.

Utah juniper is the most important of its genus in the pinyon-juniper woodland (Lanner 1975). Its range extends over Utah, Nevada, northern Arizona, and parts of California. One-seed juniper extends from central Colorado to central Arizona and it is widely distributed throughout most of New Mexico. Some one-seed juniper is found in the Palo Duro Canyon of west Texas. Alligator juniper, a sprouting species, is also widespread in the Southwest. It is abundant in Mexico and reaches its northern limits in north-central Arizona and New Mexico. Rocky Mountain juniper occurs only at the highest elevations of the pinyon-juniper woodland (Lanner 1975).

Outside the pinyon-juniper woodland, western juniper *(Juniperis occidentalis)* is very abundant in eastern Oregon (Fig. 9.4). East of New Mexico, additional species of juniper, such as redberry juniper *(J. pinchoti)*, a sprouting species, Ashe juniper *(J. ashei)*, and eastern redcedar *(J. virginiana)*, are abundant. The latter junipers are usually associated with rocky slopes such as escarpments, ridges, or rimrocks throughout the Great Plains (Wells 1970) and alkaline soils such as the Edwards Plateau in central Texas. From the rocky areas where juniper is considered to be the climax in the plains, trees have spread rapidly into the surrounding grasslands in the absence of fire. Eastern redcedar is the most extensive species and occurs from Texas northeastward throughout most of the eastern United States.

Herbaceous species vary considerably throughout the pinyon-juniper zone. On the the eastern flank of the Rocky Mountains in New Mexico, blue grama *(Bouteloua gracilis)* is the dominant herbaceous understory (Fig. 9.3) with some sideoats grama *(B. curtipendula)* and wolftail *(Lycurus phleoides)* and a few forbs often intermixed (Pieper et al. 1971). In north-central Arizona blue grama and sideoats grama remain as dominant grasses but junegrass *(Koeleria cristata)*, bottlebrush squirreltail *(Sitanion hystrix)*, muttongrass *(Poa fendleriana)*, and black dropseed *(Sporobolus interruptus)* are important components (Clary 1971). Other sites contain desert needlegrass *(Stipa speciosa)* (Thatcher and Hart 1974), red threeawn *(Aristida longiseta)*, and ring muhly *(Muhlenbergia torreyi)* (Jameson and Reid 1965), and western wheatgrass *(Agropyron smithii)* (Clary and Morrison 1973). Forbs and half-shrubs make up about 50 percent of the herbaceous composition in north-central Arizona. Broom snakeweed *(Xanthocephalum sarothrae)*, sulfur eriogonum *(Eriogonum cognatum)*, plumweed birdbeak *(Cordylanthus wrightii)*, and goldeneye *(Viguiera* sp.) are dominants (Clary 1971).

Fig. 9.4. Western juniper is very abundant in eastern Oregon. (Photo courtesy of Oregon State University.)

Northward, warm-season grasses drop out of the understory. In southwestern Colorado on Mesa Verde National Park, the meadow stage that develops four years after burning is dominated by Indian ricegrass *(Oryzopsis hymenoides)*, bottlebrush squirreltail, and muttongrass (Erdman 1970). Forbs and shrubs are also present at various stages of succession after burning. In Utah bluebunch wheatgrass *(Agropyron spicatum)* and western wheatgrass are the most abundant grasses with lesser amounts of Sandberg bluegrass *(Poa sandbergii)*, bottlebrush squirreltail, and Indian ricegrass (Barney and Frischknecht 1974). Forbs are a minor component.

In Nevada several sources (Blackburn et al. 1969; Jensen 1972; Klebenow et al. 1976) indicate that common grasses include bottlebrush squirreltail, Sandberg bluegrass, needle-and-thread *(Stipa comata)*, Great Basin wildrye *(Elymus cinereus)*, and cheatgrass *(Bromus tectorum)*. Other grasses include western wheatgrass, Indian ricegrass, and Thurber needlegrass *(S. thurberiana)*. Forbs vary widely in species and abundance but can include *Lupinus* spp., *Phlox* spp., *Aster* spp., *Senecio* spp., *Eriogonum* spp., *Ranunculus* spp., *Cryptantha* spp., *Mentzelia* sp., *Eriastrum* spp., *Castilleja* spp., *Machaeranthera* sp., *Argemone* sp., *Sphaeralcea* spp., *Nicotiana* sp., *Lygodesmia* sp., and *Chenopodium* spp.

Further north in pure stands of western juniper in eastern Oregon, bluebunch wheatgrass and Idaho fescue *(Festuca idahoensis)* are the dominant grasses (Franklin and Dyrness 1969). Sandberg bluegrass and Thurber needlegrass are common. Other grasses include bottlebrush squirreltail, needle-and-thread, cheatgrass, annual fescue *(Vulpia octoflora)*, and junegrass. Forbs are not very abundant, but the most common perennial forbs are dandelion *(Agoseris* sp.), western yarrow *(Achillea millefolium)*, woolly-yarrow *(Eriophyllum lanatum)*, locoweed *(Astragalus* sp.), fleabane *(Erigeron linearis)*, and lupine *(Lupinus* spp.).

Herbage yields of pinyon-juniper stands can vary considerably depending on surface texture of soil and stage of succession (Thatcher and Hart 1974). Soils with a vesicular, massive, or platy surface layer may have very little grass regardless of the stage of plant succession (Thatcher and Hart 1974). However a reasonable average herbage yield for plant communities with moderate amounts of pinyon-juniper seems to be 675 kg/ha (600 lb/acre) (Jameson and Reid 1965; Clary 1971; Pieper et al. 1971; Jensen 1972). Yields on good soils with good precipitation can be as high as 1575 kg/ha (1400 lb/acre) (Pieper et al. 1971; Springfield 1976).

Shrubs that dominate the understory of pinyon-juniper include big sagebrush *(Artemisia tridentata)*, black sagebrush *(A. nova)*, antelope bitterbrush *(Purshia tridentata)*, shrub live oak *(Quercus turbinella)*, Mexican cliffrose *(Cowania mexicana* var. *mexicana)*, Gambel oak, serviceberry *(Amelanchier* sp.), and true mountain mahogany *(Cercocarpus montanus)*. Other shrubs associated with pinyon-juniper include fringed sagebrush *(A. frigida)*, Wright silktassel *(Garrya wrightii)*, currant *(Ribes cereum)*, desert peach *(Prunus andersonii)*, mountain lover *(Pachistima myrsinites)*, skunkbush sumac *(Rhus trilobata)*, ephedra *(Ephedra* spp.), curlleaf mountain mahogany *(C. ledifolius)*, chokecherry *(Prunus virginiana)*, mockorange *(Philadelphus lewisii)*, pointleaf manzanita *(Arctostaphylos pungens)*, winterfat *(Ceratoides lanata)*, snowberry *(Symphoricarpos vaccinoides)*, algerita

(Berberis trifoliata), Wright eriogonum *(Eriogonum wrightii)*, rabbitbrush *(Chrysothamnus* spp.), Apache plume *(Fallugia paradoxa)*, desert blackbrush *(Coleogyne ramosissima)*, fourwing saltbush *(Atriplex canescens)*, broom snakeweed, dalea *(Dalea* sp.), horsebrush *(Tetradymia canescens)*, and yucca *(Yucca* spp.). This wide variety of shrubs reflects the many different plant communities with which pinyon-juniper associates and has been taken from various references (Arnold et al. 1964; Dwyer and Pieper 1967; Blackburn et al. 1969; O'Rourke and Ogden 1969; Blackburn and Tueller 1970; Erdman 1970; Aro 1971; Clary 1971; Barney and Frischknecht 1974; West et al. 1975; Klebenow et al. 1976; Springfield 1976; Young and Evans 1976).

EFFECTS OF FIRE

Pinyon and Juniper

The effect of fire on pinyon and nonsprouting juniper trees depends largely upon the height and spatial distribution of trees, herbaceous fuel, weather conditions, and season. In open pinyon-juniper stands (Fig. 9.3) with an understory of 675 to 1125 kg/ha (600 to 1000 lb/acre) of fine fuel, Jameson (1962) and Dwyer and Pieper (1967) found that pinyon and juniper were easily killed by spring fires if trees were less than 1.2 m (4 ft) tall when air temperature was 21° to 23°C (70° to 74°F), relative humidity 20 to 40 percent, and wind speed 16 to 32 km/hr (10 to 20 mi/hr). Lower air temperatures in January [9° to 12°C (49° to 54°F)], a relative humidity of 44 percent, and a wind speed of 10 to 13 km/hr (6 to 8 mi/hr) caused a very spotty burn in which crown kill varied from 30 to 70 percent for trees 0.6 to 1.2 m (2 to 4 ft) tall. Mortality two years after the burn, however, was 70 percent (Jameson 1962). A wildfire in June when air temperature was 36°C (97°F), wind was 16 to 24 km/hr (10 to 15 mi/hr), and relative humidity was 17 to 25 percent resulted in a 100-percent kill of all trees less than 1.2 m (4 ft) tall, but was no more effective on taller trees than when air temperatures were 21° to 23°C (70° to 74°F) (Jameson 1962).

   Trees taller than 1.2 m (4 ft) in open pinyon-juniper stands are difficult to kill unless there are heavy accumulations of fine fuel beneath the trees. On the wildfire studied by Dwyer and Pieper (1967), only 24 percent of the pinyon and 13.5 percent of the juniper which exceeded 1.2 m (4 ft) died. Jameson (1962) found that most juniper over 1.2 m (4 ft) tall had only a 30- to 40-percent crown kill unless tumbleweeds had accumulated at the base of the trees. As a result of this added fuel, 60 to 90 percent of the crowns were killed, particularly those of trees 2.4 to 3.0 m (8 to 10 ft) tall. East of the pinyon-juniper zone in Texas, juniper trees up to 3.7 m (12 ft) tall were easily killed when fine fuel was 2245 kg/ha (2000 lb/acre) or higher (Wink and Wright 1973).

   Trees in closed stands of pinyon-juniper (Fig. 9.5) with no grass or sagebrush in the understory are difficult to kill because fires do not carry easily (Arnold et al. 1964; Blackburn and Bruner 1975). In 35- to 45-cm (14- to 18-in.) rainfall areas,

Fig. 9.5. Closed stand of pinyon and juniper with no understory, in Nevada. (Photo courtesy of University of Nevada.)

Fig. 9.6. Mixture of big sagebrush and pinyon-juniper in Nevada. (Photo courtesy of University of Nevada.)

dense stands [1223 to 2440 trees/ha (495 to 988/acre)] of mixed pinyon and juniper can be burned on hot days (Hester 1952), but pure stands of juniper are almost impossible to burn (Blackburn and Bruner 1975). In the Great Basin it is commonly believed that winds over 55 km/hr (35 mi/hr) would be required to burn pure stands of juniper in the 25- to 38-cm (10- to 15-in.) precipitation zone. Thus many attempts to burn such stands have failed (Arnold et al. 1964; Aro 1971). As the proportion of pinyon to juniper increases and the density increases, the stands are easier to burn (Truesdell 1969; Blackburn and Bruner 1975).

Mixtures of sagebrush and pinyon-juniper (Fig. 9.6) are common throughout the Great Basin and it is feasible to burn and kill large pinyon and juniper trees in these communities (Bruner and Klebenow 1978). Most of the work done by Bruner and Klebenow has been done in pinyon-juniper stands with 45- to 60-percent shrub and tree cover when air temperature was 16° to 24°C (60° to 75°F), relative humidity was below 25 percent, and maximum wind speed was 8 to 40 km/hr (5 to 25 mi/hr).

Alligator juniper and redberry juniper are the only sprouting species. Smith et al. (1975) found that if the tops of redberry juniper trees were removed before they reached 12 years of age, 99 percent mortality could be expected. Older trees were not studied. This data implies that if subjected to fire every 10 years or so, sprouting species of juniper may have a difficult time invading grasslands. Schroeder (1956) found that about 40 percent of alligator juniper trees less than 4.6 m (15 ft) tall could be killed by burning individual plants.

Forbs and Grasses

In Utah the most abundant annual forbs during the first stage of succession are pale alyssum (*Alyssum alyssoides*), flixweed tansymustard (*Descurainia sophia*), annual sunflower (*Helianthus annuus*), coyote tobacco (*Nicotiana attenuata*), and Russian thistle (*Salsola Kali*) (Barney and Frischknecht 1974). Generally, however, none of these forbs constitute a large amount of cover on pinyon-juniper burns (Arnold et al. 1964; Barney and Frischknecht 1974). Cheatgrass is usually the most abundant annual and has a cover value as high as 12.6 percent on three-year-old burns. Thereafter it gradually declines to 0.9 percent over a period of 20 years (Barney and Frischknecht 1974). On some sites, however, cheatgrass may never show up (Klebenow et al. 1976) which might be used as a guide to determine whether some areas can be reclaimed to grasses. In Nevada tapertip hawksbeard (*Crepis acuminata*) and lupine increase abundantly after fire (Klebenow et al. 1976). *Balsamorhiza* species and *Castilleja* species also come back reasonably well (Klebenow et al. 1976).

Composition of perennial grasses varies depending on the location, as discussed earlier. In the northern latitudes west of the Rocky Mountains, cool-season grasses dominate with a gradual transition to the dominance of blue grama eastward in shortgrass plains and southward in central Arizona and New Mexico. Vallentine (1971) has cited a number of authors as to the tolerance of various grasses to fire. Species that are only slightly damaged by fall fires include bluebunch wheatgrass, Indian ricegrass, galleta grass *(Hilaria jamesii)*, bottlebrush squirreltail, Great Basin wildrye, and blue grama. Those moderately affected by fire include junegrass, needle-and-thread, Thurber needlegrass, and threeawns. Species severely affected by fire include ring muhly, sideoats grama, and Idaho fescue. Sandberg bluegrass, cheatgrass, western wheatgrass, and crested wheatgrass are unaffected by fall fires.

Shrubs

Vigorous sprouters after fire include serviceberry, Wright silktassel, shrub live oak, skunkbush sumac, true mountain mahogany, chokecherry, mockorange, winterfat, snowberry, algerita, rabbitbrush, fourwing saltbush, and horsebrush (Wright et al. 1979). Weak sprouters include antelope bitterbrush, desert bitterbrush, curlleaf mountain mahogany, broom snakeweed, mountain lover, yucca, fringed sagebrush, and Wright eriogonum (Klebenow et al. 1976; Wright et al. 1979). Antelope bitterbrush and mountain lover are extremely slow in recovering after fire (Mckell 1950; Nord 1965). Mexican cliffrose is eliminated on burns in Nevada (Klebenow et al. 1976). Nonsprouting species include big sagebrush, black sagebrush, and desert blackbrush. Nevertheless, these nonsprouting species, except for desert blackbrush, have the ability to reestablish themselves quickly from seed. Desert blackbrush reestablishes itself very slowly (Jenson et al. 1960, Bowns and West 1976).

Succession After Fire

The general successional pattern after fire in pinyon-juniper of the southern Rocky Mountains has been worked out by Arnold et al. (1964) and most recently by Erdman (1970) and Barney and Frischknecht (1974). The order of vegetational changes in juniper woodland after fire as reported by Barney and Frischknecht is as follows: juniper woodland → fire → skeleton forest (dead trees) and bare soil → annual stage → perennial grass-forb stage → perennial grass-forb-shrub stage → perennial grass-forb-shrub-young juniper stage → shrub-juniper stage → juniper woodland.

Mature stands of juniper (100 or more years) consist primarily of 35 percent bare ground, 19 to 60 percent canopy cover, which is dominated by the tree overstory that inhibits grass production (Johnsen 1962), and a few scattered shrubs and perennial and annual grasses (Arnold et al. 1964; Clary 1971; Barney and Frischknecht 1974). Erosion is frequently a problem from bare soils between trees in mature stands (Plummer 1958). After a fire juniper seedlings and annuals begin to invade and reach maximum abundance in the first three to four years (Arnold et al. 1964; Barney and Frischknecht 1974). Where partial shade is present, pinyon seedlings can also be abundant (Erdman 1970).

The perennial-grass-forb stage usually follows in the fourth to sixth year. Little rabbitbrush *(Chrysothamnus viscidiflorus)* resprouts in the first year or two and shrubs such as sagebrush and broom snakeweed begin appearing after the sixth year in plant communities in and around the Great Basin. In southeastern Colorado Gambel oak, serviceberry, true mountain mahogany, and antelope bitterbrush are the dominant shrubs (Erdman 1970). After 40 years the shrubs begin to die out and the cover and density of juniper increases dramatically (Barney and Frischknecht 1974). Barney and Frischknecht found that Utah juniper begins to bear fruit for the first time at 33 years of age, which accounts for the ability of many juniper stands at 40 years of age or older to establish new trees and increase the number of trees dramatically (Erdman 1970).

Plant succession according to Erdman continues in three more stages. The open shrub stage becomes a thicket in about 100 years. As the sere progresses toward climax, the trees begin to overtop the shrubs and gradually suppress the shrubs as the forest matures. After several centuries the understory is composed mainly of a sparse shrub component, some grass, and several forbs. In the absence of disturbances, a climax pinyon-juniper forest occurs in about 300 years.

MANAGEMENT IMPLICATIONS

In open stands of pinyon-juniper in the Southwest, fire can be used effectively to kill pinyon and juniper trees less than 1.2 m (4 ft) tall. Taller trees are very difficult to kill, even with hot fires, unless tumbleweeds have accumulated at the tree bases. Thus open stands of tall pinyon and juniper trees can be eliminated only by chaining or dozing followed by burning, to render the microclimate unfavorable for tree seedlings.

Several management agencies have tried various techniques (Arnold et al. 1964; Aro 1971; Clary 1971; Blackburn and Bruner 1975) to reclaim closed pinyon-juniper stands (with no understory of grasses or shrubs). Prescribed burning, or some combination of burning with other treatments (followed by artificial seeding when necessary) is the most effective procedure to reclaim closed stands of pinyon-juniper (Aro 1971; Springfield 1976). Without any prior treatment, burning must be done on hot days [35° to 38°C (95° to 100°F)] with low relative humidity and 13- to 32-km/hr (8- to 20-mi/hr) winds, conditions considered too hazardous by most land managers (Arnold et al. 1964). Thus mechanical treatment followed by burning is probably the most acceptable technique to reclaim dense stands of pinyon-juniper, even though it is expensive. Burning should be delayed two to three years after chaining to assure that most of the pinyon and juniper seeds have germinated (Meagher 1943).

Grasses will increase dramatically following burning and seeding treatments in closed stands of pinyon-juniper. Herbage yields on the Hualapi Indian Reservation in northern Arizona, seeded with crested wheatgrass, western wheatgrass, weeping lovegrass, and yellow sweetclover (*Melilotus officinalis*), produced 1865 kg/ha (1660 lb/acre) compared to 65 kg/ha (60 lb/acre) for the unburned control (Aro 1971). On another large-scale burning and seeding program in pinyon-juniper woodland, Aro (1971) reported that forage production increased 560 kg/ha (500 lb/acre). Pinyon-juniper communities in northern Arizona that were chained and seeded but not burned produced 1100 kg/ha (980 lb/acre) of grasses, forbs, and shrubs five to eleven years after treatment, compared to 250 kg/ha (220 lb/acre) on control plots

(Clary 1971). Where native grasses were present in the understory, reseeding was not necessary (Aro 1971).

Mixtures of sagebrush and pinyon-juniper can be burned without prior treatment. Generally thick stands with 45- to 60-percent cover are selected for burning and burned into areas with less shrub cover. Some areas are left to reseed naturally, but aerial seeding is usually considered desirable.

Pinyon-juniper stands converted to grassland should be reburned about every 20 to 40 years. A definite time is difficult to set because reinvasion of pinyon and juniper is dependent on the kind of initial treatment, time span between treatments, intensity of the burn, and grazing intensity after the burn. A better guide would be to reburn when the tallest pinyon or juniper tree reaches a height of 1.2 m (4 ft).

REFERENCES

Arend, J. L. 1950. Influence of fire and soil on distribution of eastern red cedar in the Ozarks. *J. For.* **48**:120–130.

Arnold, J. F., D. A. Jameson, and E. H. Reid. 1964. The pinyon-juniper type of Arizona: Effect of grazing, fire, and tree control. USDA Prod. Res. Rep. No. 84. Washington, D. C.

Aro, R. S. 1971. Evaluation of pinyon-juniper conversion to grassland. *J. Range Manage.* **24**:188–197.

Barney, M. A., and N. C. Frischknecht. 1974. Vegetation changes following fire in the pinyon-juniper type of west-central Utah. *J. Range Manage.* **27**:91–96.

Billings, W. D. 1952. The environmental complexion related to plant growth and distribution. *Quart. Rev. Biol.* **27**:257–265.

Blackburn, W. H., and A. D. Bruner. 1975. Use of fire in manipulation of the pinyon-juniper ecosystem. In *The Pinyon-Juniper Ecosystem: A Symposium.* Utah State Univ., Logan, pp. 91–96.

Blackburn, W. H., R. E. Eckert, Jr., and P. T. Tueller. 1969. Vegetation and soils of the Crane Springs Watershed. Nevada Agric. Exp. Stn. Paper R-55. Reno.

Blackburn, W. H., and P. T. Tueller. 1970. Pinyon and juniper invasion in black sagebrush communities in east-central Nevada. *Ecology* **51**:841–848.

Bowns, J. E., and N. E. West. 1976. Blackbrush (*Coleognyne ramosissima* Torr.) on southwestern Utah rangelands. Utah Agric. Exp. Stn. Res. Rep. 27. Logan.

Bruner, A. D., and D. A. Klebenow. 1978. A technique to burn pinyon-juniper woodlands in Nevada. USDA For. Serv. Gen. Tech. Rep. INT-219. Intermt. For. and Range Exp. Stn., Ogden, Utah.

Burkhardt, J. W., and E. W. Tisdale. 1969. Nature and successional status of western juniper vegetation in Idaho. *J. Range Manage.* **22**:264–270.

Burkhardt, J. W., and E. W. Tisdale. 1976. Causes of juniper invasion in southwestern Idaho. *Ecology* **57**:472–484.

Clapp, E. H. 1936. The western range: A great but neglected natural resource. *Senate Doc.* **199**:1–620.

Clary, W. P. 1971. Effects of Utah juniper removal on herbage yields from Springerville soils. *J. Range Manage.* **24**:373–378.

Clary, W. P., and D. C. Morrison. 1973. Large alligator junipers benefit early-spring forage. *J. Range Manage.* **26**:70–71.

Daubenmire, R. 1947. *Plants and Enrironment.* Wiley, New York.

Dwyer, D. D., and R. D. Pieper. 1967. Fire effects on blue grama-pinyon-juniper rangeland in New Mexico. *J. Range Manage.* **20**:359–362.

Emerson, F. W. 1932. The tension zone between the grama grass and pinyon-juniper associations in northeastern New Mexico. *Ecology* **13**:347–358.

Erdman, J. A. 1970. Pinyon-juniper succession after natural fires on residual soils of Mesa Verde, Colorado. BYU Sci. Bull. Biol. Series 11(2).

Franklin, J. F., and C. T. Dyrness. 1969. Vegetation of Oregon and Washington. USDA For. Serv. Res. Paper PNW-80. Pac. Northwest For. and Range Exp. Stn., Portland, Ore.

Hester, D. A. 1952. The pinyon-juniper fuel type can really burn. *Fire Contr. Notes* **13**(1):26–29.

Howell, J., Jr. 1941. Pinyon and juniper woodland of the Southwest. *J. For.* **39**:542–545.

Jameson, D. A. 1962. Effects of burning on a galleta-black grama range invaded by juniper. *Ecology* **43**:760–763.

Jameson, D. A., and E. H. Reid. 1965. The pinyon-juniper type of Arizona. *J. Range Manage.* **18**:152–154.

Jensen, N. E. 1972. Pinyon-juniper woodland management for multiple use benefits. *J. Range Manage.* **25**:231–234.

Jensen, D. E., M. W. Butan, and D. E. Dimock. 1960. Blackbrush burns. Report on field examinations. Las Vegas Grazing Dist., Nevada.

Johnsen, T. N., Jr. 1962. One-seed juniper invasion of northern Arizona grasslands. *Ecol. Monogr.* **32**:187–207.

Klebenow, D. A., R. Beall, A. Bruner, R. Mason, B. Roundy, W. Stager, and K. Ward. 1976. Controlled fire as a management tool in the pinyon-juniper woodland, Nevada. Annual Prog. Rep., Univ. of Nevada, Reno.

Küchler, A. W. 1964. Potential natural vegetation of the conterminous United States. Manual to accompany the map. Am. Geogr. Soc. Spec. Pub. 36. (With map, rev. ed., 1965, 1966.)

Lanner, R. M. 1975. Pinyon pines and junipers of the southwestern woodlands. In *The Pinyon-Juniper Ecosystem: A Symposium*. Utah State Univ., Logan.

Leopold, A. 1924. Grass, brush, timber and fire in southern Arizona. *J. For.* **22**(6):1–10.

McKell, C. M. 1950. A study of plant succession in the oak brush *(Quercus gambelii)* zone after fire. M. S. Thesis. Univ. Utah, Salt Lake City.

Meagher, G. S. 1943. Reaction of piñon and juniper seedlings to artificial shade and supplemental watering. *J. For.* **41**:480–482.

Nabi, A. A. 1978. Variation in successional status of pinyon-juniper woodlands of the Great Basin. M.S. Thesis. Utah State Univ., Logan.

Nord, E. C. 1965. Autecology of bitterbrush in California. *Ecol. Monogr.* **35**:307–334.

O'Rourke, J. T., ad P. R. Ogden. 1969. Vegetative response following pinyon-juniper control in Arizona. *J. Range Manage.* **22**:416–418.

Pearson, G. A. 1920. Factors controlling the distribution of forest types, Part II. *Ecology* **1**:289–309.

Pearson, G. A. 1931. Forest types in the southwest as determined by climate and soil. USDA Tech. Bull. 247. Washington, D.C.

Phillips, F. J. 1909. A study of piñon pine. *Bot. Gaz.* **48**:216–223.

Pieper, R. D., J. R. Montoya, and V. L. Groce. 1971. Site characteristics on pinyon-juniper and blue grama ranges in south-central New Mexico. New Mexico Agric. Exp. Stn. Bull. 573. Las Cruces.

Plummer, A. P. 1958. Restoration of pinyon-juniper ranges in Utah. In Proc. Soc. Amer. For. Soc. Amer. For., Mills Bldg., Washington, D.C., pp. 207–211.

Reveal, J. L. 1944. Single-leaf pinyon and Utah juniper woodlands of western Nevada. *J. For.* **42**:276–278.

Schroeder, W. L. 1956. Juniper and brush control. Bur. Indian Affairs, USDI. (mimeo)

Smith, M. A., H. A. Wright, and J. L. Schuster. 1975. Reproductive characteristics of redberry juniper. *J. Range Manage.* **27**:126–128.

Springfield, H. W. 1976. Characteristics and management of southwestern pinyon-juniper ranges: The status of our knowledge. USDA For. Serv. Res. Paper RM-160. Rocky Mtn. For. and Range Exp. Stn., Fort Collins.

St Andre, G., H. A. Mooney, and R. D. Wright. 1965. The pinyon woodland zone in the White Mountains of California. *Amer. Midl. Natur.* **73:**225–239.

Thatcher, A. P., and V. L. Hart. 1974. Spy Mesa yields better understanding of pinyon-juniper range ecosystem. *J. Range Manage.* **27:**354–357.

Truesdell, P. S. 1969. Postulates of the prescribed burning program of the Bureau of Indian Affairs. *Proc. Tall Timbers Fire Ecol. Conf.* **9:**235–240.

Vallentine, J. F. 1971. *Range Development and Improvements.* Brigham Young Univ. Press, Provo, Utah.

Wells, P. V. 1970. Postglacial vegetational history of the Great Plains. *Science* **167:**1574–1582.

West, N. E., K. H. Rea, and R. J. Tausch. 1975. Basic synecological relationships in pinyon-juniper woodlands. In *The Pinyon-Juniper Ecosystem: A Symposium.* Utah State Univ., Logan, pp.41–52.

Wink, R. L., and H. A. Wright. 1973. Effects of fire on an Ashe juniper community. *J. Range Manage.* **26:**326–329.

Woodbury, A. M. 1947. Distribution of pigmy conifers in Utah and northeastern Arizona. *Ecology* **28:**113–126.

Wright, H. A., L. F. Neuenschwander, and C. M. Britton. 1979. The role and use of fire in sagebrush-grass and pinyon-juniper plant communities. USDA For. Serv. Gen. Tech. Rep. INT-58. Intermt. For. and Range. Exp. Stn., Ogden, Utah.

Young, J. A., and R. A. Evans. 1976. Control of pinyon saplings with picloram or karbutilate. *J. Range Manage.* **29:**144–147.

CHAPTER

10

PONDEROSA PINE

H istorical evidence indicates that fires have always been an ecological force in ponderosa pine *(Pinus ponderosa)* communities, regardless of whether they were seral or climax (Dutton 1887; Daubenmire 1943; Arnold 1950; Weaver 1951a, 1951b, 1959; Cooper 1960; Biswell 1963, 1967; Gartner and Thompson 1972; van Wagtendonk 1974; Hall 1976). These fires thinned the stands, eliminated young pines and/or climax mixed-conifer species including thickets, and kept the ponderosa pine forests open and parklike (Fig. 10.1) with an understory of herbs and shrubs (Arnold 1950; Beale 1958; Weaver 1947, 1951a, 1967b; Cooper 1960; Gartner and Thompson 1972; Biswell 1972; Progulske 1974; Hall 1976). Only in the Sierra Nevada mixed conifer type were there relatively few grasses and shrubs in the understory, but the forest was more open with fewer incense cedar *(Calocedrus decurrens)* and white fir *(Abies concolor)* thickets (van Wagtendonk 1974). The forest floor was carpeted with needles, bear clover *(Chamaebatia foliolosa)*, and some forbs and grasses (van Wagtendonk 1974).

Successful prescribed burns not only have shown the potential to thin stands of ponderosa pine, allow crop trees to increase in diameter, eliminate thickets (Weaver 1967a), and increase forage yields (Cooper 1960; Biswell 1972; Hall 1976; Severson and Boldt 1977), but also to reduce wildfire hazards for five to seven years (Truesdell 1969). The ability of ponderosa pine to survive severe wildfires following natural fires at 5- to 10-year frequencies has been confirmed in Arizona by a 32,390-hectare (80,000-acre) wildfire in 1903-1904 (Kallander 1969) and by a 23,075-hectare (57,000-acre) wildfire in 1971 (Wagle and Eakle 1977). On these burns few trees were killed that had a recent fire history, but on untreated areas the fire destroyed ev-

Fig. 10.1.   An open stand of ponderosa pine in Arizona with a grassy understory, giving a parklike appearance. Note the absence of young conifer regeneration. Historical documents in this region indicate that fires every seven to nine years maintained stands of ponderosa pine in this condition.

erything in its path, except for those burns which occurred during late evening and early morning when the fires were cool (Wagle and Eakle 1977).

Truesdell (1969) reported that no fire larger than 4 ha (10 acres) in size had ever occurred in ponderosa pine on the Fort Apache Reservation within seven years following a controlled burn. Knorr (1963), working in the same area, made a special study that showed wildfires were only one fourteenth as large as fires on untreated areas. For three fire seasons following controlled burning of 23,315 ha (64,000 acres) on the Fort Apache Reservation, Kallander et al. (1955) showed an 82-percent reduction in the number of wildfires, a 94-percent reduction in area burned, and a 65-percent reduction in the average size of a fire. Similarly, on the Colville Indian Reservation in Washington, Weaver (1957) showed a 90-percent reduction in the number of acres burned, a 94-percent reduction in damage and a 79-percent reduction in the cost of control per wildfire on previously burned communities. To maintain such low fire hazards, Biswell et al. (1973) recommended burning ponderosa pine communities every five to seven years. Where fuel buildups are high, several successive burns of low intensity must occur to reduce the surface fuels to acceptable levels (Weaver 1957; Biswell 1963; van Wagtendonk 1974), if it can be done at all (Gordon 1967).

On the other hand, prescribed fires have killed quality ponderosa pine trees on the Fort Apache Indian Reservation in eastern Arizona, especially were crown fires occurred (Lindenmuth 1960). Light surface fires, hot surface fires, and crown fires killed 0.8, 9.2, and 47.3 percent respectively, of the potential crop trees in the 10- to 18-cm (4- to 7-in.) dbh class. The same fires killed 21.3, 57.5, and 73.7 percent of the trees in the 0- to 1.4-m (0- to 4.5-ft) height class. Thus it appears that, depending on one's objectives, prescribed burns have the potential to clean out understories of potential crop trees and to thin stands of ponderosa pine if fire prescriptions are written on a unit-objective basis. Some losses of wood products will occur and the thinning will not be uniform (Wooldridge and Weaver 1965), but if weighed against the benefits of fuel reduction and the potential for a variety of uses that are compatible with the carrying capacity of the land, the loss of wood products may be justified (Meyers 1974).

Most managers have preferred to protect their forests from intentional fires because they have deep feelings of wariness and uncertainty as to whether fire can be used safely and effectively. There is too much evidence of impressive conflagrations in ''pristine'' forests—especially ponderosa pine—to believe that hot fires did not also play a major role along with the gentle burns which undoubtedly occurred (C. E. Boldt, pers. comm.). Most managers are uneasy about potential ''benefits'' of using prescribed fire to reduce fuel buildups and to precommercially thin forests. The net result has been a high incidence of crown fires and overstocked stands on poor sites that grow slowly and are too expensive to thin by hand (Weaver 1943; Biswell et al. 1973; Hall 1976).

Managers realize the strong tendency of overstocked ponderosa pine stands to stagnate (Biswell et al. 1973; Meyers 1974; Hall 1976) (Fig. 10.2). The need for precommercial thinning and release of stagnant trees is common knowledge. Most

Fig. 10.2. A stagnated stand of ponderosa pine in Arizona 40 to 50 years old with 2000 to 3000 stems per acre. (Photo courtesy of USDA Forest Service.)

managers, however, have not been convinced that they can safely substitute fire for conventional thinning. There have been few studies to document the burning prescriptions for results such as those claimed by Weaver (1947) and Biswell et al. (1973). Moreover there have been few economic studies such as the one by Clary et al. (1975) for the southern Rockies to show how to manage ponderosa pine communities considering all the resources.

FIRE HISTORY

Frequency of natural fires in ponderosa pine varied considerably depending on the regional area and site. In Arizona and New Mexico the frequency for climax and seral communities of ponderosa pine was between 4.8 and 11.9 years (Weaver 1951a). Show and Kotok (1924), Wagener (1961), and Kilgore and Taylor (1979) found that it averaged eight to ten years for seral communities of ponderosa pine in California, as did Hall (1976) for seral ponderosa pine in the Blue Mountains of eastern Oregon. Prior to 1875, climax ponderosa pine communities in the Sierra Nevada mixed conifer had an average fire frequency of six to nine years (range 2 to 23 years) (Kilgore and Taylor 1979). Weaver (1959; 1967a), working with a wide range of climax and seral ponderosa pine community types in eastern Washington, showed that fire frequency varied from 6 to 47 years.

In the Bitterroot National Forest of eastern Idaho and western Montana fire frequency averaged from 6 to 11 years (range 2 to 20 years) for climax ponderosa pine and 7 to 19 years (range 2 to 48 years) for ponderosa pine that was seral to Douglas fir *(Pseudotsuga menziesii)* (Arno 1976). Eastward in South Dakota, historical documents from 1874 to 1880 show that lightning fires were prevalent in the Black Hills of South Dakota (Gartner and Thompson 1972). Photographs taken in 1874 and a century later *(Progulske 1974)* confirm the significant role of fire and Progulske estimated the fire frequency to be 15 to 20 years. Hendrickson (1972) estimated the fire frequency in Colorado and Wyoming to be 12 to 25 years. In this area there are many poor, dry sites where litter buildup is slow and the grass understory is thin.

DISTRIBUTION, CLIMATE, ELEVATION, AND SOILS

Ponderosa pine is widely distributed, both as a climax and seral species. It spans a north-south distance (southern British Columbia to Durango, Mexico) of 3710 km (2300 miles) and an east-west distance (California to north-central Nebraska) of 2030 km (1260 miles) (Fig. 10.3). Climax communities include the interior ponderosa and Pacific ponderosa pine types (Fowells 1965). They are usually characterized by shrubby or grassy understories (Daubenmire 1969). The interior ponderosa pine type forms a narrow belt along the east side of the Sierra Nevada-Cascade Ranges, southern British Columbia, eastern Oregon and Washington, central and northern Idaho, western Montana, South Dakota, Utah, western Colorado, Arizona, New Mexico,

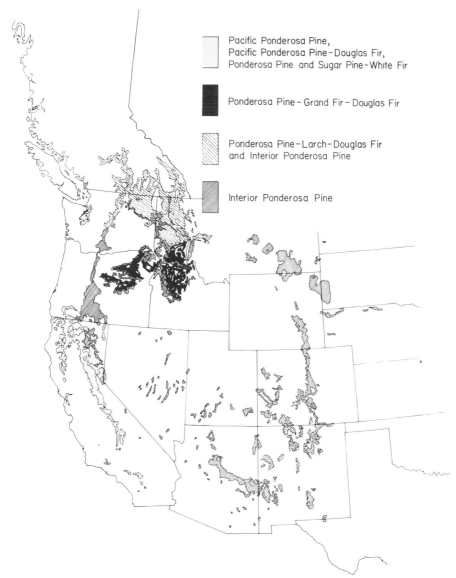

Pacific Ponderosa Pine,
Pacific Ponderosa Pine-Douglas Fir,
Ponderosa Pine and Sugar Pine-White Fir

Ponderosa Pine- Grand Fir- Douglas Fir

Ponderosa Pine-Larch-Douglas Fir
and Interior Ponderosa Pine

Interior Ponderosa Pine

Fig. 10.3. Distribution of ponderosa pine in the United States and Canada. (Modified from Küchler 1964. Reproduced with permission of the American Geographical Society.)

and central Mexico. The Pacific ponderosa pine type occurs on the lowermost west slopes of the Sierra Nevada and southern Cascade ranges and cross ranges of northern California and southern Oregon (Fowells 1965).

In other forest-habitat types (climax associations of species that show the strongest evidence of self-perpetuation on a site), namely, ponderosa pine-larch-Douglas fir, ponderosa pine-grand fir-Douglas fir, ponderosa pine-sugar pine-fir, and Pacific ponderosa pine-Douglas fir, ponderosa pine is a component of the climax cover, but never predominates unless the community is disturbed. The first habitat type is typical of western Montana, northern Idaho, northeastern Washington, and southern British Columbia. The ponderosa pine-grand fir-Douglas fir type, mixed-conifer, occurs primarily in eastern Oregon and central Idaho. The ponderosa pine-sugar pine-fir type is often called the Sierra Nevada mixed-conifer in California, where it is most extensive on the west slope of the Sierra Nevada Mountains (Fowells 1965). The pacific ponderosa pine-Douglas fir type is typically found on the east slope of the north coast ranges of California and Oregon, the Siskiyou Mountains, and southern Cascades.

Annual precipitation of climax ponderosa pine plant communities averages 51 cm (20 in.), but generally varies from 25 to 28 cm (10 or 11 in.) to 71 cm (28 in.) throughout western North America (Weaver 1951b; Brayshaw 1955, 1965; Fowells 1965; Foiles and Curtis 1973; Boldt and Van Deusen 1974; Currie 1975; Clary 1975; Skovlin et al. 1976). This variation is related to latitude, slope, aspect, soil type, and precipitation-to-evaporation ratio. In northern California, where ponderosa pine may be a seral or climax species, precipitation generally varies from 75 to 125 cm (30 to 50 in.) per year, but can be as high as 240 cm (95 in.) per year (P. M. McDonald, pers. comm.). Seral ponderosa pine communities in eastern Oregon and Washington, Idaho, and western Montana grow where precipitation varies from 38 to 76 cm (15 to 30 in.) per year (Hall 1976).

Elevation varies according to the geographical region. In British Columbia elevation for ponderosa pine ranges from 305 to 1310 m (1000 to 4300 ft) (Brayshaw 1965). Along the eastern flank of the Cascades and the Blue Mountains ponderosa pine grows at elevations from 490 to 1525 m (1600 to 5000 ft) (Weaver 1951b; Harris 1954; Biswell et al. 1955; Franklin and Dyrness 1973; Skovlin et al. 1976). In California it is most commonly found at elevations from 150 to 1065 m (500 to 3500 ft) in the north and from 1615 to 2225 m (5300 to 7300 ft) in the south (Fowells 1965). Ponderosa pine in the central and southern Rocky Mountains varies in elevation from 1525 to 2895 m (5000 to 9500 ft) (Currie 1975; Clary 1975), but is generally most common at elevations from 1980 to 2590 m (6500 to 8500 ft) (Cooper 1960; Schubert 1974; Clary 1975). In the Black Hills elevation for ponderosa pine ranges from 975 to 2195 m (3200 to 7200 ft) (Gartner and Thompson 1972; Boldt and Van Deusen 1974). Although exceptions can be found, the best developed stands are at elevations of 1220 to 2440 m (4000 to 8000 ft) on benches, plateaus, and west and south aspects (Fowells 1965).

Soils may be derived from a variety of igneous and sedimentary rocks (Schubert 1974). Generally, however, soils of the ponderosa pine type have developed primarily from quartzite, argillite, schist, basalt, andesite, granite, or shale and, to a lesser extent, from limestone and sandstone (Foiles and Curtis 1973; Schubert 1974; Currie 1975). Soil textures generally vary from sands to clay loam and occasionally to clays (Arnold 1950; Ogilvie 1955). Trees grow best on loam soils and on moderately sandy or gravelly soils (Schubert 1974). Clay soils often inhibit germination of pine seeds and do not favor good tree development (Schubert 1974) because they yield less of their total water for plant growth and are better suited to the more drought-tolerant plants such as grasses (Clary 1975; Skovlin et al. 1976).

NATURAL SUCCESSION WITH FIRE

Natural succession with fire in ponderosa pine forests can be envisioned by the following sequence of events: windfall, insect attacks, mortality, and then fire. Because of these events openings are created naturally (Cooper 1961; Weaver 1943, 1959, 1964). Ponderosa pine seed is blown on to the bare mineral soil of these openings from adjacent areas. Seedlings of ponderosa pine become established most easily on a bare mineral soil surface (Schultz and Biswell 1959) in open sunlight (Schubert 1974). Shade from larger trees, charred remains, stumps, and fallen logs help to provide the necessary environment for seedling survival (Biswell et al. 1973). Later, grasses compete with the pine seedlings and partially thin them (Pearson 1942; Pieper and Biswell 1961). In grassy areas it will take healthy pine trees about five years to overcome competition from grasses, during which time they are susceptible to fire. It usually takes eight years before the pine needles alone beneath young trees will be adequate to carry a fire (Cooper 1960).

While the openings are still predominantly grass, a subsequent fire will act as a natural thinning agent. The fire will be of a lower intensity than in the thick mats of pine needles, killing the smaller, thin-barked saplings and leaving the healthy, more vigorous trees (Biswell et al. 1973). Dense thickets of young ponderosa pine trees will not escape fire, causing the cycle to begin from mineral soil again (Morris and Mowat 1958; Hall 1976).

This type of regeneration in ponderosa pine forest resulted in a pristine forest that was represented by many age classes arranged in distinct groups, rather than in admixtures of all ages as found in climax forests (Biswell et al. 1973). Thus a pattern of disjunct groups or clusters of even-aged and even-sized trees, saplings, or seedlings existed, often less than 0.4 ha (1 acre) in size. Each class was separate; a healthy stand seldom contained older trees with younger reproduction growing under them (Biswell et al. 1973).

As passing surface fires helped to kill the younger, weaker trees and dense thickets in a young pine stand (Fig. 10.4), competition and stand stagnation were minimized. Nutrients in the litter were recycled (Vlamis et al. 1955; Moir 1966; Wollum and Schubert 1975). Lower branches and foilage of the remaining saplings and trees were pruned and thinned (Biswell et al. 1973), although pruning of the lower branches apparently does not stimulate leader growth (Barrett 1968) because height is controlled genetically (Wollum and Schubert 1975).

Repeated fires checked encroachment of the less fire-resistant species associates and pine seedlings in both seral and climax stands of ponderosa pine. This burning permitted the development of mature ponderosa pine trees with expanded canopies, sometimes nearly closed, thereby reducing herbaceous growth and increasing the

Fig. 10.4.   Entire thickets such as the one in the foreground (top photo) rarely escaped surface fires, although individual trees might be spared by the fire leaving scattered trees similar to the condition illustrated in the bottom photo. Dense stands of stagnated ponderosa pine seldom occurred in pristine forests.

thickness of pine litter on the forest floor. As the trees reached greater heights in isolated groups, they became more vulnerable to lightning. The largest, and often the oldest, trees were most commonly and even repeatedly struck by lightning (Biswell et al. 1973). These trees served as conductors which could ignite surface fires and burn dead and diseased trees, thickets, and heavy accumulations of fuels. Each time the fire produced an opening that included charred remains of former trees, grasses and shrubs invaded along with the pine seedlings. Then the cycle repeated itself with a mosaic of even-aged groups of pines being maintained. This is a simplified picture of what happens in the "typical" ponderosa pine forest type. A typical forest rarely exists, but all of these successional phases can be found in forests, as well as variations of these successional phases.

VEGETATION AND FIRE EFFECTS

Since ponderosa pine is an integral component of six forest-habitat types in the West, each with different understory species, different associations, and somewhat different fire relationships, each type will be discussed separately in the sections to follow. Yields of the understory, however, will be discussed in this section.

Forage, particularly grasses and shrubs, is an important component of seral and climax ponderosa pine communities. In the Black Hills, where ponderosa pine is a climax species, Pase (1958) found that understory compositions were approximately as follows: grasses and grass-like plants, 66 percent; forbs, 15 percent; and shrubs, 19 percent. Herbaceous yields generally range from 845 to 1910 kg/ha (750 to 1700 lb/acre) in openings of pine forests to less than 55 kg/ha (50 lb/acre) under 70 to 100-percent canopies (Pase 1958; Cooper 1960; Biswell 1972; Hall 1976; Severson and Boldt 1977). In the central and southern Rocky Mountains and the Black Hills of South Dakota open ponderosa pine pole stands produced yields of 280 to 450 kg/ha (250 to 400 lb/acre) if thinned to a 10.4- to 13.8-m^2 basal area per hectare (45- to 60-ft/acre) (Smith 1967; Clary et al. 1975; Severson and Boldt 1977). In eastern Washington seral ponderosa pine communities in the ponderosa pine-grand fir-Douglas fir habitat type produced 560 to 675 kg/ha (500 to 600 lb/acre) of herbage (Hall 1976). Shading and litter are the primary factors that depress plant yields beneath ponderosa pine trees. Pase (1958) and Biswell (1972) showed exponential relationships between litter and herbage yields. Herbage yields decline rapidly as pine needles build up to 6740 kg/ha (6000 lb/acre). Thereafter, yields drop to 55 kg/ha (50 lb/acre).

The second year after burning in northern Arizona, forage yields increased from 80 to 690 kg/ha (70 to 610 lb/acre) on unthinned stands and from 550 to 715 kg/ha (490 to 640 lb/acre) on thinned stands (Pearson et al. 1972). The increases were due primarily to forbs. Grasses held steady or decreased slightly. First year grass-yields had increased on thinned stands, but decreased slightly on unthinned stands.

Climax Ponderosa Pine Communities

EAST SLOPE OF CASCADES, BLUE MOUNTAINS, AND NORTHERN ROCKIES

Vegetation. This region includes parts of central and eastern Oregon and Washington, Idaho, western Montana, and southern British Columbia. The understory of the ponderosa pine communities are dominated by grasses or shrubs (Brayshaw 1965; Daubenmire 1969) (Fig. 10.5). Common grasses include

Fig. 10.5. A ponderosa pine-Idaho fescue habitat type with an understory dominated by bluebunch wheatgrass and Idaho fescue. Both grasses are shade-intolerant and will decline as canopy cover increases and pine needles accumulate. Photo taken in northern Idaho. (Photo courtesy of University of Idaho.)

bluebunch wheatgrass *(Agropyron spicatum)*, needle-and-thread *(Stipa comata)*, Idaho fescue *(Festuca idahoensis)*, and rough fescue *(Festuca scabrella)*. Idaho fescue usually dominates the wettest sites and bluebunch wheatgrass dominates many of the drier sites in Washington, Oregon, and Idaho. Needle-and-thread is usually associated with sandy soils but is also found in the driest areas. Rough fescue is most abundant east of the summit of the northern Rockies. It frequently grows in combination with Idaho fescue, but is more tolerant of cold weather. Other grass species present usually include Sandberg bluegrass *(Poa sandbergii)*, bottlebrush squirreltail *(Sitanion hystrix)*, junegrass *(Koeleria cristata)*, onespike danthonia *(Danthonia unispicata)*, pinegrass *(Calamagrostis rubescens)*, elk sedge *(Carex geyeri)*, and letterman needlegrass *(Stipa lettermanii)*. Heavily grazed areas are often dominated by cheatgrass *(Bromus tectorum)* (Weaver 1967a).

Mixed with the grasses are small amounts of numerous forbs including arrowleaf balsamroot *(Balsamorhiza sagittata)*, woodland star *(Lithophragma inflatum)*, Collinsia *(Collinsia parviflora)*, whitlow grass *(Draba verna)*, springbeauty *(Montia linearis)*, forget-me-not *(Myosotis micrantha)*, low pussytoes *(Antennaria dimorpha)*, and silky lupine *(Lupinus sericeus)* (Daubenmire and Daubenmire 1968). Other vegetation includes wild strawberry *(Fragaria* spp.), bedstraw *(Galium aparine)*, bigleaf sandwort *(Arenaria macrophylla)*, fleabane *(Erigeron compositus)*, and heartleaf arnica *(Arnica cordifolia)* (Brayshaw 1955; Hall 1973; Skovlin et al. 1976; Volland 1976). Fireweed *(Epilobium angustifolium)* is common on hotspots (Weaver 1951b). Overgrazed areas contain dalmation toadflax *(Linnaria dalmatica)* and common St. Johnswort *(Hypericum perforatum)* (Daubenmire and Daubenmire 1968).

Dominant shrubs are typically snowberry *(Symphoricarpos albus)*, ninebark *(Physocarpus malvaceus)*, oceanspray *(Holodiscus discolor)*, smoothleaf sumac *(Rhus glabra)*, and antelope bitterbrush *(Purshia tridentata)*. Less abundant shrubs include snowbrush *(Ceanothus velutinus)*, birchleaf spiraea *(Spiraea betulifolia)*, redstem ceanothus *(Ceanothus sanguineus)*, willow *(Salix* spp.), rose *(Rosa* spp.), mockorange *(Philadelphus lewisii)*, bittercherry *(Prunus emarginata)*, and service-berry *(Amelanchier alnifolia)*.

Fire Effects.   In general fire is beneficial to grasses in ponderosa pine associations of eastern Oregon and Washington because it removes pine needle mats (Weaver 1951b). Because of the precipitation zone and size of most plants, bunchgrasses will recover from fire in one to three years, whereas most forbs will not be harmed for more than a year (Wright et al. 1979). All of the shrubs, except for antelope bitterbrush, are vigorous sprouters and very tolerant of fire (Wright 1971). For most of these communities, Weaver (1967a) estimated a fire frequency of 6 to 22 years, and Arno (1976), with limited data, documented an average fire frequency of 6 to 11 years with a rage of 2 to 20 years.

By contrast, fire frequency in the ponderosa pine-bitterbrush habitat type (a widely distributed community in west-central North America) was probably burned every 50 years or so. This hypothesis is based on observations and current knowledge of the suceptibility of antelope bitterbrush to fire (Nord 1965; Weaver 1967a; Wright et al. 1979). Following one fire in such a community, Weaver (1967a) noted that the ground was still comparatively free of fuel accumulations and brush after 20 years. After 27 years bitterbrush seedlings became frequent. In general, ponderosa pine communities with shrub understories apparently had lower fire frequencies than communities with grass understories (Arno 1976).

CENTRAL ROCKIES

Vegetation.   Currie (1975) has described vegetation of ponderosa pine-bunchgrass ranges of the central Rockies. Generally six forage species—Arizona fescue *(Festuca arizonica)*, mountain muhly *(Muhlenbergia montana)*, blue grama *(Bouteloua gracilis)*, sun sedge *(Carex heliophila)*, little bluestem *(Schizachyrium scoparium)*, and fringed sagebrush *(Artemisia frigida)*—make up 95 percent of the herbage composition by weight. Arizona fescue and mountain muhly dominate the southern two-thirds of the zone. Northward at the northern border of Colorado and into Wyoming, Idaho fescue replaces Arizona fescue. At the upper limit of the ponderosa pine zone, Thurber fescue *(Festuca thurberi)* will replace Arizona fescue and may be associated with Idaho fescue. Other desirable grasses include parry oatgrass *(Danthonia parryi)*, junegrass, little bluestem, blue grama, western wheatgrass *(Agropyron smithii)*, and bottlebrush squirreltail. Less desirable grasses include sleepygrass *(Stipa robusta)*, pullup muhly *(Muhlenbergia filiformis)*, and tumble windmillgrass

(Schedonnardus paniculatus). The most conspicuous forbs are geraniums *(Geranium* spp.), purple milkvetch *(Astragalus striatus)*, western yarrow *(Achillea millefolium)*, Lambert locoweed *(Oxytropis lambertii)*, pusssytoes *(Antennaria rosea)*, trailing fleabane *(Erigeron flagellaris)*, cinquefoils *(Potentilla* spp.), asters *(Aster* spp.), and bluebells *(Mertensia* spp.). Shrubs are not very prevalent, but in local areas bearberry *(Arctostaphylos uva-ursi)*, true mountain mahogany *(Cercocarpus montanus)*, Gambel oak *(Quercus gambelii)*, and ceanothus *(Ceanothus* spp.) are present.

Fire Effects. Daubenmire (1943) indicates that fire probably had a significant role in ponderosa pine forests of this region. Based on the predominantly grass understory, this assumption could easily be true. Certainly the grasses could tolerate fire well. However since litter is usually meager and the tree canopy is generally no more than 25 percent (Daubenmire 1943), one could also reason that fire is less frequent in ponderosa pine communities of this region than in most other areas where it grows. There seems to be less fine fuel for potential fires and fire frequency may be close to the estimated 12- to 25-year interval suggested by Hendrickson (1972).

SOUTHERN ROCKIES

Vegetation. Ponderosa pine is generally the climax tree (Fig. 10.6) where it occurs in pure stands in this region. However, litter builds up quickly and this area has a high incidence of lightning strikes (Komarek 1966). Principal grass species in the pine understory include mountain muhly, muttongrass *(Poa fendleriana)*, Arizona fescue, black dropseed *(Sporobolus interruptus)*, pine dropseed *(Blepharoneuron tricholepis)*, bottlebrush squirreltail, sedge *(Carex geophila)*, and blue grama (Arnold 1950), although mountain muhly and Arizona fescue are the dominant grasses (Clary 1975). The most abundant forbs include western yarrow, pussytoes, sandwort *(Arenaria fendleri)*, senecio *(Senecio* spp.), fleabane, thistle *(Cirsium* spp.), asters, penstemons *(Penstemon* spp.), clovers (Trifolium spp.), and lupine *(Lupinus* spp.).

Fire Effects. Grazing and, presumably, burning are most harmful to mountain muhly, Arizona fescue, muttongrass, and pine dropseed (Arnold 1950). Mountain muhly takes at least three years to fully recover from fire (Gaines et al. 1958); it dropped from 70 to 60 percent of the total grass composition two years after burning. Bottlebrush squirreltail, an inconspicuous seral species, will increase fivefold by the end of the second year after burning (Gaines et al. 1958). Forbs are generally harmed very little because their foliage has disintegrated by the time of the burn. Fendler ceanothus *(Ceanothus fendleri)* is the only browse species over most of this area (Kruse 1972) that produces numerous seedlings after fire (Pearson et al. 1972.). Alligator juniper *(Juniperus deppeana)* and Gambel oak, both sprouting species, invade stands of ponderosa pine at the lower elevations, but fire, along with competition and shade from pines, will keep these species in check (Leopold 1924) (Fig. 10.7).

Fig. 10.6. A climax stand of ponderosa pine in Arizona, as indicated by the young ponderosa pine trees in the understory. An average fire frequency of seven to nine years would not have let the young trees become established. (Photo courtesy of USDA Forest Service.)

Fig. 10.7. Ponderosa pine with an understory of Gambel oak.

Fig. 10.8. Effect of fire exclusion on ponderosa pine is shown by these two photographs. The top photograph (courtesy of South Dakota State Historical Society, Pierre, S. D.) was taken in 1874 when natural fires took their course. The bottom photograph (courtesy of South Dakota State University) was taken a century later following complete fire protection during the last 75 years.

BLACK HILLS

Comparison of photographs taken in 1874 and a century later (Fig. 10.8) from the same camera points in the Black Hills clearly illustrates how forests can change under prolonged fire protection and intensive land use (Progulske 1974). Many pine stands are much denser than they were before human settlement and the aggressive pine has occupied many of the smaller grassy openings. Ponderosa pine reproduces more prolifically in the Black Hills, year in and year out, than almost anywhere else in the species' vast range (C. E. Boldt, pers. comm.). Fire today is being tested as a tool to check encroachment of pine into grasslands (Gartner and Thompson 1972) (Fig. 10.9).

Vegetation. Vegetation of commercial timberland in the Black Hills area has been described by Pase (1958). Ponderosa pine occupies 95 percent of the area. White spruce *(Picea glauca)*, an extension of the boreal forest, accounts for the remaining 5 percent on moist northerly slopes at higher elevations. Paper birch *(Betula papyrifera)* and aspen *(Populus tremuloides)* occupy cool, moist sites throughout the northern half of the Black Hills, often forming dense stands on old burns. Bur oak *(Quercus macrocarpa)* can be found on the northern fringe of the Black Hills. Other species mentioned by Gartner and Thompson (1972) include willows, redosier dogwood *(Cornus stolonifera)*, water birch *(Betula occidentalis)*, American elm *(Ulmus americana)*, boxelder *(Acer negundo), cottonwoods (Populus* spp.), and mountain ash *(Sorbus scopulina)*. True mountain mahogany is the most extensive shrub community in the Black Hills. Other understory shrubs include common juniper

Fig. 10.9. Ponderosa pine encroaching on mixed prairie grassland near the Black Hills.

(Juniperus communis), bearberry, russet buffaloberry *(Sheperdia canadensis)*, red raspberry *(Rubus strigosus)*, chokecherry *(Prunus virginiana)*, American plum *(P. americana)*, pincherry *(P. pensylvanica)*, western snowberry *(Symphoricarpos occidentalis)*, eastern hop-hornbeam *(Ostrya virginiana)*, beaked hazel *(Corylus cornuta)*, serviceberry, rose, and creeping barberry *(Berberis repens)* (Thilenius 1971).

Principal understory herbaceous species in open stands of ponderosa pine include Pennsylvania sedge *(Carex pensylvanica)*, Kentucky bluegrass *(Poa pratensis)*, roughleaf ricegrass *(Oryzopsis asperifolia)*, poverty oatgrass *(Danthonia spicata)*, fuzzyspike wildrye *(Elymus innovatus)*, prairie dropseed *(Sporobolus heterolepis)*, little bluestem, western yarrow, Missouri goldenrod *(Solidago missouriensis)*, American strawberry *(Fragaria vesca)*, pussytoes, white clover *(Trifolium repens)*, and cream peavine *(Lathyrus ochroleucus)*. Under dense canopies, Pennsylvania sedge, Missouri goldenrod, peavine, common juniper, and creeping barberry are the most persistent plants. Poverty oatgrass, bearberry, western snowberry, and serviceberry are moderately tolerant of shade. A tree canopy cover of 40 percent is the breaking point where many species begin to disappear (Pase 1958).

Fire Effects.   The response of understory species in this plant asssociation to fire has not been studied. Based on other studies, however, most of these species are tolerant to fire. Gartner and Thompson (1972) have studied the effect of fire on plants in grasslands adjacent to the Black Hills. Their data show that sedges, prairie dropseed, Missouri goldenrod, cream peavine, and chokecherry tolerate fire very well. Erdman (1970) found that true mountain mahogany was a sprouter in southern Colorado.

The season of a fire would have the greatest effect on herbaceous plants. According to Orr (1959), severe thunderstorms are common during summer months. During this time of the year fires could spread in heavy grass (a mixture of dead and green grass) or needle litter and severely harm warm-season species and favor cool-season species such as Kentucky bluegrass, an introduced species. Spring burning, on the other hand, would favor warm-season grasses and harm cool-season species.

PACIFIC PONDEROSA PINE TYPE

The Pacific ponderosa pine type occupies the lowermost slopes of the Sierra Nevada Mountains and southern Cascade Range and cross ranges of northern California and southern Oregon (Fowells 1965). It is found mostly on sites of high quality where it tends to form dense stands (P. M. McDonald, pers. comm.). It is a climax type, although the area it occupies shifts with disturbance. At low elevation, ponderosa pine can be found on outliers of higher site quality where it outgrows the manzanitas *(Arctostaphylos* spp.) and oaks *(Quercus* spp.). Lack of disturbance aids species extension; disturbance causes its retreat. At the upper elevational limits of the type, the more tolerant conifers invade and, without disturbance, capture the area. Disturbance favors extension of the Pacific ponderosa pine type into the Sierra Nevada mixed-conifer zone. Stand structure reflects the magnitude and timing of disturbance; hence vegetation pattern tends toward even-aged groups in an uneven-aged mosaic.

Vegetation.   Shrub and herbaceous vegetation in the Pacific ponderosa pine type is varied and abundant. However it is sparse beneath dense stands but increasingly abundant as stand density decreases. Because of the high site quality, disturbed areas do not remain bare for long. Resprouts of many shrub species as well as dormant shrub seeds in the soil burst forth to pioneer succession. Numerous invader species (annual grasses and forbs) also are blown in by wind. Together the pioneer and invader species constitute a formidable challenge to conifer regeneration. Manzanita, deerbrush *(Ceanothus integerrimus)*, coffeeberry *(Rhamnus californica)*, a variety of oaks, poison oak *(Rhus diversiloba)*, snowberry, and other ceanothus species plus a host of annual forbs and grasses constitute the understory at lower elevations in this forest type.

Fire Effects.   Except for some of the manzanita species, all of the shrubs are vigorous sprouters following fire and some reproduce from seed. Annual plant seeds are also abundant, so these species come back quickly after fire. The abundance of shrubs and grasses after a prescribed burn is often difficult to deal with. They are very competitive with pine seedlings. However, slash reduction is inescapable and if done properly, prescribed burning shows positive potential (McDonald and Schimke 1966).

Prescribed burning trials in stands of the Pacific ponderosa pine type were conducted along roadsides in the winter (P. M. McDonald, pers. comm.). These were limited trials involving only 0.4 to 0.8 ha (1 or 2 acres) of land. Nevertheless, conifer and hardwood seedlings and saplings under 100-year-old ponderosa pines were reduced greatly, extending the scenic view and reducing the hazard of roadside fires.

Seral Ponderosa Pine Communities

PONDEROSA PINE–LARCH–DOUGLAS FIR

Vegetation.   This habitat type occurs primarily in western Montana, northern Idaho, northeastern Washington, and southern British Columbia (Fig. 10.10). It occupies areas that are intermediate between the dry ponderosa pine lands and the more moist larch-Douglas fir type. It is presumed to be a subclimax type successionally, often having a definite fire origin and tending to pass slowly into a Douglas fir climax with significant amounts of ponderosa pine and western larch *(Larix occidentalis)*. Other species occurring in limited amounts are grand fir *(Abies grandis)* and lodgepole pine *(Pinus contorta)*.

Common grass and forb associates include rough fescue, pinegrass, elk sedge, Idaho fescue, bluebunch wheatgrass, arrowleaf balsamroot, and heartleaf arnica. The more common shrubs associated with ponderosa pine in this cover type are the blue huckleberry *(Vaccinium globulare)*, dwarf huckleberry *(V. caespitosum)*, nine-

Fig. 10.10. A climax stand of ponderosa pine-larch-Douglas fir in Missoula, Montana. Seral communities of this habitat type are predominantly ponderosa pine with significant amounts of western larch.

bark, bearberry, chokecherry *(Prunus virginiana, P. Pensylvanica, P. virginiana* var. *melanocarpa)*, serviceberry, Wood rose *(Rosa woodsii)*, antelope bitterbrush, oceanspray, redstem ceanothus, snowberry, birchleaf spiraea, mockorange, and snowbrush.

Fire Effects.   Because ponderosa pine is more fire-resistant than most associated tree species (except western larch), past fires have had a profound effect on its distribution. Although young seedlings are readily killed by fire, older trees are protected from fire damage by their thick bark. Associated species such as Douglas fir and grand fir are considerably less fire-tolerant, especially in the sapling and pole-size classes. Thus periodic fires have maintained ponderosa pine in this cover type, where without fire disturbance, climax species would have attained dominance.

Fire has also influenced understory vegetation. Several workers have observed that burning reduces shrub cover substantially and increases grass cover, especially on the more xeric sites (Brayshaw 1965; Daubenmire and Daubenmire 1968). In seral ponderosa pine stands maintained by fire, pinegrass, a very fire-tolerant species, dominates the herbaceous layer along with other grasses on some sites (Brayshaw 1965; Hall 1976). As litter accumulates and crown cover closes (as results from long fire-free intervals), heartleaf arnica becomes dominant (Brayshaw 1965).

Of the shrubs associated with this cover type antelope bitterbrush is the most readily eliminated by fire, although on some sites fire exclusion and consequent competition from increases in canopy density may have the same effect (Sherman 1966). Rodent caches of seed are also significant in the establishment of antelope bitterbrush and the cache placement pattern is strongly correlated with sites free of pine litter (Sherman 1966). As a result of fire exclusion, litter-free sites are becoming less abundant and a decrease in antelope bitterbrush populations may be anticipated.

The majority of shrubs associated with ponderosa pine in this cover type are well adapted to fire and resprout vigorously. In western Montana Miller (1977) documented the sprouting ability of blue huckleberry, big huckleberry *(Vaccinium membranaceum)*, snowberry, and birchleaf spiraea. Redstem ceanothus will resprout, but it also comes back as a vigorous seedling following fire (Orme and Leege 1976). Heat causes the hilar fissure to permanently open and allow water to enter the seed. Snowbrush also resprouts and germinates from seed following fire (Lyon and Stickney 1976). Other species that easily regenerate from root crowns include Scouler willow *(Salix scouleriana)*, Sierra maple *(Acer glabrum)*, and serviceberry (Lyon and Stickney 1976). Additional research has been summarized by Wright (1971) which shows that chokecherry, ninebark, oceanspray, Wood rose, and snowberry are also vigorous sprouters. The grasses and forbs best adapted to fire are pinegrass, arnica *(Arnica latifolia)*, and bracken fern *(Pteridium aquilinium)* (Lyon and Stickney 1976).

PONDEROSA PINE–GRAND FIR–DOUGLAS FIR

Plant communities of ponderosa pine in eastern Washington, eastern Oregon, and central Idaho are usually seral to climax mixtures of the ponderosa pine-grand fir-Douglas fir habitat type (Hall 1976). This type is also referred to as mixed-conifer. Forests of this habitat type vary in their composition but the primary type of concern here contains ponderosa pine, Douglas fir, grand fir, and sometimes western larch.

For the seral communities of ponderosa pine which have been maintained by fire, pinegrass, a very fire-tolerant species, dominates the herbaceous layer with elk as a codominant (Hall 1976) (Fig. 10.11). As litter accumulates and the crown cover closes, heartleaf arnica become dominant (Hall 1976). Shrubs are rare in seral communities of mixed-conifer but usually very abundant in seral communities of ponderosa pine-Douglas fir. The most common shrubs are snowberry, ninebark, oceanspray, and birchleaf spiraea — all vigorous sprouters following fire (Weaver 1967a).

Ponderosa pine is the most fire-tolerant species of all trees in the mixed-conifer zone because it has a fire-resistant bark containing a 0.3- to 0.6-cm (0.12- to 0.25 in.) thick dead layer at 5 cm (2 in.) diameter (Hall 1976). Grand fir bark, on the other hand, remains green and photosynthetically active up to 10 cm (4 in.) in diameter (Hall 1976). Thus it is easily killed by surface fires; it can even be killed by sunscald (Hall 1976). Douglas fir is intermediate in suceptibility to fire.

PONDEROSA PINE–SUGAR PINE–FIR

Vegetation. In the Sierra Nevada mixed-conifer forest type (Fig. 10.12), ponderosa pine is part of a complex six-species mixture, which as a whole is consid-

Fig. 10.11. Seral communities of ponderosa pine in the ponderosa pine-grand fir-Douglas fir habitat type in eastern Washington have an understory that is dominated by pinegrass and elk sedge. (Photo courtesy USDA Forest Service.)

Fig. 10.12. Sierra Nevada mixed-conifer type with an overstory of ponderosa pine and incense cedar. Understory consists of large and small incense cedar trees. Note heavy accumulations of fuel on forest floor. (Photo courtesy of National Park Service.)

ered climax. Early travelers recorded that these forests were dominated by ponderosa pine with some incense cedar and California black oak *(Quercus kelloggii)* at the lower elevations and increasing numbers of sugar pine *(Pinus lambertiana)* and white fir at the upper end of the zone (van Wagtendonk 1974). Douglas fir was more prevalent in the northern portion of the region. Today ponderosa pine occurs both as individual trees and in small stands or groves. Occasionally it is found occupying an entire hillside, especially on south or west aspects. Site quality generally is high and, as might be expected, the stands tend to be dense on the south and west aspects. The different tree species in the Sierra Nevada mixed-conifer vary in shade tolerance; thus all strata in the stands have become occupied because of protection from fire.

Lesser species commonly consist of less than 10 species beneath dense stands, but in the type as a whole over 100 species of grasses, forbs, and shrubs have been identified (P.M. McDonald, pers. comm.). The understory, where present, includes young tree reproduction, Mariposa manzanita *(Arctostaphylos manzanita)*, wedgeleaf ceanothus *(Ceanothus cuneatus)* deerbrush, and gooseberry *(Ribes* sp.). The first two are nonsprouters and depend on seed for survival. Beneath the larger trees the forest floor is carpeted with needles, bear clover *(Chamaebatia foliolosa)*, and various grasses and forbs (van Wagtendonk 1974).

Fire Effects.   The less shade-tolerant ponderosa and sugar pines are best adapted to a regime of periodic surface fires and would be favored with prescribed burning (van Wagtendonk 1974). However protection from natural surface fires has permitted the forest floor to become a tangle of understory vegetation and accumulated debris. Shade-tolerant incense cedar and white fir thickets often predominate in the understory (Fig. 10.12). Thus a wildfire in this forest, as it exists today, could easily reach catastrophic proportions. For those who wish to reintroduce fire into the Sierra Nevada mixed-conifer forest, it will have to be done carefully with refined prescriptions (van Wagtendonk 1974).

PONDEROSA PINE–DOUGLAS FIR

Vegetation.   This type is widely developed in northern California and southern Oregon, particularly in the north coast ranges and Siskiyou Mountains. Ponderosa pine and Douglas fir together predominate, although neither species is present to the extent of 80 percent. Incense cedar, sugar pine, and a variety of other conifers and oaks are often present in small amounts.

Many annual grasses and forbs, including legumes, as well as perennial plants such as mountain brome *(Bromus carinatus)* and blue wildrye *(Elymus glaucus)* are common in this area. Major forb genera include *Madia, Lotus, Epilobium, Cirsium, Clarkia,* and *Trifolium.* Common annual grass genera are *Bromus, Festuca,* and *Hordeum.*

Common and white-leaved manzanita *(Arctostaphylos manzanita* and *A. vicsida)*, both nonsprouting shrubs, and deerbrush, a sprouting species, are the most prevalent shrubs (Biswell and Schultz 1958; McDonald 1976a). Occasional stands of western mountain mahogany *(Cercocarpus betuloides)*, mountain whitethorn *(Ceanothus cordulatus)*, California hazel *(Corylus rostrata* var. *californica)*, and flowering dogwood *(Cornus floridana)*, all sprouting species, are also present. Greenleaf manzanita *(A. patula)*, snowbrush, Sierra plum *(Prunus subcordata)*, and bush chinkapin *(Castanopsis sempervirens)* also occur in ponderosa pine communities of northern California (Bentley et al. 1971). Prevalent tree species include Pacific madrone *(Arbutus menziesii)* and various oaks.

Fire Effects.   After logging, slash volumes ranged from 215 to 270 metric tons/ha (85 to 110 tons/acre) in this forest type (Sundahl 1966). The need for slash reduction is inescapable. If site preparation also can be accomplished simultaneously, costs to restock a forest can be reduced. Controlled burning of 14 different compartments totaling 370 ha (330 acres) over four years indicated the positive potential of this technique (Wright 1978). Specific planning, prescriptions, and care, however, must be used (McDonald and Schimke 1966). Drawbacks are dependent on rather specific weather and the propensity for breaking dormancy of copious brush seeds in the soil.

Shrubs can be very prevalent after a burn because of their ability to sprout and germinate from seed. However, other than general observations, no specific fire effects data is available for the shrub and herbaceous species.

MANAGEMENT IMPLICATIONS

How should a ponderosa pine forest be managed? A widely applied management scheme is a two-stage shelterwood cutting method (Schubert 1974; Boldt and Van Deusen 1974; McDonald 1976b). This regeneration method seems to work well in climax communities of ponderosa pine and in some successional pine stands where the associated conifers are desired species. However, in successional pine stands of the Pacific Northwest where ponderosa pine is preferred over Douglas fir and grand fir, the seed-tree method is used to allow more light through the canopy (Hall 1973).

Shelterwood cutting, in most cases, involves the removal of all merchantable trees except for 30 to 44/ha (12 to 18/acre) residual, large, full-crowned, and thrifty trees that are left to produce seed and some shade. Advance reproduction, small poles, and large poles usually remain in the understory. On good sites slash is disposed of both in individual piles and in windrows (McDonald 1976a, 1976b). In the interior ponderosa pine type, however, everything except the residual shelterwood trees is usually lopped and scattered (Fig. 10.13) (Boldt and Van Deusen 1974; Dieterich 1976). After 10 years or so if the new stand is more than adequately stocked [80% or more of the milacres are occupied by at least one solidly established pine tree 0.3 to 0.9 m (1 to 3 ft) tall], stagnation may become a threat.

Thus the shelterwood and seed-tree methods, as presently used, are highly satisfactory for regenerating ponderosa pine but they may lead to overstocked seedling stands. The reason is that they provide only a low level of control over seed dispersal and seedling establishment. Seed tree cutting [10 to 30 residual seed trees/ha (4 to 12/acre)], of course, provides better control than the shelterwood cutting method, but control may be reduced by advanced reproduction or seed dispersal before harvest. Some additional control may be attained by limiting site disturbance and allowing grasses and shrubs to compete, but such vegetation must be present and competitive at time of the regeneration cut. Grazing should probably be restricted to protect seedlings when pine stocking is marginal as well as to enhance competition by ground cover when pine production is too abundant.

For those who wish to use silviculture as a tool to obtain maximum production of

Fig. 10.13. In shelterwood cutting schemes everything except the residual shelterwood trees is usually lopped and scattered. (Photo courtesy of University of Oregon.)

wood products, natural regeneration is uncertain (Rietveld and Heidman 1976) and therefore clear-cutting and planting is the preferred method to establish ponderosa pine trees (Shearer and Schmidt 1970). However, silviculture should be a means of improving the quality and diversity of the landscape and of providing as high a potential for a variety of uses as is compatible with carrying capacity of the land (Meyers 1974).

Once a new forest has been established, there should be an adequate number of trees 3 to 3.7 m (10 to 12 ft) tall before regular prescribed burning begins, although low intensity burns will leave trees 1.8 to 2.4 m (6 to 8 ft) tall unharmed (Biswell 1968). Here the primary goal is to reduce the density within a range of 310 to 640 trees/ha (125 to 260/acre) (Barrett 1965, 1973). Additional benefits will be to reduce surface fuels and regenerate desirable shrubs where applicable. Subsequent fires at five to seven year intervals can be used to maintain low fire hazard conditions and kill unnecessary reproduction so that a stand of healthy trees will develop. Burns are conducted in the fall in most areas, although Biswell (1968) recommends winter and spring burning for California.

McComb (pers. comm.) has been using a modification of the traditional shelterwood cutting method. By leaving the slash in place, packing it with a dozer, and then burning it he has obtained good regeneration of ponderosa pine as well as snowbrush and redstem ceanothus. Moreover, with this method of prescribed burning, fuel loads are reduced, and low fire intensities can be used in future years for thinning and maintenance of low fire hazards (Wright 1978). Seed production from ponderosa pine trees remains adequate as long as percent of crown scorched remains less than 65 percent (Rietveld 1976). Trees scorched less than 35 percent are the best seed producers (Rietveld 1976).

In other cases, selective cutting or thinning may be the first treatment in a ponderosa pine forest. These forests are not ready for the two-stage shelterwood or seed-tree cutting methods. Biswell (1968) suggests that in such forests the heavy debris be removed by a series of prescribed fires. Fire can be used to maintain low fire hazard conditions and kill tree reproduction until the forest is ready to be harvested by clear-cutting, a two-stage shelterwood method, group selection, or the seed-tree method, and then burned.

Dense stands of pine saplings or mature stands of ponderosa pine with dense understories of young trees pose one of the most serious management problems (Fig. 10.2). Few saw-log trees will be produced without major mechanical and hand thinning (Barrett 1973), which is expensive (Weaver 1943). Prescribed fire shows potential for thinning uneven-aged stands of trees if one is willing to accept the risk of some commercial timber losses (Weaver 1961; Lindenmuth 1962), but not for thinning dense, even-aged stands (Gordon 1967). There is no easy solution. Looking at the picture on a long-term basis, these forests should be reclaimed as soon as possible because the trees are capable of growing several times faster (Weaver 1959; Barrett 1973; Hall 1976). Destroying the stand and starting anew is one alternative (Boldt and Van Deusen 1974), but only if the area is not seeded after burning to prevent soil erosion. Hand or mechanical thinning may be another alternative. Hand or mechanical thinning can save 10 years of stand growth (Barrett 1973), but such thinning seriously increases the fire hazard for five years (Fahnestock 1968).

Seeding or planting trees after a dense cover of grass is established usually results in total failure (Wright 1978). This method is not the way nature established trees. Grasses were thin on many areas after a burn and thickened gradually (Wright 1978). Where possible, competition must be kept to a minimum during the first year or two after the establishment, and thus the seeding of grasses on timber sites should be kept to a minimum (Rietveld and Heidmann 1976).

We have stressed the management of trees in this section because it is the primary resource on timber sites. However, the other resources, forage, wildlife, water, and recreation, will be present and aesthetically pleasing if the area is managed to optimize a resource or combination of resources, including wood (Meyers 1974; Severson and Boldt 1977).

In conclusion, fire should play a much greater role in the management of our ponderosa pine forests than it does today, although it is not equally necessary in all associations that support ponderosa pine. It will take much time to get ponderosa pine forests to the stage where they can be easily managed with fire. But once that stage is achieved, fire hazards and suppression costs will be reduced and investments for cultural treatments will decline yet forests will be both more productive and attractive.

REFERENCES

Arno, S. F. 1976. The historical role of fire on the Bitterroot National Forest. USDA For. Serv. Res. Paper INT-187. Intermt. For. and Range Exp. Stn., Ogden, Utah.

Arnold, J. F. 1950. Changes in ponderosa pine-bunchgrass ranges in northern Arizona resulting from pine regeneration and grazing. *J. For.* **48**:118–126.

Barrett, J. W. 1965. Spacing and understory vegetation affect growth of ponderosa pine saplings. USDA For. Serv. Res. Note PNW-27. Pac. Northwest For. and Range Exp. Stn., Portland, Ore.

Barrett, J. W. 1968. Pruning of ponderosa pine—effect on growth. USDA For. Serv. Res. Paper PNW-68. Pac. Northwest For. and Range Exp. Stn., Portland, Ore.

Barrett, J. W. 1973. Latest results from the Pringle Falls ponderosa pine spacing study. USDA For. Serv. Res. Note PNW-209. Pac. Northwest For. and Range Exp. Stn., Portland, Ore.

Beale, E. F. 1958. Wagon road from Fort Defiance to the Colorado River. 35 Cong. 1 Sess., Sen. Exec. Doc. 124.

Bentley, J. R., S. B. Carpenter, and D. A. Blakeman. 1971. Early brush control promotes growth of ponderosa pine planted on bulldozed site. USDA For. Serv. Res. Note PSW-238. Pac. Southwest For. and Range Exp. Stn., Berkeley, Calif.

Biswell, H. H. 1963. The use of fire in wildland management in California. *Proc. Tall Timbers Fire Ecol. Conf.* **2**:62–97.

Biswell, H. H. 1967. The use of fire in wildland management in California. In S. V. Wantrup and J. J. Parsons (eds.) *Natural Resources, Quality and Quantity*. Univ. Calif. Press, Berkeley, pp. 71–86.

Biswell, H. H. 1968. Forest fire. Talk given at NSF Field School of Natural History, Asilomar. Calif.

Biswell, H. H. 1972. Fire ecology in ponderosa pine-grassland. *Proc. Tall Timbers Fire Ecol. Conf.* **12**:69–96.

Biswell, H. H., H. R. Kallander, R. Komarek, R. J. Vogl, and H. Weaver. 1973. Ponderosa fire management. Misc. Pub. No. 2, Tall Timbers Res. Stn., Tallahassee, Fla.

Biswell, H. H., and A. M. Schultz. 1958. Manzanita control in ponderosa. *Calif. Agr.* **12**(2):12. Berkeley.

Biswell, H. H., A. M. Schultz, and J. L. Launchbaugh. 1955. Brush control in ponderosa pine. *Calif. Agr.* **9**(1):3,14. Berkeley.

Boldt, C. E., and J. L. Van Deusen. 1974. Silviculture of ponderosa pine in the Black Hills: The status of our knowledge. USDA For. Serv. Res. Paper RM-124. Rocky Mtn. For. and Range Exp. Stn., Fort Collins, Colo.

Brayshaw, T. C. 1955. An ecological classification of the ponderosa pine stands in the southwestern interior of British Columbia. Ph.D. Thesis, Univ. of B. C., Vancouver.

Brayshaw, T. C. 1965. The dry forest of southern British Columbia. In V. J. Krajina (ed.) *Ecology of Western North America*. Vol. I, Dept. of Botany, Univ. of B. C., Vancouver.

Clary, W. P. 1975. Range management and its ecological basis in the ponderosa pine type of Arizona: The status of our knowledge. USDA For. Serv. Res. Paper RM-158. Rocky Mtn. For. and Range Exp. Stn., Fort Collins, Colo.

Clary, W. P., W. H. Kruse, and F. R. Larson. 1975. Cattle grazing and wood production with different basal areas of ponderosa pine. *J. Range Manage.* **28**:434–437.

Cooper, C. F. 1960. Changes in vegetation, structure, and growth of southwestern pine forests since white settlement. *Ecol. Monogr.* **30**:129–164.

Cooper, C. F. 1961. Pattern in ponderosa pine forests. *Ecology* **42**:493–499.

Currie, P.O. 1975. Grazing management of ponderosa pine-bunchgrass ranges of the central Rocky Mountains: The status of our knowledge. USDA For. Serv. Res. Paper RM-159. Rocky Mtn. For. and Range Exp. Stn., Fort Collins, Colo.

Daubenmire, R. 1943. Vegetational zonation in the Rocky Mountains. *Bot. Rev.* **9**:325–393.

Daubenmire, R. 1969. Structure and ecology of coniferous forests of the northern Rocky Mountains. In *Coniferous Forests of the Northern Rocky Mountains: Proceedings of the 1968 Symposium.* Univ. of Mont., Missoula, pp. 25–41.

Daubenmire, R., and J. B. Daubenmire. 1968. Forest vegetation in eastern Washington and northern Idaho. Wash. State Agric. Exp. Stn. Tech. Bull. 60. Pullman.

Dieterich, J. H. 1976. Prescribed burning in ponderosa pine— state of the art. Paper presented at Region 6 Eastside Prescribed Fire Workshop, Bend, Ore., May 3–7.

Dutton, D. E. 1887. Physical geology of the Grand Canyon District. In U. S. Geol. Surv. Second Annual Rep. Washington, D. C., pp. 49–166.

Erdman, J. A. 1970. Pinyon-juniper succession after natural fires on residual soils of Mesa Verde, Colorado. Brigham Young Univ. Sci. Bull. Biol. Series 11(2). Provo, Utah.

Fahnestock, G. R. 1968. Fire hazard from precommercial thinning of ponderosa pine. USDA For. Serv. Res. Paper PNW-57. Pac. Northwest For. and Range Exp. Stn., Portland, Ore.

Foiles, M. W., and J. D. Curtis. 1973. Regeneration of ponderosa pine in the Northern Rocky Mountain-Intermountain Region. USDA For. Serv. Res. Paper INT-145. Intermt. For. and Range Exp. Stn., Ogden, Utah.

Fowells, H. A. 1965. Silvics of forest trees of the United States. USDA Handb. 271. Washington, D.C.

Franklin, J. F., and C. T. Dyrness. 1973. Natural vegetation of Oregon and Washington. USDA For. Serv. Gen. Tech. Rep. PNW-8. Pac. Northwest For. and Range Exp. Stn., Portland, Ore.

Gaines, E. M., H. R. Kallander, and J. A. Wagner. 1958. Controlled burning in southwestern ponderosa pine: Results from the Blue Mountain Plots, Fort Apache Indian Reservation. *J. For.* **56**:323–327.

Gartner, F. R., and W. W. Thompson. 1972. Fire in the Black Hills forest-grass ecotone. *Proc. Tall Timbers Fire Ecol. Conf.* **12**:37–68.

Gordon, D. T. 1967. Prescribed burning in the interior ponderosa pine type of northern California—a preliminary study. USDA For. Serv. Res. Paper PSW-45. Pac. Southwest For. and Range Exp. Stn., Berkeley, Calif.

Hall, F. C. 1973. Plant communities of the Blue Mountains in eastern Oregon and southeastern Washington. USDA For. Serv. Pac. Northwest Reg., R6 Area Guide 3-1. Portland, Ore.

Hall, F. C. 1976. Fire and vegetation in the Blue Mountains—implications for land managers. *Proc. Tall Timbers Fire Ecol. Conf.* **15**:155–170.

Harris, R. W. 1954. Fluctuations in forage utilization on ponderosa pine ranges in eastern Oregon. *J. Range Manage.* **7**:250–255.

Hendrickson, W. H. 1972. Perspective on fire and ecosystems in the United States. In *Fire in the Environment Symposium Proceedings.* Denver, Colo.

Kallander, H. R. 1969. Controlled burning on the Fort Apache Indian Reservation, Arizona. *Proc. Tall Timbers Fire Ecol. Conf.* **9**:241–249.

Kallander, H. R., H. Weaver, and E. M. Gaines. 1955. Additional information on prescribed burning in virgin ponderosa pine in Arizona. *J. For.* **53**:730,731.

Kilgore, B. M., and D. Taylor. 1979. Fire history of a Sequoia-mixed conifer forest. *Ecology* **60**:129–142.

Knoor, P. N. 1963. One effect of control burning on the Fort Apache Indian Reservation. Ariz. Annual Watershed Symp., Phoenix **7**:35–37.

Komarek, E. V., Sr. 1966. The meteorological basis for fire ecology. *Proc. Tall Timbers Fire Ecol. Conf.* **5**:85–125.

Kruse, W. H. 1972. Effects of wildfire on elk and deer use of a ponderosa pine forest. USDA For. Serv. Res. Note RM-226. Rocky Mtn. For. and Range Exp. Stn., Fort Collins, Colo.

Küchler, A. W. 1964. Potential natural vegetation of the conterminous United States. Manual to accompany the map. Amer. Geogr. Soc. Spec. Pub. 36. (With map, rev. ed., 1965, 1966.)

Leopold, A. 1924. Grass, brush, timber and fire in southern Arizona. *J. For.* **22**:1–10.

Lindenmuth, A. W., Jr. 1960. A survey of effects of intentional burning fuels and timber stands of ponderosa pine. USDA For. Serv. Res. Paper RM-54. Rocky Mtn. For. and Range Exp. Stn., Fort Collins, Colo.

Lindenmuth, A. W., Jr. 1962. Effects of fuels and trees of a large intentional burn in ponderosa pine. *J. For.* **60**:804–810.

Lyon, J. L., and P. F. Stickney. 1976. Early vegetal succession following large northern Rocky Mountain wildfires. *Proc. Tall Timbers Fire Ecol. Conf.* **14**:355–375.

McDonald, P. M. 1976a. Forest regeneration and seedling growth from five major cutting methods in north-central California. USDA For. Serv. Res. Paper PSW-115. Pac. Southwest For. and Range Exp. Stn., Berkeley, Calif.

McDonald, P. M. 1976b. Shelterwood cutting in a young-growth, mixed conifer stand in north-central California. USDA For. Serv. Res. Paper PSW-117. Pac. Southwest For. and Range Exp. Stn., Berkeley, Calif.

McDonald, P. M., and H. E. Schimke. 1966. A broadcast burn in second-growth clearcuttings in the north central Sierra Nevada. USDA For. Serv. Res. Note PSW-99. Pac. Southwest For. and Range Exp. Stn., Berkeley, Calif.

Meyers, C. A. 1974. Multipurpose silviculture in ponderosa pine stands of the Montane Zone of central Colorado. USDA For. Serv. Res. Paper RM-132. Rocky Mtn. For. and Range Exp. Stn., Fort Collins, Colo.

Miller, M. 1977. Response of blue huckleberry to prescribed fires in a western Montana larch-fir forest. USDA For. Serv. Res. Paper INT-188. Intermt. For. and Range Exp. Stn., Ogden, Utah.

Moir, W. H. 1966. Influence of ponderosa pine on herbaceous vegetation. *Ecology* **47**:1045–1048.

Morris, W. G., and E. L. Mowat. 1958. Some effects of thinning a ponderosa pine thicket with a prescribed fire. *J. For.* **56**:203–209.

Nord, E. C. 1965. Autecology of bitterbrush in California. *Ecol. Monogr.* **35**:307–334.

Ogilvie, R. T. 1955. Soil texture of *Pinus ponderosa* plant communities in British Columbia. M.A. Thesis, Univ. of B. C., Vancouver.

Orme, M. L., and T. A. Leege. 1976. Emergence and survival of redstem ceanothus *(Ceanothus sangiuneus)* following prescribed burning. *Proc. Tall Timbers Fire Ecol. Conf.* **14**:391–420.

Orr, H. K. 1959. Precipitation and streamflow in the Black Hills. USDA For. Serv. Res. Paper RM-44. Rocky Mtn. For. and Range Exp. Stn., Fort Collins, Colo.

Pase, C. P. 1958. Herbage production and composition under immature ponderosa pine stand in the Black Hills. *J. Range Manage.* **11**:238–243.

Pearson, G. A. 1942. Herbaceous vegetation a factor in natural regeneration of ponderosa pine in the Southwest. *Ecol. Monogr.* **12**:313–338.

Pearson, H. A., J. R. Davis, and G. H. Schubert. 1972. Effects of wildfire on timber and forage production in Arizona. *J. Range Manage.* **25**:250–253.

Pieper, R. D., and H. H. Biswell. 1961. Relationship between trees and cattle in ponderosa pine. *Calif. Agric.* **15**(5):12. Berkeley.

Progulske, D. R. 1974. Yellow ore, yellow hair, yellow pine. A photographic study of a century of forest ecology. South Dakota State Univ., Brookings, Agric. Exp. Stn. Bull. 616.

Rietveld, W. J. 1976. Cone maturation in ponderosa pine foliage scorched by wildfire. USDA For. Serv. Res. Note RM-317. Rocky Mtn. For. and Range Exp. Stn., Fort Collins, Colo.

Rietveld, W. J., and L. J. Heidmann. 1976. Direct seeding ponderosa pine on recent burns in Arizona. USDA For. Serv. Res. Note RM-312. Rocky Mtn. For. and Range Exp. Stn., Fort Collins, Colo.

Schubert, G. H. 1974. Silviculture of southwestern ponderosa pine: The status of our knowledge. USDA For. Serv. Res. Paper RM-123. Rocky Mtn. For. and Range Exp. Stn., Fort Collins, Colo.

Schultz, A. M., and H. H. Biswell. 1959. Effect of prescribed burning and other seedbed treatments on ponderosa pine seedling emergence. *J. For.* **57**:816,817.

Severson, K. E., and C. E. Boldt. 1977. Options for Black Hills forest owners: Timber, forage, or both. *Rangeman's J.* **4**(1):13–15.

Shearer, R. C., and W. C. Schmidt. 1970. Natural regeneration in ponderosa pine forests of Western Montana. USDA For. Serv. Res. Paper INT-86. Intermt. For. and Range Exp. Stn., Ogden, Utah.

Sherman, R. J. 1966. Spatial and chronologic patterns of *Purshia tridentata* as influenced by *Pinus ponderosa* overstory. M.S. Thesis. Ore. State Univ., Corvallis.

Show, S. B., and E. I. Kotok. 1924. The role of fire in the California pine forests. USDA Bull. 1294. Washington, D.C.

Skovlin, J. M., R. W. Harris, G. S. Strickler, and G. A. Garrison. 1976. Effects of cattle grazing methods on ponderosa pine-bunchgrass range in the Pacific Northwest. USDA Tech. Bull. 1531. Washington, D.C.

Smith, D. R. 1967. Effects of cattle grazing on a ponderosa pine-bunchgrass range in Colorado. USDA Tech. Bull. 1371. Washington, D.C.

Sundahl, W. E. 1966. Slash and litter weight after clearcut logging in two young-growth timber stands. USDA For. Serv. Res. Note PSW-124. Pac. Southwest For. and Range Exp. Stn., Berkeley, Calif.

Thilenius, J. F. 1971. Vascular plants of the Black Hills of South Dakota and adjacent Wyoming. USDA For. Serv. Res. Paper RM-71. Rocky Mtn. For. and Range Exp. Stn., Fort Collins, Colo.

Truesdell, P. S. 1969. Postulates of the prescribed burning program of the Bureau of Indian Affairs. *Proc. Tall Timbers Fire Ecol. Conf.* **9**:235–240.

van Wagtendonk, J. W. 1974. Refined burning prescriptions for Yosemite National Park. Nat. Park Serv. Occas. Paper No. 2.

Vlamis, J., H. H. Biswell, and A. M. Schultz. 1955. Effects of prescribed burning on soil fertility in second-growth ponderosa pine. *J. For.* **53**:905–909.

Volland, L. A. 1976. Plant communities of the central Oregon pumice zone. USDA For. Serv., Pac. Northwest Reg., R6 Area Guide 4-2, 110. Portland, Ore.

Wagener, W. 1961. Past fire incidence in the Sierra Nevada forests. *J. For.* **59**:739–747.

Wagle, R. F., and T. W. Eakle. 1977. Effect of a controlled burn on damage caused by wildfire. Final Rep. for Cooperative Agreement 16-364-CA, Res. work unit RM-2108, Tempe, Ariz.

Weaver, H. 1943. Fire as an ecological and silvicultural factor in the ponderosa pine region of the Pacific slope. *J. For.* **41**:7–15.

Weaver, H. 1947. Fire—nature's thinning agent in ponderosa pine stands. *J. For.* **45**:437–444.

Weaver, H. 1951a. Fire as an ecological factor in the southwestern ponderosa pine forests. *J. For.* **49**:93–98.

Weaver, H. 1951b. Observed effects of prescribed burning on perennial grasses in the ponderosa pine forests. *J. For.* **49**:267–271.

Weaver, H. 1957. Effects of prescribed burning in ponderosa pine. *J. For.* **55**:133–138.

Weaver, H. 1959. Ecological changes in the ponderosa pine forest of the Warm Springs Indian Reservation in Oregon. *J. For.* **57**:15–20.

Weaver, H. 1961. Ecological changes in the ponderosa pine forest of Cedar Valley in southern Washington. *Ecology* **42**:416–420.

Weaver, H. 1964. Fire management problems in ponderosa pine. *Proc. Tall Timbers Fire Ecol. Conf.* **3**:60–79.

Weaver, H. 1967a. Fire and its relationship to ponderosa pine. *Proc. Tall Timbers Fire Ecol. Conf.* **7**:127–149.

Weaver, H. 1967b. Some effects of prescribed burning on the Coyote Creek Test Area Colville Indian Reservation. *J. For.* **65**:552–558.

Wollum, A. G., II, and G. H. Schubert. 1975. Effect of thinning on the foliage and forest floor properties of ponderosa pine stands. *Proc. Soil Sci. Soc. of Amer.* **39**(5):968–972.

Wooldridge, D. D., and Weaver, H. 1965. Some effects of thinning a ponderosa pine thicket with prescribed fire. *J. For.* **63**:92–95.

Wright, H. A. 1971. Shrub response to fire. In Wildland shrubs—their biology and utilization: Symposium proc. July 1971, Logan, Utah. USDA For. Serv. Gen. Tech. Rep. INT-1. Intermt. For. and Range Exp. Stn., Ogden, Utah.

Wright, H. A. 1978. The effect of fire on vegetation in ponderosa pine forests: A state-of-the-art review. Texas Tech Univ. Range and wildl. Information Series No. 2. Lubbock.

Wright, H. A., L. F. Neuenschwander, and C. M. Britton. 1979. The role and use of fire in sagebrush-grass and pinyon-juniper plant communities: A state-of-the-art review. USDA For. Serv. Gen. Tech. Rep. INT-58. Intermt. For. and Range Exp. Stn., Ogden, Utah.

CHAPTER

11

DOUGLAS FIR AND ASSOCIATED COMMUNITIES

D ouglas fir* *(Pseudotsuga menziesii)* extends over a wide range in western North America, from parallel 55°N in northern British Columbia, south to parallel 19°N near Mexico City, and from the Pacific Ocean to the east slope of the Rocky Mountains (Fig. 11.1) (Isaac 1963). West of the Cascade Range from British Columbia through Washington and Oregon and west of the Sierra Nevada Mountains it is classified as *P. menziesii* var. *menziesii* (Little 1979) or coastal Douglas fir (Rowe 1972). East of these mountain ranges, it is *P. menziesii* var. *glauca* or interior Douglas fir.

This chapter discusses all the vegetative zones where Douglas fir is a dominant species, regardless of its successional status. To facilitate the presentation, Douglas fir forests will be categorized into three sections: the West Coast, the northern Interior and Rocky Mountains, and the central and southern Rocky Mountains.

---

*Douglas fir should be hyphenated since it is not a true fir. It was not done so in this text to avoid expensive last-minute corrections.

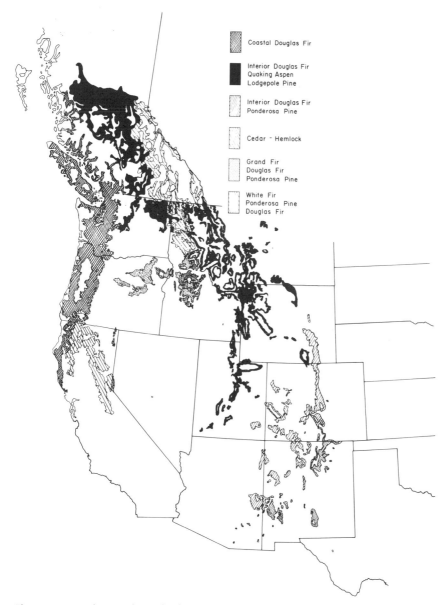

Fig. 11.1. Distribution of Douglas fir and associated forests. (Modified from Küchler 1964 and Fremlin 1974. Reproduced by permission of the American Geographical Society and the Minister of Supply and Services Canada from the publication "The National Atlas of Canada" published by the Macmillan Co. of Canada Limited, Toronto, Ontario in association with the Department of Energy, Mines and Resources Canada and Information Canada, Ottawa, Canada 1974.)

FIRE IN DOUGLAS FIR FORESTS

Various terms will be used in this chapter that have specific meanings. Several kinds of fire are common in these forests. The most spectacular is the crown fire which burns along the ground and in the tree canopy spreading from treetop to treetop. It is the most common type of stand-replacing fire in coniferous forests since everything from the ground upward is killed. Stand-replacing fire is a burn which kills all or most of the living trees in a forest and initiates forest succession or regrowth (Romme 1980). The surface fire sweeps over the ground consuming litter, dead fuels, living herbs, and shrubs and scorching tree trunks (Daubenmire 1959). Trees may or may not be killed depending on fire severity and bark and foliage characteristics. Before Europeans arrived surface fires were more frequent in Douglas fir forests than crown fires, but crown fires were the most extensive. The ground fire smolders for long periods in thick accumulations of organic matter. These fires are usually flameless and may burn for days or months after the surface or crown fire is out.

Other terms are used relative to fires. Fire regime is the characteristic frequency, extent, intensity, severity, seasonality, and effects of fires on an ecosystem during some designated time period. Fire intensity is the rate of energy transmission or heat release at the fire front (Albini 1976). Fire severity is the effect of the fire on the ecosystem.

Trees differ in their resistance to burning. Thick bark and lack of branches and foliage close to the ground are common adaptations of mature trees to surface fire. The relative tolerance of conifers in the Douglas fir region of the Pacific Northwest (Starker 1934) to surface fire is:

Most resistant	Western larch *(Larix occidentalis)*
Very resistant	Ponderosa pine *(Pinus ponderosa)*, Douglas fir
Medium	Grand fir *(Abies grandis)*, lodgepole pine *(Pinus contorta)*, western white pine *(Pinus monticola)*
Low	Western redcedar *(Thuja plicata),* western hemlock *(Tsuga heterophylla)* noble fir *(Abies procera)*
Very low	Subalpine fir *(Abies lasiocarpa)* Pacific silver fir *(Abies amabilis)*

The three tree species most resistant to surface fires have very thick bark, whereas subalpine fir has very thin bark (Starker 1934). The remaining species have moderate bark thickness except lodgepole pine and western redcedar which have thin bark. The susceptibility of trees to fire injury is illustrated in Table 11.1. Western larch is clearly the most resistant tree and subalpine fir is least resistant. Western redcedar survives surface fires far better than western hemlock (Habeck and Mutch 1973). Although fires often burn through the bark and cambium of western redcedar and char areas of sapwood, the trees continue to live; similar damage usually kills other conifers (Arno and Davis 1980). The wet valley bottoms and north-facing slopes where western redcedar frequently grows also tend to protect the trees.

Table 11.1. Susceptibility of Trees to an August 1926 Wildfire in the Cedar-Hemlock and Spruce-Fir Zones of Northern Idaho.

	Tree Mortality (%)					
	1926 and 1927	1928	1929	1930	1931	Five-Year Total
Western larch	2	1	7	—	3	13
Douglas fir	43	—	3	2	2	49
Western redcedar[a]	73	3	2	—	1	79
Grand fir	44	11	22	5	—	82
Western white pine	60	14	9	1	1	82
Englemann spruce	49	31	4	—	4	88
Western hemlock	67	6	9	5	4	90
Lodgepole pine	6	60	17	—	9	91
Subalpine fir	54	31	15	—	—	100

Adapted from Starker (1934). Reproduced by permission of the *Journal of Forestry*.
[a]Most trees were in wet valley bottoms where site conditions (moister environment) provided some protection from fire.

Mutch (1970) reasoned that if species have developed reproductive mechanisms (underground rhizomes, root sprouting, serotinous cones) and anatomical mechanisms (thick bark, epicormic sprouting) to survive periodic fires, then fire-dependent plants might also possess characteristics obtained through natural selection which actually enhances the flammability of the communities. Although communities may be ignited accidentally or randomly, the character of burning is not random. Mutch developed the following hypothesis which attempts to deal with the interaction between fire and the ecosystem: fire-dependent plant communities burn more readily than non–fire-dependent communities because natural selection has favored development of characteristics that make them more flammable. Fire occurrence may be more than the commonly accepted fire climate-fuel moisture basis. For example, dry bracken is highly flammable material. Isaac (1940) stated ''not only do fires favor bracken, but bracken favors fire, creating a vicious circle that tends to perpetuate the fern patch and eliminate other weed species, brush, and the coniferous seedlings''. Natural selection may have enhanced the flammability of such plants as bracken fern *(Pteridium aquilinum)* and lodgepole pine but it has not enhanced the flammability of all fire-dependent plants. Western larch is not very flammable but repeated burning favors its dominance. Mature western larch are resistant to fire because of thick bark and branching high above ground.

Only some plants and plant communities show evidence of a possible evolutionary link between fire and their flammability. This relationship appears to work both ways, some plants and plant communities may have become more flammable with fire, others became less flammable. It is reasonable to propose the following revisions to the hypothesis of Mutch (1970): fire-dependent plants and plant communities burn more readily or less readily than non–fire-dependent plants and plant communities because natural selection has favored the development of characteristics that make them more flammable or less flammable.

Shrubs and herbs are generally well adapted to fire. They are forest remnant species with vegetative parts that survive and normally sprout from surviving plant parts after fire, recolonizing the area. Seedlings emerge on new burns from seed stored in the soil and duff as well as from seed of the ubiquitous pioneers.

Plants that have reproductive capabilities from within the mineral soil, whether by seed, root crown, or rhizomes, are best able to survive forest fire. Plants with reproductive organs within the duff layer or above it will be reduced or eliminated by an intense fire (McLean 1969).

The adaptation of plants to fire is outlined below (Rowe 1981).

Disseminules-Based, Propagating Primarily by Diaspores

1. **Invaders.** Highly dispersive, pioneering fugitives with short-lived disseminules.
2. **Evaders.** Species with relatively long-lived propagules that are stored in the soil or canopy.
3. **Avoiders.** Shade-tolerant species that slowly reinvade burned areas, late successional, often with symbiotic requirements.

Vegetative-Based, Propagating Primarily by Horizontal and Vertical Extension

4. **Resisters.** Intolerant species whose adult stages can survive low-severity fires.
5. **Endurers.** Resprouting species, intolerant or tolerant, with shallow or deep buried perennating buds.

There are examples of each type in the Douglas fir forests. Hawkweeds *(Hieracium* spp.) are invaders. Fireweed *(epilobium angustifolium)* is an invader because of its highly dispersive seed and an endurer because of its buried perennating buds. Snowbrush *(Ceanothus velutinus)* and lodgepole pine are evaders because the former emerges after fire from seed stored in the soil while the latter perpetuates itself because of the seed stored in serotinous cones. Western hemlock, western redcedar, grand fir, and Pacific silver fir are avoiders. Ponderosa pine, western larch, and Douglas fir are resisters. Many shrubs and herbs are endurers (Fig. 11.2). Blue huckleberry *(Vaccinium globulare)* is an endurer that has its rhizomes below the duff layer.

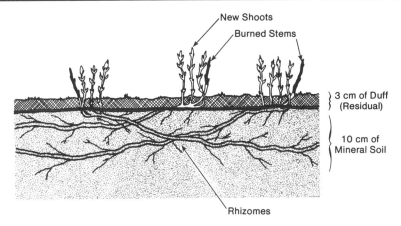

New Shoots

Burned Stems

} 3 cm of Duff
  (Residual)

10 cm of
Mineral Soil

Rhizomes

Fig. 11.2.   Sprouting of blue huckleberry from burned rhizomes. (From Miller 1977.)

Fire damage to living plants is influenced by many factors. A major one is fire severity, which is closely related to weather, fuel loading, size and distribution of fuel,
and moisture content of fuel and soil.

Survival of below-ground vegetative parts is affected by temperatures experienced
in the duff and upper portions of the mineral soil. Reduction in duff depth affects reproductive organs such as the rhizomes of blue huckleberry; duff reduction is usually
greatest when duff and mineral soil are dry and there are high fuel loadings above
ground (Norum 1976). Duff and soils are wet in spring because of snowmelt or winter
rains. Summer and early autumn fires usually reduce duff levels more than spring
fires because of lower moisture levels in soil and lower duff (Table 11.2). In the example, fire intensity, heat output during burning, was higher in spring because of the
higher fuel loadings of rotten logs, but the effect on blue huckleberry (fire severity)
was less; the higher moisture in lower duff and mineral soil dissipated heat, reducing
damage to the rhizomes.

Fire during the growing season will usually reduce sprouting from shrubs and
herbs more than fire during the dormant season. This relationship is compounded and
intensified during summer burns by the greater heat penetration and duff removal
when soils are dry.

Small diameter [0 to 0.6 cm (0 to 0.25 in.)] fuels and rotten logs influence fire intensity positively whereas abundant live stems of some shrubs reduce fire intensity
(Norum 1976; Miller 1977). Fire rarely carries in duff alone. Duff reduction depends
on those factors influencing fire intensity and residence time, particularly dry fine
fuels, smaller diameter dead woody fuels, and dry rotten logs. The longer duration of
fire in heavy fuels under dry conditions not only harms the rhizomes of endurer
shrubs but it also causes a greater kill of residual conifers.

Weather affects fire intensity and kind of fire. Hot dry winds are often effective in
changing a gentle surface fire into a fast moving crown fire. Wind speeds usually
peak during afternoons. This is normally the time when crown fires burn vast areas of
conifer forest. Wind speeds slow as evening approaches, fire intensity declines, and

Table 11.2. A Comparison of the Effect of Spring Versus Fall Fires in a *Pseudotsuga Menziesii/Vaccinium globulare* Habitat Type, Montana

Character	Prescribed Burn	
	Spring	Fall
Number of blue huckleberry stems (no./m^2) before fire	30	22
Number of blue huckleberry stems (no./m^2) first year after fire	42	28
Number of blue huckleberry stems (no./m^2) second year after fire	49	37
Rotten, 7.6-cm (3-in.) and larger preburn fuel weight (kg/m^2)	5.8	4.1
Total preburn fuel weight (kg/m^2)	7.6	5.6
Total fuel weight reduction (kg/m^2)	4.6	3.8
Fire intensity (kcal/sec/m^2)	103	72
Preburn duff depth (cm)	7.2	7.1
Duff depth reduction (%)	24	53
Upper duff moisture content (%)	41	71
Lower duff moisture content (%)	90	79
Soil moisture content (%)	29	13
Average duff surface temperature (°F)	252	359
Average mineral soil surface temperature (°F)	144	233

Adapted from Miller (1977).

sometimes the crown fires are not able to be sustained changing to a surface fire. As winds increase the next day, the fire intensity may also pick up and, if there is a ladder of fuels into the treetops, another crown fire will develop.

Topography and elevation affect fire frequency, severity, intensity, and kind of fire. North-facing slopes and higher-elevation forests are generally cool, wet, and have longer fire-free intervals. Extensive areas of level or gently undulating terrain favor fire. Dissected topography causes complex wind currents that often result in conditions unfavorable to fire. Fire burns upslope more readily than downslope. Back eddies on the downslope side of the hill slow the passage of fire. Some dissected topography may cause such wind movements that one head fire moves into another resulting in the fires going out for lack of fuel. Upper slopes and ridgetops may be burned more frequently because of dry fuel conditions and a higher frequency of lightning strikes. The patterns of vegetation development after centuries of wildfires are related to topographic relief.

The development of a conflagration, a large stand-replacing fire, is dependent, among other factors, upon weather, terrain, and large accumulations of fuel (Martin and Brackebusch 1974). Burns at frequent intervals have a lower intensity than those at longer intervals. Long time intervals between fires foster the accumulation of large quantities of fuel in the tree, understory, and forest floor layers (Bray and Gorham 1964; Turner and Long 1975).

Most fire history studies have been relatively recent in western North America and more work is in progress. Terminology is often confusing in a developing discipline as it is at present in fire history. The following terminology is from Romme (1980). Fire-return interval refers to a point (i.e., a tree) or a very small stand, whereas fire-free interval is for an area larger than a small stand. Mean fire-return interval is the arithmetic average of all fire intervals at a point or a very small stand. Mean fire-free interval is the mean of all fire intervals in an area larger than a small stand (the size is specified). Fire rotation is the length of time necessary for a specified area to burn.

Fire scars are usually used to document fire occurrence (Fig. 11.3). Stand-replacing fires frequently leave few survivors whereas fires of low intensity result in little or no scarring of trees (Lewis 1981). Indian burning was common in or adjacent to certain Douglas fir forests. These would probably have been frequent, low intensity fires that left fewer scarred trees than the less frequent lightning-caused fires.

Fire-return intervals are highly variable from region to region, zone to zone, and community to community. For example, mean fire-return intervals ranged from 15 to 30 years in small stands in the Douglas fir zone of the northern Rocky Mountains with maximum intervals of 35 to 60 years (Arno 1976) and minimum intervals of 1 to 10 years (Tande 1979).

Fig. 11.3.   Fire scarred ponderosa pine cross section from a climax ponderosa pine forest in Arizona. Thirty-one fire scars are identifiable between the years 1540 and 1876. (Photo courtesy of USDA Forest Service.)

The mean fire-return interval even varies depending upon whether one point or several points are sampled in a small stand. Arno (1976) gave an example of a mean fire-return interval of 7 years in a stand and 11 years on a single tree scarred by 15 fires between 1735 and 1900. One should base mean fire-return interval per stand rather than per tree. It is unlikely that every fire would burn hot enough at the base of any given tree in a stand to inflict a new wound.

Serious misunderstandings can arise if one confuses terminology. For example Tande (1979) gave a mean fire-free interval for the entire study area of 43,200 ha (95,000 acres) and a different value for the mean fire-return interval for 50-ha (100-acre) stands.

Clear-cut logging with or without subsequent slash burning is now the common substitute for stand-replacing wildfire as the primary cause of forest destruction. Both types of perturbation cause destruction of the tree overstory but they are different treatments. Wildfire leaves much more wood on the site than does clear-cutting, so clear-cutting has a more severe effect on nutrient cycling and probably also on microclimate.

Individual trees or groups of trees often survive single wildfires and act as a seed source to reforest the burn (Franklin and Dyrness 1973). Repeated wildfires usually eliminate surviving trees and kill regenerating conifers. Elimination of the seed source within large burns may leave extensive areas without trees for decades, as happened in the Tillamook fires in northwestern Oregon (Fig. 11.4). Clear-cut logging of large areas is equally effective in removing conifer seed sources, preventing natural regeneration. Crown fires kill all conifers, but the thick bark of Douglas fir, western larch, and ponderosa pine makes them resistant to low- to medium-intensity surface fire. This characteristic gives the species a competitive edge in reestablishing a new forest. Conifer cones not destroyed by wildfire may ripen after the fire, even on dead trees, and seed the surrounding area (Isaac 1943).

Secondary succession following destruction of the forest is influenced by numerous factors. The presence or absence of wild fire, number of times repeated, fire intensity, and fire severity are important. The kind of logging and the presence or absence of slash burning will influence succession. The topography, fuels, fuel moisture, and weather conditions before, during, and after burning are also influential. Other major factors include geographic location, forest zone, habitat type, plant species presence and abundance before burning, seed stored in the ground, seed sources available, and degree of soil surface disturbance during logging or burning.

Isaac (1940) indicated that there are recognizable stages in plant succession after fire. The pioneer sere, which he called the "weed-brush" stage, is most subject to further burning. Successive fires favor herbaceous species, retard brush species, and eliminate conifer seedlings and saplings, thus prolonging or perpetuating the "weed-brush" stage. The general successional pattern after wildfire or after clear-cut logging with or without slash burning is illustrated in Fig. 11.5. The pioneer stage lasts about five years or longer depending on the circumstances (Fig. 11.6). It is composed of shade-intolerant, herbaceous pioneer species, forest remnant herbs and shrubs that survived burning, conifer remnants, and postfire conifer seedlings. The semitolerant

Fig. 11.4. A bracken fern community 17 years after the third Tillamook fire and 29 years after the first one, northwest Oregon. There is no conifer regeneration, due to lack of a seed source. Stake is 1 m (3.3 ft) tall.

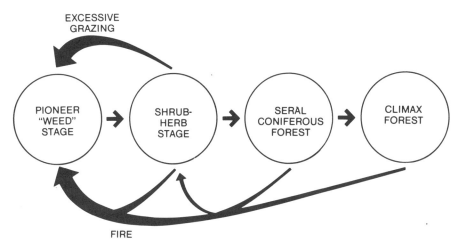

Fig. 11.5. General successional pattern in Douglas fir forests after logging and burning or after wildfires.

Fig. 11.6. A logged and slash burned plot (a) first year, (b) second year (woodland groundsel), (c) third year (fireweed), and (d) fifth year after burning (fireweed and snowbrush) in western hemlock zone of the Oregon Cascade Range. (Photos courtesy of USDA Forest Service.)

forest remnant species and/or perennial herbs and shrubs that establish after fire dominate the shrub-herb stage (Fig. 11.7). Conifer seedlings and saplings are more abundant if a seed source is available, and shade-intolerant pioneer species are still present in small amounts. The shrub-herb stage normally lasts about 5 to 15 years, but at higher elevations it may take 50 years or more before conifers assume dominance. This stage may last indefinitely if there is no conifer seed source (Fig. 11.8) or if shrubs are so dense that conifers cannot become established. Overgrazing by domestic or wild ungulates may destroy forest remnant shrubs and herbs and conifer seedlings maintaining a zootic climax of herbaceous perennial exotics and/or unpalatable shrubs and herbs. Rapidly growing shade-intolerant confiers dominate the seral forest, overtop shade-tolerant conifers, and suppress their growth. After several centuries the shade-intolerant conifers mature and die. Gradually the shade-tolerant conifers take over dominance as the others die. Eventually all shade-intolerant (seral) conifers have died and uneven-aged stands of shade-intolerant conifers are left. This climax forest will perpetuate itself until the next perturbation when secondary succession begins again. On certain dry and/or infertile sites shade-intolerant trees such as Douglas fir in certain west coast forests and lodgepole pine or aspen *(Populus tremuloides)* in some Rocky Mountain forests dominate both the seral and the climax forests.

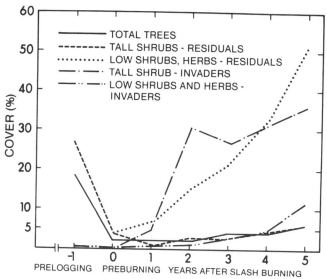

Fig. 11.7. Canopy cover (%) of selected plant groups before and after clear-cut logging and slash burning in the western hemlock zone, Oregon Cascade Range. (From Dyrness 1973.) Copyright 1973, the Eological Society of America.

Fig. 11.8. Douglas fir tree in foreground was the single survivor in this area of the 1933, 1939, and 1945 Tillamook fires. By 1962 some of its progeny were established in a shrub-herb stage dominated by vine maple, big-leaf maple, trailing blackberry, and swordfern.

249

WEST COAST FORESTS

Vegetation, Climate, and Soils

Douglas fir (Fig. 11.9) reaches its maximum development, produces its maximum yield, and is of greatest commercial importance west of the Cascade Range in southwestern British Columbia, western Washington, and western Oregon (McArdle et al. 1961). It dominates this region but is not the climax species on many sites.

In northern sections, soils have developed from glacial till, glaciomarine sediments, and various outwash deposits (Forest Soils Committee of the Douglas fir Region 1957). The northern Cascades are largely composed of ancient sedimentary rock with intrusions of granite batholiths (Franklin and Dyrness 1973). Volcanism is common in the Cascade Range with soils being derived from andesite, basalt, and pumice. The Coast Range has Eocene-age basalts, shales, sandstones, and mudstones. The Siskiyou Mountains have rocks of the Triassic and Jurassic periods including lava flows, volcanic ash beds of andesite and basalt, sandstones, mudstones, limestones, and chert.

Soils in southwestern British Columbia and the Puget Sound area have developed mostly from unconsolidated (surficial) parent materials of Pleistocene and recent ages. Soil profiles are poorly developed and the texture and composition of parent materials is still dominant.

Fig. 11.9.   A west coast Douglas fir forest. (Photo courtesy of USDA Forest Service.)

Colluvium of various textures is common on steeper slopes throughout western Washington and Oregon. On flatter areas with more benchlike topography, deeper, less stony soils have developed from basalt, finely weathered sandstone, or shale. Many deep soils are derived from water, gravity, or airborne deposits.

The soils are leached, having been formed under high rainfall conditions and pH ranges from 4.5 to 6.5. Outside of the glaciated regions, most soils are well supplied with a clay fraction. Cation exchange capacity varies widely from 6 to 60 m.e./100 g usually averaging 20 to 60 in nonglaciated soils and 6 to 35 in glaciated areas. Most nonglaciated soils have nitrogen contents of 0.20 to 0.60 percent in the 0- to 15-cm (0- to 6-in.) depth whereas glaciated soils range from 0.05 to 0.25 percent; Siskiyou Mountain soils have only 0.05 to 0.09 percent nitrogen.

There are three major forest zones: western hemlock at the lower elevations, Pacific silver fir at higher elevations, and the Coastal Valley Zone in the rainshadow areas (Krajina 1965; Franklin and Dyrness 1973).

WESTERN HEMLOCK ZONE

This is the largest zone in the coastal forest extending from southeastern Alaska (56°N latitude) to northern California (38°N latitude) (Krajina 1965; Rowe 1972; Franklin and Dyrness 1973). The western hemlock zone extends from the Sitka spruce *(Picea sitchensis)* zone (fog belt forest) bordering the Pacific Ocean to the higher elevation Pacific silver fir zone in the Coast and Cascade Ranges. The range varies in elevation from sea level to 900 m (0 to 3000 ft) at 45°N latitude in southern areas to sea level to 300 m (0 to 1000 ft) at 55°N latitude. Precipitation varies from about 150 to 300 cm (60 to 120 in.).

Douglas fir often is the dominant tree in even-aged stands because of past catastrophic fires or logging. Western hemlock and western redcedar are the climax-dominant trees. Deciduous trees are not common but red alder *(Alnus rubra)* and bigleaf maple *(Acer macrophyllum)* do frequent specialized habitats such as riparian sites.

There are four common communities representative of the spectrum of variation in this zone. The *Tsuga heterophylla/Berberis nervosa* community is the climatic climax (Franklin and Dyrness 1973). Orloci (1965) refers to this as the *Tsuga heterophylla/Gaultheria shallon* community. Old-growth stands are predominantly Douglas fir, western hemlock, and western redcedar with succession favoring the hemlock and cedar. Major shrubs are Oregongrape *(Berberis nervosa)*, salal *(Gaultheria shallon)*, vine maple *(Acer circinatum)*, Pacific rhododendron *(Rhododendron macrophyllum)*, and red huckleberry *(Vaccinium parvifolium)* (Krajina 1965; Franklin and Dyrness 1973). Common herbs include twinflower *(Linnaea borealis)*, cutleaf goldthread *(Coptis laciniata)*, rattlesnake plantain *(Goodyera oblongifolia)*, and swordfern *(Polystichum munitum)*.

The *Tsuga heterophylla/Polystichum munitum* community is found in moister habitats where summer drought is rare. Lateral seepage from upper slope positions is often the source of the extra moisture. Overstory trees include the seral Douglas fir

Fig. 11.10.   A 500-year-old *Tsuga*/*Polystichum* stand with one large, old Douglas fir and every size class of western hemlock, vine maple, and swordfern in the understory.

and climax western hemlock and western redcedar (Fig. 11.10). There are few shrubs with vine maple being the most frequent. The understory is dominated by a lush growth of swordfern. Other herbs include Oregon oxalis *(Oxalis oregana)*, three-leaved coolwort *(Tiarella trifoliata)*, deer fern *(Blechnum spicant)*, and white inside-out flower *(Vancouveria hexandra)*.

Western redcedar is the dominant tree in wet to very wet habitats though Douglas fir and western hemlock may be present. Understory dominants include devil's club *(Oplopanax horridum)*, swordfern, lady fern *(Athyrium filix-femina)*, deer fern, and maidenhair fern *(Adiantum pedatum)*.

The *Pseudotsuga menziesii/Holodiscus discolor* community occurs on dry sites, particularly on ridges. Douglas fir is usually present in all age classes. Western hemlock and western redcedar are usually absent. Common shrubs include oceanspray *(Holodiscus discolor)*, beaked hazel *(Corylus cornuta)*, snowberry *(Symphoricarpos mollis)*, and salal.

There are many variations in these generalized communities, particularly from south to north (Franklin and Dyrness 1973). Douglas fir decreases in importance from south to north. Pacific rhododendron is frequently the shrub layer dominant in Oregon but occurs only sporadically further north. Swordfern-dominated communities are more common in the northern half of the zone. Pacific silver fir, Alaska huckleberry *(Vaccinium alaskaense)*, and rustyleaf *(Menziesia ferruginea)* occur in the western hemlock zone in British Columbia (Orloci 1965) but rarely further south.

PACIFIC SILVER FIR ZONE

This zone lies above the western hemlock and below the subalpine forest zones (Franklin and Dyrness 1973). Krajina (1965) recognized this as a wetter subzone of the western hemlock zone. The Pacific silver fir zone extends from about 44°N latitude in the Cascade Range northwards to the southern tip of the Alaska panhandle at 56°N latitude. This is the most extensive forest zone in the Olympic Mountains (Fonda and Bliss 1969). The elevational range is about 900 to 1500 m (3000 to 5000 ft) in Oregon to 450 to 1100 m (1500 to 3500 ft) in British Columbia. This zone is wetter and cooler than the western hemlock zone with annual precipitation ranging from 200 to 660 cm (80 to 260 in.) (Krajina 1965; Franklin and Dyrness 1973).

The zone varies widely depending upon stand age, history, and locale (Franklin 1965). Pacific silver fir, western hemlock, noble fir, Douglas fir, western redcedar, and western white pine are the common tree species; yellow cedar *(Chamaecyparis nootkatensis)* is a minor tree species.

Common shrubs are Alaska huckleberry, ovalleaf huckleberry *(Vaccinium ovalifolium)*, red huckleberry, rustyleaf, salal, Oregongrape, vine maple, and Pacific rhododendron, with the latter three species most common in southern parts of the range. Major herbs include deer fern, three-leaved coolwort, Oregon oxalis, queencup beadlily *(Clintonia uniflora)*, deerfoot vanillaleaf *(Achlys triphylla)*, western coolwort *(Tiarella unifoliata)*, twinflower, and common beargrass *(Xerophyllum tenax)* (Orloci 1965; Fonda and Bliss 1969; Franklin and Dyrness 1973).

COASTAL VALLEY ZONE

The interior valleys of western Oregon to the Strait of Georgia and southeastern Vancouver Island are in the rainshadow of coastal mountains to the west. This zone is drier, and warmer and has a more extensive summer drought than the adjacent western hemlock zone (Krajina 1965; Franklin and Dyrness 1973). Variation in climate, vegetation, and soils is great from south to north. Elevation ranges from sea level to 1400 m (0 to 4500 ft). Annual precipitation ranges from 50 to 130 cm (20 to 50 in.).

The zone has stands of Douglas fir, Oregon white oak *(Quercus garryana)*, natural prairie, chaparral, lodgepole pine, western white pine, and mixed conifers (Krajina 1965; Thilenius 1968; Franklin and Dyrness 1973). One of the most common communities is *Pseudotsuga menziesii/Holodiscus discolor*. Douglas fir is the major seral tree as well as the climax dominant (Fig. 11.11). Other trees include incense cedar *(Calocedrus decurrens)*, Pacific madrone *(Arbutus menziesii)*, big-leaf maple, chinkapin *(Chrysolepis chrysophylla)*, and tanoak *(Lithocarpus densiflorus)*. Shrubs include oceanspray, beaked hazel, snowberry, and salal. Whipplevine *(Whipplea modesta)* is one of the few important herbs (Bailey 1966). Thilenius (1968) recognized oak communities dominated by Oregon white oak, with the *Quer-*

Fig. 11.11. All ages of Douglas fir are present in this 475-year-old *Pseudotsuga/Holodiscus* stand. Surface fire has scorched 5 m (15 ft) of tree trunk.

cus garryana/Rhus diversiloba community most widespread and occupying the most xeric sites. Oregon white oak is often the sole dominant, but some stands are seral to Douglas fir or grand fir (Sprague and Hansen 1946). Pacific madrone and big-leaf maple may also be present. In southern Oregon and northern California, California black oak *(Q. kelloggii)* is also present in the oak woodlands. Whittaker (1960) recognized oak woodland as the driest forested formation in the Siskiyou Mountains of southwestern Oregon.

MIXED EVERGREEN ZONE

In the Klamath Mountains of southwestern Oregon and northern California is a forest of conifers and sclerophyllous broad-leaved trees. The zone is relatively warm and wet in winter and hot and dry in summer (Franklin and Dryness 1973). Annual precipitation is 60 to 170 cm (24 to 70 in.) with less than 20 percent falling during the growing season (Gratkowski 1961a). Whittaker (1960) found that geologic parent material had an important influence on vegetation. A common community is dominated by Douglas fir with sugar pine *(Pinus lambertiana)*, tanoak, Pacific madrone, and chinkapin in the tree canopy. Common shrubs are canyon live oak *(Q. chrysolepis)*, Oregongrape, Pygmy Oregongrape *(Berberis pumila)*, trailing blackberry *(Rubus ursinus)*, baldhip rose *(Rosa gymnocarpa)*, poison oak *(Rhus diversiloba)*, and salal. The herbaceous layer is not well developed.

On xeric diorite sites there are chaparral-Douglas fir communities characterized by less than 50 percent coverage of Douglas fir and closed canopies of schlerophylls (Whittaker 1960). Tanoak is the dominant scherophyll but Pacific madrone and canyon live oak are abundant.

MIXED CONIFER ZONE

Forests of Douglas fir, sugar pine, ponderosa pine, incense cedar, grand fir, and white fir *(Abies concolor)* occur at mid-elevation in the Cascade Range in southern Oregon and in the Sierra Nevada Mountains of northern California (Küchler 1964; Franklin and Dyrness 1973). Douglas fir is probably the most abundant tree but it tends to decrease from north to south in favor of the pines (Franklin and Dyrness 1973). Major understory species include Pacific rhododendron, beaked hazel, chinkapin, oceanspray, Oregongrape, and trailing blackberry.

Fire History

Low frequency, high intensity crown fires were the norm before settlement in west coast coniferous forests. Fires of 50,000 to 1 million hectares (100,000 to 2 million acres) were common (Martin et al. 1976). Lower intensity surface fires were frequent, but generally burned over relatively small areas.

Douglas fir and western redcedar can attain ages of 1000 years (Fowells 1965) but few old stands lived beyond 500 years because of wildfires. The causes of fires prior to settlement were lightning (Schmidt 1957; Schroeder and Buck 1970) and Indians (Kirkwood 1902; Johannessen et al. 1971). Lightning and people continue to be the major causes of fire. Of fires started in the west-central Cascade Range of Oregon from 1910 to 1977, 53 percent were caused by lightning and 45 percent were caused by humans (Burke 1979).

Extensive fires burned on Vancouver Island 150, 230, 310, 360, 410, 560, 760, and 870 years ago (Schmidt 1970). Eight major fires, and numerous minor ones, occurred in the Nimpkish Valley on Vancouver Island between 150 and 1100 years ago for a mean fire-free interval of about 135 years for the entire valley. In the coastal valley zone, major fires occurred in the University of British Columbia experimental forest in 1860, 1770, 1660, and 1550 (Eis 1962) for a mean fire-free interval of about 100 years. There were also numerous small fires between the major fire events. Hemstrom (1979) found a fire rotation of 382 years for forests below 970 m (3175 ft) in Mount Rainier National Park in Washington's Cascade Range. The mean fire-free interval for large fires (greater than 5% of the area) is 78 years.

Fire frequency in west coast forests is not well understood and detailed fire history work still needs to be done (Heinzelmann 1978). Fire-free intervals for the Douglas fir forest type have been estimated at from 150 to over 500 years (Franklin et al. 1981). The fire-free interval in the western hemlock forest has been estimated at greater than 150 years (Martin et al. 1976). In a higher elevation zone, Pacific silver fir requires a fire-return interval of 700 to 800 years to maintain dominance in the stand (Schmidt 1957). In Mount Rainier National Park, the fire rotation is 450 years for forests above 970 m (3175 ft).

Fire regime varies from vegetation zone to vegetation zone. There was apparently a shorter fire-return interval in the warmer, drier areas such as the coastal valley zone than in the cooler, moister western hemlock zone or in the even cooler Pacific silver fir zone. Fire was more frequent on the east side of Vancouver Island (Schmidt 1970) and on the east side of the Olympic Mountains where Douglas fir is more often the climax dominant and where lightning ignition is more common (Pickford et al. 1980). Within a vegetation zone the warmer, drier exposures have shorter fire-return intervals than cooler, moister areas such as north-facing slopes or stream-side corridors.

The most widespread forest fires frequently owe their origins to a coincidence of lightning storms and pronounced summer drought. Since the year 1300 all but two major fires at Mount Rainier coincided with a major regional drought (Hemstrom 1979). Hot, dry winds often contribute to the development of a massive fire (Cramer 1957; Countryman 1974). Easterly "foehn" winds can be exceptionally dry and blow continuously for 24 to 48 hr, sometimes at gale force (Cramer 1957). Strong east winds were blowing when the Tillamook fire, which started August 14, 1933, blew up 11 days later and burned 102,000 ha (225,000 acres) or 85 percent of the area burned, in a 30-hr period (Morris 1934).

Crown fires in west coast forests covered the greatest areas but were infrequent. Surface fires were more frequent. They were probably most common and covered greater areas in the coastal valley zone because of the drier summer fuel conditions and comparatively light fuel loadings. Surface fires were probably less common in the more moist western hemlock and Pacific silver fir zones. Low intensity surface fires accompany high intensity crown fires. Certain areas are only burned by surface fires because of weather, fuel, or topographic conditions. Bailey (1966) studied relic old-growth stands which survived within a huge 190-year-old burn. In the western hemlock zone, 63 percent of the relic stands had evidence of surface fire while in the coastal valley zone 100 percent of the stands had been burned by surface fire. On riparian habitat, in the middle of the Tillamook burn, red alder, a fire-sensitive species, survived the 1933 holocaust and subsequent 1939 and 1945 burns (Bailey and Poulton 1968). Surface fires in the stands caused fire scars during the 1933 and 1939 fires. The moist environment of riparian habitats and north-facing slopes are the main defense that western redcedar has against fire (Sharpe 1974) (Fig. 11.12).

The west coast Douglas fir forests grow in a relatively narrow band from the Alaska panhandle to northern California. The more northerly parts are cooler and moister and probably have less frequent summer droughts. Fire-return intervals are probably longer in northerly parts, intermediate in central areas, and shorter in southerly parts of each vegetation zone.

The best guess for a fire-free interval in the coastal valley zone is about 40 years, in the western hemlock zone about 200 years, and in the Pacific silver fir zone about

Fig. 11.12. Stream-side corridor of western redcedar and Douglas fir in western hemlock zone near Hope, B. C. The trees were not killed in a stand-replacing fire that burned slope in the background.

500 years. Some stands in the Pacific silver fir zone show no evidence of ever having been burned (Hines 1971). The fire-free interval will vary considerably within each zone due to topography and other factors.

Ecological Effects of Fire

The cycle from forest destruction through pioneer stages to another forest varies among the forest zones. The western hemlock zone is most widespread and has been extensively studied.

WESTERN HEMLOCK ZONE

The outline of the cycle from Douglas fir to hemlock presented in the next two paragraphs is mostly from Munger (1940). Fire or clear-cutting usually precede the initiation of a stand of Douglas fir. Many seeds in cones survive the severe crown fires of late summer. The seedlings establish despite adverse factors and gradually overtake both brush and other species of trees. As the Douglas fir continues its growth, hemlocks, cedars, and firs appear in the understory. These shade-tolerant species often occupy spaces in the understory where Douglas fir reproduction could not survive. After 100 to 200 years a conspicuous understory of shade-tolerant species is present.

During the second century, Douglas fir begins to thin out. Trees die from competition, wind, snow-breakage, insects, or disease and this allows the shade-tolerant species to expand in number and size. By the end of the second century the invading shade-tolerant conifers are conspicuous. After 300 to 400 years the shade-tolerant species may outnumber Douglas fir trees. The latter are mature and are no longer making significant growth. In 500 to 800 years the forest converts to the climax western hemlock community, barring disturbance by fire, logging, or windthrow. Western redcedar dominates wet areas along stream courses on poorly drained sites (Fig. 11.12), whereas western hemlock dominates moist but well-drained sites (Neiland 1958). Even though western hemlock establishes on the forest floor, it seems to germinate and grow best on decaying wood, either on rotting logs and stumps or on the bases of dead or living standing trees.

Douglas fir loses dominance as succession proceeds in most communities in the western hemlock zone, but it may not be completely removed from climax stands in southern parts of this zone. Douglas fir will live for more than 1000 years (Schmidt 1957; Franklin and Dyrness 1973), so only periodic regeneration is necessary to keep the species as a minor component. Bailey (1966) found Douglas fir in smaller tree size classes in five of six habitat types studied. Douglas fir probably remains a subordinate species in some climax stands depending upon windthrow sites and other forms of tree mortality to provide enough light for successful reproduction.

The pioneer stage of secondary succession has been studied by Isaac (1940), Bailey (1966), West and Chilcote (1968), Morris (1958, 1970), Dyrness (1973), and Miller and Miller (1976). After three wildfires in the Cascade Range of northern Washington there was a decline in coverage and frequency of residual trees, since mortality continued for several years (Miller and Miller 1976). The greatest postfire mortality was in western hemlock; western redcedar was intermediate and Douglas fir experienced the least postfire mortality. Seedlings of the three species emerged in all three of these comparatively small burns. Total plant cover was lowest the year after the fire. Disseminules of weedy pioneer species and of semitolerant or shade-tolerant shrubs and herbs were the source of many plants. These seedlings establish and increase rapidly during the first three years. Forest remnant species that sprout from surviving underground parts increase at a slower rate. Abundant invader species after fire are fireweed, grasses, and varied-leaved collomia *(Collomia heterophylla)*. Forest species establishing from seed included Pacific willow *(Salix lasiandra)*, trailing blackberry, Oregongrape, oceanspray, and red huckleberry. Common forest remnant species include Oregongrape, salal, and twinflower. Mosses and liverworts, especially mosses, increase rapidly. The increase in moss cover after burning is apparently common in the northern part of the western hemlock zone but uncommon in southern Washington and Oregon.

Some invader species last only a year. Woodland groundsel *(Senecio sylvaticus)* commonly dominates clear-cuts the year after slash burning (West and Chilcote 1968). It has high nutrient requirements and these are generally satisfied only during the first year after burning. Over the next few years there is an increase in the shade-intolerant herbs: fireweed, autumn willowweed *(Epilobium paniculatum)*, velvet-grass *(Holcus lanatus)*, pearly everlasting *(Anaphalis margaritacea)*, false dandelion *(Hypochaeris radicata)*, prickly lettuce *(Lactuca serriola)*, smooth hawksbeard *(Crepis capillaris)*, thistle *(Cirsium vulgare, C. arvense)*, and white hawkweed *(Hieracium albiflorum)* (Isaac 1940; West and Chilcote 1968; Dyrness 1973). Species composition and abundance of pioner species are closely related to seed availability. Seed of pioneer species are dispersed by wind over great distances. There is considerable heterogeneity in abundance of pioneer species from place to place.

The pioneer stage gradually gives way to the shrub-herb stage which is dominated by residual forest herbs or shrubs such as swordfern, vine maple, Pacific rhododendron, oceanspray, beaked hazel, Oregongrape, salal, and trailing blackberry or by invaders such as red alder, willows, and snowbrush (Fig. 11.6). Conifers that establish after natural seeding or artificial regeneration gradually assume dominance after forest destruction. Overgrazing or repeated wildfires will either retard (grazing) or kill (burning) the conifers, maintaining a cover of mostly exotic herbaceous perennials and forest remnant species.

Morris (1970) observed some differences in species composition of burned clear-cuts in the Oregon Coast Mountains versus the Cascade Mountains. Vine maple and snowbrush are more common in the Cascades while swordfern, bracken fern, Australian fireweed *(Erechtites prenanthoides)*, phacelia *(Phacelia* sp.), pearly everlasting, thimbleberry *(Rubus parviflorus)*, and salmonberry are more common in the Coast Range. Certain differences are to be expected within the western hemlock zone between northern California and northern British Columbia but they have not been studied.

Fire severity is important. Many intense wildfires occur when forest fuels are very dry; a greater proportion of the area is severely burned than in a slash fire conducted when soil and lower duff are moist. Slash fires are usually conducted after the first autumn rain and are not as severe as summer wildfire. Less than 6 percent of the area is severely burned in most slash fires (Tarrant 1956; Morris 1970). Prescribed burning of logging slash is conducted to reduce brush competition with tree reproduction, to reduce fire hazard, and to prepare seedbed or planting sites (Morris 1970). Any fire reduces competition from residual woody and herbaceous perennial species but a hot fire burning over dry duff and soil in summer will reduce competition for the longest period of time. However the seeds of certain woody species such as snowbrush, mountain whitethorn *(Ceanothus cordulatus)*, and greenleaf manzanita *(Arctostaphylos patula)* germinate in response to heat from the fire (Gratkowski 1961a, 1961b). Soil temperatures in excess of 120°C (248°F) for long periods will kill mountain whitethorn seed whereas germination is induced by heat treatment of 60° to 105°C (140° to 220°F) (Gratkowski 1974).

After burning, red alder sometimes dominates for many years on deep, moist soils particularly near the Pacific coast (Fig. 11.13). Since alder grows more quickly than

Fig. 11.13. Red alder has invaded after clear-cut logging of Douglas fir forests in the western hemlock zone, Oregon Coast Range.

Douglas fir it overtops the conifer and may reduce growth. Once Douglas fir overtops red alder its height and diameter growth may be promoted because the alder is a nitrogen fixer (Miller and Murray 1978).

Natural conifer regeneration after logging is about the same on burned and unburned areas during the first 16 years (Morris 1970). Burning reduces brush cover in most areas but total herbaceous cover was not greatly affected. Snowbrush is an exception to the general rule because it excludes conifer regeneration in some cases.

In the Cascade Range, after 25 years, the differences between slash burned and unburned logged areas is usually subtle, although marked differences do occur in some cases (Clarke 1976). An 80 percent stocking level of conifer regeneration occurred on 75 percent of burned plots but there was only a 61 percent stocking on unburned plots. There is better stocking on burned plots where salal and Pacific rhododendron are present prior to burning. Where snowbrush is present, some burned plots have little conifer regeneration.

Tree establishment is slow after a large stand-replacing wildfire, particularly if a second or third fire follows in a few decades; it often results in many open spaces and a many-aged stand. In a small area in the Tillamook Burn, Bailey and Poulton (1968) found Douglas fir stumps that ranged from 320 to 440 years old. This was apparently very common in the past. Individual tree growth patterns suggest essentially competition-free growing conditions for a century or more where age differences in old growth stands range from 50 to 100 years (Franklin et al. 1981).

In managed forests most stands are dominated by conifers 25 to 50 years after burning. Douglas fir is usually the sole dominant. There may be some western hemlock and western redcedar. The most common shrubs are vine maple, red huckleberry, and Pacific rhododendron. The shade-intolerant herbaceous invaders are eliminated from stands adequately stocked with conifers. Where trees have not established, the forest remnant species and/or perennial herbaceous and woody invaders continue to dominate the stand for decades.

From 80 to 200 years after burning there is a gradual increase in understory species as the Douglas fir trees grow taller, allowing more light to penetrate into the understory. The tall shrubs that expand include vine maple, Pacific rhododendron, oceanspray, huckleberries, blueberries, and beaked hazel. The shade-tolerant conifers also are common in the shrub layer. Major shade-tolerant herbs, including swordfern, Oregon oxalis, twinflower, and evergreen violet *(Viola sempervirens)*, expand as well. By 200 years after burning saprophytes begin to appear in the forest stands.

After logging and slash burning there is no Douglas fir on severely burned areas but higher coverage on lightly burned areas (Dyrness 1973). Severely burned areas have the lowest herbaceous cover and the lowest tree cover. They have the greatest proportion of invader species and the smallest proportion of forest remnant plants.

Undisturbed areas consistently have higher plant cover than those burned or disturbed by logging operations. Cover values for residual forest species on the undisturbed sites are much higher than on the disturbed soil sites or on the burned sites (Fig. 11.14). The major forest understory species that benefit are vine maple, Pacific rhododendron, salal, Oregongrape, swordfern, and Oregon oxalis. Fire alone, however, usually does not eliminate the residual forest species. Even the three wildfires that swept the Tillamook burn area over a period of 12 years did not elimi-

Fig. 11.14. Douglas fir, Pacific rhododendron, red huckleberry, and Oregongrape are prominent on this *Tsuga/Berberis* site nine years after clear-cutting; there was no burning. Stake is 1 m (3.3 ft) tall.

nate many residual forest species. Seventeen years after the third wildfire, vine maple, salal, red huckleberry, swordfern, and Oregon oxalis were important constitutents of many stands. The physical destruction of residual forest species through soil disturbance during tractor logging is more common than their destruction by prescribed burning or wildfire (Bailey and Poulton 1968).

Forest habitat type is an important factor in evaluating the effect of fire. The initial stages of secondary succession following wildfire or logging and slash burning tend to mask the effect of habitat type. Logged but unburned clear-cuts are dominated by forest remnant species on undisturbed sites and by invading herbaceous species on sites where the soil surface has been disturbed causing the elimination of forest remnant species. Forest habitat types can be recognized at anytime after clear-cutting on unburned, undisturbed sites. After light slash fire, it takes about three to five years for the forest remnant vegetation to recover sufficiently to exert control over the site. On the severely burned areas and on sites where the soil surface has been removed the herbaceous and woody perennial forest understory species may not reestablish in the first 20 years. Plant succession frequently goes directly from a weed or grass stage to dominance by Douglas fir saplings.

Much of the resistance to the use of slash burning is related to fire escapes, to difficulties experienced in establishing the new Douglas fir forest, and to smoke management problems. Does slash burning after logging benefit Douglas fir regeneration? The answer is not a straight forward "yes" or "no." Slash fire reduces the competition from forest remnant shrub and herbaceous species but it also exposes Douglas fir seedlings or transplants to higher soil surface temperatures (Silen 1960) and to browsing by wild ungulates. In the southwestern Oregon Coast Range, elk are attracted to a 12-year-old logged and slash-burned area (Fig. 11.15) because of the desirable southwest exposure and palatable herbaceous forages, especially false dandelion and silver hairgrass *(Aira caryophyllea)* (Bailey 1966). Douglas fir saplings are only 0.3 m (1 ft) tall in burned areas due to elk browsing but 2 m (6 ft) tall in adjacent undisturbed areas (Fig. 11.16). Elk do not prefer the adjacent unburned northeast exposure which is dominated by the forest remnant species swordfern and so the conifer regeneration is not browsed. In areas where Roosevelt elk *(Cervis canadensis roosevelti)*, Columbian black-tailed deer *(Odocoileus hemionus columbianus)*, or domestic livestock populations are high, the drier, more exposed south-facing slopes are frequently the ones more heavily burned (Fig. 11.15). They remain longer under herbaceous cover, are difficult to restock to conifers, and are preferentially grazed and browsed. Soil surface temperature on south-facing slopes can rise above the lethal tolerance levels of Douglas fir seedlings [about 49°C (120°F)] (Silen 1960). Elk or deer can maintain the grass or weed stage for 20 years or longer. Slash burning can benefit big game by providing tracts covered by palatable grasses and forbs. These plants are very desirable during spring, summer, and fall. In winter, when snow covers herbaceous vegetation, a supply of taller browse is required. If available, vine maple, red huckleberry, and salal provide most of the browse. If slash burning has set them back many years, the big game will consume whatever is exposed (Crouch 1968; Hines 1968). If that happens to be a conifer, then it will be used. There are dif-

Fig. 11.15. Elk favor this 12-year-old southwest-facing slash burn in southwestern Oregon and keep it in a perennial grass cover slowing conifer establishment and growth.

Fig. 11.16. Opposite slope to Fig. 11.15 on a northeast aspect that did not burn. Douglas fir and western hemlock are 2 m (6.5 ft) tall, swordfern and trailing blackberry are abundant, and there is little elk use.

ferences in palatability of conifers caused by genetics and environment. Western redcedar frequently has difficulty surviving because it is highly palatable. Some Douglas fir trees are more palatable than others (Radwan 1972).

In a comparison of unburned versus slash-burned treatments, Steen (1965) found that in one pair there was 92-percent stocking by natural conifers in unburned and 100-percent stocking in burned treatments after 13 years; the second pair had a dense cover of vine maple, a forest remnant, in the unburned plot and a dense cover of snowbrush, a woody invader, on the burned plot. There were twice as many conifers on the burned plot as on the unburned, vine maple-dominated plot. After 13 years one severely burned area was still in the weed stage.

Fire is a tool of use in some parts of the coastal Douglas fir zone (Isaac 1963). Clear-cutting followed by autumn slash burning (after a rain) is used in some areas to promote the rapid establishment of a new Douglas fir forest. Early logging practices included a planned severe slash burn to reduce fire hazard. The poor fire control procedures usually precipitated several reburns thus maintaining the pioneer stage. An opposing school developed that believed burning should be eliminated (McCulloch 1944). Over time, slash burning has become recognized as being effective in some areas to remove vast accumulations of logging slash and rotten wood, retard the growth of forest remnant brush species, eliminate less desirable shade-tolerant species, remove insect infestations or disease organisms, and expose more mineral soil, the favored seedbed for Douglas fir seed (Isaac 1963). Light to moderate burns are usually beneficial to Douglas fir in the western hemlock zone while extemely hot fires are almost always damaging. Reburns are usually detrimental because they destroy conifer regeneration, expose more mineral soil, and stimulate grass competition. Natural regeneration of Douglas fir will occur on both burned and unburned areas (Isaac 1963). In northern parts of this zone, moderate intensity burns are required to reduce duff and forest remnant species on north-facing slopes and flats to benefit planted Douglas fir. Slash burning may or may not benefit conifer regeneration, but it does temporarily reduce fire hazard and makes fire control easier (Morris 1970).

Slash burning may or may not be necessary for reforestation on drier upper slopes occupied by a *Pseudotsuga menziesii/ Holodiscus discolor* community. Douglas fir can regenerate on these sites of lower productivity without slash burning. Where there is heavy accumulation of logging slash, burning can reduce fire hazard. When thickets of tall shrubs are present, slash burning can set them back, allowing easier conifer regeneration.

Slash burning delays forest remnant species in *Tsuga heterophylla/Polystichum munitum* communities. On these high productivity sites competition from swordfern, vine maple, western hemlock advanced regeneration, or the shade-intolerant red alder may prevent or delay the establishment of a second-growth Douglas fir forest. Adequate forest regeneration practices are required to take advantage of the favorable seedbed created for Douglas fir by the slash burn.

Slash burning may promote the establishment of red alder on moister parts of the western hemlock zone. The value of this tree is vigorously debated. If Douglas fir is wanted, hand planting of two- or three-year-old stock is desirable. Western redcedar is well adapted to most of these sites.

Fire may be used to help reduce or eradicate disease organisms and insect pests or it may cause them. The Douglas fir beetle *(Dendroctonus pseudotsugae)* is the foremost pest of mature Douglas fir forests on the west coast. Outbreaks of the bark beetle are related to windthrow, ice breakage, and wildfire. Beetles associated with fire-scorched trees in the Tillamook burn of 1933 killed 200 million board feet (about 800,000 m³) of green timber (Furniss 1941). Downed or weakened trees attract beetles from surrounding stands concentrating them in a small area. If the beetle population in the woods is large, and the downed trees cannot accommodate the beetles, green trees within 11 m (35 ft) will be attacked (Johnson and Belluschi 1969).

The Douglas fir beetle has not caused serious damage in recently logged old-growth stands along the west coast. Clear-cut logging keeps slash accumulation away from the forest edge and leaves few attractive breeding sites for the beetles (Mitchell and Sartwell 1974).

Burning of logging slash may favor the root disease *Rhizina undulata*. High soil temperatures are necessary to germinate ascospores of this fungus. The disease is presently not a major problem but it may become more important in the future since it kills young trees (Nelson and Harvey 1974).

PACIFIC SILVER FIR ZONE

Early stages of succession have not been studied in detail in the higher elevation Pacific silver fir zone (Franklin and Dyrness 1973). Many of the weedy pioneers are the same as in the western hemlock zone. Succession proceeds from the pioneer stage to a perennial herb stage and then to a closed forest canopy of trees, usually Douglas fir, noble fir, Pacific silver fir, and western hemlock. The growth rate of tree seedlings is slow (Sullivan 1978). The *Abies amabilis/Achlys triphylla* habitat type, occurring on southerly exposures, has greater conifer height but fewer trees per hectare than the *Abies amabilis/Rhododendron macrophyllum/Vaccinium membranaceum* habitat type. Total shrub canopy cover in the 7- to 15-year-old slash-burned clear-cuts was only 11 to 44 percent and herb cover ranged from 7 to 34 percent (Sullivan 1978). Major shrubs are snowbrush, Pacific rhododendron, and Alaska huckleberry. Only beargrass and bracken fern are common in the herb layer.

Succession proceeds over hundreds of years towards a climax forest dominated by Pacific silver fir and western hemlock either singly or in combination (Hanzlik 1932; Fonda and Bliss 1969). Vast forest fires at too short a fire-return interval have probably limited the distribution of Pacific silver fir. This fire-sensitive species requires about 700 to 800 years to reenter the stand. Pacific silver fir has a heavy seed and will not disseminate great distances (Schmidt 1957). Hines (1971) found Pacific silver fir present in *Tsuga heterophylla-Abies procera/Vaccinium membranaceum/Crytogramma crispa* stands near the Oregon coast. The average annual precipitation is 500 cm (200 in.) and there is summer fog. He did not find any fire scars or charcoal in soil profiles. Pacific silver fir is absent in higher elevation Coast Range stands where average annual precipitation is 330 to 500 cm (130 to 200 in.) and summer fog is infrequent. The stands are only 400 to 450 years old. The mean fire-return interval for stands in the coastal fog belt is probably double those in the Coast Range Mountains.

Dwarf mistletoe *(Arceuthobium camylopodum)* infects western hemlock in many old-growth forests causing excessive defect and adding to residue volumes after logging (Ruth 1974). Broadcast burning helps eradicate residual dwarf mistletoe infected seedlings and residues.

Balsam woolly aphid *(Adelges piceae)* attacks several true fir species but is particularly serious in mid-elevation Pacific silver fir stands. Residual silver fir remaining after logging is attacked by the aphid. Prescribed burning is successful in destroying the Pacific silver fir regeneration and providing mineral soil for establishment of Douglas fir (Mitchell and Sartwell 1974).

COASTAL VALLEY ZONE

Succession after clear-cut logging and burning in the climax Douglas fir forest on Vancouver Island follows the same general patterns as that for the western hemlock and Pacific silver fir zones. Species characteristic of the undisturbed, mature forest associations are present on logged and unburned localities and some are present after burning (Mueller-Dombois 1959). The list of species growing on burned clear-cuts is much longer because of weedy invaders including fireweed, pearly everlasting, false dandelion, prickly lettuce, autumn willowweed, willow-herb *(Epilobium watsonii)*, hawksbeard, thistle, woodland groundsel, white hawkweed, sow thistle, and everlasting *(Gnaphalium microcephalum)*. These shade-intolerant weeds are ubiquitous in relation to soil moisture regime since light is the major controlling factor (Mueller-Dombois 1959).

One of the most common communities in the Coastal Valley zone is *Pseudotsuga menziesii/Holodiscus discolor*. Before settlement surface fires were relatively common in this type (Fig. 11.11); they burned fine fuels and rotten logs, killed shrub stems and smaller trees, and reduced competition for the surviving conifers. These fires maintained stocking level control of Douglas fir, preventing stagnation. All age classes of Douglas fir are normally present today in old-growth stands, probably as a consequence of the past surface fires. After logging and slash burning weedy pioneers such as woodland groundsel, fireweed, Australian fireweed, and pearly everlasting expand rapidly and occupy the site for several years. Gradually the forest remnant species become dominant. Common ones are oceanspray, beaked hazel, Pacific rhododendron, salal, Oregongrape, and trailing blackberry. Douglas fir will establish if a seed source is available and if partial shade is present on blackened surfaces.

Oregon white oak woodland is common in interior valleys (Thilenius 1968). Some of the communities are seral to climax Douglas fir, big-leaf maple, and grand fir forests (Sprague and Hanson 1946; Habeck 1962; Thilenius 1968). In the past grazing and Indian burning kept much of the interior valley area of Oregon and northern California in grassland, shrubland, and oak woodland (Kirkwood 1902; Johannessen et al. 1971; Franklin and Dyrness 1973). The vegetation is well adapted to fire. Some of the common plants are Columbia brome *(Bromus vulgaris)*, beaked hazel, poison oak, oceanspray, snowberry, Oregongrape, salal, and baldhip rose. It is hypothesized that open oak savannahs were maintained by fire and that fire control practices, implemented by settlers and continued to the present, are responsible for the major successional changes presently taking place (Franklin and Dyrness 1973).

The successional status of vegetation in the interior valleys of southwestern Oregon is not well known (Franklin and Dyrness 1973) nor is that of adjacent northern California. Both repeated burning and heavy grazing will favor grasses, forbs, poison oak, and some chaparral species such as wedgeleaf ceanothus *(Ceanothus cuneatus)* and white-leaved manzanita *(Arctostaphylos viscida)*. If burning is followed by grazing of goats for a period of years in parts of northern California, poison oak will dominate for at least 50 years.

Burning is not required to regenerate Douglas fir. It may actually be detrimental to tree establishment by creating conditions where soil surface temperatures are lethal to Douglas fir seedlings (Hallin 1968) and by attracting grazing animals to the succulent green vegetation that grows on a recent burn. Prescribed burning may have a place in perpetuating grazing and recreation resources. It may also be used to reduce fuel hazards associated with logging and urban and recreational developments.

MIXED EVERGREEN AND MIXED CONIFER ZONES

These zones have very complex vegetation patterns in southwestern Oregon and adjacent California because of complex geological, edaphic, and climatic patterns. Superimposed is another complex pattern of heavy grazing, burning, periodic logging, and other forms of human disturbance. Consequently successional sequences are not well understood. The chaparral and forest communities appear to be successionally related in many cases (Franklin and Dyrness 1973). Chaparral is probably climax on drier and more exposed sites because of inadequate summer moisture for tree growth (Gratkowski 1961a). On more mesic sites chaparral stands are often fire-induced seral types. Brushfields often develop after wildfire or logging. Snowbrush is one of the important invaders after logging and slash burning in the mixed conifer zone. It germinates after burning from seed stored in the soil. It may produce a favorable microclimate for conifer seedlings under some conditions (Zavitkovski and Newton 1968). It can fix nitrogen (Wollum et al. 1968) and it can shade Douglas fir seedlings preventing lethal soil surface temperatures (Hallin 1968). On other sites it may hinder conifer establishment (Franklin and Dyrness 1973).

NORTHERN INTERIOR AND ROCKY MOUNTAIN FORESTS

Vegetation, Climate, and Soils

There are extensive interior Douglas fir *(P. menziesii* var. *glauca)* forests east of the Cascade Range from central British Columbia to Washington, Oregon, Idaho, and Montana. Douglas fir is a climax dominant in a drier zone and is a major seral tree in a wetter forest zone. Generally precipitation is less, summers are warmer, and winters are colder than in coastal forests.

CEDAR–HEMLOCK ZONE

The cedar-hemlock zone is in the interior of British Columbia, northern Idaho, and northwestern Montana where the storm track of the westerlies carries oceanic influence as far inland as the continental divide (Daubenmire 1978). The forced rise of eastward moving Pacific air results in what Rowe (1972) describes as the "Interior Wet Belt" where annual precipitation ranges from 80 to 170 cm (32 to 67 in.) (Krajina 1965; Daubenmire and Daubenmire 1968). This is a high snowfall area with amounts ranging from 160 to 420 cm (75 to 265 in.) annually (Krajina 1965). The zone lies at intermediate elevations 400 to 1500 m (1200 to 5000 ft) where soil drought is not severely limiting and where there is adequate summer heat (Daubenmire and Daubenmire 1968).

Western hemlock and western redcedar are the climax tree dominants and are the largest trees in the Rocky Mountains. Bell (1965) considers western hemlock the climatic climax tree while western redcedar is the edaphic climax tree on wetter sites. Grand fir is also a climax dominant in southern parts; the *Abies grandis/Pachistima myrsinites* habitat type is adapted to the warmest and driest sites in the southern half of the zone (Daubenmire and Daubenmire 1968).

The *Tsuga heterophylla/Pachistima myrsinites* habitat type is the climatic climax community in northern parts of the zone where soils are relatively deep and drainage is good (Bell 1965; Daubenmire and Daubenmire 1968). Western hemlock reproduces effectively in the shade. Western redcedar and grand fir should be considered late seral trees in this habitat type (Daubenmire and Daubenmire 1968). Other seral trees include Douglas fir and western white pine. The understory is a rich growth of shrubs and herbs including Queencup beadlily, twinflower, mountain lover, western coolwort, big huckleberry *(Vaccinium membranaceum)* and western prince's pine *(Chimaphila umbellata)*.

Thuja/Oplopanax or *Thuja/Athyrium* are edaphic climax communities in wet lowlands. Most commonly western redcedar is the climax dominant in northern Idaho. In other stands western hemlock shares this role while in still others western hemlock alone reproduces (Daubenmire and Daubenmire 1968). The understory is usually a dense cover of devil's club or lady fern. Common understory constituents include queencup beadlily, western coolwort, sweetscented bedstraw *(Galium triflorum)*, twistedstalk *(Streptopus amplexifolius)*, Wood violet *(Viola glabella)*, and rattlesnake plantain.

GRAND FIR ZONE

There is a grand fir zone as the mid-elevation forest on the eastern side of the southern Washington and Oregon Cascade Range in the Blue and Ochoco Mountains (Franklin and Dyrness 1973) and in central Idaho. The zone is bounded by the subalpine forest at the upper limit and Douglas fir or ponderosa pine zones at the lower limit.

Major trees are grand fir or white fir, ponderosa pine, lodgepole pine, western larch, and Douglas fir (Franklin and Dyrness 1973). Ponderosa pine, Douglas fir, and western larch attain optimum growth in this zone (Daubenmire 1961; Roe 1967). Common communities are *Abies grandis/Pachistima myrsinites, Abies grandis/Vaccinium membranaceum,* and *Abies grandis/ Calamagrostis rubescens* (Hall 1967; Daubenmire and Daubenmire 1968).

DOUGLAS FIR ZONE

The most extensive forest zone ranges from near Prince George in central British Columbia southwards to northwestern Wyoming. In British Columbia this zone extends from the subalpine forest of the Cascade range to the arid grasslands in river valleys, eastward to the cedar-hemlock zone, and then occurs again in the Rocky Mountains (Rowe 1972). The Douglas fir zone occurs between ponderosa pine and subalpine forest zones in the Washington Cascade Range and occupies extensive areas in eastern Washington, Idaho, western Montana, and northwestern Wyoming (Franklin and Dyrness 1973).

Annual precipitation in the interior Douglas fir zone ranges from about 38 to 56 cm (15 to 20 in.) (Krajina 1965; Daubenmire and Daubenmire 1968; Pfister et al. 1977). Annual snowfall ranges from about 76 to 250 cm (30 to 100 in.) (Krajina 1965; Pfister et al. 1977). The zone ranges from 300 to 1200 m (1000 to 4000 ft) in British Columbia (Krajina 1965) and from grassland to 2380 m (7800 ft) in southwestern Montana (Pfister et al. 1977).

Douglas fir is the climax tree species of the zone. Lodgepole pine and aspen are common seral trees in central British Columbia. Farther south ponderosa pine, western larch and lodgepole pine are the most prevalent seral trees (Daubenmire and Daubenmire 1968).

The common representative community groups in this zone are *Pseudotsuga*/bunchgrass, *Pseudotsuga/ Calamagrostis,* and the *Pseudotsuga*/shrub group (Tisdale and McLean 1957; McLean and Holland 1958; Daubenmire and Daubenmire 1968; Brayshaw 1970; Beals 1974; Pfister et al. 1977).

The *Pseudotsuga menziesii*/bunchgrass group of communities occurs on the most xeric sites and has an understory dominated by bluebunch wheatgrass, Idaho fescue, or rough fescue *(Festuca scabrella).* The *Pseudotsuga menziesii/Agropyron spicatum* habitat type is often a topoedaphic climax on steep, exposed south- or west-facing slopes (McLean 1970; Pfister et al. 1977). The *Pseudotsuga menziesii/Festuca idahoensis* habitat type occurs at higher elevations [1700 to 2260 m (5600 to 7400 ft)] and has ponderosa pine as the most common seral tree. Douglas fir is the only tree in the *Pseudotsuga*/bunchgrass community in northern parts of the range and at higher elevations where it is too cold for ponderosa pine. Lodgepole pine is absent in these communities (McLean 1970; Beals 1974).

The *Pseudotsuga menziesii/Calamagrostis rubescens* habitat type is the most widely distributed in the interior Douglas fir zone extending from central B.C. to southern Montana (Beals 1974; Pfister et al. 1977) and from the east slope of the Cascade Range to the Grand Tetons of Wyoming (Daubenmire and Daubenmire 1968). It is the climatic climax community for the zone (Brayshaw 1970; McLean 1970; Beals

1974). Douglas fir is the only climax tree and there is a sparse shrub layer. A brilliant green sward of pinegrass dominates the understory (Daubenmire and Daubenmire 1968). Northwestern sedge *(Carex concinnoides)* has a high frequency throughout the range. Heartleaf arnica *(Arnica cordifolia)* is the most dominant forb (Tisdale and McLean 1957; Daubenmire and Daubenmire 1968; Brayshaw 1970) but is replaced by showy aster *(Aster conspicuus)* in central B. C. (Beals 1974). Elk sedge *(Carex geyeri)* may be a codominant with pinegrass *(Calamagrostis rubescens)* in Washington and Idaho (Daubenmire and Daubenmire 1968).

Several variants of the *Pseudotsuga/Calamagrostis* habitat type have been found. Daubenmire and Daubenmire (1968) proposed a bearberry *(Arctostaphylos uva-ursi)* phase to the community to include those stands which have the consistent occurrence of bearberry, high amounts of wild strawberry *(Fragaria* spp.), and one or more species of blueberry (Vaccinium spp.). Pfister et al. (1977) recognized four phases, ranging from xeric to mesic they are bluebunch wheatgrass, ponderosa pine, bearberry, and the pinegrass phase of the *Pseudotsuga/Calamagrostis* habitat type.

Ponderosa pine reaches its northern limit within the *Pseudotsuga/Calamagrostis* habitat type near Kamloops, B.C. (Tisdale and McLean 1957) while western larch reaches its northern limits in southeastern B.C. (McLean and Holland 1958). In southern parts of the range certain dry phases do not have lodgepole pine or western larch as seral trees (Pfister et al. 1977).

The *Pseudotsuga*/shrub type includes a group of habitat types with understories dominated by snowberry or ninebark *(Physocarpus malvaceus)* (Daubenmire and Daubenmire 1968; Pfister et al. 1977) and extends from central B.C. to Montana. Most extensive is the *Pseudotsuga menziesii/Symphoricarpos albus* habitat type. The understory is dominated by common snowberry, shinyleaf spiraea *(Spiraea lucida)*, and Wood and Nootka rose *(Rosa woodsii, R. nutkana)*. Seral trees are mostly ponderosa pine (Daubenmire and Daubenmire 1968) with some aspen (Brayshaw 1970). The *Pseudotsuga menziesii/Physocarpus malvaceus* habitat type is found in southern parts of the zone (Daubenmire and Daubenmire 1968). The understory consists of ninebark and oceanspray. Seral trees are ponderosa pine and western larch.

The *Pseudotsuga menziesii/Arctostaphylos uva-ursi* habitat type occurs in the Chilcotin of central B.C. (Beals 1974), southern B.C. (Brayshaw 1970), southwestern Alberta (Ogilvie 1963; Stringer and La Roi 1970) and in central Montana (Pfister et al. 1977). The stands occupy infertile, coarse sandy, or gravelly outwash soils in B.C. and warm, dry, gravelly loam to silt loam soils in Montana. The understory has an abundance of bearberry, Rocky Mountain juniper *(Juniperus scopulorum)*, common juniper *(J. communis)*, russet buffaloberry *(Shepherdia canadensis)*, wild strawberry, Missouri goldenrod, and wild onion *(Allium cernuum)* (Brayshaw 1970).

North of the Chilcotin in the Nechako region of B.C. Douglas fir gradually becomes more sparse while aspen, lodgepole pine, white spruce *(Picea glauca)*, Engelmann spruce *(P. engelmannii)*, and subalpine fir become more abundant (Rowe 1972). These forests are transitional between the interior Douglas fir zone and the subalpine and boreal forest zones. Douglas fir loses its position of prominence except in a drier open forest community. Few ecological studies have been done. Wild-fires have been frequent in the past permitting the development of extensive seral stands of lodgepole pine and aspen. The climax status of many sites is uncertain.

LODGEPOLE PINE ZONE

Lodgepole pine is a species with wide ecological amplitude that is distributed throughout western North America from the western half of the United States to the Yukon Territory (Fowells 1965). It is the ecological equivalent of jack pine *(Pinus banksiana)* in the eastern half of North America. These two species converge and intergrade in northern Alberta.

Elevation of lodgepole pine varies from sea level to 600 m (2000 ft) along the west coast, 500 to 1000 m (1500 to 3000 ft) in the Yukon Territory, from 1500 to 3000 m (4500 to 9000 ft) in Montana east of the Continental Divide, and between 2100 to 3500 m (7000 to 11,500 ft) in the central Rocky Mountains. Lodgepole pine is associated with the lower spruce-fir, cedar-hemlock, and Douglas fir zones.

Ecologically lodgepole pine is an extremely adapable species. It grows on a wide variety of soils, but grows best on moderately acid, sandy, or gravelly loams that are moist, light, and well drained. However it will grow on soils that are poorly drained or in frost pockets (Franklin and Dyrness 1973), on soils that are too cold for Douglas fir or ponderosa pine, or on soils that are too dry for spruce-fir (Pfister and Daubenmire 1975). Although the tree is usually seral to more shade-tolerant conifers, it is the apparent climax dominant in several widely separated areas of western North America. Lodgepole pine appears to be the climax dominant on dry, cold sites having coarse-textured soils on the Chilcotin Plateau in British Columbia (Muraro 1978b) as well as in frost pockets on pumice soil east of the Oregon Cascade Range (Youngberg and Dyrness 1959) and east of the Rocky Mountains in Montana and Wyoming (Fig. 11.17) (Loope and Gruell 1973; Hoffman and Alexander 1976; Pfister et al. 1977) and in Colorado (Moir 1969).

Lodgepole pine is the only tree in a climax *Pinus contorta/Calamagrostis rubescens* community on the Chilcotin Plateau. Various age, size, density, and height classes of trees are interspersed with small areas of dense, even-aged stands (Fig. 11.18) (Muraro 1978b). This is apparently the result of a history of generally low-severity surface fires at frequent intervals. Even-aged stands result from periodic stand-replacing crown fires when weather and fuel conditions allow. These stands have lower tree densities and are on less fertile sites than most seral lodgepole stands growing in the adjacent Douglas fir zone. Soils under the pine stands of the Chilcotin Plateau are mostly coarse-textured glacial tills. The elevation ranges from 900 to 1200 m (3000 to 4000 ft) (Holland 1964) and annual precipitation averages 32 cm (12 in.). Lodgepole pine occupies sites too dry for Douglas fir and too cold for ponderosa pine (Watt et al. 1979).

Lodgepole pine grows as both a climax tree and as seral to ponderosa pine and other conifers on pumice soils in Oregon east of the Cascade Range (Youngberg and Dyrness 1959; Volland 1976). Elevation ranges from 1200 to 1800 m (4000 to 6000

Fig. 11.17.   Lodgepole pine forest with a grass understory in southern Wyoming.

ft) and annual precipitation is from 38 to 76 cm (15 to 30 in.) with much occurring as snowfall. Lodgepole pine is climax on seasonably wet soils and on pumice with or without a perched water table where ponderosa pine is apparently excluded. Common understory shrubs are dwarf huckleberry *(Vaccinium caespitosum)*, western bog blueberry *(V. occidentale)*, bearberry, and antelope bitterbrush *(Purshia tridentata)* (Volland 1976).

East of the Continental Divide in Montana Pfister et al. (1977) recognized five habitat types where lodgepole pine is either climax or persistent. Stands are usually at 1500 to 2300 m (5000 to 7500 ft) on well drained upland sites in areas similar to colder parts of the Douglas fir zone and warmer parts of the subalpine zone. Understories are dominated by antelope bitterbrush on obsidian sand benchland, which is too frosty for Douglas fir and too dry for subalpine conifers; dwarf huckleberry and pinegrass; twinflower, dwarf huckleberry, blue ' huckleberry, and pinegrass; grouse huckleberry *(Vaccinium scoparium)*; and pinegrass. Soils are mostly a variety of noncalcareous parent materials with very acidic surface horizons. The *Pinus contorta/Calamagrostis rubescens* habitat type is on igneous parent material.

Patten (1969) studied lodgepole pine invading big sagebrush range *(Artemisia tridentata)* in Yellowstone National Park and considers it seral to spruce-fir. Pfister

Fig. 11.18.   Low intensity fires killed patches of lodgepole pine in fireline and background resulting in pine regeneration; thinning the stand caused frequent fire scars like the catface in center. John Muraro is skilled at prescribed burning for control of stand stocking, mountain pine beetle, and dwarf mistletoe damage on this climax lodgepole pine site, Riske Creek, B. C.

and Daubenmire (1975) interpret it as being climax lodgepole pine. There is a distinctive shrub and herb composition and the soil at 10 cm is considerably drier than under spruce-fir.

The Bighorn Mountains in north-central Wyoming have two climax lodgepole pine habitat types (Hoffman and Alexander 1976). The communities are at 2300 to 2600 m (7700 to 8600 ft) on sandy loam to silt loam soils. The *Pinus contorta/ Arctostaphylos uva-ursi* habitat type is confined to low fertility soils of granitic origin. It is the warmest, driest, and most frequently burned of the two climax lodgepole pine habitat types. Tree reproduction is likely to be sporadic due to the xeric nature of the environment. The *Pinus contorta/Vaccinium scoparium* habitat type occurs on granitic soils that are more mesic.

Lodgepole pine appears to be a fire climax, edaphic, or topoedaphic tree where it reproduces successfully. The sites are too dry, too wet, too cold, too infertile, or in some other way beyond the ecological amplitude of other coniferous species.

Fire History

The wildfire history of the northern Rocky Mountains and adjacent plateaus east of the Cascade Range summits is a study in contrasts. Fire-return interval and fire intensity generally have an inverse relationship. The vegetation having the most frequent fire-return intervals usually had the lowest fire intensity and vice versa. Fire-return intervals varied from greater than 500 years in certain moist subalpine forests to about 6 years in dry forest or grassland communities of valleys and lower exposed slopes (Arno 1976; Tande 1979; Strang and Parminter 1980). Both crown fires and surface fires occurred. Fire severity varied with vegetation type, fuel, weather, and topography. The more moist vegetation, cedar-hemlock, grand fir, and subalpine zones, commonly had catastrophic crown fires and some surface fires (Habeck 1973; Habeck and Mutch 1973; Arno 1976; Tande 1979). The drier forest zones, Douglas fir, lodgepole pine, and ponderosa pine, also had some crown fires under drought conditions, strong winds, and high fuel accumulations. Low to medium intensity surface fires, however, were most common in these forests (Loope and Gruell 1973; Arno 1976; Tande 1979; Strang and Parminter 1980). Often major fires burned at several intensities in reaction to changes in stand structure, fuel loadings, topography, and, especially, weather (Arno 1980). Under strong winds and drought conditions fires would crown where there was adequate fuel. Then, as conditions moderated, fires would creep along the ground with only occasional flareups (Muraro 1978b; Arno 1980).

Lightning is a major ignition source in this region (Habeck and Mutch 1973). Nowhere in the world are lighting fires more important than in the western forests of the United States and Canada (Taylor 1971). Most fires are small with a few fires during drought years burning large areas (Wellner 1970; Habeck and Mutch 1973). Six of 39 fires burned more than half of the study area in the Bitterroot Valley in western Montana between 1734 and 1900 (Arno 1976); only three of 72 fires burned more than half the 43,200-ha (95,000-acres) study area in Jasper National Park (Alberta) between 1665 and 1975 (Tande 1979). Tande also found that since fire suppression started in 1913, it has been very successful, putting most fires out before they reached appreciable size.

The time since the last burning affects fuel loading and fire severity. After wildfire fuel loadings are frequently dangerously high (Fig. 11.19, curve I) because of standing and fallen dead trees. As these trees decay, fuel loadings decline and again rise as the stand becomes older. In a subalpine *Abies/Pachistima* habitat type in the Cascade Range of Washington, as the stand age increased from 2 to 400 years, fuel loading increased from 60 to 380 metric tons/ha (27 to 170 tons/acre) (Fahnestock 1976). Similar fuel loadings are present in the cedar-hemlock zone. A recently logged cedar-hemlock forest had 295 to 365 metric tons/ha (132 to 163 tons/acre) (Muraro 1971b) (Fig. 11.20). Forest associations vary in their fuel loadings. There is generally more fuel on productive sites. Habeck (1976) found lower fuel loadings in mixed ponderosa pine-Douglas fir stands than in grand fir-western redcedar stands.

Fire-return intervals are highly variable. Arno (1980) has reviewed recent studies

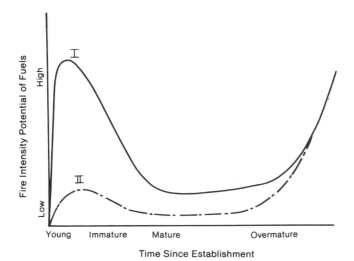

Fig. 11.19.   Fuel cycles and fire intensity potential in lodgepole pine. (Adapted from Brown 1975.)

Fig. 11.20.   A recently logged cedar-hemlock forest having heavy slash loadings.

on the forest fire history for the northern Rocky Mountains. Mean fire-return intervals ranged from 15 to 30 years in the Douglas fir zone with maximum intervals from about 35 to 60 years (Arno 1976) and minimum intervals from 1 to 10 years (Tande 1979). Warm, dry *Pseudotsuga*/bunchgrass communities generally have a fire-return interval averaging 10 years in the Chilcotin region of B.C. (Strang and Parminter 1980), 6 to 11 years in the Bitterroot Valley of Montana (Arno 1976), and 20 to 25 years in northern Yellowstone National Park (Houston 1973). The *Pseudotsuga menziesii/Calamagrostis rubescens* habitat type is very common and has a fire-return interval of about 8 to 30 years (Arno 1980; Strang and Johnson 1980). In seral ponderosa pine stands of an *Abies grandis/Calamagrostis rubescens* habitat type in eastern Oregon, the time between underburnings is about 10 years (Hall 1976).

Lodgepole pine is the climax dominant in the lodgepole pine zone and it dominates seral forests in Douglas fir, grand fir, cedar-hemlock, and subalpine forest zones. Fires are more frequent and less intense in areas having dry summers; the mean fire-return interval is 25 to 50 years (Arno 1980). Muraro (1978b) estimates the mean fire-return interval for climax lodgepole pine forests at 20 to 40 years in the Chilcotin plateau of central B.C. Lodgepole pine is not known for its ability to survive burning because its bark is not thick, but fire scars are common on trees growing in these climax stands. The tree in Fig. 11.21 survived fires at 12, 16, 19, 27, and 59 years of age for a mean fire-return interval of 9 years. Mean fire-return intervals in mostly seral lodgepole pine forests average 27 years in Jasper National Park (Tande 1979) and 90 years in lower parts of the spruce-fir zone in Kananaskis Provincial Park (Alberta) (Hawkes 1980); it may be as high as 300 years in high elevation sites where fuel accumulation is slow (Romme 1979). North-facing slopes have a longer mean fire-return interval than other aspects (Hawkes 1980).

The grand fir zone has quite a variable fire regime. Dry sites, especially steep south- and west-facing slopes, have mean fire-return intervals of 25 to 50 years, whereas cooler and moister sites have mean fire-return intervals of about 70 to 250 years (Arno 1980).

In the cedar-hemlock zone about one in fifty 400-year-old western redcedar trees, growing along streams, have been extensively burned by lightning, yet these igni-

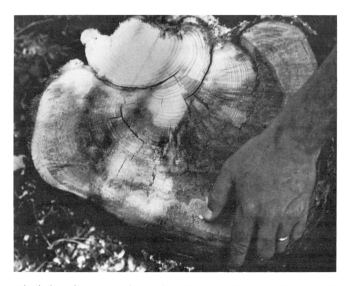

Fig. 11.21.   This lodgepole pine in Riske Creek, B. C. survived five fires at between 12 and 59 years of age.

tions failed to spread elsewhere in the grove (Habeck 1976). The mean fire-free interval for stream-side *Thuja/Athyrium* and *Thuja/Oplopanax* habitat types is greater than 200 years and fire intensities are low (Arno and Davis 1980). The fires usually have little effect on the stands. Extreme summer droughts that occur every few years, high accumulations of fuel in these productive forests, lightning or people as an ignition source, and high winds during dry periods predispose upland forests in this zone to spectacular crown fires. Uplands in the *Tsuga/Pachistima* and *Thuja/Pachistima* habitat types have mean fire-free intervals of 50 to 150 years (Arno and Davis 1980). Rugged topography and northerly exposures have the longer fire-free intervals. Preliminary findings in Idaho's Lolo National Forest indicate that mean fire-return intervals for moderate to light surface fires are from 50 to 100 years (Davis et al. 1980).

Low to medium intensity surface fire is the normal situation for mature stands in the Douglas fir, grand fir, and cedar-hemlock zones where 60 to 100 percent of the stands show evidence of surface fire after stand establishment (Arno 1980). Even under western hemlock, western redcedar, and grand fir forests, surface fire has a higher frequency than stand-replacing fire (Davis et al. 1980). Much of the fuel within the interior western hemlock zone is too wet to burn most of the time. Periodic droughts are the only times when large quantities of fuel are dry enough to burn and spectacular fires do occur due to ignition by lightning or people.

Ecological Effects of Fire

There are differences among trees in their response to fire depending on the vegetation zones and ecosystems occupied. Surviving trees are generally those with thick bark. A smaller percentage of less resistant trees will survive low intensity fires. These survivors were probably favored by some combination of reduced fuel loadings, light winds, or localized moist fuel conditions which caused a reduction in flame length and burning time.

Succession after fire depends upon many factors: vegetation zone, habitat type, topography, soil, degree of exposure, climatic variables, season burned, intensity of fire, soil moisture status during and after the burn, and propagules available. Succession follows the same general patterns outlined earlier (Fig. 11.5).

All trees within dissemination distance may seed into the bare area left after a stand-replacing fire so that within a decade or so trees are well represented in the seral vegetation (Daubenmire and Daubenmire 1968). If a second fire occurs before the seral conifers reach cone-bearing age, the coniferous tree component will be eliminated and seral herb or shrub stages will remain for decades (Fig. 11.22).

CEDAR–HEMLOCK ZONE

After small fires western redcedar and western hemlock usually establish because of the readily available seed source. On severely burned and denuded areas or where burns have been repeated lodgepole pine and western larch are abundant. Pure stands

Fig. 11.22. A seral brush field in northern Idaho caused by multiple fires in *Thuja/Pachistima* and *Thuja/Athyrium* habitat types on southwestern exposures, northern exposures, and stream bottoms. (Photo courtesy of University of Idaho.)

of lodgepole pine are common on gravelly sites. Western larch is usually more abundant on north and east exposures where there is less evapotranspiration stress.

Western white pine and Douglas fir often reproduce well after single fires, whereas lodgepole pine and western larch are most abundant after repeated fires. In the absence of fires, Douglas fir will establish itself under the seral stands of lodgepole pine and western larch and gradually overtake them.

Even-aged stands of western white pine thrive best on the well drained, fertile soil of protected eastern or northern gradients, lower slopes, bottoms, and benches (Larsen 1929). In these locations the soil remains moist throughout the summer. Seedlings of western white pine have a prodigious taproot which is supplemented in later life by a deep, wide-spreading system of lateral roots (Fowells 1965). This makes the trees unusually windfirm. Western white pine grows rapidly and generally overtops its competitors (Larsen 1929).

Western larch and lodgepole pine are generally found as seral species at the upper elevations of the cedar-hemlock zone and are extremely well adapted to fire; lodgepole pine frequently overlaps into the spruce-fir zone (Fig. 11.23). Lodgepole pine can produce some seed at 10 years of age and after 25 years it can produce heavy seed crops every two to three years (Fowells 1965). Seeds are often held in serotinous cones until heated by a fire. Both of these characteristics enable lodgepole pine to regenerate quickly after fire. Western larch, on the other hand, is favored by a remarkably long life and a very high degree of resistance to injury from fire. It possesses an unusually thick bark and sparse foliage that is less flammable than lodgepole pine; larch crowns are high and narrow. Mature larch trees will usually survive several

Fig. 11.23.   Lodgepole pine is the major tree regenerating after wildfire in this high elevation spruce-fir site in the Canadian Rockies.

fires. The trees mature in 300 to 400 years and may live for as long as 900 years (Fowells 1965). They produce heavy seed crops every five to seven years.

The climax trees (western hemlock, western redcedar, and grand fir) and seral trees (western white pine, Douglas fir, western larch, and lodgepole pine) invade a new burn simultaneously provided seed sources are available (Daubenmire and Daubenmire 1968). The seral trees grow faster and soon overtop the slower growing climax tree species.

The climax species respond to local differences in habitat. Western redcedar requires wet, well aerated soils; hemlock needs moist soil without free water, although sexual reproduction in an established forest appears to be somewhat dependent upon decaying logs and stumps for germination substrates (Habeck and Mutch 1973); grand fir thrives with much less moisture than either western redcedar or western hemlock. Therefore these species in a climax forest are grouped according to variations in soil moisture. Western redcedar occurs in greatest abundance near the streams on well watered soil. Western hemlock is usually found on moist flats and on lower slopes, while grand fir prefers the rolling lowland and favorable south slopes.

Fire control in this forest zone has reduced the production of younger seral communities and is contributing to an increase of intermediate- to old-aged stages of succession (Habeck and Mutch 1973). These successional developments will produce high accumulations of fuel in the future, which means that diligent fire control will be required. It is doubtful whether such control can continue to be effective, because of the predictable occurrence of lighning in these hazardous fuels (Habeck and Mutch 1973) during the periodic summer droughts.

The largest areas of fire in the cedar-hemlock zone have been catastrophic, stand-replacing crown fires during unusually dry summers. These fires are intense and can consume the duff when it is dry, leaving an exposed mineral soil. In spite of the high fire intensity, the floristic composition of postburn communities is predominantly derived from plants or seeds already present at the time of the fire (Lyon and Stickney 1976). Much of the understory vegetation is adapted to fire; many species sprout or germinate from seed stored in soil or duff and establish quickly after fire. In the catastrophic Sundance fire in northern Idaho, only 16 percent of the species forming the first-year vegetation were from off-site sources (Lyon and Stickney 1976). Of the on-site sources, 33 percent were from rhizomes, 24 percent from seed or fruit, and 23 percent from root crowns. Root crowns are the most important fire survival mechanism for tall shrubs while rhizomes are the most important for low shrubs and herbaceous plants. Of the tall shrubs, Scouler willow *(Salix scouleriana)*, Sierra maple *(Acer glabrum)*, and serviceberry were most prominent. Low shrubs that survive as rhizomes include spiraea, snowberry, and thimbleberry. Adapted herbaceous species include lupine, arnica, and bracken fern.

On mesic sites there is essentially complete coverage by the fourth year after the burn (Lyon and Stickney 1976). Species which establish from windblown seed include hawkweed, fireweed, and scouler willow. Heat treatment stimulates germina-

Table 11.3.   Effect of Fire on Shrub Frequency (%) in the Cedar-Hemlock Zone.

Species	None, Closed Stands	Logged, No Burn	Logged, Piled and Burned	Single Broadcast Burn	Multiple Broadcast Burn
Decreasers					
Lonicera utahensis	52	60	56	37	25
Menziesia glabella	19	16	22	17	9
No Change					
Vaccinium spp.	69	71	67	55	57
Pachistima myrsinites	54	88	90	85	67
Rosa gymnocarpa	40	45	45	38	40
Amelanchier alnifolia	22	26	43	27	34
Acer glabrum	38	28	26	36	33
Increasers					
Ceanothus velutinus	0	0	8	24	18
Prunus emarginata	0	6	9	30	27
Alnus sinuata	1	7	6	19	30
Ribes spp.	4	6	22	12	40
Salix scouleriana	2	16	44	75	80
Spiraea lucida	8	17	49	50	50
Ceanothus sanguineus	6	12	19	32	26
Rubus parviflorus	32	61	61	83	73

Adapted from Mueggler (1965). Copyright 1965, the Ecological Society of America.

tion of on-site seed sources of snowbrush, redstem ceanothus *(Ceanothus sanguineus)*, elderberry *(Sambucus caerulea, S. racemosa)*, and gooseberry *(Ribes lacustre, R. viscosissimum)* (Table 11.3) (Daubenmire and Daubenmire 1968; Lyon and Stickney 1976).

Shrubs in the cedar-hemlock zone sprout prolifically following fire (Mueggler 1965; Lyon and Stickney 1976). These include Scouler willow, Sierra maple, snowbrush, serviceberry, oceanspray, alder *(Alnus sinuata)*, spiraea, mountain lover, thimbleberry, and huckleberry. Redstem ceanothus, bittercherry *(Prunus emarginata)*, syringa, and cascara *(Rhamnus purshiana)* are less active sprouters. Disturbance by fire harms more herbs than it benefits. All decreasers are forest species that decline in frequency and cover after logging and burning or after burning only. Trail plant *(Adenocaulon bicolor)* declines greatly because of increased light as well as because of burning. Pinegrass is a particularly effective invader after multiple fires.

Slash burning (Fig. 11.24) is a common practice after clear-cut logging to remove fine fuels and duff. The depth of duff removal is controlled to a degree through skilled manipulation of fuel, duff and soil moisture, and weather conditions.

After repeated burns some areas in the cedar-hemlock zone remain in the seral shrub stage for 50 years or so (Mueggler 1965; Lyon 1971) (Fig. 11.22). Repeated burns are detrimental to the reproduction of conifers but provide excellent wildlife habitat. This forest zone has many shrubs palatable to elk and deer, such as serviceberry, redstem ceanothus, snowbrush, willow, and mountain lover.

Fig. 11.24.   Slash burning is conducted in the cedar-hemlock zone to reduce fire hazard from high fuel loadings and to reduce duff depth promoting the regeneration of desired conifers. At least moderately intense fires are required to reduce depth of duff, although larger logs remain.

As the tree canopy increases willow, redstem ceanothus, snowbrush, and bittercherry decrease (Mueggler 1965). Mountain lover, huckleberry, honeysuckle, thimbleberry, and mountain maple remain as common shrubs.

GRAND FIR ZONE

Fire has been of high enough frequency in many stands to limit grand fir to the status of a minor subordinate. After decades of vigorous fire suppression, grand fir and Douglas fir are becoming of greater importance under moist sites dominated by ponderosa pine (West 1969; Habeck 1976). In the Selway-Bitterroot wilderness of northern Idaho Habeck (1976) found that grand fir, or a grand fir-white fir hybrid, is expanding into warmer, dry sites, higher, cooler sites, and even wetter stream-side sites dominated by western redcedar.

Either ponderosa pine, lodgepole pine, western larch, or Douglas fir may dominate seral stands. The shade-tolerant grand fir seeds in gradually and establishes in the understory. The most common seral tree is Douglas fir (Daubenmire and Daubenmire 1968). The successional sequence of grand fir sites in eastern Washington and Idaho is very similar to that of western hemlock sites (Daubenmire and Daubenmire 1968). Logging alone has a negligible effect on shrub cover but logging followed by burning reduces shrub cover about 50 percent the first year after fire (Pengelly 1963). Most shrubs sprout the first year after fire. Redstem ceanothus germinated from seed stored in the soil and reached 23 percent cover the first growing season after fire.

Repeated wildfires in the early 1900s created massive brushfields that serve as winter range for elk. Wildlife managers have studied the effect of prescribed burning on seral brushfields of the *Abies grandis/Pachistima myrsinites* habitat type in Idaho (Leege and Hickey 1971) (Fig. 11.25). Principal shrubs include scouler willow, redstem ceanothus, mountain maple, serviceberry, cascara, bittercherry, syringa *(Philadelphus lewisii)*, and oceanspray. Most shrubs sprout prolifically from root crowns or rhizomes. Spring burning is conducted soon after snowmelt so the soil is moist. Spring burning promotes more sprouting than fall burning. Spring burning after snowmelt is easier to control, less expensive, and produces more sprouts but fewer shrub seedlings; fall burning stimulates more seed germination of redstem ceanothus and bittercherry.

Low intensity, frequent surface fires were the rule in the *Abies grandis/ Calamagrostis rubescens* habitat type in the Blue Mountains of eastern Oregon (Hall 1976). Pinegrass is a continuous sward wherever light is adequate. When dry this fuel ignites readily during a dry lightning storm. The fire burns over the ground to the base of a tree where there is an accumulation of needles. If it is a fire-resistant tree such as ponderosa pine without a large accumulation of dry needles and branches, the tree will survive. Fire burned about once every 10 years, maintaining an open stand of trees because of periodic thinning. Ponderosa pine would have been most favored by fire. It is hypothesized that thickets of ponderosa or lodgepole pine or Douglas fir would have burned as a crown fire. In the open savannahs these crown fires would

Fig. 11.25. Three months after spring burn on a southwest-facing slope of *Abies grandis/Pachistima myrsinites* habitat type near Selway River, Idaho. Sprouting shrubs are mostly thimbleberry, paper birch, serviceberry, bittercherry, willow, and redstem ceanothus. (Photo courtesy of University of Idaho.)

change to surface fires burning the dry pinegrass fuels. Fire selectively killed grand fir and Douglas fir favoring ponderosa pine. Ponderosa pine develops fire-resistant bark when saplings are about 5 cm (2 in.) in diameter while grand fir and Douglas fir bark remain green and photosynthetically alive up to 10 cm (4 in.) in diameter (Hall 1976).

Fire suppression is favoring grand fir and Douglas fir over ponderosa pine. Thickets of conifers have developed. Fuels are accumulating setting the stage for stand-replacing fires. Without underburning the *Pinus ponderosa/Calamagrostis rubescens* community is changing to the *Abies grandis/Calamagrostis rubescens-Arnica cordifolia* community. Lack of underburning is apparently retarding growth of ponderosa pine because of too high a stocking of trees and saplings. With a denser tree overstory, cover of pinegrass, elk sedge, wild strawberry, arnica, balsam woolleyweed *(Hieraceum scouleri)*, and snowbrush are declining. The productivity of the community for timber, wildlife, and livestock is declining under a no-burn policy (Hall 1976).

DOUGLAS FIR ZONE

Historical fires kept forests open providing much herbage and browse for grazing animals. Fire suppression has favored tree regeneration at the expense of shrubs, grasses, and rapid tree growth (Fig. 11.26).

Fires delayed successional replacement by Douglas fir (Habeck and Mutch 1973) and perpetuated the dominance of the seral trees ponderosa and lodgepole pine, western larch, and aspen. Fire protection and selective logging are accelerating natural trends toward elimination of ponderosa pine (Franklin and Dyrness 1973). Ponderosa pine is often the most preferred and productive timber species.

Fig. 11.26. Snow King Mountain, Jackson, Wyoming in February 1918 (top), 39 years after the 1879 fire, and in February 1978 (bottom), illustrating the marked increase in forest cover. (Photo courtesy of USDA Forest Service.)

Fig. 11.27. In Kootenay National Park, B. C., the larger Douglas fir trees (center) survived a stand-replacing fire that swept uphill on the left. Steeper, more dissected topography and or weather conditions probably contributed to the patchy burning by stand-replacing fire (right) which left numerous survivors.

Topography, fuels, and weather influence burning conditions and subsequent forest succession (Fig. 11.27). More dissected topography and avalanche tracks do not favor widespread stand-replacing fire; the many patches of survivors act as seed sources for burned areas.

The *Pseudotsuga*/bunchgrass group of communities usually has an incomplete cover of herbaceous species. Ponderosa pine is the common seral tree in the southern stands. The xeric sites are not conducive to the development of a dense tree cover and frequent fire will favor grass over trees for decades. Surface fires were frequent wherever fuels were adequate; there were probably few crown fires. Now, dense stands of young conifers are very common, a ladder of fuels of live and dead trees reach from the ground to treetops (Fig. 11.28). Most wildfires will be stand-replacing crown fires. Douglas fir is invading grasslands throughout the Douglas fir zone (Fig. 11.29).

The *Pseudotsuga*/*Calamagrostis* communities have had frequent fires. It is hypothesized that surface fires kept the understory open in a parklike appearance, burning thickets as crown fires, burning lodgepole pine stands killed by mountain pine beetles in stand-replacing fires, and promoting the establishment of another dense lodgepole pine forest. Lightning-caused fires probably often smoldered for several months. Light surface fires and moderate to severe crown fires were often experienced in a single large burn, depending upon weather and fuel conditions (Muraro 1971a, 1978b; Arno 1980) (Fig. 11.27). Common shrubs, such as baldhip rose, spiraea, russet buffaloberry, and serviceberry are well adapted to fire and occur in all seral stages as well as in climax stands (Tisdale and McLean 1957). Pinegrass is al-

Fig. 11.28. A Douglas fir/bunchgrass stand in Kootenay National Park, B. C. that has too high a stocking of young trees for good tree growth, for most wildlife, range, or aesthetic purposes.

Fig. 11.29. Douglas fir is invading this bluebunch wheatgrass grassland in Radium, B. C.

ways present; bearberry, Richardson's sedge *(Carex richardsonii)*, and twinflower increase in older stands because they are more fire susceptible.

The effect of fire on 30 common shrubs and herbs in the *Pseudotsuga/ Calamagrostis* habitat type has been studied by McLean (1969) in 1-, 4-, 12- and 19-year-old wildfires compared with 90-year-old lodgepole pine and 250-year-old Douglas fir stands. Plants can be rated for relative fire resistance based on rooting characteristics (Table 11.4). Generally plants with reproductive organs above the duff layer are most susceptible (i.e., white hawkweed, bearberry, twinflower) while

Table 11.4. Relative Fire Resistance of Thirty Species in the Douglas Fir Zone Classified by Root System.

Group	Description and Species	
A	Fibrous roots, no rhizomes	
	White hawkweed *(Hieracium albiflorum)*	s
	Yellow hawkweed *(H. umbellatum)*	s
B	Fibrous roots and stolons	
	Bearberry *(Arctostaphylos uva-ursi)*	s
	Strawberry *(Fragaria glauca)*	s
	Twinflower *(Linnaea borealis)*	s
· C	Fibrous roots and rhizomes growing mostly in duff or between duff and mineral soil	
	Northern sedge *(Carex concinnoides)*	s
	Rattlesnake plantain *(Goodyera oblongifolia)*	s
	Sidebells pyrola *(Pyrola secunda)*	s
	Grouse whortleberry *(Vaccinium scoparium)*	m
	Yarrow *(Achillea millefolium)*	m
D	Fibrous roots and rhizomes growing mostly 1.5–5 cm below mineral soil surface	
	Heartleaf arnica *(Arnica cordifolia)*	s
	American vetch *(Vicia americana)*	m
	Creamy peavine *(Lathyrus ochroleucus)*	m
	Fireweed (Epilobium angustifolium)	m
E	Fibrous roots and rhizomes mostly 5–13 cm below mineral soil surface and showing signs of being able to regenerate from those depths	
	Western Prince's Pine *(Chimaphila umbellata)*	s
	Bunchberry *(Cornus canadensis)*	m
	Snowberry *(Symphoricarpos albus)*	r
	Spiraea *(Spiraea betulifolia)*	r
F	Species without rhizomes but which have taproots	
	Buffaloberry *(Sheperdia canadensis)*	m
	Indian paintbrush *(Castilleja miniata)*	m
	Mountain lover *(Pachistima myrsinites)*	m
	Lupine *(Lupinus arcticus)*	r
	Timber milkvetch *(Astragalus miser* var. *serotinus)*	r

From McLean (1969).
Key: s—susceptible; m—moderately resistant; r—resistant.

plants with reproductive organs buried in the mineral soil that could sprout after fire from that depth are resistant to fire. Western prince's pine is an exception, however. Extent of duff reduction affects conifer seedling establishment. In Montana, western larch is favored where little duff is left (Shearer 1975), whereas Douglas fir germinates and survives better on moderate amounts of duff (Norum 1976).

Ponderosa pine and western larch are normally the only seral trees in the *Pseudotsuga menziesii/Physocarpus malvaceus* habitat type in eastern Washington and northern Idaho (Daubenmire and Daubenmire 1968). Snowbrush invades this site after burning (Pengelly 1963) wherever a seed source is stored in the soil. The *Pinus ponderosa/Physocarpus malvaceus* and *Pseudotsuga/Physocarpus* habitat types have similar floristic and edaphic characteristics except for the climax tree dominants (Daubenmire and Daubenmire 1968). It may be that past fire history played a role in removing Douglas fir seed sources from stands now classified as *Pinus/Physocarpus*. Repeated intense fires in this most mesic ponderosa pine habitat type in eastern Washington and northern Idaho may have eliminated Douglas fir centuries ago.

LODGEPOLE PINE FORESTS

The importance of lodgepole pine in seral and climax stands merits a special section dealing with its relationship to fire.

In the classical sense we think of lodgepole pine as a seral species that follows fire in the Douglas fir, cedar-hemlock, and spruce-fir vegetation zones. However it is a very complex species that can behave as a minor seral, dominant seral, prolonged seral, persistent, or climax species (Fig. 11.30). Its ability to grow on sites unsuitable for other trees, in cold climates where fires are frequent, or where cone serotiny persists plays an important role in its distribution (Lotan 1975a).

Where fires are relatively infrequent, occurring less than every 200 years, lodgepole pine is a minor seral species (100 years) on moist sites in the Douglas fir and cedar-hemlock zones (Pfister and Daubenmire 1975). It is a major seral species (100 to 200 years) in the spruce-fir zone where succession progresses slower. In these cases lodgepole pine and other conifers establish themselves concurrently as mixed stands of species. Lodgepole pine grows rapidly and gains early dominance. Then the shade-tolerant associates persist and assume dominance as individual lodgepole pines die from mountain pine beetle, root fungi, dwarf mistletoe, commandra blister rust, or competition from shade-tolerant species.

In the prolonged seral, persistent, and climax stages lodgepole pine is clearly the dominant species and there is little evidence to indicate that the tree will be succeeded by other species (Clagg 1975). They may all be climax stands because of soils, microclimate, or fire. Since many of these stands occur on rolling topography with slopes less than 30 percent (Moir 1969) and most are less than 100 years old (Lotan 1975b), we are of the opinion that many stands, though not all, have been maintained by fire. Level to rolling topography lends itself to wide ranging, sweeping fires much more frequently than steep, dissected topography. Lightning fires are frequent in the western United States and Canada (Taylor 1971), which makes this theory very plau-

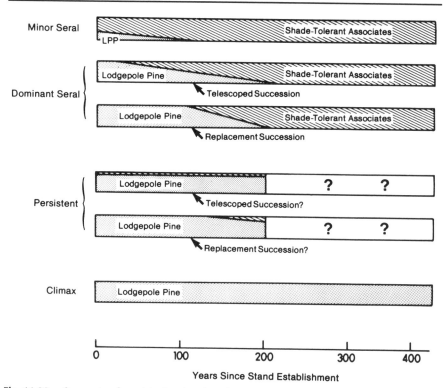

Fig. 11.30. Successional models showing the relative proportion of lodgepole pine and shade-tolerant conifer associates over time. (From Pfister and Daubenmire 1975.)

sible. Fire recurred in most lodgepole pine stands that originated with fire (Brown 1975). Frequent fires over time (100 years) would keep the climax species suppressed and gradually eliminate them from the stand. The high incidence of cone serotiny in the Intermountain Region (Lotan 1975a) further supports this theory.

The persistent stands may or may not be fire-induced. Since these stands are dominated by lodgepole pine possessing nonserotinous cones, one can reason that fires have not occurred for a long time or that the stands have only been subjected to repeated fires of a very low intensity. In our judgment these stands could easily be considered climax.

The extensions of lodgepole pine on cold sites or frost pockets in the Douglas fir and ponderosa pine zones, on dry sites in the spruce-fir zone, and on poorly drained sites are edaphic or microclimatic climaxes.

Lodgepole pine is known to have serotinous cones, but cone serotiny varies widely from one geographic area to another (Critchfield 1957; Crossley 1956; Tackle 1961), and more recent evidence has shown that serotiny also varies greatly within local regions of the Rocky Mountains (Lotan 1967, 1968, 1975a). Critchfield (1957) found that most cones of lodgepole pine along the Pacific Coast and in the Sierra Nevada and southern Cascade Ranges are nonserotinous. In the northern Cascades and the Blue Mountains, the habitat is variable (Mowat 1960). Obviously none of the lodgepole pine stands of the Rocky Mountains appear to have only serotinous cones and many have less than 50 percent serotinous cones. Lotan (1975a) concluded that young trees a few decades old have a greater tendency to bear open cones than older trees. This is a unique adaptation of the tree to fire. Dominant trees have a greater percentage of closed cones than suppressed trees. Where fire is absent for a long time in stands of pure lodgepole pine there is a shift from a dominance by trees with serotinous cones to one by those with nonserotinous cones, since a greater proportion of the new trees come from seed of nonserotinous cones. However, serotinous cones will also release seed without the aid of fire when the cones eventually reach the ground and direct sunlight melts the resinous bond. The bond will melt when temperatures reach 45° to 50°C (113° to 122°F) (Cameron 1953). Thus there always seems to be a certain percentage of lodgepole pine trees with serotinous cones in stands but the percentage can shift back and forth over time. This phenotype seems to be highly selective by fire regime (Perry and Lotan 1979).

Lodgepole pine produces abundant seed crops every two to three years. Serotinous cones remain on the trees for 15 to 20 years, enabling the trees to accumulate large quantities of seed. These seeds remain viable for 30 years (Lotan 1975a).

Under natural conditions heat from a forest fire starts the cone opening. After the passage of fire, seed is released on the exposed mineral soil. Competition is minimal and new seedlings quickly establish themselves the next spring. Reproduction is often so dense that the stand soon stagnates. Growth is usually slow, but persistent, and maturity is attained in about 80 years in the cedar-hemlock zone but takes much longer in cold, dry habitats; maximum age is about 450 years (Critchfield 1957). Insects and diseases cause high mortality in 100- to 200-year-old trees resulting in an accumulation of dead material on the forest floor making the stands highly susceptible to another fire.

A wide variety of burning conditions exist during most large wildfires in lodgepole pine (Muraro 1971a). Fire behavior during damp weather or night burning has a different effect on the regenerative capacity of the site compared to hot, dry, windy daytime conditions. For example, at night winds usually decrease, humidity rises, and the tendency for the fire to burn as a crown fire will decrease. The fire will creep and smoulder on the surface in an erratic fashion, exposing mineral soil on shallow, dry sites and leaving a residual duff layer on moister sites. Mortality is primarily from root and stem damage and a thinned stand may result. Temperature requirements for serotinous cones are not met. Good seedbed conditions for lodgepole pine are spotty and the fire has made no pine seed available. With the onset of daytime conditions, flame heights will increase, surface fire will cover the ground more uniformly, and crowning of individual trees will become more general, dependent on the availability of ground fuels and the proximity of the crowns to the ground fuels. Flame heights will increase with increasing wind or fuel until all the crowns are involved. A fully developed crown fire will develop unless the crowns are too widely spaced. Tree mortality will be complete in the crown fire and mineral soil exposure will be uniform except as influenced by differing microsites. Temperatures in the crown fire will be high enough to open serotinous cones without neccessarily becoming lethal to the seed. Lodgepole pine seedbed requirements are generally satisfied and an abundance of seed is available. If burning conditions continue to improve, fire behavior is characterized by a greater combustion of aerial fuels, needles, and branches up to 1 cm (⅓ in.) and occasionally cones are consumed. Temperatures lethal to seed are general, the organic layer is completely removed except for local wet sites, and physical alteration of the surface mineral soil may occur on dry sites. Areas subjected to this type of fire generally present a favorable seedbed except for the most severely burned sites where viable lodgepole pine seed has been destroyed. The result of these three levels of fire behavior is, in the first case, a thinned stand, in the second, an extremely dense stand of regeneration, and in the third, a low density stand of regeneration.

The effect of variable fire intensities on understory species in the lodgepole pine zone is not as well known. In general shrubs and perennial herbs will not be as affected by lower intensity fires whereas the severe crown fires will kill all but the most resistant species if the soil is dry. Annual and perennial herbs may occupy severely burned sites for prolonged periods of time if no tree seed source or live shrub root crowns remain.

Fire behavior influences the fuel complex for many years after the burn (Fig. 11.19). In Curve I, a moderate crown fire, the initial blowdown, and a high protracted rate of lodgepole pine downfall as roots decompose ensure that these areas will experience heavy ground fuel loadings for many years (Muraro 1971b). Fuel loadings will decline as the forests mature and then build up sharply as trees become overmature and die. The area of moderate intensity is the first to regenerate but presents a fuel complex most conducive to destruction by another fire. If a second wildfire occurs during drought conditions within 15 to 20 years, chances are excellent that a severe site virtually devoid of tree growth will result due to total destruction of the limited seed available in stands of this age (Muraro 1971b).

In the area of high intensity fire (Fig. 11.19, Curve II) greater fuel depletion ensures that blowdown is complete within a relatively short time. Decomposition will be more rapid because of proximity of branchless material to the ground.

Dwarf mistletoe *(Arceuthobium* spp.) can damage lodgepole pine, ponderosa pine, Douglas fir, western larch, and western hemlock (Alexander and Hawksworth 1975) in Douglas fir and associated forests. Wildfire is probably the primary factor governing the distribution and abundance of dwarf mistletoes. Relatively complete burns tend to sanitize the infested stands because trees usually reinvade burned areas faster than the mistletoe (Jones 1974). Thinning, selective cutting, windthrow, insects, diseases, and light and moderate intensity burns tend to create favorable conditions for dwarf mistletoe intensification. Consequently, the disease has reached high levels in many second-growth stands (Alexander and Hawksworth 1975). The disease has also reached high levels in many very old stands of lodgepole pine.

Fire has played an important role in determining the present distribution of dwarf mistletoe in lodgepole pine. Stand-replacing fires that killed all trees also killed the mistletoe whereas wildfires that left infected surviving trees were not effective in sanitizing the stand. The tree mortality, stunted trees, witches' broom, resin-infiltrated stem cankers and accumulated dead fuels tend to increase fire behavior and flammability potential. The mistletoe enhances vertical fuel continuity and the likelihood that ground fires will burn out individual tree crowns (Brown 1975).

The maintenance of seral lodgepole pine stands by intermittent fire has insured that dwarf mistletoe seed sources are not eliminated, that climax stands of nonsusceptible trees are not developed, and that genetic sources of resistant lodgepole pine are not enhanced (Alexander and Hawksworth 1975). The control of wildfires has probably resulted in an increase of dwarf mistletoe in lodgepole pine forests.

If there is an adequate supply of serotinous cones, clear-cutting followed by slash burning is the most practical method to keep many lodgepole pine stands healthy and productive (Lotan 1975b). Uneven aged management or partial cutting is seldom practical because lodgepole pine trees are not windfirm and often blow down after being thinned (Lotan 1975b). Moreover, in the Rocky Mountain area partial cutting encourages the spread and intensification of dwarf mistletoe (Alexander and Hawksworth 1975; Lotan 1975b). Thus Lotan suggests that "where management considerations preclude clearcutting, the choice may be limited to leaving the area uncut!"

On the Chilcotin Plateau of central B.C., all multistoried lodgepole pine stands are infected with dwarf mistletoe. The extensive low intensity fires that have occurred there periodically for centuries thin the stand, prepare a seedbed in places, and permit regeneration where there is sufficient light. The young trees are in turn infected with dwarf mistletoe from overstory remnants (Muraro 1978a). Where intense fires have burned, killing all trees and regenerating even-aged, denser lodgepole pine stands, there is little dwarf mistletoe. Clear-cutting, limbing before forwarding to insure adequate distribution of fuel, then burning with prescribed fire all slash and killing all advanced regeneration, followed by seeding if necessary, will insure that the next generation of pine is free of dwarf mistletoe.

Fire, lodgepole pine, and its insect and disease organisms are intimately related. The two major damaging agents are mountain pine beetle *(Dendroctonus ponderose)* and dwarf mistletoe *(Arceuthobium americanum)* (Alexander 1974).

Mountain pine beetles have been active in lodgepole pine ecosystems for as long as these trees have existed (Roe and Amman 1970). Beetles show strong preference for large-diameter trees which are usually over 80 years of age (Safranyik et al. 1974; Amman et al. 1977). They favor large, vigorous trees having a wide phloem layer; it comprises their feeding and breeding habitat. Beetles may kill only a few trees or in major outbreaks, decimate entire stands. Outbreaks usually last eight or nine years. Greater losses usually occur in pure lodgepole pine stands and at mid-elevations where conditions are favorable for brood rearing.

In the northern Rocky Mountains lodgepole pine has some commercial value at the higher elevations where 40-cm (16-in.) trees may only have a 35-percent insect attack. It has very low commercial value in the Douglas fir zone because the mountain pine beetle attacks at an early age and only 25 percent of the older stands can be harvested. Usually at these lower elevations all trees over 30 cm (12 in.) in diameter have been attacked by beetles.

Mountain pine beetles overshadow all other insects as a cause of fuel in lodgepole pine forests (Brown 1975). On medium sites after a beetle epidemic about 130 to 200 metric tons/ha (60 to 90 tons/acre) of beetle-killed boles and crowns accumulate. These great quantities of dry fuel increase the probability of intense stand-replacing fires. Lodgepole would normally regenerate after such a fire from seed stored in any surviving serotinous cones. The relationship of lodgepole pine to mountain pine beetle and fire appears to favor perpetuation of lodgepole pine (Brown 1975) and so also perpetuate a habitat for the mountain pine beetle. In the cooler, higher-elevation subalpine forests where fire frequency is much lower, beetle kill of lodgepole pine favors acceleration of succession towards a spruce-fir climax.

Habitat types reflect differences in environment; beetle behavior and survival differs among habitat types (Roe and Amman 1970). In the upper-elevation *Abies lasiocarpa/Vaccinium scoparium* habitat type, 64 percent or more of the 40-cm (16-in.) trees can be expected to survive whereas in other mid-elevation *Abies lasiocarpa* habitat types studied by Roe and Amman (1970) only 25 percent or less will survive.

Throughout the lodgepole pine forests from British Columbia to Colorado there will be continued beetle kill in mature stands of lodgepole pine. Decades of fire suppression have enabled many lodgepole pine stands to reach near-maturity and diameter classes attractive to both beetles and loggers. A century or more ago most stands would have burned by now. Outbreaks of the beetle can be expected to be more serious in the near future because there has probably never been as much mature, even-aged lodgepole pine as there is at present.

Prescribed fire is an effective tool in achieving land management objectives in multi-aged, climax lodgepole pine stands on the Chilcotin Plateau of central B.C. (Muraro 1978b). Fire has been used to control mountain pine beetle, for sanitation or stand rehabilitation in areas of dwarf mistletoe infection, and for precommercial thinning. All applications of fire take advantage of the perpetual seed supply available in lodgepole pine stands and the ability to release varying amounts of seed through fire manipulation.

In general, most of the seral species in the lodgepole pine zone are not palatable to game, and this vegetation type does not seem to have much potential for game except as hiding areas for elk during the hunting season. Huckleberries are the primary shrubs palatable to mule deer *(Odocoileus hemionus hemionus)* (Wallmo et al. 1962). On clear-cuts in Idaho, forage reaches its maximum production of about 1100 kg/ha (1000 lb/acre) in about 10 years and declines to that of the control [330 kg/ha (300 lb/acre)] in about 20 years (Basile and Jensen 1971).

CENTRAL AND SOUTHERN ROCKY MOUNTAIN FORESTS

Douglas fir does not achieve the magnificent proportions of its coastal relative in the central and southern Rocky Mountains (Fig. 11.31). It has a number of shortcomings as a timber species. The Douglas fir zone occupies a significant place sandwiched between the ponderosa pine and spruce-fir zones. There is no cedar-hemlock zone in

Fig. 11.31.   A Douglas fir forest at 2570 m (8400 ft) in Yellowstone National Park that is aesthetically pleasing but has little commercial value as mechantable timber. (Photo courtesy of USDI National Park Service.)

this region and the lodgepole pine zone is restricted to Wyoming and northern Colorado.

The central Rocky Mountains include mountainous regions in southern Idaho and Wyoming, Utah, and most of Colorado. The southern Rocky Mountains are in southern Colorado, New Mexico, and Arizona.

Vegetation, Climate, And Soils

The Douglas fir zone and the montane zone are essentially synonymous. Pure stands of climax Douglas fir are fairly rare. Much of the zone is occupied by seral brush-dominated communities, aspen, and lodgepole pine. Climax stands of aspen and lodgepole pine are also common in the central Rocky Mountains.

The geology of the Rocky Mountains is extremely complex. Igneous, volcanic, glaciated, and sedimentary parent materials are present. Soils are as variable as the parent material. The coniferous forests tend to be on podzolic soils (Turner and Paulsen 1976).

Douglas fir is characteristically the climatic climax species that lies above the ponderosa pine zone and below the spruce-fir zone in the central Rockies. At lower elevations of 1800 to 2350 m (6000 to 7700 ft) climax stands of Douglas fir are usually on mesic sites of steep north-facing slopes with deep soils (Marr 1967). Douglas fir intergrades with ponderosa pine at the lowest edges and on dry sites within the zone but it is the climatic climax type at 2440 to 2750 m (8000 to 9000 ft) on all aspects (Marr 1967). At upper elevations Douglas fir intergrades with the subalpine forest and it is an important seral tree in lower parts of the spruce-fir zone. Elevation of the Douglas fir zone varies from 2300 to 2750 m (7500 to 9000 ft) and precipitation ranges from 50 to 75 cm (20 to 30 in.) (Marr 1967).

Lodgepole pine extends southward through Wyoming into central Colorado in climax forests and as seral constituents of Douglas fir, aspen, and spruce-fir climax communities. There is a lodgepole pine zone in Colorado above the Douglas fir zone and below the spruce-fir zone at about 2440 to 2870 m (8000 to 9400 ft) (Moir 1969). Lodgepole pine was discussed in the previous section and that information essentially applies to the central Rocky Mountains.

ASPEN FORESTS

Aspen occupies much of the Douglas fir zone in the central and southern Rocky Mountains. It varies from scrubby trees at low elevations along creeks flowing through sagebrush or ponderosa pine to large trees on flats and moist slopes in the Douglas fir zone. Aspen may also be a seral tree in lower parts of the spruce-fir zone (Fig. 11.32).

Aspen grows under a wide variety of environmental conditions but its range is related to relatively cool, dry summers and winters with deep snow packs. Annual precipitation varies from 40 to 100 cm (16 to 40 in.) and elevation ranges from 2000 to 3200 m (6500 to 10,500 ft) (Mueggler 1976). Aspen grows on a wide variety of soils ranging from rocky talus slopes to deep, heavy clays. Better stands are usually on deep, moist loamy soils. Aspen usually occupies soils higher in organic matter, moisture, and pH than adjacent conifers (Hoff 1957). Conifers are more competitive than aspen on rocky sites but they have a difficult time becoming established on clay sites.

Fig. 11.32.   Aspen will last one generation and be replaced by Engelmann Spruce in the spruce-fir zone of southern Wyoming.

Temperature seems to be the main factor limiting the distribution of aspen. The upper elevational limit of aspen is associated with a mean annual temperature of 2°C (35°F); the lower elevational limit is associated with a mean annual temperature of 7°C (45°F) (Baker 1925).

Aspen has generally been considered a seral species able to dominate after fire for one generation until replaced by less fire-enduring but more shade-tolerant conifers. This is true in many areas throughout the southern, central and northern Rocky Mountains. However there are areas of optimum aspen development in Colorado and Utah where conifer invasion is either nonexistent or so slow that over 1000 years of fire-free conditions are required for aspen stands to progress to a conifer climax (Fig. 11.33) (Mueggler 1976). Uneven age distribution of the aspen trees reveals that they are able to perpetuate under certain conditions in the absence of fire or cutting. Stable aspen stands can be found from British Columbia to New Mexico.

The understories of stable aspen stands are usually dominated by a rich mixture of forbs and grasses. A common habitat type in Colorado and Wyoming is *Populus tremuloides/Thalictrum fendleri* (Hoffman and Alexander 1980). Major understory species include hairy brome *(Bromus ciliatus)*, elk sedge, blue wildrye *(Elymus glaucus)*, meadow rue *(Thalictrum fendleri)*, American vetch *(Vicia americana)*, vetchling *(Lathyrus leucanthus)*, sweet-anise *(Osmorhiza* spp.), Richardson geranium *(Geranium richardsonii)*, licoriceroot *(Ligusticum porteri)*, western yarrow, larkspur *(Delphinium* spp.), Engelmann aster *(Aster engelmannii)*, and wild strawberry. Thurber fescue *(Festuca thurberi)*, Kentucky bluegrass, mountain brome *(Bromus marginatus)*, Wheeler bluegrass *(Poa nervosa)*, slender wheatgrass, bluebells *(Mertensia leonardi)*, lupine *(Lupinus argenteus)*, and cow parsnip *(Heracleum lanatum)*

Fig. 11.33.   Stable aspen stands such as this one in southern Colorado are common in parts of Utah and Colorado.

are important on other aspen sites (Costello 1944). One habitat type with a shrub understory is *Populus tremuloides/Symphoricarpos oreophilus*. Two other common shrubs are serviceberry and Wood rose.

Productive aspen ranges produce from 670 to 1570 kg/ha (600 to 1400 lb/acre) of forage depending on openness of the stand. Releasing competition from dense stands of aspen can increase forage from 450 kg/ha (400 lb/acre) to 1350 to 2000 kg/ha (1200 to 1800 lb/acre) (Ellison and Houston 1958). These communities are very important watersheds and summer ranges for deer, elk, moose, and domestic livestock.

MIXED CONIFER FORESTS

Mixed conifer forests occupy sites that are moister than the ponderosa pine zone but are drier and usually at lower elevation than the spruce-fir zone in Arizona, New Mexico, and southern Colorado. The species are a blend from the Douglas fir and spruce-fir zones (Fig. 11.34) (Alexander 1974; Jones 1974). The elevational boundaries of the mixed conifer and spruce-fir forests shift with geographic location. Most mixed conifer forests occur from 2440 to 3050 m (8000 to 10,000 ft) where annual precipitation varies from 50 to 76 cm (20 to 30 in.). Common overstory species include Douglas fir, ponderosa pine, white fir, Engelmann spruce, aspen, southwestern white pine *(Pinus strobiformis)*, blue spruce *(Picea pungens)*, and corkbark fir *(Abies lasiocarpa* var. *arizonica)*, often more or less in that order of abundance (Alexander 1974). The order of shade-tolerance is Engelmann spruce, corkbark fir > white fir > Douglas fir > blue spruce > southwestern pine > ponderosa pine > aspen (Jones 1974). The predominant overstory species are Douglas fir and white fir, but some mixed conifer forests are seral to spruce-fir forest (Alexander

Fig. 11.34. Mixed conifer forest near Cloudcroft, New Mexico.

1974). At lower elevations and drier sites Douglas fir, ponderosa pine, and southwestern pine are climax tree species, but with increase in elevation and soil moisture status dominance shifts towards combinations of Douglas fir, blue spruce, white fir and Engelmann spruce (Moir and Ludwig 1979). There are differences in tree dominance among mountain ranges.

There are a number of habitat types in mixed conifer forests. The three described below are representative examples and are major ecosystems in Arizona and Colorado (Moir and Ludwig 1979).

The *Picea pungens/Carex foena* habitat type is common at 2620 to 2770 m (8600 to 9100 ft) in central and northern Arizona. Blue spruce is the climatic climax tree. Douglas fir is a strong codominant in one phase with an understory of sedge. There is also a ponderosa pine phase that has an understory of mountain muhly *(Muhlenbergia montana)*. Aspen is common in the pine phase. Fire has played a major role in this habitat type.

The *Abies concolor-Pseudotsuga menziesii/Quercus gambelii* habitat type is widespread in New Mexico and Arizona and southern Colorado at elevations of 2340 to 2900 m (7700 to 9500 ft), generally on western, southern, and eastern exposures. Either or both white and Douglas fir dominate forest regeneration. Gambel oak is the common understory tree or shrub. Various grasses are present in the three phases. Ponderosa pine, a seral tree, is the most important timber species. The *Abies concolor-Pseudotsuga menziesii/Acer glabrum* habitat type is common in cool, moist canyons and on uplands for much of the mixed conifer forest in New Mexico and adjacent Arizona. It is one of the wettest and coolest habitat types and is found at 2400 to 3000 m (7900 to 10,000 ft). Either or both white and Douglas fir may dominate the forest regeneration after fire. Ponderosa pine is minor but Sierra maple is always present. Aspen is an important seral tree.

Fire History

Fire has been of major importance in forests of the central and southern Rocky Mountains. Most forests established after fire disturbance.

Before settlement fire maintained an open condition in climax Douglas fir stands of the central Rockies through ground fires that occurred every 25 to 100 years (Loope and Gruell 1973).

There was plenty of evidence of past fires in aspen forests in Utah in the 1920s (Baker 1925). On flats where grasses intermingled with aspen, recurrent surface fires spread with ease. Light fires occurred every 7 to 10 years before the arrival of Europeans. Soon after settlement the fires were frequent and large. With the advent of grazing the quantity of fine fuel was reduced and fires became fewer and fewer and have been of little consequence from about 1920 to the present.

Most mixed conifer stands in the southern Rocky Mountains were established after fire (Moir and Ludwig 1979). Arizona and New Mexico have the highest frequency of lightning fires in the United States and Canada (Schroeder and Buck 1970). Indians started many fires (Cooper 1960). The drier, lower elevation habitat types dominated in climax by white fir, Douglas fir, and ponderosa pine and in seral stages by ponderosa pine burned about every 5 to 12 years (Weaver 1951). On cooler, moister sites such as the *Abies concolor-Pseudotsuga menziesii/Acer glabrum* habitat type, the mean fire-return interval would be longer. Fires would be either light and erratic due to wet fuels or intense stand-replacing fires during dry years, because of the heavier fuel loadings. In the White Mountains of Arizona the mean fire-return interval is 36 years (Dieterich 1980). Further investigation has revealed more fire events, reducing the mean fire-return interval to 22 years (J. H. Dieterich, pers. comm.). Considerably more research is required before one can make reasonable extrapolations for the area as a whole.

Ecological Effects of Fire

Historically, ground fire was most common in the central Rockies with the occasional crown fire where there was a combination of heavy fuels and extreme fire weather conditions. Today many stands have not been burned for over 100 years. In dry, windy weather crown fires will kill most trees because of heavy fuel accumulations and the continuous ladder of fuel from ground to treetop. If fire suppression continues for another 50 to 100 years the wildfires will be devastating and the Douglas fir seed source will be inadequate in some areas for a long time (Loope and Gruell 1973).

For best establishment and growth Douglas fir needs a minimum of competition and some shade (30 to 40%) during the day (Ryker 1975). The shade relieves temperature and moisture extremes. Thus it would seem that a burned area with logging debris and a nearby seed source would favor establishment of Douglas fir. Aspen resprouts vigorously after fire from underground rootstocks and can serve as a nurse crop for Douglas fir, ponderosa pine, and lodgepole pine if conifer seed is available (Marr 1967; Morgan 1969). Lodgepole pine will succeed aspen and Douglas fir will succeed lodgepole pine in classical ecological situations where all three species are found together. However this does not always happen. Many ecotypes of aspen are adapted to moist situations, heavy soils, or drought sites and will remain as pure stands.

Very few forage plants grow beneath old stands of Douglas fir. Heartleaf arnica may be one of the most prevalent species, but the dense shade and the thick duff layer minimizes the chance for many forb or shrub species to become established.

The seral communities of lodgepole pine and aspen are very important in the Douglas fir zone and each merits special attention. The lodgepole pine community was discussed previously. Aspen is discussed below.

ASPEN FORESTS

Human interference with the natural fire cycles threatens the existence of aspen in many parts of the Douglas fir zone. Douglas fir is invading aspen groves such as the one in Fig. 11.35 and will continue to do so since the fire-return interval is now much

Fig. 11.35. Continuous Douglas fir forests on north-facing slopes and grasslands and aspen groves on south-facing slopes are common scenes from British Columbia to Colorado as in this scene in Aspen Grove, B. C.

longer because of fire suppression. Aspen and Douglas fir will continue to invade these wheatgrass and fescue grasslands wherever soils and moisture are adequate. Historically in the Rocky Mountains, aspen forests were burned at frequent intervals (Baker 1925). Grazing by domestic livestock and big game are keeping the quantities of fine fuels at a low level. Aspen leaf fall alone is a poor fuel.

Aspen stands are deteriorating in many parts of the central and southern Rocky Mountains. Aspen is plagued with diseases (Schier 1975) and is overbrowsed by game and livestock (Baker 1925; Krebill 1972; Gruell and Loope 1974). It is seral in many stands and is gradually being replaced by conifers. Some aspen stands are deteriorating and reverting to grassland (Morgan 1969; Jones 1974). Overbrowsing can hasten this conversion to grassland (Schier 1975).

Aspen responds vigorously after fire because of the combined effect of removal of apical dominance and increased soil temperatures (Maini and Horton 1966; Steneker 1974). The minimum, optimum, and maximum temperatures for sucker formation after the breaking of apical dominance are 15°, 24°, and 35 °C (60°, 75°, and 95 °F), respectively (Maini and Horton 1966). Moderate fires which kill the tree canopy and eliminate part of the litter and duff are most effective in stimulating suckering (Horton and Hopkins 1965).

At high elevations or high latitudes where summer growing temperatures are marginal, soil temperature may be the limiting growth factor. In these areas increased soil temperature with or without fire, not canopy kill, may be the most important factor which stimulates new aspen suckers (Bailey and Wroe 1974). At high elevations in Colorado the absence of fire is apparently favoring aspen groves. They are expanding into Thurber fescue grasslands (Langenheim 1962).

Prescribed burning stimulates sucker production in decadent aspen (Bartos and Mueggler 1979). The density of aspen suckers increases one and two years after fire to about 30,000/ha (14,000/acre) and then natural mortality starts to thin suckers by about the third year. Fire may temporarily lower herbage and browse production the first year following burning but there is greater production in subsequent years. Fire favors forbs and grasses over woody species for several years. Fireweed, lupine *(Lupinus parviflorus)*, and slender wheatgrass *(Agropyron trachycaulum)* are some of the major increasers after burning in northwestern Wyoming.

Fire can easily be used to maintain healthy stands of aspen for watersheds, aesthetic values, or game management. However, we do not know much about the effect of fire on grasses and forbs.

Prescription techniques have not been worked out for burning aspen stands in the Rocky Mountains, although we know that aspen is more difficult to burn than most conifer species (Fechner and Barrows 1976). To regain the original prevalence of aspen in these forests it appears that better grazing management and the reintroduction of fire will be necessary.

MIXED CONIFER FORESTS

Wildfires have played an important role in the composition and structure of mixed conifer forests (Alexander 1974; Jones 1974). Generally primeval fires were light, varying in intensity depending on fuel buildup and weather conditions. This created a situation whereby some areas did not burn, some burned intensely as crown fires, and most burned lightly leaving the large resistant trees, killing shrub topgrowth, and removing dead fuels.

Ponderosa pine and Douglas fir are the most resistant to fire. White fir, southwestern white pine, and aspen are intermediate in susceptibility to fire. Least resistant and easily killed are corkbark fir, Engelmann spruce, and blue spruce. Thus Douglas fir and ponderosa pine usually predominate in the overstory while spruce and corkbark fir may predominate in the understory. White fir and southwestern white pine are often fire-scarred and defective, while patches of even-aged aspen are common, often marking hot spots. The variability of fire severity always insures a conifer seed source.

Following stand-replacing fires, initial stages of succession are usually forbs, grass, aspen, or oak scrub (Alexander 1974; Hanks and Dick-Peddie 1974). Aspen is the most common seral tree after fire on most habitat types (Moir and Ludwig 1979). Gambel oak regenerates after fire on some sites. Sometimes ponderosa pine restocks burns directly. Grasses frequently dominate drier sites after logging or burning and conifer establishment is slow; grass is very competitive with trees. Where seed is available Douglas fir and southwestern white pine will invade pioneer ponderosa pine stands. Aspen and Gambel oak are invaded by shade-tolerant conifers if there is a seed source.

Natural fires affected the abundance of ponderosa pine and aspen in the *Picea pungens/Carex foena* habitat type. Most were surface fires but crown fires developed during dry years where there were high fuel loads and canopies were closed. Aspen suckers rapidly developed and achieved dominance over pioneer forbs in a few years.

There were frequent natural fires in the *Abies concolor-Pseudotsuga menziesii/ Quercus gambelii* habitat type (Hanks and Dick-Peddie 1974). After a hot fire, forbs last a year or more before new oak sprouts assume dominance. Herbaceous sage *(Artemisia ludoviciana)* is the most common forb. Natural succession from oaks to conifers is very slow. The oak stage maintains dominance for about 50 to 100 years before there is substantial reforestation with conifers. Ponderosa and southwestern white pine are important seral trees although both Douglas fir and white fir may establish within oak thickets. Any recurring light fires will kill young Douglas and white fir. This habitat type is important for wildlife because of the favorable supply of browse and herbage. Fire can probably be used to add to the wildlife habitat values.

Natural fires were probably light, erratic, and infrequent in the *Abies concolor-Pseudotsuga menziesii/Acer glabrum* habitat type (Moir and Ludwig 1979). The normally wet, cool conditions do not favor fire. A few stand-replacing crown fires would occur during exceptionally dry years.

The erratic intensities of past fires may account for the variable structure in today's forests—a cluster of aspen growing on a localized hotspot, mixed aspen and conifers, an open canopy of old Douglas fir, and sapling thickets of mixed aspen and Douglas fir all on the same slope (Jones 1974). Light, erratic fires probably favor several diseases and insect pests. Dwarf mistletoes are the most difficult disease problem in mixed conifer forests (Jones 1974). The most serious is Douglas fir dwarf mistletoe *(Arceuthobium douglasii)*, which infects Douglas and corkbark fir, and southwestern dwarf mistletoe *(A. vaginatum)*, which infects ponderosa pine. Douglas fir beetles kill many mature and overmature Douglas fir trees. Mistletoe-infected trees become more susceptible to bark beetles. The most damaging outbreaks of bark beetles are associated with drought.

From a theoretical standpoint burning following clear-cutting should not be necessary in the mixed conifer forests because of the climax mixture of trees. In many cases it is not necessary. However, keeping the forest healthy and productive, and the fire hazard low, requires that a resource manager use the complete arsenal of management techniques. Dwarf mistletoe and bark beetle infestations may be common in overmature forests. Not letting the forest become overmature can solve some problems. Most diseases are species-specific and can be controlled by harvesting the infected trees. In other cases spot clear-cutting and burning may be necessary to minimize a disease or bark beetle infestation.

The future uses of prescribed fire in mixed conifer forests are summarized as follows: as a follow-up to clear-cutting where aspen dominance is to be established or perpetuated; as a means of rejuvenating aspen stands for aesthetic and wildlife purposes where clear-cutting is not feasible; as a sanitation measure against major pests (bark beetles, defoliators, western budworm) and diseases (dwarf mistletoe and trunk rot) following clear-cutting of parts of stands; where an overstory predominance of ponderosa pine or Douglas fir is to be maintained against aggressive invasion by more shade-tolerant species; as a means of reducing fine fuels in stands resistant to fire using moderate burning conditions; for seedbed preparation during development of a cone crop after seed cutting in the shelterwood system; and as a seminatural process in wilderness areas (Jones 1974).

The role of prescribed burning in fire hazard abatement is unclear. Periodic controlled burning is an effective means of reducing wildfire hazard in ponderosa pine forests (Weaver 1951; Kallander 1969; Wagle and Eakle 1979). Mixed conifer forests lie at higher elevations where critical fire weather is infrequent. Fuel loadings are higher in mixed conifer forests averaging 98 metric tons/ha (44 tons/acre) versus 49 metric tons/ha (22 tons/acre) in ponderosa pine forests (Sackett 1979). Although primeval wildfires were light and fire intensity patchy in mixed conifer forests, more stand-replacing crown fires can be expected in the future. Decades of fire exclusion have resulted in increased fuel loadings. Well developed understories of shade-tolerant conifers are common (Jones 1974) and will act as fuel ladders from ground to treetop. It is not certain that forest management be aimed toward these species. If it is, fire will be excluded. Hazard abatement burns have been conducted in some mixed conifer stands (Weaver 1951; Kallander 1969). Considerable skill is required in conducting these burns (Buck 1971). Tests have shown that controlled burning is a crude tool for fuel reduction but it can be refined considerably (Jones 1974). Hazard abatement burns will probably become an integral part of forest management in at least the drier parts of the mixed conifer forests.

REFERENCES

Albini, F. A. 1976. Estimating wildfire behavior and effects. USDA For. Serv. Gen. Tech. Rep. INT-30. Intermt. For. and Range Exp. Stn., Ogden, Utah.

Alexander, R. R. 1974. Silviculture of central and southern Rocky Mountain forests: A summary of the status of our knowledge of timber types. USDA For. Serv. Res. Paper RM-120. Rocky Mtn. For. and Range Exp. Stn., Fort Collins, Colo.

Alexander, M. E., and F. G. Hawksworth. 1975. Wildland fires and dwarf mistletoes: A literature review of ecology and prescribed burning. USDA For. Serv. Gen. Tech. Rep. RM-14. Rocky Mtn. For. and Range Exp. Stn., Fort Collins, Colo.

Amman, G. D., M. D. McGregor, D. B. Cahill, and W. H. Klein. 1977. Guidelines for reducing losses of lodgepole pine to the mountain pine beetle in unmanaged stands in the Rocky Mountains. USDA For. Serv. Gen. Tech. Rep. INT-36. Intermt. For. and Range Exp. Stn., Ogden, Utah.

Arno, S. F. 1976. The historical role of fire on the Bitterroot National Forest. USDA For. Serv. Res. Paper INT-187. Intermt. For. and Range Exp. Stn., Ogden, Utah.

Arno, S. F. 1980. Forest fire history in the Northern Rockies. *J. For.* **78**:460–465.

Arno, S. F., and D. H. Davis. 1980. Fire history of western red cedar/hemlock forests in northern Idaho. In Proc. Fire History Workshop, Oct. 20–24, 1980, Univ. Ariz., Tucson. USDA For. Serv. Gen. Tech. Rep. RM-81. Rocky Mtn. For. and Range Exp. Stn., Fort Collins, Colo., pp. 21–26.

Bailey, A. W. 1966. Forest associations and secondary plant succession in the southern Oregon Coast Range. Ph.D. Diss., Oregon State Univ., Corvallis.

Bailey, A. W., and C. E. Poulton. 1968. Plant communities and environment interrelationships in a portion of the Tillamook Burn, northwestern Oregon. *Ecology* **49**:1–13.

Bailey, A. W., and R. A. Wroe. 1974. Aspen invasion in a portion of the Alberta parklands. *J. Range Manage.* **27**:263–266.

Baker, F. S. 1925. Aspen in the Central Rocky Mountain region. USDA Bull. No. 1291. Washington, D.C.

Bartos, D. L., and W. F. Mueggler. 1979. Influence of fire on vegetation production in the aspen ecosystem in western Wyoming. In M. S. Boyce and L. D. Hayden-Wint (eds.) *North American Elk: Ecology, Behavior and Management.* Univ. Wyoming, Laramie, pp. 75–78.

Basile, J. V., and C. E. Jensen. 1971. Grazing potential on lodgepole pine clearcuts. USDA For. Serv. Res. Paper INT-98. Intermt. For. and Range Exp. Stn., Ogden, Utah.

Beals, C. E. 1974. Forest associations of the southern Caribou zone, British Columbia. *Syesis* 7:201–233.

Bell, M. A. M. 1965. The dry subzone of the interior western hemlock zone. Part 1. Phytocenoses. In V. J. Krajina (ed.) *Ecology of Western North America*. Univ. British Columbia, Dept. of Botany, Vancouver, pp. 42–56.

Bray, J. R., and E. Gorham. 1964. Litter production in forests of the world. In J. B. Cragg (ed.) *Advances in Ecological Research 2*. Academic Press, London, pp. 101–151.

Brayshaw, T. C. 1970. The dry forests of southern British Columbia. *Syesis* 3:17–43.

Brown, J. K. 1975. Fire cycles and community dynamics in lodgepole pine forests. In D. M. Baumgartner (ed.) *Management of Lodgepole Pine Ecosystems Symposium Proceedings*. Wash. State Univ. Coop. Ext. Serv. Pullman, pp. 429–456.

Buck, B. 1971. Prescribed burning on the Apache National Forest. In Proc. planning for fire management symposium Dec. 9–11, 1971. Southwestern Interagency Fire Council. Univ. Ariz., Tucson, pp. 165–172.

Burke, C. J. 1979. Historic fires in the central western Cascades, Oregon. M. S. Thesis. Oregon State Univ., Corvallis.

Cameron, H. 1953. Melting point of the bonding material in lodgepole pine and jack pine cones. Can. Dept. Resour. and Dev., For. Res. Div. Silvicult. Leafl. 86. Ottawa, Ont.

Clagg, H. B. 1975. Fire ecology in high-elevation forests in Colorado. M. S. Thesis. Colorado State Univ., Fort Collins.

Clarke, E. H. 1976. Residue research—recent developments. In Proc. Northwest Forest Fire Council Annual Meeting, Olympia, Wash., pp. 80–86.

Cooper, C. F. 1960. Changes in vegetation, structure, and growth of southwestern pine forests since white settlement. *Ecol. Monogr.* **30**:129–164.

Costello, D. 1944. Important species of the major forage types in Colorado and Wyoming. *Ecol. Mongr.* **14**:108–134.

Countryman, C. M. 1974. Can southern California wildland conflagrations be stopped? USDA For. Serv. Gen. Tech. Rep. PSW-7. Pac. Southwest For. and Range Exp. Stn., Berkeley, Calif.

Cramer, O. P. 1957. Frequency of dry east winds over northwest Oregon and southwest Washington. USDA For. Serv. Res. Paper No. PNW-24. Pac. Northwest For. and Range Exp. Stn., Portland, Ore.

Critchfield, W. B. 1957. Geographic variation in *Pinus contorta*. Maria Moors Cabot Found. Pub. 3.

Crossley, D. I. 1956. Fruiting habits of lodgepole pine. Can. Dept. North Aff. and Nat. Resourc., For. Res. Div. Tech. Note 35. Edmonton, Alberta.

Crouch, G. L. 1968. Forage availability in relation to browsing of Douglas-fir seedlings by black-tailed deer. *J. Wildl. Manage.* **32**: 542–553.

Daubenmire, R. 1959. *Plants and Environment*. 2nd ed. Wiley, New York.

Daubenmire, R. 1961. Vegetative indicators of rate of height growth in ponderosa pine. *For. Sci.* **7**:24–34.

Daubenmire, R. 1978. *Plant Geography*. Academic Press, New York.

Daubenmire, R., and J. B. Daubenmire. 1968. Forest vegetation in eastern Washington and northern Idaho. Wash. State Agric. Exp. Sta. Tech. Bull. 60. Pullman.

Davis, K. M., B. D. Clayton, and W. C. Fischer. 1980. Fire ecology of Lolo National forest habitat types. USDA For. Serv. Gen. Tech. Rep. INT-79. Intermt. For. and Range Exp. Stn., Ogden, Utah.

Dieterich, J. H. 1980. The composite fire interval—a tool for accurate interpretation of fire history. In Proc. Fire History Workshop Oct. 20–24, 1980, Univ. Arizona, Tucson. USDA For. Serv. Gen. Tech. Rep RM-81. Rocky Mtn. For. and Range Exp. Stn., Fort Collins, Colo., pp. 8–14.

Dyrness, C. T. 1973. Early stages of plant succession following logging and burning in the western Cascades of Oregon. *Ecology* **54**:57–69.

Eis, S. 1962. Statistical analysis of several methods for estimation of forest habitats and tree growth near Vancouver, B.C. Univ. of British Columbia, Faculty of For. Bull. No. 4.

Ellison, L., and W. R. Houston. 1958. Production of herbaceous vegetation in openings and under canopies of western aspen. *Ecology* **39**:337–345.

Fahnestock, G. R. 1976. Fires, fuels and flora as factors in wilderness management, the Pasayten case. *Proc. Tall Timbers Fire Ecol. Conf.* **15**:33–69.

Fechner, G. H., and J. S. Barrows 1976. Aspen stands as wildfire fuel breaks. Eisenhower Consortium Bull. No. 4.

Fonda, R. W., and L. C. Bliss. 1969. Forest vegetation of the montane and subalpine zones, Olympic Mountains, Washington. *Ecol. Monogr.* **39**:271–301.

Forest Soils Committee of the Douglas-fir Region. 1957. An introduction to forest soils of the Douglas-fir region of the Pacific Northwest. Western Forestry and Cons. Assn., Portland, Ore.

Fowells, H. A. 1965. Silvics of forest trees of the United States. USDA For. Serv. Agric. Handb. No. 271. Washington, D.C.

Franklin, J. F. 1965. Ecology and silviculture of the true fir-hemlock forests of the Pacific Northwest. *Soc. Amer. For. Proc.* **1964**:28–32.

Franklin, J. F., K. Cromack, Jr., W. Denison, A. McKee, C. Maser, J. Sedell, F. Swanson, and H. Juday. 1981. Ecological characteristics of old-growth forest ecosystems in the Douglas-fir region. USDA For. Serv. Gen. Tech. Rep. PNW-(In Press). Pac. Northwest For. and Range Exp. Stn., Portland, Ore.

Franklin, J. F., and C. T. Dyrness. 1973. Natural vegetation of Oregon and Washington. USDA For. Serv. Gen. Tech. Rep. PNW-8. Pac. Northwest For. and Range Exp. Stn., Portland, Ore.

Fremlin, G.(ed.) 1974. Vegetation. In *The National Atlas of Canada*. Macmillan Co. Can. Ltd., Toronto, Ont. and Dept. Energy, Mines, Res., Inf. Can., Ottawa, pp. 45, 46.

Furniss, R. L. 1941. Fire and insects in the Douglas-fir region. *Fire Contr. Notes* **5**:211–213.

Gratkowski, H. 1961a. Brush problems in southwestern Oregon. USDA For. Serv. Pac. Northwest For. and Range Exp. Stn., Portland, Ore.

Gratkowski, H. 1961b. Brush seedlings after controlled burning of brushlands in southwestern Oregon. *J. For.* **59**:885–888.

Gratkowski, H. 1974. Origin of mountain whitethorn brush fields on burns and cuttings in Pacific Northwest Forests. *Proc. Western Soc. Weed Sci.* **27**:5–8.

Gruell, G. E., and L. L. Loope. 1974. Relationships among aspen, fire, and ungulate browsing in Jackson Hole, Wyoming. USDA For. Serv. and USDI Park Serv., U.S. Gov. Printing Off., Washington, D.C.

Habeck, J. R. 1962. Forest succession in Mammouth Township, Polk County, Oregon since 1850. *Mont. Acad. Sci. Proc.* **21**:7–17.

Habeck, J. R. 1973. A phytosociological analysis of forests, fuels, and fire in the Moose Creek drainage, Selway-Bitterroot Wilderness. Univ. Montana-USDA For. Serv. Pub. R1-73-022.

Habeck, J. R. 1976. Forests, fuels, and fire in the Selway-Bitterroot Wilderness, Idaho. *Proc. Tall Timbers Fire Ecol. Conf.* **14**:305–375.

Habeck, J. R., and R. W. Mutch. 1973. Fire-dependent forests in the Northern Rocky Mountains. *Quat. Res.* **3**:408–424.

Hall, F. C. 1967. Vegetation-soil relations as a basis for resource management on the Ochoco National Forest of central Oregon. Ph.D. Diss., Oregon State Univ., Corvallis.

Hall, F. C. 1976. Fire and vegetation in the Blue Mountains—implications for land management. *Proc. Tall Timbers Fire Ecol. Conf.* **15**:155–170.

Hallin, W. E. 1968. Soil surface temperature on cutovers in southwest Oregon. USDA For Serv. Res. Paper PNW-78. Pac. Northwest For. and Range Exp. Stn., Portland, Ore.

Hanks, J. P., and W. A. Dick-Peddie. 1974. Vegetational patterns of the White Mountains, New Mexico. *Southwestern Natur.* **18**:371–382.

Hanzlik, E. J. 1932. Type successions in the Olympic Mountains. *J. For.* **30:** 91, 92.

Hawkes, B. C. 1980. Fire history of Kananaskis Provincial Park—mean fire return intervals. In Proc. Fire History Workshop, Oct. 20–24, 1980, Univ. Ariz., Tucson. USDA For. Serv. Gen. Tech. Rep. RM-81. Intermt. For. and Range Exp. Stn., Ogden, Utah., pp. 42–45.

Heinzelmann, M. L. 1978. Fire in wilderness ecosystems. In J. C. Hendee, G. H. Stankey, R. C. Lucas (eds.) *Wilderness Management.* USDA For. Serv. Misc. Pub. No. 1365. Washington, D.C., pp. 249–278.

Hemstrom, M. A. 1979. A recent disturbance history of forest ecosystems at Mount Rainier National Park. Ph.D. Diss., Oregon State Univ., Corvallis.

Hines, W. W. 1968. Trees and deer, a progress report on the Cedar Creek Deer Study. Oregon Game Comm., Portland.

Hines, W. W. 1971. Plant communities in the old-growth forests of north coastal Oregon. M. S. Thesis. Oregon State Univ., Corvallis.

Hoff, C. C. 1957. A comparison of soil, climate, and biota of conifer and aspen communities in the central Rocky Mountains. *Amer. Midl. Natur.* **58**:115–140.

Hoffman, G. R., and R. R. Alexander. 1976. Forest vegetation of the Bighorn Mountains, Wyoming: A habitat type classification. USDA For. Serv. Res. Paper RM-170. Rocky Mtn. For. and Range Exp. Stn., Fort Collins, Colo.

Hoffman, G. R., and R. R. Alexander. 1980. Forest vegetation of the Routt National Forest in northwestern Colorado: A habitat type classification. USDA For. Serv. Res. Paper RM-221. Rocky Mtn. For. and Range Exp. Stn., Fort Collins, Colo.

Holland, S. S. 1964. Landforms of British Columbia: A physiographic outline. British Columbia Dep. of Mines and Petroleum Resour. Bull. No. 48.

Horton, K. W., and E. J. Hopkins. 1965. Influence of fire on aspen suckering. Can. Dept. For., Dept. Northern Off. Nat. Resour., For. Br. Pub. 1095. Ottawa, Ont.

Houston, D. B. 1973. Wildfires in northern Yellowstone National Park. *Ecology* **54**:1111–1117.

Isaac, L. A. 1940. Vegetative succession following logging in the Douglas-fir region with special reference to fire. *J. For.* **38**:716–721.

Isaac, L. A. 1943. *Reproductive Habits of Douglas-Fir.* Charles Lathrop Pack Forestry Foundation, Washington, D.C.

Isaac, L. A. 1963. Fire—a tool not a blanket rule in Douglas-fir ecology. *Proc. Tall Timbers Fire Ecol. Conf.* **2**:1–17.

Johannessen, C. L., W. A. Davenport, A. Millet, and S. McWilliams. 1971. The vegetation of the Willamette Valley. *Ann. Assoc. Amer. Geogr.* **61**:286–302.

Johnson, N. E., and P. G. Belluschi. 1969. Host-finding behavior of the Douglas-fir beetle. *J. For.* **67**:290–295.

Jones, J. R. 1974. Silviculture of southwestern mixed conifers and aspen: The status of our knowledge. USDA For. Serv. Res. Paper RM-122. Rocky Mtn. For. and Range Exp. Stn., Fort Collins, Colo.

Kallander, H. 1969. Controlled burning on the Fort Apache Indian Reservation, Arizona. *Proc. Tall Timbers Fire Ecol. Conf.* **9**:241–249.

Kirkwood, J. E. 1902. The vegetation of northwestern Oregon. *Torreya* **2**:129–134.

Krajina, V. J. (ed.) 1965. Ecology of western North America. Vol. 1. Univ. British Columbia, Dept. of Botany. Vancouver.

Krebill, R. G. 1972. Mortality of aspen on the Gros Ventre elk winter range. USDA For. Serv. Res. Paper INT-129. Intermt. For. and Range Exp. Stn., Ogden, Utah.

Küchler, A. W. 1964. Potential natural vegetation of the conterminous United States. Manual to accompany the map. Amer. Geogr. Soc. Spec. Pub. 36. (With map, rev. ed., 1965, 1966.)

Langenheim, J. H. 1962. Vegetation and environmental patterns in the Crested Butte area, Gunnison County, Colorado. *Ecol. Monogr.* **32**:249–285.

Larsen, J. A. 1929. Fires and forest succession in the Bitterroot Mountains of northern Idaho. *Ecology* **10**:67–76.

Leege, T. A., and W. O. Hickey. 1971. Sprouting of northern Idaho shrubs after prescribed burning. *J. Wildl. Manage.* **35**:508–515.

Lewis, H. T. 1981. Hunter-gatherers and problems for fire history. In Proc. Fire History Workshop, Oct. 20–24, 1980, Univ. Ariz., Tucson. USDA For. Serv. Gen. Tech. Rep. RM-81. Rocky Mtn. For. and Range Exp. Stn., Ogden, Utah, pp. 115–119.

Little, E. L., Jr. 1979. Checklist of United States trees (native and naturalized). USDA Handb. 541. Washington, D.C.

Loope, L. L., and G. E. Gruell. 1973. The ecological role of fire in the Jackson Hole Area, northwestern Wyoming. *Quat. Res.* **3**:425–443.

Lotan, J. E. 1967. Cone serotiny of lodgepole pine near West Yellowstone, Montana. *For Sci.* **13**:55–59.

Lotan, J. E. 1968. Cone serotiny of lodgepole pine near Island Park, Idaho. USDA For. Serv. Res. Paper INT-52. Intermt. For. and Range Exp. Stn., Ogden, Utah.

Lotan, J. E. 1975a. The role of cone serotiny in lodgepole pine forests. In D. M. Baumgartner (ed.) *Management of Lodgepole Pine Ecosystems Symposium Proceedings*. Wash. State Univ. Coop. Ext. Serv., Pullman, pp. 471–495.

Lotan, J. E. 1975b. Regeneration of lodgepole pine forests in the northern Rocky Mountains. In D. M. Baumgartner (ed.) *Management of Lodgepole Pine Ecosystems Symposium Proceedings*. Wash. State Univ. Coop. Ext. Serv., Pullman, pp. 516–535.

Lyon L. J. 1971. Vegetal development following prescribed burning of Douglas-fir in south-central Idaho. USDA For. Serv. Res. Paper INT-105. Intermt. For. and Range Exp. Stn., Ogden, Utah.

Lyon, J. L., and P. F. Stickney. 1976. Early vegetal succession following large northern Rocky Mountain wildfires. *Proc. Tall Timbers Fire Ecol. Conf.* **14**:355–375.

Maini, J. S., and K. W. Horton. 1966. Vegetative propagation of *Populus* spp. I. Influence of temperature on formation and initial growth of aspen suckers. *Can. J. Bot.* **44**:1183–1189.

Marr, J. 1967. Ecosystems of the east slope of the Front Range in Colorado. Univ. Colorado Studies, Series in Biol., No. 8. Fort Collins, Colo.

Martin, R. E., and A. P. Brackebusch. 1974. Fire hazard and conflagration prevention. In O. P. Cramer (ed.) Environmental effects of forest residues management in the Pacific Northwest. USDA For. Serv. Gen. Tech. Rep. PNW-24. Pac. Northwest For. and Range Exp. Stn., Portland, Ore., pp. 610–630.

Martin, R. E., D. D. Robinson, and W. H. Schaeffer. 1976. Fire in the Pacific Northwest—perspectives and problems. *Proc. Tall Timbers Fire Ecol. Conf.* **15**:1–23.

McArdle, R. E., W. H. Meyer, and D. Bruce. 1961. The yield of Douglas-fir in the Pacific Northwest. USDA Tech. Bull. No. 201. Washington, D.C.

McCulloch, W. F. 1944. Slash burning. *For. Chron.* **20**:111–118.

McLean, A. 1969. Fire resistance of forest species as influenced by root systems. *J. Range Manage.* **22**:120–122.

McLean, A. 1970. Plant communities of the Similkameen Valley, British Columbia, and their relationships to soils. *Ecol. Monogr.* **40**:403–423.

McLean, A., and W. D. Holland. 1958. Vegetation zones and their relationship to the soils and climate of the upper Columbia Valley. *Can. J. Plant Sci.* **38**:328–345.

Miller, M. 1977. Response of blue huckleberry to prescribed fires in a western Montana larch-fir forest. USDA For. Serv. Res. Paper INT-188. Intermt. For. and Range Exp. Stn., Ogden, Utah.

Miller, M. M., and J. W. Miller. 1976. Succession after wildlife in the North Cascades National Park complex. *Proc. Tall Timbers Fire Ecol. Conf.* **15**:71–83.

Miller, R. E., and M. D. Murray. 1978. The effects of red alder on growth of Douglas-fir. In D. G. Briggs, D. S. DeBell and W. A. Atkinson (eds.) Utilization and management of alder. USDA For. Serv. Gen. Tech. Rep. PNW-70. Pac. Northwest For. and Range Exp. Stn., Portland, Ore., pp. 283–306.

Mitchell, R. G., and C. Sartwell. 1974. Insects and other arthropods. In O. P. Cramer (ed.) Environmental effects of forest residues management in the Pacific Northwest. USDA For. Serv. Gen. Tech. Rep. PNW-24. Pac. Northwest For. and Range Exp. Stn., Portland, Ore., pp. R1–R22.

Moir, W. H. 1969. The lodgepole pine zone in Colorado. *Amer. Mid. Natur.* **81**:87–98.

Moir, W. H., and J. A. Ludwig. 1979. A classification of spruce-fir and mixed conifer habitat types in Arizona and New Mexico. USDA For. Serv. Res. Paper RM-207. Rocky Mtn. For. and Range Exp. Stn., Fort Collins, Colo.

Morgan, M. D. 1969. Ecology of aspen in Gunnison County, Colorado. *Amer. Mid. Natur.* **82**(2):204–228.

Morris, W. G. 1934. Forest fires in Washington and Oregon. *Oreg. Hist. Quart.* **35**:313–339.

Morris, W. G. 1958. Influence of slash burning regeneration, other plant cover, and fire hazard in the Douglas-fir region. USDA For. Serv. Res. Paper 29. Pac. Northwest For. and Range Exp. Stn., Portland, Ore.

Morris, W. G. 1970. Effects of slash burning in overmature stands in the Douglas-fir region. *For Sci.* **16**:259–270.

Mowat, E. L. 1960. No serotinous cones on central Oregon lodgepole pine. *J. For.* **58**:118–119.

Mueggler, W. F. 1965. Ecology of seral shrub communities in the cedar-hemlock zone of northern Idaho. *Ecol. Monogr.* **35**:165–185.

Mueggler, W. F. 1976. Type variability and succession in Rocky Mountain aspen. In Utilization and marketing as tools for aspen management in the Rocky Mountains. USDA For. Serv. Gen. Tech. Rep. RM-29. Rocky Mtn. For. and Range Exp. Stn., Fort Collins, Colo., pp. 16–19.

Mueller-Dombois, D. 1959. The Douglas-fir forest associations on Vancouver Island in their initial stage of secondary succession. Ph.D. Diss. Univ. of British Columbia, Vancouver.

Munger, T. T. 1940. The cycle from Douglas-fir to hemlock. *Ecology* **21**:451–459.

Muraro, S. J. 1971a. The lodgepole pine fuel complex. Can. For. Serv. Inf. Rep. BC-X-53. Victoria, B.C.

Muraro, S. J. 1971b. Prescribed fire impact in cedar-hemlock logging slash. Can. Dept. Envir., Can. For. Serv. Pub. 1295. Ottawa, Ont.

Muraro, S. J. 1978a. Prescribed fire—a tool for the control of dwarf mistletoe in lodgepole pine. In Proc. of the symp. on dwarf mistletoe control through forest management. USDA For. Serv. Gen. Tech. Rep. PSW-31. Pac. Southwest For. and Range Exp. Stn., Berkeley, Calif., pp. 124–127.

Muraro, S. J. 1978b. The use of prescribed fire in the management of lodgepole pine. In D. E. Dube (ed.) Fire ecology in resource management. Can. For. Serv. Inf. Rep. NOR-X-210. Northern For. Res. Cent., Edmonton, Alberta., pp. 82–89.

Mutch, R. W. 1970. Wildland fires and ecosystems—A hypothesis. Ecology **51**:1046–1051.

Neiland, B. J. 1958. Forest and adjacent burn in the Tillamook Burn area of northwestern Oregon. *Ecology* **39**:660–671.

Nelson, E. E., and G. M. Harvey. 1974. Diseases. In O. P. Cramer (ed.) Environmental effects of forest residues management in the Pacific Northwest. USDA For. Serv. Gen. Tech. Rep. PNW-24. Pac. Northwest For. and Range Exp. Stn., Portland, Ore., pp. S1–S11.

Norum, R. A. 1976. Fire intensity—fuel reduction relationships associated with underburning in larch/Douglas-fir stands. *Proc. Tall Timbers Fire Ecol. Conf.* **14:**559–572.

Ogilvie, R. T. 1963. Ecology of the forests in the Rocky Mountains of Alberta. Can. Dept. For., For. Res. Br. Calgary.

Orloci, L. 1965. The coastal Western Hemlock Zone on the southwestern British Columbia mainland. In V. J. Krajina (ed.) *Ecology of Western North America.* Vol. 1. Univ. British Columbia, Dept. of Bot., Vancouver, pp. 18–34.

Patten, D. T. 1969. Succession from sagebrush to mixed conifer forest in the Northern Rocky Mountains. *Amer. Midl. Natur.* **82:** 229–240.

Pengelly, W. L. 1963. Timberland and deer in the Northern Rockies. *J. For.* **61:**734–740.

Perry, D. A., and J. E. Lotan. 1979. A model of fire selection for serotiny in lodgepole pine. *Evolution* **33**(3) 958–968.

Pfister, R. D., and R. Daubenmire. 1975. Ecology of lodgepole pine. In D. M. Baumgartner (ed.) *Management of Lodgepole Pine Ecosystems Symposium Proceedings.* Wash. State Univ. Coop. Ext. Serv., pp. 27–46.

Pfister, R. D., B. L. Koralchik, S. F. Arno, and R. C. Presby. 1977. Forest habitat types of Montana. USDA For. Serv. Gen. Tech. Rep. INT-34. Intermt. For. and Range Exp. Stn., Ogden, Utah.

Pickford, S. G., G. R. Fahnestock, and R. Ottserar. 1980. Weather, fuel and lightning fires in Olympic National Park. *Northwest Sci.* **54:**92–105.

Radwan, M. A. 1972. Differences between Douglas-fir genotypes in relation to browsing preference by black-tailed deer. *Can. J. For. Res.* **2:**250–255.

Roe, A. L. 1967. Productivity indicators in western larch forests. USDA For. Serv. Res. Note INT-59. Intermt. For. and Range Exp. Stn., Ogden, Utah.

Roe, A. L., and G. D. Amman. 1970. The mountain pine beetle in lodgepole pine forests. USDA For. Serv. Res. Paper INT-71. Intermt. For. and Range Exp. Stn., Ogden, Utah.

Romme, W. H. 1979. Fire and landscape diversity in subalpine forests of Yellowstone National Park. Ph.D. Diss., Univ. Wyoming, Laramie.

Romme, W. H. 1980. Fire history terminology: Report of the Ad Hoc Committee. In Proc. Fire History Workshop, Oct. 20–24, 1980, Univ. Ariz., Tucson. USDA For. Serv. Gen. Tech. Rep. RM-81, Rocky Mtn. For. and Range Exp. Stn., Fort Collins, Colo.

Rowe, J. S. 1972. Forest regions of Canada. Dept. Environ. Can. For. Serv. Pub. No. 1300. Ottawa, Ont.

Rowe, J. S. 1981. Concepts of fire effects on plant individuals and species. In R. W. Wein and D. A. MacLean (eds.) *The Role of Fire in Northern Circumpolar Ecosystems.* Wiley, New York. (In press.)

Ruth, R. H. 1974. Regeneration and growth of west-side mixed conifers. In O. P. Cramer (ed.) Environmental effects of forest residues management in the Pacific Northwest. USDA For. Serv. Gen. Tech. Rep. PNW-24. Pac. Northwest For. and Range Exp. Stn., Portland, Ore., pp. K1–K18.

Ryker, R. A. 1975. A survey of factors affecting regeneration of Rocky Mountain Douglas-fir. USDA For. Serv. Res. Paper INT-174. Intermt. For. and Range Exp. Stn., Ogden, Utah.

Sackett, S. S. 1979. Natural fuel loadings in ponderosa pine and mixed conifer forests of the Southwest. USDA For. Serv. Res. Paper RM-213. Rocky Mtn. For. and Range Stn., Fort Collins, Colo.

Safranyik, L., D. M. Shrimpton, and H. S. Whitney. 1974. Management of lodgepole pine to reduce losses from the mountain pine beetle. Envir. Can. For. Serv. Tech. Rep. 1.

Schier, G. A. 1975. Deterioration of aspen clones in the middle Rocky Mountains. USDA For. Serv. Res. Paper INT-170. Intermt. For. and Range Exp. Stn., Ogden, Utah.

Schmidt, R. L. 1957. The silvics and plant geography of the genus *Abies* in the costal forests of British Columbia. B.C. For. Serv. Tech. Pub. T. 46.

Schmidt, R. L. 1970. A history of pre-settlement fires on Vancouver Island as determined from Douglas-fir ages. In J. H. G. Smith and J. Worrall (eds.) Tree ring analysis with special reference to Northwest America. Univ. British Columbia, Faculty of For. Bull. No. 7. Vancouver, pp. 107–108.

Schroeder, M. J., and C. C. Buck. 1970. Fire weather. USDA Handb. No. 360. Washington, D.C.

Sharpe, G. W. 1974. Western red cedar. Univ. Wash., College of For., Seattle, Wash.

Shearer, R. C. 1975. Seedbed characteristics in western larch forests after prescribed burning. USDA For. Serv. Res. Paper INT-167. Intermt. For. and Range Exp. Stn., Ogden, Utah.

Silen, R. R. 1960. Lethal surface temperatures and their interpretation for Douglas-fir. Ph.D. Diss., Oregon State Univ., Corvallis.

Sprague, F. L., and H. P. Hansen. 1946. Forest succession in the McDonald Forest, Willamette Valley, Oregon. *Northwest Sci.* **20**:89–98.

Starker, T. J. 1934. Fire resistance in the forest. *J. For.* **32**:462–467.

Steen, H. K. 1965. Variation in vegetation following slash fires near Oakridge, Oregon. USDA For. Serv. Res. Paper PNW-25. Pac. Northwest For. and Range Exp. Stn., Portland, Ore.

Steneker, G. A. 1974. Factors affecting suckering of trembling aspen. *For Chron.* **50**:32–34.

Strang, R. M., and A. H. Johnson 1980. Prescribed forest rangeland burning (Dewdrop project), Annual report. 1979. B.C. For. Serv., Victoria.

Strang, R. M., and J. V. Parminter. 1980. Conifer encroachment on the Chilcotin grasslands of British Columbia. *For. Chron.* **56**:13–18.

Stringer, P. W., and G. H. La Roi. 1970. The Douglas-fir forests of Banff and Jasper National Parks, Canada. *Can. J. Bot.* **48**:1703–1726.

Sullivan, M. J. 1978. Regeneration of tree seedlings after clearcutting on some upper-slope habitat types in the Oregon Cascade Range. USDA For. Serv. Res. Paper PNW-245. Pac. Northwest For. and Range Exp. Stn., Portland, Ore.

Tackle, D. 1961. Silvics of lodgepole pine. USDA For. Serv. Misc. Pub. 19. Intermt. For. and Range Exp. Stn., Ogden, Utah.

Tande, G. F. 1979. Fire history and vegetation pattern of coniferous forests in Jasper National Park, Alberta. *Can. J. Bot.* **57**:1912–1931.

Tarrant, R. F. 1956. Effects of slash burning on some soils of the Douglas-fir region. *Proc. Soil Sci. Soc. Amer.* **20**:408–411.

Taylor, A. R. 1971. Lightning—agent of change in forest ecosystems. *J. For.* **68**:477–480.

Thilenius, J. F. 1968. The *Quercus garryana* forests of the Willamette Valley, Oregon. *Ecology* **49**:1124–1133.

Tisdale, E. W., and A. McLean. 1957. The Douglas-fir zone of southern British Columbia. *Ecol. Monogr.* **27**:247–266.

Turner, G. T., and H. A. Paulsen, Jr. 1976. Management of mountain grasslands in the central Rockies: The status of our knowledge. USDA For. Serv. Res. Paper RM-161. Rocky Mtn. For. and Range Exp. Stn., Fort Collins, Colo.

Turner, J., and J. N. Long. 1975. Accumulation of organic matter in a series of Douglas-fir stands. *Can. J. For. Res.* **5**:681–690.

Volland, L. A. 1976. Plant communities of the central Oregon pumice zone. USDA For. Serv. R6 Area Guide 4-2.

Wagle, R. F., and T. W. Eakle. 1979. A controlled burn reduces the impact of a subsequent wildfire in a ponderosa pine vegetation type. *For. Sci.* **25**:123–129.

Wallmo, O. C., W. L. Regelin, and D. W. Reichert. 1972. Forage use by mule deer relative to logging in Colorado. *J. Wildl. Manage.* **36**:1025–1033.

Watt, W., R. Chilton, R. M. Annas, and R. Coupe. 1979. The Caribou forest region. In R. M. Annas and R. Coupe (eds.) *Biogeoclimatic Zones and Subzones of the Caribou Forest Region.* B.C. Min. of For., pp. 18–95.

Weaver, H. 1951. Fire as an ecological factor in the southwestern ponderosa pine forests. *J. For.* **49**:93–98.

Wellner, C. A. 1970. Fire history in the northern Rocky Mountains. In *The Role of Fire in the Intermountain West, Symposium.* Intermt. Fire Res. Coun., Missoula, Mont., pp. 42–64.

West, N. E. 1969. Successional changes in the montane forest of the central Oregon Cascades. *Amer. Midl. Natur.* **81**:265–271.

West, N. E., and W. W. Chilcote. 1968. *Senecio sylvaticus* in relation to Douglas-fir clearcut succession in the Oregon Coast Range. *Ecology* **49**:1101–1107.

Whittaker, R. H. 1960. Vegetation of the Siskiyou Mountains, Oregon and California. *Ecol. Monogr.* **30**:279–338.

Wollum, A. G., C. T. Youngberg, and F. W. Chichester. 1968. Relation of previous timber stand age to nodulation of *Ceanothus velutinus. For. Sci.* **14**:114–118.

Youngberg, C. T., and C. T. Dyrness. 1959. The influence of soils and topography on the occurrence of lodgepole pine in central Oregon. *Northwest Sci.* **33**:111–120.

Zavitkovski, J., and M. Newton. 1968. Ecological importance of snowbrush *(Ceanothus velutinus)* in the Oregon Cascades. *Ecology* **49**:1134–1145.

12

SPRUCE–FIR

F ires are rare in spruce and spruce-fir forests because of climatic conditions. Nevertheless fire is a natural force that can enhance regeneration of all spruce species. Along the Pacific Coast lightning fires rarely occur because thunderstorms are rare and fog is often present. In the Rocky Mountains and along the east coast frequent summer showers keep the possibility of lightning fires to a minimum. The understories of all spruce communities stay moist and fires are rare, except during extreme droughts, despite the high flammability of spruce leaves. During droughts fires can ravage Rocky Mountain spruce forests because limbs are close to the ground and are highly flammable (LeBarron and Jemison 1953; Ruth 1965; Kallander 1969). Historically in western North America, most stands of spruce originated after fire, blowdown, erosion, and, more recently, clear-cutting (Ruth 1965). Today fire's major role in spruce forests is as a silvicultural tool to clean up slash, sanitize infected forests, and aid in preparing mineral soil seedbeds for establishment of new trees.

SITKA SPRUCE

Distribution, Climate, Soils, and Vegetation

Sitka spruce *(Picea sitchensis)* occurs in a narrow zone, generally from a few to 50 km (30 miles) wide, along the Pacific Coast of North America (Fig. 12.1) from northwestern California to Kodiak Island, Alaska (Woodfin 1973). Climate is wet and mild with an annual precipitation of 100 to 380 cm (40 to 150 in.) per year (Ruth 1965). Summer fogs are frequent and fog drip adds as much as 26 percent to the an-

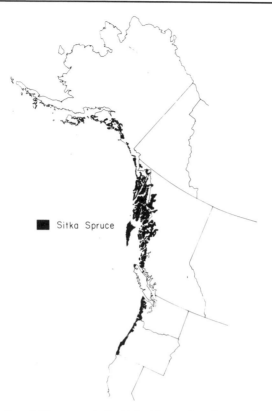

Fig. 12.1. Distribution of Sitka spruce in the United States, Canada, and Alaska. (Modified from Küchler 1964 and Fremlin 1974. Reproduced with permission of the American Geographical Society and the Minister of Supply and Services Canada from the publication "The National Atlas of Canada" published by the Macmillan Co. of Canada Limited, Toronto, Ontario in association with the Department of Energy, Mines and Resources Canada and Information Canada, Ottawa, Canada 1974.)

nual precipitation (Ruth 1954). Soils are deep, fertile, and fine-textured (Franklin and Dyrness 1969). Sitka spruce is generally found at elevations below 180 m (500 ft) but can be found as high as 610 m (2000 ft) (Fowells 1965).

The Sitka spruce zone is a variant of the cedar-hemlock zone, but is distinguished by Sitka spruce (Fig. 12.2) (Franklin and Dyrness 1969). In Alaska low temperature limits growth and Sitka spruce is usually found on southern slopes (Fowells 1965). At the southern extreme in southwestern Oregon and northwestern California Sitka spruce mixes with coastal redwood *(Sequoia sempervirens)* and moisture is the limiting factor. Sitka spruce is usually found on northern slopes, but coastal redwood occupies the somewhat warmer sites.

The most common tree species in climax communities of the Sitka spruce zone include Sitka spruce, western hemlock *(Tsuga heterophylla)*, and western redcedar *(Thuja plicata)*. Less common species include Douglas fir *(Pseudotsuga menziesii)*,

Fig. 12.2. Old-growth Sitka spruce, showing buttressed and swollen base. Swordfern and Oregon oxalis dominate the understory. (Photo courtesy of USDA Forest Service.)

grand fir *(Abies grandis)*, and Pacific silver fir *(A. amabilis)*. Mature forests have lush understories with dense growths of shrubs, dicotyledonous herbs, ferns, and cryptograms (Franklin and Dyrness 1969). Sites with average moisture contain swordfern *(Polystichum munitum)*, Oregon oxalis *(Oxalis oregana)*, false lily-of-the-valley *(Maianthemum bifolium* var. *kamschaticum)*, western springbeauty *(Montia sibirica)*, three-leaved coolwart *(Tiarella trifoliata)*, evergreen violet *(Viola semper-virens)*, Wood violet *(V. glabella)*, Smith fairybells *(Disporum smithii)*, red huckleberry *(Vaccinium parvifolium)*, and rustyleaf *(Menziesia ferruginea)*. The species composition varies slightly at drier or wetter sites.

Fire Effects

Initial succession after fire usually falls into three categories: where red alder *(Alnus rubra)* is very abundant; where Sitka spruce and western hemlock become established immediately after fire; and where other shrubs dominate while conifers are being established (Franklin and Dyrness 1969). Red alder reproduces abundantly and grows fast following disturbance by logging or fires on many sites (Franklin and Dyrness 1969). It is a nonleguminous nitrogen-fixing species (Tarrant et al. 1969) and can significantly improve soil properties. Often it grows in pure stands and is very competitive with conifers. Natural reestablishment of Sitka spruce, western hemlock, and western redcedar in red alder stands is slow or nonexistent. Occasionally reestablishment takes place on downed, rotting logs on some sites within older stands of red alder. Following clear-cutting chemicals are sometimes used to suppress alder and permit the conifers to compete successfully (Ruth 1965).

In conifer mixtures succession after fire varies depending on the site. Shrub competition from salmonberry *(Rubus spectabilis)*, thimbleberry *(R. parviflorus)*, Pacific red elder *(Sambucus callicarpa)*, black elderberry *(Sambucus racemosa* var. *melanocarpa)*, huckleberries (*Vaccinium* spp.), rustyleaf, devilsclub *(Oplopanax horridum)*, salal *(Gaultheria shallon)*, and others is very serious on bottomland sites (Fowells 1965; Franklin and Dyrness 1969). On these sites Sitka spruce and western hemlock will gradually become established among the shrubs depending on seed supply, seedbed conditions, and moisture. Other sites may become established with Sitka spruce and western hemlock immediately after a fire and produce even-aged stands, as at Cascade Head on the Oregon Coast (D. Minore, pers. comm.).

Moisture must be abundant in order for Sitka spruce and western hemlock seedlings to become established (Fowells 1965). Then establishment of all species is best on mineral soils with side shade, but Sitka spruce is the most competitive of all conifer species on mineral soils (Ruth 1965). Conifer seedlings also become established on organic seedbeds such as rotten wood or moss. On these sites Sitka spruce still appears more competitive than western hemlock and western red cedar (Minore 1972), although the lack of light will favor western hemlock and western red cedar (Ruth 1965).

To maintain commercial timber stands with a predominance of Sitka spruce, clear-cutting and broadcast burning (where there is heavy slash) seem most desirable to provide a seedbed that will favor Sitka spruce. Once established, Sitka spruce is a vigorous, fast-growing tree that readily overtops western hemlock and cedars to occupy the dominant position in the stand (Ruth 1965; Fowells 1965). As the stand matures Sitka spruce maintains its dominance with some hemlock. Sitka spruce dominates for a long time because it will live for 700 to 800 years, whereas western hemlock lives for about 500 years (Fowells 1965). As the forest ages the old western hemlock begins to die and is succeeded by younger hemlock trees which will eventually replace the spruce forest in the absence of fire, although complete elimination of Sitka spruce seldom occurs.

Management Implications

The most commonly used silvicultural system in the Sitka spruce-western hemlock type is clear-cutting with the reliance on natural seeding (Ruth 1965). Much clear-cutting is done in patches of 25 ha (100 acres) or less to break up slash into smaller units and insure adequate seed from surrounding trees. The seed is light and will travel at least 300 m (1000 ft) (Ruth 1965). This form of natural regeneration of Sitka spruce and western hemlock is usually a combination of advanced western hemlock reproduction that survived the logging operation and new seedlings cast from nearby trees (Ruth 1965). These stands ultimately contain 30 percent Sitka spruce and 70 percent western hemlock (Harris 1966).

Prescribed fire is seldom used in the spruce-hemlock type because burning conditions seldom exist (wet climate) and western hemlock and Sitka spruce slash deteriorate rapidly (Ruth 1965). However, on old growth stands, slash accumulations are huge and critical weather conditions can and do occur. Thus fire is often used on these sites to reduce fire hazard and to suppress competition from shrubs (Isaac 1940; Ruth 1965). Fire enhances the establishment of Sitka spruce and survival is many times greater than for western hemlock seedlings (Harris 1966). Seedlings of Sitka spruce, 15 cm (6 in.) tall, were twice as abundant as western hemlock seven years after a slash burn in southeastern Alaska (Harris 1966). Sitka spruce establishes best on neutral soils where moisture is adequate whereas western hemlock establishes best on acid soils that are unburned (Taylor 1929).

Other silvicultural practices involve harvesting small trees (prelogging) from old growth stands in advance of the main logging operation (Ruth 1965) as well as thinning (Farr and Harris 1971). Prelogging permits a more efficient harvest of all timber (Ruth 1965) whereas thinning of young stands permits the remaining trees to grow faster (Farr and Harris 1971). Shelterwood cutting has been tried to a limited extent, but this method has resulted in overstocked regeneration (Williamson and Ruth 1976).

Today's fastest growing stands of spruce-hemlock on the west coast have originated from early logging areas, extensive blowdowns, and fire (Ruth 1965). Fire is not a necessary management tool in this region and has not been tested extensively. It appears, however, that it can be used to remove excessive debris, temporarily suppress undesirable brush, and shift the stand composition to a higher percentage of Sitka spruce, depending on site, soil disturbance by logging practices, and disease problems.

ENGLEMANN SPRUCE—SUBALPINE FIR

Distribution, Climate, Soils, and Vegetation

Engelmann spruce *(Picea engelmannii)* is associated with subalpine fir *(Abies lasiocarpa)* (Fig. 12.3) and occurs throughout the Rockies from southern New Mexico to central British Columbia and to some extent along the east slope of the Coast Range in Oregon and Washington (Fig. 12.4). In the central and southern Rockies, the spruce-fir zone is generally associated with an elevation of 2685 to 3505 m (9000 to 11,500 ft). Northward the elevation of spruce-fir sites ranges from 1500 to 2440 m (5000 to 8000 ft) and can be found as low as 455 m (1500 ft) in British Columbia. At intermediate elevations the structure of old growth Engelmann spruce-subalpine fir forest is nearly pure spruce in the overstory with fir predominating in the understory (Fowells 1965; Alexander 1974a; Clagg 1975).

The climate is cold and humid, characterized by cool summers, frequent summer showers, very cold winters, and heavy snowfall (Sudworth 1916). Precipitation is generally from 75 to 100 cm (30 to 40 in.), but may be as low as 60 cm (24 in.)

Fig. 12.3.   Spruce-fir in the northern Rocky Mountains. (Photo courtesy of USDA Forest Service.)

(Alexander 1974a). Soils are variable, often shallow, but moist, and have a loam texture. Grasses can compete severely with tree seedlings, especially on heavier soils, whereas coarse-textured sand and gravelly soils may dry too fast for seedling establishment (Alexander 1974b).

Composition of the Engelmann spruce forest varies with elevation, exposure, and latitude (Fowells 1965). Subalpine fir is a common associate throughout the Rocky Mountains and Cascades. In Alberta the spruce-fir zone mixes with white spruce *(Picea glauca)*, balsam poplar *(Populus balsamifera)*, and paper birch *(Betula papyrifera)*. Southward, in the northern Rocky Mountains at the lower limits of its elevation, Engelmann spruce mixes with western white pine *(Pinus monticola)*, western redcedar, and western hemlock; at middle elevations it mixes with interior Douglas fir *(Pseudotsuga menziesii* var. *glauca)*, western larch *(Larix occidentalis)*, and grand fir; and at the highest elevations where Engelmann spruce extends to timberline, it mixes with mountain hemlock *(Tsuga mertensiana)*, whitebark pine *(Pinus albicaulis)*, and subalpline larch *(Larix lyallii)*. In the central Rocky Mountains common associates at its lower limits include lodgepole pine *(Pinus contorta)*, Douglas fir, blue spruce *(Picea pungens)*, and quaking aspen *(Populus tremuloides)*; at its middle and upper limits whitebark pine, bristlecone pine *(Pinus aristata)*, and limber pine *(P. flexilis)* are included. In the southern Rocky Mountains in New Mexico and Arizona, white fir *(Abies concolor)*, Douglas fir, blue spruce, aspen, and occasionally, ponderosa pine *(Pinus ponderosa)* are associates at its lower limits; and corkbark fir *(Abies lasiocarpa* var. *arizonica)* and bristlecone pine at the middle and higher elevations.

Understory species for the northern Rocky Mountains has been described in the

■ Engelman Spruce – Alpine Fir

Fig. 12.4. Distribution of the spruce-fir zone in the United States and Canada. (Modified from Küchler 1964 and Fremlin 1974. Reproduced with permission of the American Geographical Society and the Minister of Supply and Services Canada from the publication "The National Atlas of Canada" published by the Macmillan Co. of Canada Limited, Toronto, Ontario in association with the Department of Energy, Mines and Resources Canada and Information Canada, Ottawa, Canada 1974.)

greatest detail by Daubenmire and Daubenmire (1968). Common shrubs include big huckleberry *(Vaccinium membranaceum)*, grouse huckleberry *(V. scoparium)*, rustyleaf, and common beargrass *(Xerophyllum tenax)*. Less common are alder *(Alnus sinuata)*, mountain lover *(Pachistima myrsinites)*, and honeysuckle *(Lonicera utahensis)*. Herbs usually include arnica *(Arnica* sp.*)*, anemone *(Anemone piperi)*, queencup beadlily *(Clintonia uniflora)*, western meadow rue *(Thalictrum occidentale)*, violet *(Viola* sp.*)*, elk sedge *(Carex geyeri)*, and pinegrass *(Calamagrostis rubescens)*.

In the central Rocky Mountains, understory vegetation has been described by Mogren and Barth (1974), Clagg (1975), and Hanley et al. (1975). Major shrubs include grouse huckleberry, currant *(Ribes* spp.), elderberry *(Sambucus* sp.), creeping barberry *(Berberis repens)*, rose *(Rosa* spp.), *Rubus* spp., and *Clematis* sp. Indicative herbs of old stands are arnica, aster *(Aster* spp.), goldenweed *(Aplopappus parryi)*, purple reedgrass *(Calamagrostis purpurascens)*, and Ross sedge *(Carex brevipes)*. Little data is available for the southern Rocky Mountains, but Fowells (1965) indicates the presence of common juniper *(Juniperus communis)* and creeping hollygrape *(Odostemon repens)*. Arnica should also be present.

Fire Effects

Thin bark, persistence of dead lower limbs, and high flammability of green material, makes Engelmann spruce especially susceptible to severe damage by fire during drought years (Fig. 12.5). However folowing destruction on northern exposures, Engelmann spruce and subalpine fir usually return at once (Fowells 1965). On southern exposures such species as aspen, Douglas fir, western larch, lodgepole pine, and western white pine may become established initially, but are eventually replaced by the more shade-tolerant spruce and subalpine fir (LeBarron and Jemison 1953). Under the seral species Engelmann spruce may be suppressed for 50 to 100 years but will respond quickly to release and soon outgrow its common associates (Garmon 1957).

The effect of fire on understory species has only been described for the central Rocky Mountains by Mogren and Barth (1974) and Clagg (1975). *Clematis* sp., creeping barberry, rose, elderberry, *Rubus* sp., and huckleberries responded vigorously to fire and persisted in old stands. Dominant forbs in early successional stages

Fig. 12.5. Wildfire in a high elevation spruce-fir forest in Banff National Park. Lower right area in photo was not burned.

included arnica *(Arnica cordifolia; A. latifolia)*, fireweed *(Epilobium angustifolium)*, goldenpea *(Thermopsis divaricarpa)*, and crown closure approached in 74 to 100 years. Several years after fire goldenweed appeared and later aster appeared. These species remained in mature stands.

Grasses and sedges after fire included pinegrass, *Carex siccata*, Ross sedge and elk sedge as the dominants. Bentgrass *(Agrostis scabra)*, spike trisetum *(Trisetum spicatum)*, and purple reedgrass appeared in the 18-year-old burn. Spike trisetum passed out of the stand in 74 to 100 years and purple reedgrass and Ross sedge remained in mature stands.

Interspersed throughout the spruce-fir zone are mountain forb-grasslands (described in Chapter 5) that have a rich mixture of forbs and grasses. These areas have significant summer grazing value but usually have quite a different mixture of species compared to the understory of spruce-fir. Little is known about the effect of fire on these species.

Management Implications

As a silvicultural tool fire is desirable for a few purposes in spruce-fir forests but should be used wisely. Prescribed fires can be used to reduce available food supply for the spruce beetle *(Dendroctonus rufipennis)*, which may occupy cull logs or windthrown trees (Schmid and Beckwith 1972) and expose mineral soil which is the best seedbed for Engelmann spruce because soil moisture is more stable (Alexander 1974a). However, fire is not always necessary and all management schemes should provide adequate seed trees (Noble and Ronco 1978) and leave adequate debris on the mineral soil to provide shade for the seedlings (Alexander 1977; Noble and Alexander 1977). Hot fires are not desirable because they sterilize the mineral soil and leave it unprotected (Roe and Schmidt 1964; Noble and Alexander 1977; Noble and Ronco 1978).

The present recommended cutting practices for Engelmann spruce are to use either some form of partial cutting ($< 30\%$) or a combination of partial cutting and small cleared openings without complete cleanup of slash and other logging debris (Alexander 1973). The latter method adds flexibility, for some pile burning in cleared openings may be desirable to control insects or diseases. In no case, however, should the clear-cut and burn areas on north slopes be more than 135 m (450 ft) wide in the central Rocky Mountains (Alexander 1974a; Noble and Ronco 1978) or 185 m (660 ft) wide in the northern Rocky Mountains (Boyd and Deitschman 1969; Roe et al. 1970). On southern slopes the width of clear-cuts should not be more than 60 m (200 ft) (Roe et al. 1970; Alexander 1974b). Large blocks of uncut trees need to be left for seed and to minimize windthrows.

Seedling establishment is usually best on silt and clay loam soils because they have a more stable supply of moisture than sandy soils and are less quickly invaded by grasses than clay soils (Roe et al. 1970). Shrubs and bare soil provide good sites for regeneration (Alexander 1974a), but establishment is best where there is shade from dead materials and a minimum of competition (Roe and Schmidt 1964). Shade lowers soil temperature and conserves moisture. Both of these factors improve the microenvironment for seedling establishment and survival (Alexander 1974a; Noble and Alexander 1977).

Burning and scarification have been tested to achieve similar objectives. In northern Idaho, Boyd and Deitschman (1969) found that spruce stocking on seedbeds five years after prescribed burning was as good as on scarified seedbeds where 40 percent or more of the area had exposed mineral soil. Burning had the advantages of being cheaper, being usable over a wider range of topographic conditions, and creating conditions more favorable for obtaining acceptable stocking with fewer seedlings. This reduces future expenditures for cleaning and thinning.

RED SPRUCE

Distribution, Climate, Soils, and Vegetation

Red spruce *(Picea rubens)* grows from sea level to 1370 m (4500 ft) in the Northeast (New England States, New York, New Brunswick, Nova Scotia, and southern Quebec and Ontario) and from 1070 to 1830 m (3500 to 6000 ft) in the Appalachian Mountains (Fig. 12.6) (Fowells 1965; Harlow and Harrar 1968). Annual precipitation varies from 90 to 200 cm (35 to 80 in.). Acid, sandy loam soils with considerable moisture support the best spruce (Harlow and Harrar 1968) but soils supporting red spruce are often shallow and rocky where competition from hardwood species is minimal (Westveld 1953). At the higher elevations in mountains red spruce often grows on organic soils overlying rock (Koristan 1937; Oosting and Billings 1951).

Many tree species are associated with red spruce (Fowells 1965). Throughout its range yellow birch *(Betula lutea)*, sugar maple *(Acer saccharum)*, American beech *(Fagus grandifolia)*, balsam fir *(Abies balsamea)*, and paper birch are common. In the northern part of its range additional associates include white spruce, eastern hemlock *(Tsuga canadensis)*, eastern white pine *(Pinus strobus)*, northern white cedar *(Thuja occidentalis)*, tamarack *(Larix laricinia)*, red maple *(Acer rubrum)*, white ash *(Fraxinus americana)*, quaking aspen, and bigtooth aspen *(Populus grandidentata)*. In the Appalachian Mountains yellow buckeye *(Aesculus octandra)*, sweet birch *(Betula lenta)*, and black cherry *(Prunus serotina)* are present.

Small trees and shrubs associated with the various mixtures of red spruce forest include witch hobble *(Viburnum alnifolium)*, striped maple *(Acer pensylvanicum)*, pincherry *(Prunus pensylvanica)*, mountain maple *(Acer spicatum)*, mountain ash *(Sorbus americana* and *S. decora)*, beaked hazel *(Corylus cornuta)*, hazel alder *(Alnus serrulata)*, speckled alder *(A. rugosa)*, redosier dogwood *(Cornus stolonifera)*, lowbush blueberry *(Vaccinium angustifolium)*, red raspberry *(Rubus strigosus)*, lambkill *(Kalmia angustifolia)*, black huckleberry *(Gaylussacia baccata)*, and rhododendrons *(Rhododendron* spp.). Herbs are common, but of low volume.

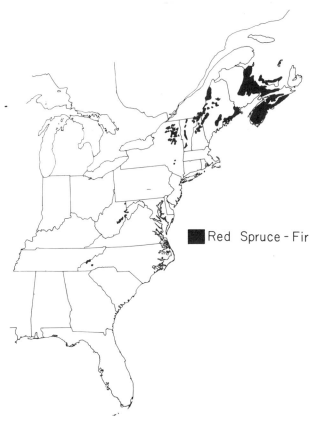

Fig. 12.6. Distribution of red spruce in the United States and Canada. (Modified from Küchler 1964 and Fremlin 1974. Reproduced with permission of the American Geographical Society and the Minister of Supply and Services Canada from the publication "The National Atlas of Canada" published by the Macmillan Co. of Canada Limited, Toronto, Ontario in association with the Department of Energy, Mines and Resources Canada and Information Canada, Ottawa, Canada 1974.)

Fire Effects

Fires occur in red spruce but are rare, generally considered destructive, and of little silvicultural value. Possibly this is because many areas containing red spruce are rooted in shallow organic soil that overlies rock (Koristan 1937). In these areas fire would destroy mature trees by exposing the roots. Hot, deep-burning fires may destroy the organic seedbed. However, on loam soils a seedbed of mineral soil with shade is desirable for red spruce and balsam fir establishment (Fowells 1965; Frank and Bjorkbom 1973). Work by Randall (1976) indicated that prescribed burning holds promise for establishment of spruce and fir on some sites. Moisture must be plentiful, soil temperatures moderate, and competition minimal. Initially red spruce seedlings can become established with as little as 10 percent of full sunlight (Vezina and Peck 1964), but then need 50 percent of full sunlight for optimal growth as the seedling develops (Shirley 1943). On burns, no seedlings become established for the first two years, but 50 percent stocking can be attained after nine years (Randall 1976). Heavy soils that contain clay generally have good herbaceous cover and are poor seedbeds for conifers.

Very few studies of seral vegetation following a burn have been conducted in red spruce forest mixtures. An intensive study by Martin (1956) in southwestern Nova Scotia indicated that mature spruce forests, except for openings, have few herbs and shrubs. After burning forbs and grasses increase dramatically during the first 6 years and maintain a dominant aspect for at least 40 years while the conifers are slowly becoming established. Some of the most abundant forbs and shrubs on 1- to 40-year-old burns were bracken fern *(Pteridium aquilinum)*, teaberry *(Gaultheria procumbens)*, starflower *(Trientalis borealis)*, bunchberry *(Cornus canadensis)*, black huckleberry, blueberry, lambkill, and red raspberry. In this region it appears that once the humus layer under red spruce is destroyed (either directly by burning or by rapid decay after burning when exposed to the sun) on poor soils, bracken fern and the heath plants *(Kalmia, Gaultheria, Arctostaphylos, Gaylussacia, Vaccinium,* and *Rhododendron)* replace the shade-loving species of the closed forest rapidly and create a poor microenvironment for conifers and hardwoods (Martin 1956). The roots of the heath plants grow almost entirely in the humus and competition for available moisture is intense (Martin 1956). Martin found that much of the regeneration of eastern white pine took place on sphagnum moss *(Sphagnum* sp.), indicating that moisture was a limiting factor for germination of conifers over most of the burned areas. Thus at this time there does not seem to be a place for fire, at least not on a large scale, in the management of red spruce.

In the Appalachian Mountains vegetation within a year after burning is frequently a dense, rank growth of trailing blackberry *(Rubus ursinus)* and red raspberry briers and is promptly followed by an abundance of pincherry and yellow birch *(Betula lutea)* (Koristan 1937). Pincherry is dominant for 15 to 20 years and is replaced by yellow birch and less tolerant hardwoods. Several tall shrubs, such as rhododendron, mountain laurel *(Kalmia latifolia)*, serviceberry *(Amelanchier canadensis)*, and black huckleberry contribute to stand density but occupy subordinate positions. Conifers will gradually become established under the trees and shrubs and become dominant after a long period of time.

In the Appalachian Mountains blackberries and raspberries are too competitive for red spruce and balsam fir (Koristan 1937) and must be shaded out by hardwoods before conifers can become established. Second burns on large clear-cuts can annihilate conifer reproduction and prompt the establishment of hardwood cover. However, establishment of a spruce-fir forest after clear-cut harvesting followed by fire depends on the distribution of seed trees and the survival of conifer regeneration on the ground at the time of clearcutting.

Management Implications

Since red spruce is more tolerant of shade than balsam fir and the associated hardwoods a selection system of management can be recommended on the east coast (Frank and Bjorkbom 1973; Frank and Blum 1978). This method of management (harvest about 30 percent of the trees every 10 to 15 years) permits young spruce trees to become established, discourages hardwood growth, minimizes the chances for fire from slash accumulations, permits slash to decompose rapidly in a moist environment, minimizes the chances for insect and disease buildups, and provides a more attractive forest (Fig. 12.7) (Frank and Bjorkbom 1973). Mature or overmature stands (Fig. 12.8) are usually clear-cut, strip cleared, or cut in a shelterwood pattern. These methods of harvesting, while not necessarily less desirable, can be recommended because partial cutting may result in considerable damage to the remaining trees. Windfall may be a serious threat after partial cutting and the risk of insect and disease infestation is high (Frank and Bjorkbom 1973).

Following any form of clear-cutting, scarification (mixing the top 2 to 3 in. of soil) is desirable after logging to expose mineral soil and optimize the chances for germination of red spruce and balsam fir seedlings (Fowells 1965; Frank and Bjorkbom 1973). Establishment is usually best on the shallow, less fertile soils which discourage production of the highly competitive species such as American beech, sugar ma-

Fig. 12.7. A well stocked stand of well formed trees being perpetuated by the selection method of silviculture. (Photo courtesy of USDA Forest Service.)

Fig. 12.8. An unmanaged, uneven-aged spruce-fir stand with a large component of hemlock. (Photo courtesy of USDA Forest Service.)

ple, bracken fern, and red raspberry (Koristan 1937; Westveld 1953; Frank and Bjorkbom 1973). These shrub and hardwood species usually grow on the most fertile soils where red spruce also does well without competition. Less competitive species include yellow birch, red maple, aspen, and paper birch (Westveld 1963). Scattered sprouts of trees and shrubs without herbaceous vegetation usually provide desirable microenvironments for red spruce regeneration.

Balsam fir grows faster and regenerates more easily than red spruce (Koristan 1937; Westveld 1953); therefore selection management is more desirable than clear-cutting. Balsam fir is also more susceptible to insects and diseases than red spruce (Frank and Bjorkbom 1973). Nevertheless balsam fir is a marketable tree species as long as it is harvested before it begins to deteriorate. This species can be harvested during the initial stand entries under the selection system. The result is a forest that is predominantly red spruce which remains relatively free of insects and diseases until maturity (Frank and Bjorkbom 1973).

WHITE AND BLACK SPRUCE

White spruce and black spruce *(P. mariana)* are common throughout the boreal forest zone, but will not be discussed here. Their ecology and relation to fire will be discussed in the next chapter.

REFERENCES

Alexander, R. R. 1973. Partial cutting in old-growth spruce-fir. USDA For. Serv. Res. Paper RM-110. Rocky Mtn. For. and Range Exp. Stn., Fort Collins, Colo.

Alexander, R. R. 1974a. Silviculture of central and southern Rocky Mountain forests: A summary of the status of our knowledge by timber types. USDA For. Serv. Res. Paper RM-120. Rocky Mtn. For. and Range Exp. Stn., Fort Collins, Colo.

Alexander, R. R. 1974b. Silviculture of subalpine forests in the central and northern Rocky Mountains: The status of our knowledge. USDA For. Serv. Res. Paper RM-121. Rocky Mtn. For. and Range Exp. Stn., Fort Collins, Colo.

Alexander, R. R. 1977. Cutting methods in relation to resource use in central Rocky Mountain spruce-fir forests. *J. For.* **75**:395–400.

Boyd, R. J., and G. H. Deitschman. 1969. Site preparation aids natural regeneration in western larch-Engelmann spruce strip clearcuttings. USDA For. Serv. Res. Paper INT-64. Intermt. For. and Range Exp. Stn., Ogden, Utah.

Clagg, H. B. 1975. Fire ecology in high-elevation forests in Colorado. M.S. Thesis. Colorado State Univ., Fort Collins.

Daubenmire, R., and J. B. Daubenmire. 1968. Forest vegetation of eastern Washington and northern Idaho. Wash. State Agric. Exp. Stn. Tech. Bull. 60. Pullman.

Farr, W. A., and A. S. Harris. 1971. Partial cutting of western hemlock and Stika spruce in southeast Alaska. USDA For. Serv. Res. Paper PNW-124. Pac. Northwest For. and Range Exp. Stn., Portland, Ore.

Fowells, H. A. 1965. Silvics of forest trees of the United States. USDA Agric. Handb. 271. Washington, D.C.

Frank, R. M., and J. C. Bjorkbom. 1973. A silvicultural guide for spruce-fir in the Northeast. USDA For. Serv. Gen. Tech. Rep. NE-6. Northeastern For. Exp. Stn.,Upper Darby, Pa.

Frank, R. M., and B. M. Blum. 1978. The selection system of silviculture in spruce-fir stands—procedures, early results, and comparisons with unmanaged stands. USDA For. Serv. Res. Paper NE-425. Northeastern For. Exp. Stn., Upper Darby, Pa.

Franklin, J. F., and C. T. Dyrness. 1969. Vegetation of Oregon and Washington. USDA For. Serv. Res. Paper PNW-80. Pac. Northwest For. and Range Exp. Stn., Portland, Ore.

Fremlin, G. (ed.) 1974. Vegetation. In *The National Atlas of Canada*. Macmillan Co. Can. Ltd., Toronto, Ont. and Dept. Energy, Mines, Res., Inf. Can., Ottawa, Can., pp. 45,46.

Garmon, E. H. 1957. The occurrence of spruce in the interior of British Columbia. Dept. of Lands and Forest, B.C. For. Serv. Tech. Pub. T-49.

Hanley, D. P., W. C. Schmidt, and G. M. Blake. 1975. Stand structure and successional status of two spruce-fir forests in southern Utah. USDA For. Serv. Res. Paper INT-176. Intermt. For. and Range Exp. Stn., Ogden, Utah.

Harlow, W. M., and E. S. Harrar. 1968. *Textbook of Dendrology*. McGraw-Hill, New York.

Harris, A. S. 1966. Effects of slash burning on conifer regeneration in southeast Alaska. USDA For. Serv. Res. Note NOR-18. Northern For. Exp. Stn., Juneau, Alaska.

Isaac, L. A. 1940. Vegetation succession following logging in the Douglas-fir region with special reference to fire. *J. For.* **38**:716–721.

Kallander, H. 1969. Controlled burning on the Fort Apache Indian Reservation, Arizona. *Proc. Tall Timbers Fire Ecol. Conf.* **9**:241–249.

Koristan, C. F. 1937. Perpetuation of spruce on cut-over and burned lands in the higher southern Appalachian Mountains. *Ecol. Monogr.* **7**:126–167.

Küchler, A. W. 1964. Potential natural vegetation of the conterminous United States. Manual to accompany map. Amer. Geogr. Soc. Spec. Pub. 36. (With map, rev. ed., 1965, 1966.)

LeBarron, R. K., and G. M. Jemison. 1953. Ecology and silviculture of the Engelmann spruce-alpine fir type. *J. For.* **51**:349–355.

Martin, J. L. 1956. An ecological survey of burned-over forest land in southwestern Nova Scotia. *For. Chron.* **31**(4):313–336.

Minore, D. 1972. Germination and early growth of coastal tree species on organic seedbeds. USDA For. Serv. Res. Paper PNW-135. Pac. Northwest For. and Range Exp. Stn., Portland, Ore.

Mogren, E. W., and R. C. Barth. 1974. Early revegetation of a burned subalpine forest in the Colorado Front Range. Dept. For. Wood Sci., Colorado State Univ., Fort Collins.

Noble, D. L., and R. R. Alexander. 1977. Environmental factors affecting natural regeneration of Engelmann spruce in the central Rocky Mountains. *For. Sci.* **23**:420-429.

Noble, D. L., and F. Ronco, Jr. 1978. Seedfall and establishment of Engelmann spruce and subalpine fir in clearcut openings in Colorado. USDA For. Serv. Res. Paper RM-200. Rocky Mtn. For. and Range Exp. Stn., Fort Collins, Colo.

Oosting, H. J., and W. D. Billings. 1951. A comparison of virgin spruce-fir forests in the northern and southern Appalachian system. *Ecology* **32**:84–103.

Randall, A. G. 1976. Natural regeneration in two spruce-fir clearcuts in eastern Maine. *Research Life Sci. Univ. Maine.* **23**(13):1–10.

Roe, A. L., R. R. Alexander, and M. D. Andrews.1970. Engelmann spruce regeneration practices in the Rocky Mountains. USDA For. Serv. Prod. Res. Rep. 115. Washington, D.C.

Roe, A. L., and W. C. Schmidt. 1964. Factors affecting natural regeneration of spruce in the Intermountain Region. USDA For. Serv., Intermt. For. and Range Exp. Stn. (Mimeo.)

Ruth, R. H. 1954. Cascade Head climatological data 1936 to 1952. USDA For. Serv. Res. Paper. Pac. Northwest For. and Range Exp. Stn., Portland, Ore.

Ruth, R. H., 1965. Silviculture of the coastal Sitka spruce-western hemlock type. In Proc. Soc. Amer. For., Mills Bldg., Washington, D.C., pp. 32–36.

Schmid, J. M., and R. C. Beckwith. 1972. The spruce beetle. USDA For. Pest Leaflet. Washington, D.C.

Shirley, H. L. 1943. Is tolerance the capacity to endure shade? *J. For.* **41**:339–345.

Sudworth, G. B. 1916. The spruce and balsam firs of the Rocky Mountain region. USDA Bull. 327. Washington, D.C.

Tarrant, R. F., K. C. Lu, W. B. Bollen, and J. F. Franklin. 1969. Nitrogen enrichment of two forest ecosystems by red alder. USDA For. Serv. Res. Paper PNW-76. Pac. Northwest For. and Range Exp. Stn., Portland, Ore.

Taylor, R. F. 1929. The role of Sitka spruce in the development of second-growth in southeastern Alaska. *J. For.* **27**:532–534.

Vezina, P. E., and G. Y. Peck. 1964. Solar radiation beneath conifer canopies in relation to crown closure. *For. Sci.* **10**(4):443–451.

Westveld, M. 1953. Ecology and silviculture of the spruce-fir forests of eastern North America. *J. For.* **51**:422–430.

Williamson, R. L., and R. H. Ruth. 1976. Results of shelterwood cutting in western hemlock. USDA For. Serv. Res. Paper PNW-201. Pac. Northwest For. and Range Exp. Stn., Portland, Ore.

Woodfin, R. O., Jr. 1973. Sitka spruce. USDA For. Serv. Pub. FS-265. Washington, D.C.

CHAPTER

13

RED AND WHITE PINE

The historical role of fire in maintaining stands of red pine *(Pinus resinosa)* and eastern white pine *(Pinus strobus)* in southern Canada, the Lake States, New England, and southern Appalachians is prevalent throughout the literature. In southern Canada the importance of fire in the perpetuation of eastern white pine was well documented by Maissurow (1935, 1941) and Mayall (1941). Most recently fire-return intervals for red pine have been documented as an average of 29 to 37 years (Burgess and Methven 1977; Alexander et al. 1979). The lowest fire-return interval of 14 years has been reported by Cwynar (1977) for the Barron Township, Alonquin Park. Ridge tops burn more frequently than the lower lying areas, which accounts for variation in fire frequencies and differences in vegetation (Ferris 1980). Using an age-class distribution model, Van Wagner (1978) estimated the fire cycle in the Boundary Waters Canoe area to be 50 years prior to 1911.

Spurr (1953, 1954) and Frissel (1973) also showed solid evidence of the historical role of fire for the establishment of red and white pine in Itasca State Park. In northern Wisconsin Maissurow (1941) concluded that in the last five centuries, forest fires have burned through 95 percent of the virgin forest of northern Wisconsin. Vogl (1970) attributed the presence of red pine and jack pine *(Pinus banksiana)* of the Wisconsin pine barrens to past fires.

In New England Hawes (1923) was one of the first writers to document historical data relating to the abundance of white pine in New England and how these trees towered high above the surrounding vegetation, although their origin from fire was first documented by Maissurow (1935), Candy (1939), and Mayall (1941). By 1850 the lumber industry had reached its peak in southern and central Maine and most of the virgin spruce and pine trees had been harvested (Hawes 1923).

In New Hampshire Cline and Spurr (1942) attributed windfalls and fire as the major factors in preparing new seedbeds for white pine. Data from Pennsylvania showed the existence of past fires by the presence of white pine stands on upper slopes facing southeast or southwest (Hough and Forbes 1943). In southern New England white pines occurred among the oak *(Quercus* sp.) and hickory *(Carya* sp.) woods and were highly prized trees (Bromley 1935). In this area historical data indicated that the original forest was open and parklike (Fig. 13.1). Presumably it was kept like this through the use of fire by the Indians (Bromley 1935; Day 1953; Graham et al. 1963).

As one can readily see, the northern pine-hardwoods region would not have had a reference to the word pine without the historical role of fire (Day and Woods 1977). Typically, red and white pines were taller than the associated vegetation. Continuous stands of red and white pine were favored by country broken up with ridges and lakes (Van Wagner 1970) but also survived fires on level terrain if widely spaced (Vogl 1964, 1967). Their absolute height [up to 45 m (150 ft)] and thick bark enabled a number of red and white pine trees to escape occasional severe fires and provided seed over a period of years where mineral seedbeds with minumum competition may have existed.

Fig. 13.1. An open, parklike stand of eastern white pine in Belknap County, New Hampshire. Stands like this are rare today. (Photo courtesy of USDA Forest Service.)

Establishment of red and eastern white pine was usually easiest on the sandy soils where hardwoods had more difficulty getting established (Smith 1940; Mayall 1941; Horton and Bedell 1960). Without fire as a component in natural communities containing pines the communities will succeed to a spruce-fir climax on the northern boundaries and to various combinations of climax hardwoods throughout most of the remaining range within 200 to 300 years (Heinselman 1973). Except for some of the oak species, hardwoods are very susceptible to fire injury or mortality (Stickel and Marco 1936; Oosting 1942; Little 1953; Hepting and Shigo 1972).

Since the white and red pine type is a natural "fire type," the applicability of controlled burning in its regeneration is unquestionable (Horton and Bedell 1960). The primary problem is to demonstrate its controlled use to a fire-prevention-conscious public then to establish experiments for the testing of techniques and silvicultural efficiency in comparison with alternative measures, which is being done.

DISTRIBUTION, CLIMATE, SOILS, AND VEGETATION

Distribution

The red and white pine region (northern pine-hardwoods), as defined in this chapter, includes southern Canada from southeastern Manitoba to Newfoundland and the northern and eastern states from Minnesota and northeastern Iowa to the Atlantic coast, and southward along the Appalachian Mountains to northern Georgia (Horton and Bedell 1960; Fowells 1965; Rowe 1972) (Fig. 13.2). It lies south of the boreal forest in Canada and north of the central hardwoods in the United States (Horton and Bedell 1960). Topography varies from flat and rolling with numerous lakes in the Lake States with low ridges and steep scarps (Heinselman 1973) to rough and hilly in the New England states (Cline and Spurr 1942). Elevation is generally below 610 m (2000 ft) throughout the Lake States, New England, and southern Ontario, Canada, but exceeds this elevation in the southern Appalachians where species of this vegetation zone may be found at elevations as high as 1220 m (4000 ft) (Cope 1932).

Climate

Climate is cool and humid with comparable amounts of precipitation falling each month (Cline and Spurr 1942; Little 1959), except in the region of the Lake States. There about two-thirds of the precipitation falls during the warm season—April to September (Fowells 1965). Elsewhere half the precipitation falls during the warm season. Annual precipitation varies from 50 cm (20 in.) in northern Minnesota to 150 cm (60 in.) in some eastern localities (Fowells 1965). Generally it ranges from 50 cm (20 in.) to 100 cm (40 in.) over most of the region. July temperatures average 17° to 23°C (62° to 72°F). The northern limit of the northern pine-hardwood region is related to the frost-free period and closely parallels the 2°C (35°F) mean annual isotherm (Haddow 1948).

Fig. 13.2. Distribution of red and eastern white pine in southeastern Canada and northeastern United States. (Modified from Küchler 1964 and Fremlin 1974. Reproduced with the permission of the American Geographical Society and the Minister of Supply and Services Canada from the publication "The National Atlas of Canada" published by the Macmillan Co. of Canada Limited, Toronto, Ontario in association with the Department of Energy, Mines and Resources Canada and Information Canada, Ottawa, Canada 1974.)

Soils

Most of the northern pine-hardwoods region was glaciated during the late Pleistocene period (Cook et al. 1952) so the soils are young, relatively coarse-textured (sandy and gravelly) and have weakly developed profiles (U.S. Soil Conservation Service 1951), especially southern Canada, the Lake States, and northern New England.

Some of the uplands in New England have a cap of silty material over the glaciated till (Fowells 1965). From central Pennsylvania southward and in southwestern Wisconsin the soils are much older, generally are finer-textured, and have well-developed profiles. All soils within this region were derived from granites, gneisses, schists, sandstones, and less commonly, from phyllites, slates, shales, and limestones (Fowells 1965).

Vegetation

OVERSTORY

Pines.   Vegetation of this region varies widely depending on fire history, drainage, aspect, soil texture, soil fertility, roughness of topography, prevalence of lakes, and geographical area. With few exceptions the pines, which help characterize the region, owe their origin to fire and the hardwoods range from seral to climax species. Eastern white pine occurs as a seral species, and in some cases as portions of climax communities, throughout the entire northern pine-hardwoods region (Horton and Bedell 1960). It is the most widespread of all the pine species in the pine-hardwood region and grows farther south than either red pine or jack pine (Fowells 1965). Eastern white pine covers about 2,834,000 ha (7,000,000 acres) in the United States (Little et al. 1973) and frequently grows in combination with red pine in the Lake States.

Red pine was formerly most abundant and grew to the largest size in the northern half of the three Lake States and southern Ontario (Fowells 1965; Van Wagner 1970) although scattered stands could be found throughout the northern two-thirds of the northern pine-hardwoods region on glaciated soils to the Atlantic coast (Cook et al. 1952). Acreage of red pine has been expanded to 405,000 ha (1,000,000 acres) in the United States through extensive plantings, particularly in the Lake States (Benzie 1973). Jack pine overlaps the red pine zone and extends northward into the boreal forest as far as the southern Northwest Territories and as far west as the foothills of the Rocky Mountains (Fowells 1965) where lodgepole pine *(Pinus contorta)* becomes its ecological equivalent. Lodgepole pine does not spread eastward because it is intolerant to sweet-fern blister rust *(Cronartium comptoniae)* (Holst 1965).

Generally jack pine grows on the poorest sandy sites or may serve as a seral species to red and white pine on the slightly more fertile sandy soils (Kittredge 1934). It covers more than 810,000 ha (2,000,000 acres) in the Lake States (Benzie 1973). Pitch pine *(Pinus rigida)* is associated with sandy or shallow soils, primarily south of the glaciated region in southern New England and Appalachian Mountains, where fires occur too frequently for other species to remain as a dominant (Little 1959).

Climax Hardwoods.   Hardwood species are the ultimate climax of the northern pine-hardwood region on all but the poorest sites, in the absence of fire (Kittredge 1934; Hepting and Hedgcock 1937; Swan 1970; Little 1973). Climax compositions of hardwoods vary from sugar maple *(Acer saccharum)*-American basswood *(Tilia americana)* in Minnesota and Wisconsin to sugar maple-American beech *(Fagus* *grandifolia)*-eastern hemlock *(Tsuga canadensis)* in Michigan, Indiana, Ohio, and northern New York to sugar maple-yellow birch *(Betula lutea)*-American beech-eastern hemlock in the Northeast (Tubbs and Godman 1973; Filip and Leak 1973). Eastern hemlock dominates sites that have thin soils and are low in fertility (Rogers 1978). It grows well on moist sites but does not attain dominance or form large aggregates on the richest soils.

In the glaciated region black birch *(Betula nigra)* is frequently a climax associate (Cline and Spurr 1942), whereas black cherry *(Prunus serotina)* is often a climax associate on residual soils (Hough and Forbes 1943). On poorly drained sites of the glaciated region basswood, balsam fir *(Abies balsamea)*, and black ash *(Fraxinus nigra)* are the climax dominates (Buell and Bormann 1955).

Boreal Forest Species.   Balsam fir is a climax associate of white spruce *(Picea glauca)* in the southern boreal forest on well drained sites where the climate is cold and moist (Fowells 1965). It also grows with red spruce *(Picea rubens)* in New England, but red spruce is not a part of the boreal forest and is covered under another section of this text. Black spruce *(Picea mariana)* and tamarack *(Larix laricina)* are climax on boggy peat soils (Frissell 1973; Johnston 1973). Further north black spruce can grow on much drier, cold soils (Fowells 1965). All these boreal species can be found on the northern borders of the northern pine-hardwood region, depending on topographic relief and coolness of the area. Jack pine is a seral species of the boreal forest but also grows on the poorest sandy soils of upland sites in the northern pine-hardwood region. Paper birch *(Betula papyrifera)* has the same distribution as balsam fir in the boreal forest but is a seral species on well-drained sandy loams.

Other Tree Species.   Quaking aspen *(Populus tremuloides)* is the most widely distributed species in North America (Fowells 1965) and is a common seral species in the northern pine-hardwood region following fire (Stoeckler 1948; Graham et al. 1963). It is a vigorous sprouter following fire (Shirley 1931; Horton and Hopkins 1965; Sando 1972; Perala 1974). Aspen will grow on shallow rocky soils and loamy sands to heavy clays (Fowells 1965) but does best on loam soils (Stoeckler 1948; Kirby et al. 1957; Horton and Hopkins 1965). Where this species occurs with jack pine, jack pine will usually be growing on the least fertile sandy sites (Horton and Bedell 1960).

Black ash typically grows as a seral species with black spruce in bogs or where there is excess water (Fowells 1965). It grows throughout the northern pine-hardwood region. Red maple *(Acer rubrum)*, northern red oak *(Quercus rubra)*, big tooth aspen *(Populus grandidentata)*, and balsam poplar *(P. balsamifera)* are other common tree species following disturbances in the eastern United States (Heinselman 1973). American elm *(Ulmus americana)* grows best on well-drained flats or bottomlands throughout the eastern United States following fire (Maissurow 1941) but is not restricted to these sites (Putnam 1951).

Northern white cedar *(Thuja occidentalis)* is a climax dominant on nonacidic, flowing, drained bogs (Grant 1929; Little 1950). Northern white cedar grows in the southeastern boreal forest and overlaps the red pine-white pine zone. It grows on neutral or alkaline soils of limestone origin and develops best on moist peat soil where the organic matter is well decomposed (Watson 1936; Little 1950). It does not do well on extremely wet or extremely dry sites (Fowells 1965).

UNDERSTORY

The most commonly associated shrubs include blueberries *(Vaccinium angustifolium* and *V. myrtilloides)*, beaked hazel *(Corlyus cornuta)*, American hazel *(C. americana)*, sweetfern *(Comptonia peregrina)*, bearberry *(Arctostaphylos uva-ursi)*, prairie willow *(Salix humilis)*, mountain maple *(Acer spicatum)*, green alder *(Alnus crispa)*, bush honeysuckle *(Diervilla lonicera)*, New Jersey tea ceanothus *(Ceanothus americanus)*, sand cherry *(Prunus pumila)*, American fly honeysuckle *(Lonciera canadensis)*, black huckleberry *(Gaylussacia baccata)*, thimbleberry *(Rubus parviflorus)*, red raspberry *(R. strigosas)*, black raspberry *(R. occidentalis)* and smooth blackberry *(R. canadensis)*, bunchberry *(Cornus canadensis)*, elderberry *(Sambucus* sp.), serviceberry *(Amelanchier* sp.), bear oak *(Quercus ilicifolia)*, common juniper *(Juniperus communis)*, and chokecherry *(Prunus virginiana)* (Kittredge 1934; Fowells 1965; Heinselman 1973; Hansen et al. 1973).

Most fire research on shrubs has been concentrated on the hazel and blueberry species. Hazel is a vigorous sprouter and is often competitive with young red and white pine trees. Blueberries can be managed commercially and for wildlife (Trevett 1962). Throughout the entire northern forest region grasslands are common on clay loam soils that are often maintained as grasslands with recurrent fires (Swan 1970). Big bluestem *(Andropogon gerardi)* and little bluestem *(Schizachyrium scoparium)* are dominant grasses following fires in forest openings (Neiring et al. 1970; Swan 1970), although many grass and forb species are present (Swan 1970).

EFFECTS OF FIRE ON VEGETATION

Lake States and Southern Ontario

OVERSTORY

The physiography, preponderance of lakes, variations in soils, understory, and variation in drainage have historically enabled past fires to create a mosaic of vegetation in the Lake States region with a preponderance of red and white pine (Hansen et al. 1973; Heinselman 1973, 1981). Each of these pine species has a lot of common requirements that are generated by fires, but white pine is less restrictive in its requirements (Horton and Bedell 1960; Van Wagner and Methven 1978). Both species do well on glaciated soils where intense fires occur at 50- to 200-year intervals (Van Wagner 1978; Heinselman 1981), although moderate surface fires (Fig. 13.3) may occur more frequently (Burgess and Methven 1977; Alexander et al. 1979). Thus in

Fig. 13.3. Moderate surface fire in hazel understory beneath red pine. (Photo courtesy of USDA Forest Service.)

natural areas both species, and particularly red pine, do well in rough topography dotted with streams and lakes where there are natural topographic features to cause variation in fire intensity so that some areas burn gently or not at all, even though the pines may be killed over the area at large (Van Wagner 1970). Virgin stands of white pine were confined exclusively to previously burned-over areas (Maissurow 1935).

For optimum germination and establishment red pine needs bare mineral sandy soil, 35 percent full sunlight, acid soils (pH 5.0 to 6.5), good soil moisture, a seed source, and minimum competition from balsam fir and hazel (Methven and Murray 1974) as well as from competing hardwoods (Horton and Bedell 1960). If the fire burns sufficiently deep to consume duff to the mineral soil (Chrosciewicz 1968; Van Wagner 1972; Methven 1978) competition from beaked hazel and American hazel will be adequately reduced (Buckman 1962, 1964; Van Wagner 1963; Methven and Murray 1974). Balsam fir will also be killed if surface fires are at least of moderate intensity (C. E. Van Wagner, pers. comm.). High fire intensities are needed in aspen and white birch to enable red pine and white pine to become established if a seed source is available (Methven 1978). Often moss *(Polytrichum* sp.), a fire plant, will cover the bare soil after fire (Cline and Steed 1933; Ahlgren 1976a), keeping it moist and relatively free from competition, which permits red pine seedlings to establish up to six years or longer after a fire (Ahlgren 1976a, 1976b). Good seed years only occur every three to seven years (Horton and Bedell 1960) so moss plays a unique role in holding the area until a good crop of red pine seed is produced. Shelterwood-strip harvesting in a mature red pine stand provided favorable growing conditions for red pine nursery stock (Benzie and Alm 1977).

White pine can germinate under the same conditions as red pine but needs less light (20 to 25 percent for optimum germination and establishment), prefers finer sands or loams, can compete under hardwoods (Horton and Bedell 1960), and can germinate over a few centimeters of duff (Smith 1968; Van Wagner and Methven 1978). White pine also has good seed years every three to five years, more frequently than red pine, and produces more seed than red pine, which increases its chances of getting established over red pine (Ahlgren 1976a). Litter surfaces have higher temperatures [20°C (36°F) higher] than mineral soil surfaces and can cause severe seedling mortality (Smith 1951). Sandy soils are two to five times as efficient in conducting heat as organic material (Buoyoucos 1913).

In competition with each other, red pine grows more rapidly than white pine. This gives red pine an edge if a seedbed is available at the time a seed crop is produced. Roots of red pine are capable of penetrating difficult materials, whereas the roots of white pine will spread laterally and search for pockets of deeper soils (Horton and Bedell 1960). Red pine is relatively disease-free, except for occasional infestations of scleroderris canker (Skilling and O'Brien 1973), whereas white pine is subject to damage by weevils and white pine blister rust (Mayall 1941; Ahlgren 1976a). Red pine and white pine are common on moist sites, but red pine is most competitive on dry sites and white pine is most competitive on wet sites following fire on glaciated soil (Horton and Bedell 1960).

Once established, white pine is more wind resistant than red pine up to 25 years of age (Curtis 1939), although both species are windfirm on most soils (Horton and Bedell 1960). Neither species can tolerate fire until 50 years or so after germination (McConkey and Gedney 1951; Van Wagner 1970). At this time red pine is usually 18 m (60 ft) tall and the bark is thick enough to tolerate most fires. Younger trees have limbs too close to the understory vegetation that can easily ignite the extremely flammable red pine leaves and destroy the trees (Van Wagner 1970). The older trees are quite tolerant of fire, provided they are moderately to widely spaced (Vogl 1964; Van Wagner 1970). White pine is less resistant to stand-replacing fires, for it is more susceptible to root injury (McConkey and Gedney 1951) and white pine is extremely susceptible to rot after injury by fire (Frissell 1973). However large trees easily tolerate light to moderate surface fires about every 20 to 30 years.

If crown fires occur more often than every 50 years, jack pine, aspen, and white birch will be the dominant species (Heinselman 1973, 1981). Jack pine usually dominates the poorest sandy sites; aspen will be most prevalent on sandy-loam and loam soils; and white birch will be most prevalent on well-drained cold sites that are best suited to the boreal forest species. Black spruce can dominate peaty bog sites if burned every 25 to 40 years. Jack pine and black spruce produce seed at 10 to 15 years of age whereas aspen and birch are prolific sprouters (Heinselman 1973).

Jack pine produces serotinous cones that remain on trees with live seed for at least 25 years (Ellis 1911). Temperatures in excess of 60°C (140°F) will melt the semiresinous bonds and those trees that remain standing after fire will drop seed on the freshly burned upland site (Cameron 1953; Beaufait 1960). Similarly, black spruce produces semiserotinous cones (Vincent 1965) which also drop seed on the boggy peat soil from standing trees after the fire has passed. Thus both of these species easily perpetuate themselves (jack pine easier than black spruce) after frequent fires (Methven 1978).

Jack pine usually originates after fire in uncut stands (Cayford et al. 1967). It not only germinates best but shows best initial survival on a bare mineral soil (Eyre 1938; Chrosciewicz 1974) and it requires high light intensity for best growth (Beaufait 1962). However, for establishment, young seedlings need only minimal exposure to light, warm temperatures and ample moisture (Beaufait 1962). Seed in the cones of jack pine are very resistant to heat. They can tolerate 150°C (300°F) for 30 to 45 seconds and 370°C (700°F) for 10 to 15 seconds (Beaufait 1960). Temperatures in slash burns containing 37 to 50 seed trees per hectare (15 to 20 trees per acre) rarely exceeded 315°C (600°F) at a height of 5.2 m (17 ft) (Beaufait 1961). For natural regeneration following harvesting seed trees are the only successful means of providing seeds since cones high in the crowns are not consumed by burning (McRae 1979). Prescribed burns to achieve successful regeneration should be conducted in the spring (Beaufait 1962; McRae 1979).

If left unburned, the boreal forest sites will probably succeed to white spruce-fir-paper birch-white cedar-black spruce (Heinselman 1973). The warmer sites will succeed to sugar maple-yellow birch-basswood and eastern hemlock in Michigan (Maissurow 1941). Some white pine and elm may remain as codominants (Maissurow 1941). Sugar maple easily invades aspen sites, as do the boreal forest species on white birch and jack pine sites.

Jack pine can easily be retained with frequent fires or managed with prescribed burns and the seed tree method (McRae 1979). However, to retain red pine and white pine in the Lake States region, intense surface fires must occur about every 150 to 350 years (Heinselman 1973). Most stands cannot tolerate intense surface fires until they are at least 50 years of age (Van Wagner 1970, 1978). Rough topography that is dotted with lakes and streams help to restrict the area of intense fires and retain a seed source. The intense fires are necessary to get a near-mineral soil, suppress competing shrubs, and provide a microenvironment for moss which will hold a seedbed ready until a seed crop is produced. In flat open terrain, such as the pine barrens in Wisconsin, red pine must be widely spaced to protect itself from intense surface fires and continuous crown fires (Vogl 1964, 1967). Neither red pine nor white pine can become established in areas with a quick flush of herbaceous growth, whereas jack pine seedlings are not impeded by herbs and rapidly grow above them (Ahlgren 1976a).

UNDERSTORY

Shrubs are common in the understory of red and white pine communities (Horton and Bedell 1960). Hazel species are severe competitors with forest species (Ahlgren 1973). However, rootstocks grow in humus near the soil surface and several consecutive light summer fires or one intense surface fire (humus moisture down to 10 to 15%) can kill 90 percent or more of the stems (Buckman 1962, 1964). Sando (1969) and Axelrod and Irving (1978) also found that successive fires eliminated some clones, but always reduced height and proportion of area dominated by hazel species in the understory.

Light fires can induce blueberry sprouting (Sharp 1970; Trevett 1962), moderate fires slow recovery (Swan 1970; Ahlgren 1973), and severe fires seriously retard sprouting (Hall 1955; Vogl 1964; Smith 1968). Depending on the degree of damage, optimum fruiting may occur the second year after a light fire (Trevett 1962) or during the third or fourth year after moderate to intense fires (Swan 1970; Ahlgren 1973). Use of buds by ruffed grouse *(Bonasa umbellus)* during winter months may delay maximum fruiting until the fifth year (Sharp 1970). Bearberry tolerates fire well and is abundant on some forest sites (Cayford et al. 1967).

Along with hazel species, balsam fir saplings are severe competitors of red pine on sandy sites (Fig. 13.4) but can be easily killed with gentle understory burns (Methven and Murray 1974). Other shrubs severely harmed include creeping snowberry *(Gaultheria hispidula)*, trailing arbutus *(Epigaea repens)*, starflower *(Trientalis borealis)*, and Wood anemone *(Anemone quinquetolia)* (Stocks and Alexander 1980). Where white pine is prevalent on loam soils hardwood tree species are severe competitors (Fig. 13.5).

Bush honeysuckle can be temporarily harmed by fire (Swan 1970) but is generally unaffected by fire unless the intensity is severe (Ahlgren 1973). By contrast, serviceberry, flameleaf sumac *(Rhus copallina)*, prairie willow, red raspberry, sarsaparilla *(Aralia nudicaulis)*, teaberry *(Gaultheria procumbens)*, and pincherry *(Prunus pensylvanica)* do well after fire (Cayford et al. 1967; Vogl 1967; Ahlgren 1973; Stocks and Alexander 1980). Alders *(Alnus crispa* and *A. rugosa)* and sweetfern reach peak abundance about 10 years after fire and then decline (Ahlgren 1973).

Numerous grasses and forbs can be abundant the first few years after a burn (Ahlgren 1960; Methven 1973). Species on fresh burn sites include twinflower *(Linnaea borealis)*, wild lily-of-the-valley *(Maianthemum canadense)*, bunchberry, and bracken fern *(Pteridium aquilinum)* (Cayford et al. 1967; Sidhu 1973); other prevalent species include fireweed *(Epilobium angustifolium)*, Bicknell geranium *(Geranium bicknellii)*, Aster *(Aster macrophyllus)*, marchantia *(Marchantia polymorpha)*, *Aralia nudicaulis,* climbing buckwheat *(Polygonum cilinode)*, Pennsylvania sedge *(Carex pensylvanica)*, cornlily *(Clintonia borealis)*, roughleaf ricegrass *(Oryzopis asperifolia)*, woolly panic-grass *(Panicum lanuginosum)*, and meadow pinegrass *(Calamagrostis canadensis)* (Ahlgren 1960; Methven 1973).

Some of the most prevalent species in fire-maintained openings include northern bedstraw *(Galium boreale)*, big bluestem, azure aster *(Aster azureus)*, sunflower

Fig. 13.4. Balsam fir invading the understory of red pine on a sandy site in northern Minnesota. (Photo courtesy of USDA Forest Service.)

Fig. 13.5. Hardwood trees invading the understory of white pine and red pine on loam soil in northern Minnesota. (Photo courtesy of USDA Forest Service.)

(Helianthus rigidus), Pennsylvania sedge, little bluestem, and western yarrow *(Achillea millifolium)* (Vogl 1964).

Mosses and lichens are prevalent on burned soils but often disappear 5 to 10 years later (Ahlgren 1973). They have low nitrogen requirements (Daubenmire 1947) and are very important seedbed areas for the establishment of red and white pine (Ahlgren 1976a).

New England

OVERSTORY

Throughout the northeastern United States white pine generally occurred as scattered individuals in mixture with eastern hemlock and hardwoods but occasionally formed pure or nearly pure stands (Lutz and McComb 1935; Cline and Spurr 1942; Hough and Forbes 1943). It thrived best in rough and hilly topography, along valleys (warmer and drier aspects), in sandy soils, on dry exposures, and following fire on recent blowdowns (Lutz and McComb 1935; Cline and Spurr 1942; Hough and Forbes 1943). Today most pure stands of white pine have been severely damaged by the native white pine weevil (Cline and Spurr 1942).

Requirements for germination are similar to those mentioned for the Lake States region. It establishes itself best on bare mineral soil and second best on sites with *Polytrichum* or hypnum moss (Cline and Steed 1933). Areas with a sparse cover of shrubs, pioneer tree species, or herbaceous plants are also suitable (Smith 1951), but not on sites where grass is dominant (Cline and Steed 1933). A shelterwood cutting method would help to promote white pine (Smith 1951).

Once established, white pine competes successfully with hardwoods on sandy soil because it is better adapted (Smith 1940). Shade is important in initial establishment but full sunlight is necessary for best growth after the trees are a few years old (Smith 1940; Clements 1957; Horton and Bedell 1960; Shonk 1975).

White pine does not compete well with advanced growth of eastern hemlock and hardwoods on loam soils (Cline and Spurr 1942). On sandy soils, however, white pine competed well following single fires and was usually a codominant with eastern hemlock—a very aggressive climax species (Cline and Spurr 1942). At lower elevations white pine was usually confined to southern exposures. Eastern hemlock is found on all exposures (Cline and Spurr 1942) but is increasingly less abundant as the depth of soil and fertility of a site increases (Rogers 1978). Surface fires in old stands of white pine may be beneficial if they are of low intensity (Little 1973), but deepburning ground fires may cause root injuries that are more serious than crown injuries (McConkey and Gedney 1951). Where 75 percent of the surface roots were damaged, McConkey and Gedney found that trees 5 to 15 cm (2 to 6 in.) in diameter were all killed. Large trees [30 cm (12 in.) or larger] were more resistant and only 40 percent died. With less than 25-percent root damage and two-thirds crown kill survival was higher for all size classes of trees. Mortality was 80, 46, and 14 percent for the small, medium, and large trees, respectively. Bark of eastern hemlock is very fire-resistant but the roots are shallow and also very susceptible to damage by fire (Cline and Spurr 1942).

In addition to white pine, pioneer species include short-lived paper birch, pincherry, and aspen. Climax species include eastern hemlock, American beech, and yellow birch at the higher elevations and sugar maple, white ash, and basswood at the lower elevations.

In Pennsylvania and southward black cherry is a fugitive species of high commercial value (Hough and Forbes 1943). It is common in virgin forests of high plateaus and lives for 100 to 250 years. Seedling mortality is high except in forest openings. In full sunlight black cherry usually outgrows all of its associates. Thus it is very abundant in secondary successions following windthrow, single fires, or logging (Hough and Forbes 1943).

Pitch pine is common on sandy soils in southern New England and will dominate these sites where fires are frequent (Little 1973). In the absence of fire, however, there is a tendency for white pine and eastern hemlock to supplant pitch pine in the northern area and for oak or pine and oak to dominate on these plains in the southern part of its range (Bromley 1935; Cary 1936; Little 1973).

Removal of white pine and eastern hemlock in swampy areas generally led to a dominance of red maple and yellow birch in New England (Bromley 1935), or sugar maple, birch, and red spruce in Minnesota (Rogers 1978). Red pine is a rare species in the New England region, although it is present as a rare individual or group of trees (Fowells 1965). It is seldom mentioned as a major species by the various ecologists who have studied trees and fire succession in New England.

UNDERSTORY

In the New England area blueberries are common and recover quickly following light fires (Trevett 1962) but more slowly following moderate to severe fires (Swan 1970). Black huckleberry, black raspberry, smooth blackberry, teaberry, pinxter-bloom *(Rhododendron nudiflorum)* respond to fire very similarly to the various blueberry species (Swan 1970; Flinn and Wein 1977). Bracken fern and bush honeysuckle were temporary decreasers (Swan 1970). Witch hazel *(Hamamelis virginiana)* is severely harmed by fire (Brown 1960).

Shrubs and trees favored by fire included mountain laurel *(Kalmia latifolia)*, bear oak, prairie willow, eastern hop-hornbeam *(Ostrya virginiana)*, bluebuck *(Carpinus caroliniana)*, wax myrtle *(Myrica asplenifolia)*, pincherry, and flowering dogwood *(Cornus floridana)* (Skutch 1929; Brown 1960). Most other species are relatively unaffected by fire or only make up a small part of the total plant composition. Fireweed, goldenrod *(Solidago* sp.), and bunchberry are common forbs following fires in forest communities (Skutch 1929; Hansen et al. 1973).

Forbs and grasses are most prevalent on fire-maintained openings (Neiring et al. 1970; Swan 1970). Common species include big bluestem, little bluestem, redtop *(Agrostis alba)*, Pennsylvania sedge, goldenrod *(Solidago rugosa)*, butterfly milkweed *(Asclepias tuberosa)*, tickclover *(Desmodium canadense, D. paniculatum)*, roundhead lespedeza *(Lespedeza capitata)*, wild lupine *(Lupinus perennis)*, and Indiangrass *(Sorghastrum nutans)* (Neiring et al. 1970; Swan 1970).

Appalachian Region

White pine is occasionally present in this region where fire or other disturbance has given it the rare opportunity, in the absence of severe competition, to become established. However, this is not a prime growing area for white pine. Probably black cherry, a hardwood species, is the most important tree species of this region that is favored by fire at the regeneration stage.

For other species of this region and succession after fire, see the section on red spruce in Chapter 12.

MANAGEMENT IMPLICATIONS

Red and white pine on extensive "natural" areas can best be managed in rough topography with numerous lakes and streams. Such physiography helps to create variety in fire intensity and limit the extent of fires. Thus a seed source is retained. Occasional surface fires of low intensity every 25 to 50 years in localized areas appear to be the best way to manage for red and white pine naturally (Ahlgren 1976a; Van Wagner 1978; Heinselman 1981). Neither of these species are well suited to logging methods that involve clear-cutting and the exclusion of fire (Van Wagner and Methven 1978).

Opening the canopy, providing a mineral seedbed, minimizing shrub competition, and retaining a nearby seed source are all requirements for establishment of red and white pine (Van Wagner and Methven 1978). The shelterwood method is a favored procedure for regeneration (Benzie 1973; Little 1973).

Once above 18m (60 ft) tall red and white pine can tolerate light fires. Intense surface fires create desirable seedbeds but they also can do much root and crown damage to white pine. Red pine is more tolerant of intense surface fires than white pine if no more than two-thirds of the needles are scorched (Sucoff and Allison 1968; Methven 1977). Moreover, the effect of fire is generally ephemeral as the area is rapidly revegetated (Methven 1974). As the bark becomes thick enough to withstand surface fires, crown damage becomes the potential danger (Fig. 13.6). The extent of crown damage depends on surface fire intensity (Van Wagner 1973). Also pines can stand more crown scorch in the dormant season after flushing because heat that kills needles may leave the buds undamaged.

White pine can become established with 20 percent light, 2 cm (0.8 in.) of duff, and a pH below 7, whereas red pine requires 35 percent light, bare mineral soil, and a pH below 6.5 (Ahlgren 1976a). A mosaic of conditions in a burn can create conditions for a variety of red and white pine mixtures. White pine will usually be most abundant after fire but white pine blister rust *(Cronartium ribicola)* and white pine weevils take their toll on white pine whereas red pine is essentially disease-free and

Fig. 13.6. Typical young stand of red pine that is very susceptible to crown fires. (Photo courtesy of USDA Forest Service.)

grows faster than white pine. These interactions can easily result in nearly pure or at least dominant stands of red pine (Heinselman 1973; Ahlgren 1976a).

Jack pine prefers a bare mineral seedbed to become established (Beaufait 1962) and the most practical method developed to establish this species is to leave 30 to 75 seed trees per hectare (12 to 30 trees per acre) and burn the slash beneath the trees in the spring (McRae 1979). The seed in the cones is very resistant to heat (Beaufait 1961). They drop on the fresh burn and establish quickly if there is good soil moisture, warm temperatures, and exposure to light (Beaufait 1962).

Establishment of white pine on the east coast may have to be done initially by planting on sandy soils and southern exposures where competition can be minimized by some means. The problems associated with white pine blister rust and white pine weevil damage and with red pine Scleroderris canker should be considered when managing these species. Mixed stand plantings help to minimize disease problems and minimize crown fire danger (R. M. Loomis, pers. comm.). The older pine trees are the most resistant to disease and insect problems.

REFERENCES

Ahlgren, C. E. 1960. Some effect of fire on reproduction and growth of vegetation in northeastern Minnesota. *Ecology* **41**:431–445.

Ahlgren, C. E. 1973. Effects of fires on temperate forests: North-central United States. In T. T. Kozlowski and C. E. Ahlgren (eds.) *Fire and Ecosystems*. Academic Press, New York, pp. 195–223.

Ahlgren, C. E. 1976a. Regeneration of red pine and white pine following wildfire and logging in northeastern Minnesota. *J. For.* **74:**135–140.

Ahlgren, C. E. 1976b. Is fire the answer? *Timber Prod. Assoc. Bull.* **31**(Oct.–Nov.):18–19.

Alexander, M. E., J. A. Mason, and B. J. Stocks. 1979. Two and a half centuries of recorded forest fire history. Can. For. Serv., Great Lakes For. Res. Cent. Leafl. Sault Ste. Marie, Ont.

Axelrod, A. N., and F. D. Irving. 1978. Some effects of prescribed fire at Cedar Creek Natural History Area. *J. Minn. Acad. Sci.* **44**(2):9–11.

Beaufait, W. R. 1960. Some effects of high temperatures on the cones and seeds of jack pine. *For. Sci.* **6:**194–199.

Beaufait, W. R. 1961. Crown temperatures during prescribed burning in jack pine. *Mich. Acad. Sci., Arts, and Letters* **46:**251–257.

Beaufait, W. R. 1962. Procedures in prescribed burning for jack pine regeneration. Mich. Coll. Mining and Technol. Tech. Bull. 9. Ford For. Cent., L'Anse, Mich.

Benzie, J. W. 1973. Red pine. In *Silvicultural Systems for the Major Forest Types of the United States.* USDA Hanb. No. 445. Washington, D. C., pp. 58–60.

Benzie, J. W., and A. A. Alm. 1977. Red pine seedling establishment after shelterwood-strip harvesting. USDA For. Serv. Res. Note NC-224. North Central For. Exp. Stn., St. Paul, Minn.

Bouyoucos, G. J. 1913. An investigation of soil temperatures and some of the most important factors influencing it. Mich. Agric. Exp. Stn. Tech. Bull. 17. Ann Arbor.

Bromley, S. W. 1935. The original forest types of southern New England. *Ecol. Monogr.* **5:**61–89.

Brown, J. H., Jr. 1960. The role of fire in altering the species composition of forests in Rhode Island. *Ecology* **41:**310–316.

Buckman, R. E. 1962. Two prescribed summer fires reduce abundance and vigor of hazel brush regrowth. USDA For. Serv. Tech. Note LS-620. Lake States Exp. Stn., St. Paul, Minn.

Buckman, R. E. 1964. Effects of prescribed burning on hazel in Minnesota. *Ecology* **45:** 626–629.

Buell, M. F., and F. H. Bormann. 1955. Deciduous forests of Ponemah Point, Red Lake Indian Reservation, Minnesota. *Ecology* **36:**646–658.

Burgess, D. M., and I. R. Methven. 1977. The historical interaction of fire, logging and pine: A case study at Chalk River, Ontario. Can. For. Serv. Inf. Rep. PS-X-66. Petawawa For. Exp. Stn., Chalk River, Ont.

Cameron, H. 1953. Melting point of the bonding material in lodgepole and jack pine cones. Can. Dept. Resour. and Develop. For. Br., For. Res. Div. Silv. Leafl. No. 86.

Candy, R. H. 1939. Discussion on the reproduction and development of white pine. *For. Chron.* **15:**88–92.

Cary, A. 1936. White pine and fire. *J. For.* **34:**62–65.

Cayford, J. H., Z. Chrosciewicz, and H. P. Sims. 1967. A review of silvicultural research in jack pine. Can. Dept. For. Br. Dept. Pub. 1173. Can. Dept. For. and Rural Develop., Ottawa, Ont.

Chrosciewicz, Z. 1968. Drought conditions for burning raw humus on clear-cut jack pine sites in central Ontario. *For. Chron.* **44:**30, 31.

Chrosciewicz, Z. 1974. Evaluation of fire-produced seedbeds for jack pine regeneration in central Ontario. *Can. J. For. Res.* **4:**455–457.

Clements, J. R. 1957. The influence of fire on the development of the pine forest in part of Section 30, T467, R36W, Iron County, Michigan. M.S. Thesis. Univ. Mich., Ann Arbor.

Cline, A. C., and S. H. Spurr. 1942. The virgin upland forest of central New England. Harvard For. Bull. 21. Harvard For., Petersham, Mass.

Cline, A. C., and A. V. Steed. 1933. A preliminary study of the effect of ground cover types on white pine regeneration in group selection cuttings. Unpubl. Ms., Harvard For. Library, Cambridge, Mass.

Cook, D. B., R. H. Smith, and E. L. Stone. 1952. The natural distribution of red pine in New York. *Ecology* **33:**500–512.

Cope, J. A. 1932. Northern white pine in the southern Appalachians. *J. For.* **30:**821–828.

Curtis, J. D. 1939. Pruning white pine. Mass. State Coll. Ext. Serv. Leafl. 170.

Cwynar, L. C. 1977. The recent fire history of Barron Township, Algonquin Park. *Can. J. Bot.* **55:**1524–1538.

Daubenmire, R. F. 1947. Plants and the Environment, 1st ed. Wiley, New York.

Day, G. M. 1953. The Indian as an ecological factor in the northeastern forest. *Ecology* **34:**329–346.

Day, R. J., and G. T. Woods. 1977. The role of wildfire in the ecology of jack and red pine forest in Quetico Provincial Park. Fire Ecol. Study Quetico Prov. Park Rep. No. 5. Min. Nat. Res., Atikohan, Ont.

Ellis, L. M. 1911. Some notes on jack pine *(Pinus divaricata)* in western Ontario. *For. Quart.* **9:**1–14.

Eyre, F. H. 1938. Can jack pine be regenerated without fire? *J. For.* **36:**1067–1072.

Ferris, J. E. 1980. The fire and logging history of Voyageur's National Park: An ecological study. M.S. Thesis. Mich. Tech. Univ., Houghton.

Filip, S. M., and W. B. Leak. 1973. Northeastern northern hardwoods. In *Silvicultural Systems for the Major Forest Types of the United States.* USDA Handb. No. 445. Washington, D.C., pp. 75–77.

Flinn, M. A., and R. W. Wein. 1977. Depth of underground plant organs and theoretical survival during fire. *Can. J. Bot.* **55:**2550–2554.

Fowells, H. A. 1965. *Silvics of Forest Trees of the United States.* USDA Handb. No. 271. Washington, D.C.

Fremlin, G. (ed.) 1974. Vegetation. In *The National Atlas of Canada.* Macmillan Co. Can. Ltd., Toronto, Ont. and Dept. Energy, Mines, Res., Inf. Can., Ottawa, pp. 45, 46.

Frissell, S. S., Jr. 1973. The importance of fire as a natural ecological factor in Itasca State Park, Minnesota. *Quat. Res.* **3:**397–407.

Graham, S. A., R. P. Harrison, Jr., and C. E. Westell, Jr. 1963. *Aspens: Phoenix Trees of the Great Lakes Region.* Univ. Mich. Press, Ann Arbor.

Grant, M. L. 1929. The burn succession in Itasca County, Minnesota. M.A. Thesis. Univ. Minn., Minneapolis.

Haddow, W. R. 1948. Distribution and occurrence of white pine *(Pinus strobus* L.) and red pine *(Pinus resinosa* Ait.) at the northern limit of their range in Ontario. *J. Arnold Arb.* **29:**217–226.

Hall, I. V. 1955. Floristic changes following the cutting and burning of a woodlot for blueberry production. *Can. J. Agric. Sci.* **35:**143–152.

Hansen, H. L., L. W. Krefting, and V. Kurmis. 1973. The forest of Isle Royale in relation to fire history and wildlife. Univ. Minn. Agric. Exp. Stn. Tech. Bull. 294 (For. Serv. 13). Minneapolis.

Hawes, A. F. 1923. New England forests in retrospect. *J. For.* **21:**209–224.

Heinselman, M. L. 1973. Fire in the virgin forests of the Boundary Waters Canoe Area, Minnesota. *Quat. Res.* **3:**329–382.

Heinselman, M. L. 1981. Fire intensity and frequency as factors in the distribution and structure of northern ecosystems. In *Proceedings of the Conference: Fire Regimes in Ecosystem Properties.* USDA For. Serv. Gen. Tech. Rep. WO-26, Washington, D.C., pp. 7–57.

Hepting, G. H., and G. G. Hedgcock. 1937. Decay in merchantable oak, yellow poplar, and basswood in the Appalachian region. USDA Tech. Bull. 570. Washington, D.C.

Hepting, G. H., and A. L. Shigo. 1972. Difference in decay rate following fire between oaks in North Carolina and Maine. *Plant Dis. Rep.* **56:**406, 407.

Holst, M. J. 1965. Research on frost tolerance and disease resistance of lodgepole pine and jack pine at the Petawawa Experiment Station. Project Summary on file for Project P-156C.

Horton, K. W., and G. H. Bedell. 1960. White and red pine: Ecology, silviculture and management. Can. Dept. For. Bull. 124. Can. Dept. North Aff. and Natl. Resour. Ottawa, Ont.

Horton, K. W., and E. J. Hopkins. 1965. Influence of fire on aspen suckering. Can. Dept. For., For. Res. Br. Pub. No. 1095. Ottawa, Ont.

Hough, A. F., and R. D. Forbes. 1943. The ecology and silvics of forests in the high plateaus of Pennsylvania. *Ecol. Monogr.* **13**:299–320.

Johnston, W. F. 1973. Black spruce. In *Silvicultural Systems for the Major Forest Types of the United States.* USDA Handb. No. 445. Washington, D.C., pp. 62, 63.

Kirby, C. L., W. S. Bailey, and J. G. Gilmour. 1957. The growth and yield of aspen in Saskatchewan. Sask. Dept. Nat. Res., For. Br. Tech. Bull. No. 3.

Kittredge, J., Jr. 1934. Evidence of the rate of forest succession on Star Island, Minnesota. *Ecology* **15**:24–35.

Küchler, A. W. 1964. Potential natural vegetation of the conterminous United States. Manual to accompany the map. Amer. Geogr. Soc. Spec. Pub. 36. (With map, rev. ed., 1965, 1966.)

Little, S., Jr. 1950. Ecology and silviculture of white cedar and associated hardwoods in southern New Jersey. Yale Univ. School For. Bull. 56. New Haven, Conn.

Little, S., Jr. 1953. Prescribed burning as a tool of forest management in the northeastern states. *J. For.* **51**:496–500.

Little, S., Jr. 1959. Silvical characteristics of pitch pine *(Pinus rigida).* USDA For. Serv. Stn. Paper NE-119. Northeast For. Exp. Stn., Upper Darby, Pa.

Little, S., Jr. 1973. Effects of fire on temperate forests: Northeastern United States. In T. T. Kozlowski and C. E. Ahlgren (eds.) *Fire and Ecosystems.* Academic Press, New York, pp. 225–250.

Little, S., Jr., D. E. Beck, and L. Della-Bianca. 1973. Eastern white pine. In *Silvicultural Systems for the Major Forest Types of the United States.* USDA Handb. No. 445. Washington, D.C., pp. 73–75.

Lutz, H. J., and A. L. McComb. 1935. Origin of white pine in virgin forest stands of northwestern Pennsylvania as indicated by stem and basal branch features. *Ecology* **16**:252–256.

Maissurow, D. K. 1935. Fire as a necessary factor in the perpetuation of white pine. *J. For.* **33**:373–378.

Maissurow, D. K. 1941. The role of fire in the perpetuation of virgin forests of northern Wisconsin. *J. For.* **39**:201–207.

Mayall, K. M. 1941. White pine succession as influenced by fire (interim report). Nat. Res. Counc. Can. Pub. N.R.C. No. 989.

McConkey, T. W., and D. R. Gedney. 1951. A guide for salvaging pine injured by forest fires. USDA For. Serv. Northeast For. Exp. Stn. Res. Note No. 11. Northeast For. Exp. Stn., Upper Darby, Pa.

McRae, D. J. 1979. Prescribed burning in jack pine logging slash: A review. Can. For. Serv. Rep. O-X-289. Great Lakes For. Res. Cent., Sault Ste. Marie, Ont.

Methven, I. R. 1973. Fire, succession and community structure in a red and white pine stand. Can. For. Serv. Inf. Rep. PS-X-43. Petawawa For. Exp. Stn., Chalk River, Ont.

Methven, I. R. 1974. Development of a numerical index to quantify the aesthetic impact of forest management practices. Can. For. Serv. Inf. Rep. PS-X-51. Petawawa For. Exp. Stn., Chalk River, Ont.

Methven, I. R. 1977. Prescribed fire, crown scorch and mortality: Field and laboratory studies on red and white pine. Can. For. Serv. Inf. Rep. PS-X-31. Petawawa For. Exp. Stn., Chalk River, Ont.

Methven, I. R. 1978. Fire research at the Petawawa Forest Experiment Station: The integration of fire behaviour and forest ecology for management purposes. In *Proceedings of the Fire Ecology in Resource Management Workshop.* Can. For. Serv. Inf. Rep. NOR-X-210. North. For. Res. Cent., Edmonton, Alberta.

Methven, I. R., and W. G. Murray. 1974. Using fire to eliminate understory balsam fir in pine management. *For. Chron.* **50**:77–79.

Neiring, W. A., R. H. Goodwin, and S. Taylor. 1970. Prescribed burning in southern New England: Introduction to long-range studies. *Proc. Tall Timbers Fire Ecol. Conf.* **10**:267–286.

Oosting, H. J. 1942. An ecological analysis of the plant communities of Piedmont, North Carolina. *Amer. Midl. Natur.* **28**:1–126.

Perala, D. A. 1974. Prescribed burning in an aspen-mixed hardwood forest. *Can. J. For. Res.* **4:**222–228.

Putnam, J. A. 1951. Management of bottomland hardwoods. USDA For. Serv. Occas. Paper 116. Southern For. Exp. Stn., New Orleans, La.

Rogers, R. S. 1978. Forests dominated by hemlock *(Tsuga canadensis):* Distribution as related to site and postsettlement history. *Can. J. Bot.* **56:**843–854.

Rowe, J. S. 1972. *Forest Regions of Canada.* Can. For. Serv. Dept. Environ., Ottawa, Ont.

Sando, R. W. 1969. The current status of prescribed burning in the Lake States. USDA For. Serv. Res. Note NC-81. North Cent. For. Exp. Stn., St. Paul, Minn.

Sando, R. W. 1972. Prescribed burning of aspen-hardwood stands for wildlife habitat improvement. Paper presented at the 34th Midwest Fish and Wildl. Conf. (Dec. 10–13, Des Moines, Iowa).

Sharp, W. M. 1970. The role of fire in ruffed grouse habitat management. *Proc. Tall Timbers Fire Ecol. Conf.* **10:**47–61.

Shirley, H. L. 1931. Does light burning stimulate aspen suckers? *J. For.* **29:**524, 525.

Shonk, B. 1975. Hurray! The woods are on fire. *Yankee Mag.* **39**(4):80–85.

Sidhu, S. S. 1973. Early effects of burning and logging in pine-mixed woods. II. Recovery in numbers of species and ground cover of minor vegetation. Can. For. Serv. Inf. Rep. PS-X-47. Petawawa For. Exp. Stn., Chalk River, Ont.

Skilling, D. D., and J. T. O'Brien. 1973. How to identify Scleroderris canker and red pine shoot blight. USDA For. Serv. North Cent. For. Exp. Stn., St. Paul, Minn.

Skutch, A. F. 1929. Early stages of plant succession following forest fires. *Ecology* **10:**177–190.

Smith, D. M. 1951. The influence of seedbed conditions on the regeneration of eastern white pine. Conn. Agric. Exp. Stn. Bull. 545. New Haven, Conn.

Smith, D. W. 1968. Surface fires in northern Ontario. *Proc. Tall Timbers Fire Ecol. Conf.* **8:**41–54.

Smith, L. F. 1940. Factors controlling the early development and survival of eastern white pine *(Pinus strobus* L.) in central New England. *Ecol. Monogr.* **10:**373–420.

Spurr, S. H. 1953. Forest fire history of Itasca State Park, Minnesota. For. Note 18. Univ. Minn., Sch. For., Minneapolis.

Spurr, S. H. 1954. The forests of Itasca in the nineteenth century as related to fire. *Ecology* **35:**21–25.

Stickel, P. W., and H. F. Marco. 1936. Forest fire damage studies in the northeast. III. Relation between fire injury and fungal infection. *J. For.* **34:**420–423.

Stocks, B. J., and M. E. Alexander. 1980. Forest fire behavior and effects research in northern Ontario: A field oriented program. In Proc. 6th Conf. Fire and For. Meteorol. Soc. Amer. For., Washington, D.C., pp. 18–24.

Stoeckler, J. H. 1948. The growth of quaking aspen as affected by soil properties and fire. *J. For.* **46:**727–737.

Sucoff, E. I., and J. H. Allison. 1968. Fire defoliation and survival in a 47-year-old red pine plantation. Minn. For. Res. Notes 187. Univ. Minn., Sch. For., Minneapolis.

Swan, F. R., Jr. 1970. Post fire response of four plant communities in south-central New York State. *Ecology* **51:**1074–1082.

Trevett, M. F. 1962. Nutrition and growth of the lowbush blueberry. Maine Agric. Exp. Stn. Bull. 605. Univ. Maine, Orono.

Tubbs, C. H., and R. M. Godman. 1973. Lake States northern hardwoods. In *Silvicultural Systems for the Major Forest Types of the United States.* USDA Handb. No. 445. Washington, D.C., pp. 55–58.

U. S. Soil Conservation Service. 1951. Problem areas in soil conservation. U.S. Soil Conserv. Serv. Northeast Reg. Ref. Upper Darby, Pa.

Van Wagner, C. E. 1963. Prescribed burning experiments: Red and white pine. Can. Dept. For. Pub. No. 1020. Ottawa, Ont.

Van Wagner, C. E. 1970. Fire and red pine. *Proc. Tall Timbers Fire Ecol. Conf.* **10**:211–219.

Van Wagner, C. E. 1972. Duff consumption by fire in eastern pine stands. *Can. J. For. Res.* **2**:34–39.

Van Wagner, C. E. 1973. Height of crown scorch in forest fires. *Can. J. For. Res.* **3**:373–378.

Van Wagner, C. E. 1978. Age-class distribution and the forest fire cycle. *Can. J. For. Res.* **8**:220–227.

Van Wagner, C. E., and I. R. Methven. 1978. Prescribed fire for site preparation in white and red pine. In Proc. White and Red Pine Symp. Can. For. Serv. Symp. Proc. O-P-6. Great Lakes For. Res. Cent., Sault Ste. Marie, Ont.

Vincent, A. B. 1965. Black spruce: A review of its silvics, ecology and silviculture. Can. Dept. For. Pub. 1100. Ottawa, Ont.

Vogl, R. J. 1964. Vegetational history of Crex Meadows, a prairie savanna in northwestern Wisconsin. *Amer. Midl. Natur.* **72**:157–175.

Vogl, R. J. 1967. Controlled burning for wildlife in Wisconsin. *Proc. Tall Timbers Fire Ecol. Conf.* **6**:47–96.

Vogl, R. J. 1970. Fire and the northern Wisconsin pine barrens. *Proc. Tall Timbers Fire Ecol. Conf.* **10**:175–209.

Watson, R. 1936. Northern white-cedar. U.S. For. Serv. Reg. 9 Misc. Pub. Milwaukee, Wisc.

14

COASTAL REDWOOD AND GIANT SEQUOIA

Coastal redwood *(Sequoia sempervirens)* and giant sequoia *(Sequoiadendron giganteum)* are the only living members of their genera which existed in the late Cretaceous period and flourished through the Northern Hemisphere in the Tertiary period (Fowells 1965). They are the world's tallest and most massive trees. Fire has been a significant force in forests of both species but the role of fire and its importance has been different for each genus.

Coastal redwood is essentially a climax forest in which redwood is sustained by low rates of reproduction in tree replacement in the canopy (Viers 1972). It does not depend upon recurrent fires for its status but is tolerant of the low intensity fires which appear to occur occasionally on sites where stands are best developed (S. D. Viers, pers. comm.). Viers estimates fire frequencies on mesic sites at 200- to 500-year intervals and at 50- to 100-year intervals inland and at higher elevations. Where fires are more frequent, Douglas fir, *(Pseudotsuga menziesii)* a redwood associate which appears to require fire (or logging), appears more frequently. Fritz (1931) has documented inland fire frequencies of 25 years in coastal redwood stands but gave no data to indicate that fire was necessary for the survival of coastal redwood. Seedlings can become established on rotten logs (Viers 1972) or bare mineral soil (Boe 1975). Moreover, it can regenerate from basal stem sprouts, but trees from sprouts are most prevalent on cut-over areas (Carranco and Labbe 1975).

Giant sequoias in the Sierra Nevada Mountains of east-central California do not sprout. They are dependent entirely on seed for regeneration. Moreover, giant sequoia is seral to white fir *(Abies concolor)*. Thus the role and importance of fire is different than for the coastal redwoods.

In habitat types containing giant sequoias, fire frequencies averaged 10 to 18 years before 1875, depending on the site (Kilgore and Taylor 1979). Southwest-facing slopes burned more frequently than southeast-facing slopes. Most groves occur on western and southwestern slopes (Hartesveldt et al. 1975). For individual giant sequoias, fire frequencies ranged from 3 to 35 years (Kilgore and Taylor 1979). Fire is especially important in this forest to provide a mineral seedbed for regeneration (Hartesveldt and Harvey 1967; Kilgore and Biswell 1971; Kilgore 1973), recycle nutrients, and remove thickets of climax species such as incense cedar *(Calocedrus decurrens)* and white fir (Kilgore 1973).

The Leopold report of 1963 (Leopold et al. 1963) pointed out some of the problems with fire protection in National Parks and recommended restoring park forests to pre-European conditions with emphasis on more openness.

> Much of the west slope [of the Sierra] is a dog-hair thicket of young pines, white fir, incense-cedar, and mature brush—a direct function of over-protection from natural ground fires. . . . A reasonable illusion of primitive America could be recreated, using the utmost in skill, judgment, and ecologic sensitivity.

This report served as a catalyst which brought together previous field observations, research, and experience with fire and led to the formulation of a new fire policy in 1963 for National Parks that recognizes fire as a natural phenomena (Kilgore and Briggs 1972).

Present National Park Service fire management policy divides all fires into either management fires or wildfires (Kilgore 1976).

> It defines management fires as those of both natural origin and prescribed burns which contribute to the attainment of the management objectives of a park through execution of predetermined prescriptions defined in detail in a portion of the approved resources management plan.

As such, this policy allows some lightning-caused fires to burn; it recognizes prescribed burning as a tool to manage areas that have been modified by prolonged exclusion of fire; and it continues fire suppression in developed areas and for all fires not classified as management fires.

Since 1968 the Kings Canyon National Park has had an active "natural environmental fire" and prescribed burning program. Prescribed burning has been adopted as the technique for restoring fire to the giant sequoia-mixed conifer forests in the Park's middle elevations (Kilgore 1970; Kilgore and Briggs 1972).

COASTAL REDWOOD

Distribution, Climate, Elevation, and Soils

Coastal redwood occupies a narrow belt about 725 km (450 miles) long and seldom more than 40 km (25 miles) wide along the Pacific Coast from extreme southwestern Oregon to southern Monterey County in central California (Fig. 14.1) (Lindquist 1975). It is a region of moderate to heavy winter rain and frequent fog. The northern portion is characterized by relatively continuous stands but in the southern half of its range [beginning 160 km (100 miles) north of San Francisco] coastal redwood more commonly grows in small, isolated, locally suitable sites along streams and alluvial flats.

Fig. 14.1. Distribution of coastal redwood and giant sequoia in California and southwestern Oregon. (Adapted from Küchler 1964 and Fowells 1965. Reproduced by permission of the American Geographical Society and the USDA Forest Service.)

Annual rainfall averages 100 to 125 cm (40 to 50 in.) over most of the range (Lindquist 1975), but can vary from 65 to 310 cm (25 to 122 in.) (Fowells 1965). Snow rarely falls in this region but summer fogs are frequent. Apparently the fog is more important than precipitation in delineating the coastal redwood region (Fowells 1965). Fog decreases water loss from evaporation and transpiration and, by condensing on and dripping from tree crowns, adds to the soil moisture supply to some degree (Cooper 1917). Mean annual temperatures vary between 10° to 16°C (50° to 60°F) and extremes seldom rise above 38°C (100°F) or fall below -1°C (30°F) (Lindquist 1975).

The coastal redwoods (Fig. 14.2) grow from sea level to 915 m (3000 ft) but are most commonly found between 30 to 760 m (100 to 2500 ft) (Fowells 1965). The best stands have developed on flats and benches along the larger streams and on moist coastal plains, river deltas, moderate westerly slopes, and valleys opening toward the sea. However, coastal redwood does not tolerate ocean winds and considerable evidence suggests that it is sensitive to ocean salts carried inland during storms (Fowells 1965). Redwoods become smaller and give way to other species as altitude, dryness, and slope increase.

Soils are usually podzolic or alluvial in nature and have formed from either consolidated or soft sedimentary rocks. Soil texture varies from sandy loam to clay. The

Fig. 14.2.   Coastal redwood (right center) and Douglas fir (left center) and associated understory of California laurel, tanoak, and various shrubs and ferns in coastal redwood forest of northern California. (Photo courtesy of USDI National Park Service.)

soils are generally moderate to deep on slopes, especially deep on flats, have high moisture retention capacity, and are moderately to strongly acidic (Fowells 1965; Lindquist 1975).

Most mature coastal redwood trees are between 60 to 90 m (200 to 300 ft) tall and 1.8 to 3.7 m (6 to 12 ft) in diameter (Lindquist 1975). They reach' maturity in 400 to 800 years (Carranco and Labbe 1975). Maximum recorded characteristics of coastal redwood are: 112 m (367 ft) tall, 6.1 m (20 ft) in diameter, and about 2200 years old (Lindquist 1975). The bark is about 0.3 m (1 ft) thick on mature trees but may be thinner if it has been subjected to frequent fires (Fritz 1931).

Vegetation

Coastal redwood grows in pure stands along moist river flats and gentle slopes (Fowells 1965) but is usually associated with other conifer and broad-leaved trees (Lindquist 1975). Douglas fir is the most important associate and is distributed throughout the coastal redwood type (Fig. 14.2). Other associates are more limited, but they include grand fir *(Abies grandis)*, western hemlock *(Tsuga heterophylla)*, Sitka spruce *(Picea sitchensis)*, tanoak *(Lithocarpus densiflora)*, and Pacific madrone *(Arbutus menziesii)*. Broad-leaved plants of lesser importance may include vine maple *(Acer circinatum)*, big-leaf maple *(A. macrophyllum)*, red alder *(Alnus rubra)*, golden chinkapin *(Castanopsis chrysophylla)*, Oregon ash *(Fraxinus latifolia)*, Pacific wax myrtle *(Myrica californica)*, Oregon white oak *(Quercus garryana)*, cascara *(Rhamnus purshiana)*, willows *(Salix* spp.), California laurel *(Umbellularia californica)*, western raspberry *(Rubus leucodermis)*, thimbleberry *(R. parviflorus)*, and poison oak *(Rhus diversiloba)*. Forbs and grasses are a minor component in occasional openings of the forest. Most shrubs, particularly the berry *(Rubus* spp.) plants are associated with openings or cut-over areas.

Fire Effects

Coastal redwood is not dependent on fire for its survival except perhaps to create a mineral seedbed for seedlings (Boe 1975) and to remove non decaying debris (Carranco and Labbe 1975). However, few of the young trees escape fires (Lindquist 1975) and most of the large trees on cut-over areas have grown from stump sprouts (Fig. 14.3) (Carranco and Labbe 1975), even though they may only provide 8 percent of full stocking (Fowells 1965). Coastal redwood is a vigorous stump sprouter on cut-over areas, especially if the stumps are less than 1.2 m (4 ft) in diameter (Fowells 1965). In one study 81 percent of the stumps less than 1.4 m (4.7 ft) in diameter sprouted, but only 36 percent of the stumps over 3.2 m (10.5 ft) sprouted (Person 1937). Another study by Boe (1975) showed that the average percentage of sprouting over all tree sizes was 50. Most trees over two years old will resprout (Mason 1924). Thus even though young trees may be burned they will resprout and eventually escape fire for a long enough period of time to avoid being top-killed. This implies that stump sprouts are important in the regeneration of coastal redwoods on cut-over areas

Fig. 14.3. Most of the large coastal redwood trees on cut-over areas have grown from stump sprouts. (Photo courtesy of Warren Kubler, Napa, California.)

as stated by Carranco and Labbe (1975) but also that a significant number of new trees can become established on mineral seedbeds as seedlings (Boe 1975). In natural ecosystems, however, fire does not appear to be a necessary environmental factor to maintain all-aged, old growth coastal redwood stands (Viers 1972).

Fire is responsible for 90 to 100 percent of the dry rot in lower portions of coastal redwood trees (Fritz 1931). Dry rot gains entry through open fire scars or what are known as "goosepens." Recurring fires and advancing decay weakens large trees and eventually causes about 2 percent of them to topple over after each fire (Fritz 1931). However, due to resistance of non charred coastal redwood to decay, most of these trees can be salvaged many years after they have fallen (Carranco and Labbe 1975).

Disturbed areas can be quickly invaded by a number of genera such as *Montia, Oxalis, Iris, Stachys, Gaultheria, Pteridium, Cirsium,* and *Rubus* (Fritz 1931; Boe 1975). Australian fireweed *(Erechtites prenanthoides)* reached its greatest density on the burned seedbed and *Stachys* seemed to be most prevalent on the unburned seedbed. The other genera grow equally well on burned or mechanically disturbed sites.

In view of competition from weeds immediately after fire, hot spots may be very important for the establishment of redwood seedlings, as Hartesveldt and Harvey (1967) reported for giant sequoia. Establishment is most successful on hot spots because the seedlings are less subject to damping-off following emergence (Davidson 1971) and are invaded more slowly by shrubs and annual and perennial herbs. However Boe (1975) found that survival of coastal redwood seedlings can be restricted on hot spots because of severe competition from shrubs and droughty conditions. This implies that soil moisture is very important for establishment of coastal redwood, as was reported for giant sequoia by Beetham (1962), and that the occasional establishment of seedlings on rotten logs in a moist environment (Viers 1972) is still an important regenerating mechanism in coastal redwood forests.

Coastal redwood seed viability has been studied by Davidson (1971). The seed has low viability, short storage life, and low survival due to a number of pathological problems that reduce germination 85 percent. Once the remaining seeds emerge, damping-off can destroy 100 percent of the seeds that germinate in humus. However only 5 percent died due to damping-off in clay loam, which largely explains the natural occurrence of coastal redwood on disturbed soil only (Davidson 1971). As a consequence soil disturbance on cut-over stands is highly desirable for optimum establishment of coastal redwood. In natural ecosystems where there are only 50 to 60 trees/ha (20 to 25/acre) and we take into account the 1000-year life span of individual trees, only 2 to 3 trees/ha (1/acre) need to become established and enter the canopy every 40 to 50 years (Viers 1972). This rate of regeneration can be achieved without fire (Viers 1972).

Once established, coastal redwoods can grow rapidly on cut-over areas. Dominant trees on good sites are 30 to 46 m (100 to 150 ft) tall at 50 years, and 50 to 65 m (165 to 220 ft) tall at 100 years (Fowells 1965). Height growth is most rapid up to the 35th year (Lindquist and Palley 1963). On the other hand, if competition is severe, the trees can be suppressed for long periods of time but accelerate growth when released from competition (Fowells 1965). Growth of coastal redwoods is much slower under natural conditions than on cut-over areas (Viers 1972).

Well-stocked stands of old growth stands of coastal redwood support about 50 to 60 trees/ha (20 to 25/acre), whereas cut-over areas may support as many as 2470 stems/ha (1000/acre) at 20 years, including 500 dominant and codominant trees (Fowells 1965). At 60 years cut-over areas of coastal redwood have a basal area 45 m^2 (486 ft^2) on the best sites (Bruce 1923). A high stocking of seedlings on cut-over areas is desirable because of the high rainfall and the high shade tolerance of coastal redwood trees (Fowells 1965). Two to ten seed trees per acre, depending on the site, are needed to provide adequate seed for regeneration (Boe 1975). Effective seeding distance is 120 m (400 ft) on level areas and 60 m (200 ft) uphill (Fowells 1965; Boe 1975).

GIANT SEQUOIA

Distribution, Climate, Elevation, and Soils

Giant sequoia grows more or less in isolated groves on western slopes of the Sierra Nevada Mountains in central California, in a narrow belt approximately 420 km (260 miles) long and no more than 25 km (15 miles) wide (Hartesveldt et al. 1975). The northernmost grove consists of six trees west of Lake Tahoe in Placer County. The southernmost grove with 100 trees, is approximately 80 km (50 miles) northeast of Bakersfield in southern Tulare County, California (Fig. 14.1). The largest groves and the largest number of groves are found in a north-south zone of central Tulare County and southern Fresno County—a distance of 120 km (75 miles). Areas of all groves in California comprises 14,416 ha (35,607 acres).

Precipitation averages 115 to 150 cm (45 to 60 in.) per year on the best sites (Fowells 1965). Most of the precipitation comes in the form of snow from September through May. Summer storms are infrequent and light, but high levels of soil moisture are maintained within the groves during the summer months by percolation of high-elevation ground water into the groves (Rundel 1972). Outside of the groves, there is little input of ground water during the summer months (Rundel 1972).

Temperature extremes vary from −24° to 38°C (−12° to 100°F) (Fowells 1965). Winter minimum temperatures are commonly cited as the limiting factor in determining the upper elevational limits of sequoia groves (Rundel 1972). The mechanisms of this limiting factor are complex, but it appears that the seedlings are particularly susceptible to frost (Rundel 1972).

Most of the groves are at elevations between 1370 and 2285 m (4500 to 7500 ft) (Shirley 1947), although some occur as low as 855 m (2800 ft) (Hartesveldt et al. 1975) and a few as high as 2715 m (8900 ft) (Fowells 1965). Fowells describes the variation in range as follows: at the northern part of its range the trees grow at 1370 to 1675 m (4500 to 5500 ft); in the central part, at 1645 to 2135 m (5400 to 7000 ft); and at the southern end, at 1830 to 2440 m (6000 to 8000 ft). The groves are usually found in canyons where soil moisture is always adequate, but also occur on or near the tops of high exposed ridges where underground water is available (Fowells 1965).

Soils are derived primarily from granitic parent materials (Storie and Weir 1951; Rundel 1972). The soils are moderately acidic with a pH of 6 (Storie and Weir 1951), which is the ideal pH for growth of giant sequoia seedlings (Beetham 1962). Soil texture ranges from sand to clay loam (Storie and Weir 1951; Hartesveldt et al. 1975). These podzolic soils are relatively thin and not nutritionally rich (Hartesveldt et al. 1975).

Giant sequoia is the world's largest tree in terms of volume (Fig. 14.4) (Hartesveldt et al. 1975). However, it is somewhat shorter in height and larger in diameter than the coastal redwood. Average height of mature trees is 75 to 85 m (250 to 275 ft) and average diameter is 3.0 to 6.1 m (10 to 20 ft) (Hartesveldt et al. 1975). Record characteristics are 89 m (291 ft) tall (Fowells 1965), 10.2 m (33.3 ft) in diameter, and 3200 years of age (Hartesveldt et al. 1975). Thickness of bark averages O.25 m (10 in.), although individual sections of bark will exceed O.6 m (24 in.) (Hartesveldt et al. 1975). Giant sequoia grows more slowly than the coastal redwood (Carranco and Labbe 1975).

Vegetation

In the groves of giant sequoia, white fir, sugar pine *(Pinus lambertiana)*, and giant sequoia dominate the forest (Fig. 14.5) (Kilgore 1972a). Incense cedar, ponderosa

Fig. 14.4. Fire scars reveal that sequoias were frequently burned in the past and there is little evidence to show the presence of crown fires. (Photo taken in 1890 by George Reichel, courtesy of Mrs. Dorothy Whitener.)

Fig. 14.5. Following 80 years of protection many thickets of white fir, a shade tolerant species, grew into the lower crowns of the giant sequoias and large amounts of debris accumulated on the forest floor. (Photo courtesy of National Park Service.)

pine *(Pinus ponderosa)*, and California black oak *(Quercus kelloggii)* are not typical associates in the mesic habitat of giant sequoia, but become increasingly abundant on the drier sites and at lower elevations (Kilgore 1972a, 1972b). Above 2000 m (6560 ft) red fir *(Abies magnifica)* may be an important associate of giant sequoia (Rundel 1971). Douglas fir is not present in most of the groves, but is a minor associate of the canopy vegetation in several of the northern giant sequoia groves (Rundel 1971).

Shrubs and herbs are present in giant sequoia groves, but they are infrequent and grasses are almost absent (Rundel 1971; Kilgore 1972a). Ground cover is usually less than 5 percent (Rundel 1971). Dominant species of the occasional forbs include fairybells *(Disporum hookeri)*, rattlesnake plantain *(Goodyera oblongifolia)*, Sierra sweet cicely *(Osmorhiza chilensis)*, wintergreen *(Pyrola picta)*, mountain bluebells *(Mertensia ciliata)*, *Nortrientalis latifolia,* northern bedstraw *(Galium triflorum)*, trail plant *(Adenocaulon bicolor)*, and pine violet *(Viola lobata)*. Below 1800 m (5900 ft), distinct species include queencup beadlily *(Clintonia uniflora)*, rattlesnake plantain, wild ginger *(Asarum harwegi)*, and Wood strawberry *(Fragaria californica)*. Grasses and sedges *(Carex* spp.) are very rare. Occasional shrubs of rose *(Rosa* sp.) and snowberry *(Symphoricarpos acutus)* are present (Rundel 1971; Kilgore 1972a). Several other shrub species are present within the groves, but they are restricted to riparian habitats (Rundel 1971).

Fire Effects

Giant sequoia is seral to white fir and is dependent on late summer fires for its survival (Hartesveldt 1964; Biswell et al. 1968; Rundel 1971; Kilgore 1972a). Repeated disturbances in the past such as fire set back the succession of other plants and favored the reproduction of giant sequoia (Hartesveldt 1964; Kilgore 1972a). Burned plots contain as many as 54,340 giant sequoia seedlings per hectare (22,000/acre) whereas the control plots often do not have a single seedling (Kilgore and Biswell 1971). "Thus fire, and apparently the hotter the better, is the prime requisite for the reproduction of sufficient seedlings so that the species may survive" (Harvey et al. 1980).

Optimum germination occurs on bare soil when seed has a chance to lodge in a crack or be covered with soft friable soil; when soil moisture is near field capacity; when soil temperature is 20°C (68°F); when sunlight is one-half full strength; and when soil is slightly acid (pH 6–7) (Stark 1968; Hartesveldt et al. 1975). Nevertheless, germination can take place on clay soils, when temperatures range from -1.6° to 34°C (30° to 92°F), when soil pH is 9, and when the seed is not covered (Stark 1968). However, the survival rate is low under these less-than-optimal conditions (Hartesveldt et al. 1975) and soil moisture must always be near field capacity (Beetham 1962). Optimum temperatures for germination occur in April, May, September, and October; best germination occurs in the spring (Stark 1968).

Cones of giant sequoia are serotinous and may remain on trees for many years after the seeds become germinable (Harvey et al. 1980). Near normal numbers of seed have been reported in cones that were over 20 years old (Buchholz 1938), but viability of seeds decline as the cones age (Hartesveldt et al. 1975). Seed falls through-out the year because the Douglas squirrel *(Tamiasciurus douglasii)* feeds on cones two to five years old and cuts cones from the trees throughout the year (Harvey et al. 1980). This causes the cones to open and disperse about 40,000 şeeds per tree. More-over, the long-horned wood-boring beetle *(Phymatodes nitidus)* feeds on vascular tissue of the cone scales that are 4 to 11 years old (Hartesveldt et al. 1975; Harvey et al. 1980). This feeding causes the cones to dry and release about 120,000 seeds per tree, primarily during late summer and fall (Hartesveldt et al. 1975; Harvey et al. 1980). Explosive reproduction follows when numerous cones are opened by the heat of fire (Harvey et al. 1980). Thus ample seed is always available for reproduction and giant sequoia appears to be well adapted to fire.

The seeds are light and the slightest movement of air will carry them as far as 180 m (600 ft) away from the parent tree (Hartesveldt et al. 1975). Once the seed has fallen on the ground, percentage of germination will quickly drop to 1 or 2 percent (Hartesveldt et al. 1975; Harvey et al. 1980). If exposed to full sunlight for 20 days, the percentage germination will drop to zero (Hartesveldt et al. 1975).

After seedfall on a fresh burn in late summer or fall, snow may play a significant role in protecting seed from sunlight and in wedging seed among soil particles. The weight of snow may be significant, for extension of the radicle 2.5 to 5 cm (1 to 2 in.) usually takes place beneath the snow (Fry and White 1930). Following snowmelt, soil moisture would be optimum and favorable soil temperatures in moderately shaded areas could provide an optimum environment for establishment. Seedling mortality is high, primarily because of dessication, but those adjacent to rocks, limbs, and other objects in heavily burned areas usually survive (Hartesveldt and Harvey 1967).

Fire scars reveal that giant sequoias were frequently burned in the past, and there is little evidence to show the presence of crown fires (Fig. 14.4) (Hartesveldt 1964). Sierra Nevada forests were once considered to be nearly immune to continuous crown fires (Show and Kotok 1924). However, following 80 years of protection, many thickets of white fir, a shade-tolerant species, grew into the lower crowns of the giant sequoias and large amounts of debris accumulated on the forest floor (Fig. 14.5) (Hartesveldt 1964; Biswell et al. 1968). This created a severe fire hazard and pre-scribed burning was introduced into the Sequoia and Kings Canyon National Parks in 1968 to restore the natural ecosystem (Kilgore 1972a). This will prevent the possibil-ity of catastrophic wildfires and provide a mechanism for the forest to regenerate it-self (Kilgore 1972b).

Following prescribed burning in a previously protected giant sequoia grove, 87 percent of the white fir and sugar pine saplings were killed; 38 percent of these trees between 15 and 30 cm (6 to 12 in.) in diameter were killed (Kilgore 1972a). This ad-justed the successional pattern by reducing the number of surviving white fir to more nearly approach the proportion found in the mature age class, a class which germi-nated under prefire suppression conditions (Kilgore 1972a). No giant sequoia seedlings were found on this November burn, but they would normally be present after an August or September fire.

Several shrubs become prevalent after fire. Those that are obviously favored include deerbrush *(Ceanothus integerrimus)*, mountain whitehorn *(C. cordulatus)*, greenleaf manzanita *(Arctostaphylos patula)*, small-leaved ceanothus *(C. parvifolius)*, bittercherry *(Prunus emarginata)*, and gooseberry *(Ribes roezlii)* (Hartesveldt et al. 1975). Lupine *(Lupinus latifolius* var. *columbianus)* is an abundant forb on burns (Rundel 1971). Small-seeded annuals increase greatly after fire, but the shade-tolerant herbs such as rattlesnake plantain continue to maintain their original populations in understory of crowns (Hartesveldt et al. 1975).

Aside from seedling establishment, we do not know much about the regeneration of giant sequoia groves. A few forests have a good range of age classes and are regenerating, whereas others are senescent and evidently have been declining over a period of time estimated at 100 to 500 years or more—a decline that began before western civilization became a factor (Rundel 1971). A somewhat drier climate that began 8000 years ago is the reason given for the decline (Hartesveldt et al. 1975). This might mean that less of the total area now occupied by giant sequoia will ultimately be suitable habitat for the species.

Since only a few new trees need to become established each year to maintain a particular forest (Rundel 1971) it appears that the reintroduction of fire into the giant sequoia groves will permit them to maintain their present populations of trees, unless the climate becomes significantly drier. Trees less than 75 years old increase in diameter at an average of 2.5 cm (1 in.) every 3 to 5 years (Fowells 1965), but ancient, overmature trees may require more than 20 years to grow 2.5 cm (1 in.) in diameter (Hartesveldt et al. 1975).

MANAGEMENT IMPLICATIONS

Commercial Forest Practices for Coastal Redwood

The importance of fire in the coastal redwood, a climax species, is not fully documented, but historically fire was present. Natural ecosystems do not need fire as a management tool, but fire appears important to establish seedlings in cut-over stands. Resprouting alone is not adequate to fully restock a cut-over coastal redwood forest (Boe 1975). Partial stocking from seedlings is necessary and can only take place with significant success following fires or soil disturbance. Disturbance by mechanical equipment will provide adequate seedbeds for seedlings (Boe 1975) but the decay-resistant debris should be removed by fire to minimize the risk of catastrophic fires (Fritz 1931). During logging operations it may be highly desirable to burn the slash, as it accumulates to accomplish both of the objectives of seedling establishment and removal of debris. Also selective cutting may be preferable to other methods of cutting for seedling establishment to minimize droughty conditions in the upper soil layers that can be caused by high soil temperatures if the canopy is too open.

Subsequent fires after the initial treatment of logging slash may or may not be necessary in cut-over stands of coastal redwoods. We have reasonably solid evidence to say that the forest can tolerate fire well, but it may not be necessary (Viers 1972). The decision as to whether fire should be used as a management tool should be left to the forest manager, depending on the apparent need for fire to suppress understory species or create mineral seedbeds and upon the overall management objectives.

Natural Ecosystem Practices for Giant Sequoia

Giant sequoia, a seral species, definitely needs fire in its ecosystem if we are to have any hope of regenerating the groves that we see today. Fire is necessary to prepare a mineral seedbed and to keep the shade-tolerant species suppressed. Without fire, a fuel ladder of shade-tolerant trees and dead debris could easily allow a crown fire in the groves. This would kill many of the existing trees and possibly eliminate the chance for seedling establishment, as shown by a November fire (Kilgore 1972a). Fires need to occur in late summer or fall so that seeds falling after the fire will have an opportunity to germinate under ideal fall or spring conditions. The other possibility, if fires are suppressed entirely, is that of white fir eventually dominating the forest and occupying the space of all present giant sequoias as they die out.

REFERENCES

Beetham, N. M. 1962. The ecological tolerance range of the seedling stage of *Sequoia gigantea*. Ph.D. Diss. Duke Univ., Durham, N. C.

Biswell, H. H., R. P. Gibbens, and H. Buchanan. 1968. Fuel reductions and fire hazard reduction costs in giant sequoia forests. *Calif. Agric.* **22**:2–4. Berkeley.

Boe, K. N. 1975. Natural seedlings and sprouts after regeneration cuttings in old-growth redwood. USDA For. Serv. Res. Paper PSW-111. Pac. Southwest For. and Range Exp. Stn., Berkeley, Calif.

Bruce, D. 1923. Preliminary yield tables for second-growth redwood. Calif. Univ. Agric. Exp. Stn. Bull. **361**:425–467. Berkeley.

Buchholz, J. T. 1938. Cone formation in *Sequoia gigantea*. *Am. J. Bot.* **25**:296–305.

Carranco, L., and J. T. Labbe. 1975. *Logging the Redwoods*. Caxton Printers, Caldwell, Idaho.

Cooper, W. S. 1917. Redwoods, rainfall and fog. *Plant World* **20**(6):179–189.

Davidson, J. G. N. 1971. Pathological problems in redwood regeneration from seed. Ph.D. Diss. Univ. Calif., Berkeley.

Fowells, H. A. 1965. Silvics of forest trees of the United States. USDA Handb. 271. Washington, D. C.

Fritz, E. 1931. The role of fire in the redwood region. *J. For.* **29**:939–950.

Fry, W., and J. R. White. 1930. *Big Trees*. Stanford Univ. Press, Palo Alto, Calif. (9th ed. 1948.)

Hartesveldt, R. J. 1964. Fire ecology of the giant sequoia. *Nat. Hist.* **73**(10):12–19.

Hartesveldt, R. J., and H. T. Harvey. 1967. The fire ecology of sequoia regeneration. *Proc. Tall Timbers Fire Ecol. Conf.* **7**:65–77.

Hartesveldt, R. J., H. T. Harvey, H. S. Shellhammer, and R. E. Stecker. 1975. The giant sequoia of the Sierra Nevada. USDI Natl. Park Serv. Pub. No. NPS 120. Washington, D.C.

Harvey, H. T., H. S. Shellhammer, and R. E. Stecker. 1980. Giant sequoia ecology: Fire and reproduction. USDI Natl. Park Serv. Sci. Monogr. Series No. 12. Washington, D.C.

Kilgore, B. M. 1970. Restoring fire to the sequoias. *Natl. Parks and Cons. Mag.* **44**(October):16–22.

Kilgore, B. M. 1972a. Impact of prescribed burning on a sequoia-mixed conifer forest. *Proc. Tall Timbers Fire Ecol. Conf.* **12**:345–375.

Kilgore, B. M. 1972b. Fire's role in a sequoia forest. *Naturalist* **23**(1):26–37.

Kilgore, B. M. 1973. The ecological role of fire in Sierran conifer forests. *Quat. Res.* **3**:496–513.

Kilgore, B. M. 1976. From fire control to fire management: An ecological basis for policies. *Trans. N. Amer. Wildl. Natur. Resources Conf.* **41**:477–493.

Kilgore, B. M., and H. H. Biswell. 1971. Seedling germination following fire in a giant sequoia forest. *Calif. Agric.* **25**(2):8–10. Berkeley.

Kilgore, B. M., and G. S. Briggs. 1972. Restoring fire in high elevation forests in California. *J. For.* **70**:266–271.

Kilgore, B. M., and D. Taylor. 1979. Fire history of a Sequoia–mixed-conifer forest. *Ecology* **60**:129–142.

Küchler, A. W. 1964. Potential natural vegetation of the conterminous United States. Manual to accompany the map. Amer. Geogr. Soc. Spec. Pub. 36. (With map, rev. ed., 1965, 1966.)

Leopold, A. S., S. A. Cain, C. M. Cottam, I. N. Gabrielson, and T. L. Kimball. 1963. Wildlife management in the national parks. *Amer. For.* **69**(4):32–35, 61–63.

Lindquist, J. L. 1975. Redwood. USDA For. Serv. FS-262. Superintendent of Documents, U. S. Govt. Printing Office. Washington, D.C.

Lindquist, J. L., and M. N. Palley. 1963. Empirical yield tables for young-growth redwood. Calif. Agric. Exp. Stn. Bull. 796. Berkeley.

Mason, D. T. 1924. Redwood for reforestation in the Douglas-fir region. *Timberman* **26**(1):130, 132.

Person, H. L. 1937. Commercial planting on redwood cut-over lands. USDA 434. Washington, D.C.

Rundel, P. H. 1971. Community structure and stability in the giant sequoia groves of the Sierra Nevada, California. *Amer. Midl. Natur.* **85**:478–492.

Rundel, P. H. 1972. Habitat restriction in giant sequoia: The environmental control of grove boundaries. *Amer. Midl. Natur.* **87**:81–99.

Shirley, J. C. 1947. *The Redwoods of Coast and Sierra.* Univ. Calif. Press, Berkeley and Los Angeles.

Show, S. B., and E. I. Kotok. 1924. The role of fire in the California pine forests. USDA Bull. 1294. Washington, D.C.

Stark, N. 1968. Seed ecology of *Sequoiadendron giganteum. Madrono* **19**(7):267–277.

Storie, E. R., and W. W. Weir. 1951. Generalized soil map of California. Calif. Agric. Exp. Stn. Ext. Serv. Man. 6. Berkeley.

Viers, S. D., Jr. 1972. Ecology of Coast Redwood. USDI Natl. Park Serv. Prog. Rep. presented to the Western Soc. of Natur., December 27, 1972. (mimeo.) (Rpt. San Francisco).

15

SOUTHEASTERN FORESTS

Historical evidence indicates that fire was a common and widespread occurrence across the South (Harper 1962). William Bartram, a botanist, wrote about fire in 1773, after traveling from South Carolina to the Mississippi. He said, "It happens almost everyday throughout the year, in some part or other, by Indians for the purpose of rousing game, as also by lightning" (Greene 1931). Charles Lyell, an eminent English geologist who made two visits to the United States in the 1840s, was told by a University professor from Tuscaloosa, Alabama that the openness of southeast pine forests was due to frequent fires. They did not hurt the pines and suppressed the more sensitive hardwoods, which might otherwise have sprung up and choked out the pines (Harper 1962). Harper also cited George V. Nash of the New York Botanical Garden in 1894 and 1895 and Mrs. Ellen Call Long of Tallahassee in 1889. Nash and Long made similar observations as to the role of fire in maintaining pine forests.

In contrast were the proponents of European philosophy who contended that fire is destructive to pines. Harper (1962) cites W. W. Ashe in 1895, a prominent botanist and prolific writer of North Carolina; Charles Mohr in 1896, a highly respected German-born botanist of Mobile; Charles S. Sargent in 1892, publisher of *Silva of North America*; Gifford Pinchot in 1899, chief of what later became the U.S. Forest Service; and President Theodore Roosevelt in 1902, a prominent conservationist. All stated that fire is destructive. Pinchot recognized that longleaf pine *(Pinus palustris)* resisted fire better than other trees, but he never advocated any beneficial effects. Roosevelt illustrated the detrimental effects of fire by referring to a picture of Stone

Mountain, Georgia. He said "The ax and fire have removed the forest; and heavy rains have removed the soil which once covered the larger part of this rocky knob" (Harper 1962). Obviously this is an exaggeration.

Roland Harper botanized extensively in the Southeast beginning in 1900 and noticed that fire was frequent in longleaf pine forests. However, he gave it little thought until 1913. In that year he observed numerous longleaf pine seedlings on recent burns as opposed to no seedlings in adjacent tall grass (Harper 1913). Herman H. Chapman (1926, 1932, 1936, 1944), from the Yale School of Forestry, was the first scientist to provide a scientific basis for burning southeast pine. His works, along with those of Greene (1931), who wrote an article entitled "The Forest That Fire Made" and published it in *American Forests;* Herbert L. Stoddard, Sr. (1931), who published a book on Bobwhite Quail *(Colinus virginianus)* and its habits; and Heyward (1936, 1937, 1938, 1939a, 1939b), who published several articles concerning the effect of fire on soils, provided the impetus for other foresters to say what they had thought for many years. John Currey, Claude Bickford, and Hux Coulter were other foresters of this time period who spoke or published on prescribed burning in the Southeast.

From 1940 to the early 1960s there were still many reactionaries who feared fire would be misused. Bonninghausen (1962) gives a rather interesting history of fire policies in the Florida Forest Service. He said the Florida Forest Service of the early 1940s was concerned about heavy fuel accumulations beneath pine forests. A policy was initiated in 1942 to permit controlled burning under conditions in which no more than one-half the tree crown would be scorched. In 1943, a drought year, Florida had many disastrous fires. This drought year revolutionized the need for prescribed burning. Following this severe fire season, C. A. Bickford and John Currey of the Southern Forest Experiment Station removed an old manuscript from their files and published their guide for prescribed burning in longleaf and slash pine *(Pinus elliottii)* forests (Bonninghausen 1962). These methods are very similar to those being used today by forest industry, state forestry agencies, and the National Forest System in the South.

Until 1950 the Florida Forest Service was primarily concerned with reduction of rough. In 1950 the service expanded its prescribed burning policy to include sanitation, preparation of seedbeds, and control of hardwoods. At the insistence of Hux Coulter, Head of Florida Division of Forestry, the Florida Forest Service published a bulletin titled "Using Fire Wisely in the Woods." Much difficulty was encountered in getting this bulletin published, and distribution was restricted for six years following publication. Even in 1962 Bonninghausen said many foresters were afraid the information would get into the wrong hands and do more harm than good.

The U. S. Forest Service was not actively involved in forest management in the Southeast until 1920 to 1930 (Riebold 1971). Initially, however, it perceived its primary responsibility to protect forests from fire. Not until 1960 was fire generally accepted as a tool, 50 years after it was first recommended as a management tool for the Southeast by Chapman in 1909 (Riebold 1971). "A Guide to Fire by Prescription" for the Southern Region was published in 1965 by Dixon. This publication was later revised and updated by Mobley et al. (1973).

FIRE FREQUENCY

Several historians had documented frequent fires in the Southeast before 1793, but the frequency varied, depending on the site and roughness of topography. According to Chapman (1926, 1944), natural fires every two to three years enabled longleaf pine (including longleaf and slash) to dominate two-thirds [35 million hectares (87 million acres)] of the original pine forests of the south Atlantic and Gulf States (Betts 1954). Slash pine was restricted to moist areas on pond margin sites that usually escaped fires for the first five or six years (Garren 1943; Chapman 1944). In the Upper Coastal Plains and Piedmont Regions, where loblolly *(Pinus taeda)* and shortleaf pine *(P. enchinata)* were dominant, loblolly pine became established and grew best on sites that escaped fires at intervals of at least 10 years, whereas shortleaf pine could tolerate fires more frequently because of its resprouting ability (Chapman 1944). On the drier sites with coarse-textured soil, however, shortleaf pine was better adapted than loblolly pine and fire did not have to occur very often.

Without the natural role of fire in the Southeast during the past 50 years, there has been a dramatic reduction in acreage of longleaf pine and an increase of slash and loblolly pine invasions and plantings on sites formerly dominated by longleaf pine (Knight and McClure 1974; Grelen 1978). Today loblolly pine is the predominating pine in much of the South (Cost 1968; Welch 1968; Knight and McClure 1974).

DISTRIBUTION, CLIMATE, ELEVATION, AND SOILS

The southern forest occupies 78 million hectares (193 million acres) from Virginia to east Texas and Oklahoma (Crafts 1958) and can be divided into the Lower Coastal Plain (flatwoods), Upper Coastal Plain, and the Piedmont regions (Fig. 15.1). The Lower Coastal Plain occupies the Florida Peninsula and adjacent flatwoods along portions of the Atlantic and Gulf Coasts. Elevation varies from 8 to 80 m (25 to 270 ft), and average rainfall varies from 117 to 163 cm (46 to 64 in.) (Henderson and Smith 1957). Rainfall is generally highest in coastal Louisiana, Mississippi, Alabama, and southeast Florida (U. S. Department of Interior 1970). The entire region is underlain by limestone with a sandy soil mantle which varies from thick to thin.

The Upper Coastal Plain includes the eastern third of Virginia and North Carolina, the eastern half of South Carolina, central Georgia, the southern and western two-thirds of Alabama, nearly all of western Mississippi, the northern half of Louisiana, the southern third of Arkansas, a fringe of southeastern Oklahoma, and a portion of east Texas (Lovvorn 1948). Precipitation averages 127 to 152 cm (50 to 60 in.) over most of the area (Pearson and Ensminger 1957) and is fairly evenly distributed throughout the year. On the western edge in Texas and Arkansas precipitation drops to 102 cm (40 in.). Elevation varies from near sea level to 185 m (600 ft). Topography is rolling and, in some instances, even hilly.

Soils in the Upper and Lower Coastal Plains have developed from marine sands

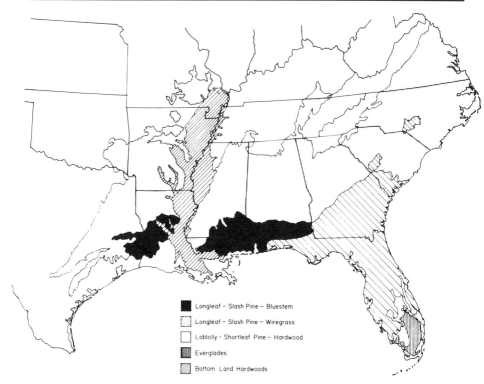

Fig. 15.1. Major vegetation types in the southeastern United States. (Modified from Küchler 1964. Reproduced with permission of the American Geographical Society.)

and clays (Pearson and Ensminger 1957). Upland soils generally have light gray to red sandy surfaces, are 13 to 25 cm (5 to 10 in.) deep, and are underlain by yellow-red, friable, sandy clay subsoils. pH values are generally around 5.O. Drainage is good. Soil organic matter and nutrients are low.

Bottomland soils along the flood plains of the Mississippi River formed from material that had worked downward from uplands. Soil textures vary from fine clays to sands (Lovvorn 1948).

Between the Appalachian Mountains and the Atlantic Coastal Plain is the Piedmont Plateau. Within the southern states it consists of a strip through middle Virginia and North Carolina, western South Carolina, and northern Georgia to central Alabama. It is a rolling to hilly region from 180 to 460 m (600 to 1500 ft) in elevation (Lovvorn 1948). Soils have been derived from complex rocks, including crystalline igneous rocks, highly metamorphosed sandstone and shale, and unmetamorphosed sandstone and shale. Soil textures range from sands to clays but were originally sandy loams and clay loams (Pearson and Ensminger 1957). Where erosion has been severe surface soils have eroded away leaving only the clay subsoil. pH values average 5.5 (Pearson and Ensminger 1957).

VEGETATION AND FIRE EFFECTS

Lower Coastal Plain

VEGETATION

Overstory. The flatwoods of Florida and parts of the Gulf and Atlantic Coasts are generally known as the longleaf-slash pine type. Longleaf pine is a deep-rooted plant that tolerates fire every two to three years (Chapman 1944). It is also drought-tolerant and occupies uplands, which are sandy sites with good to excessive drainage (Chapman 1944; Fowells 1965). In contrast, slash pine, a faster growing tree, has a wide-spreading shallow root system and is best suited to wet sites that may be seasonally flooded (Chapman 1944; Shoulders 1976) and subjected to fire every 5 to 10 years. Unburned areas in this region that are protected from fire reach the climax stage of southern mixed hardwoods (Fig. 15.2). This mixture of trees generally includes oaks *(Quercus laurifolia, Q. virginiana,* and *Q. nigra),* southern magnolia *(Magnolia grandiflora),* sweet gum *(Liquidambar styraciflua),* American holly *(Ilex opaca),* red bay *(Persea borbonia),* American beech *(Fagus grandifolia),* red maple *(Acer rubrum),* yellow-poplar *(Liriodendron tulipifera),* white ash *(Fraxinus americana),* and hickories *(Carya* spp.) (Garren 1943; Monk 1965).

Pocosins occur throughout the lower plains of North Carolina and South Carolina. They are closely related to hammocks (oaks, southern magnolia, American beech, various hickories, and other hardwoods) but consist mainly of evergreen shrubs such

Fig. 15.2. Southern mixed hardwoods that have become dominant on a slash pine site in the absence of fire. (Photo courtesy of USDA Forest Service.)

as southern wax myrtle *(Myrica cerifera)*, gallberry *(Ilex glabra)*, and pond pine *(Pinus serotina)* (Garren 1943). Poor drainage, fluctuating water tables, and infrequent fires maintain these communities. Because pond pine cones are serotinous (Little et al. 1967) they open after fire and new seed germinate. Their survival depends on infrequent fires. Sand pine *(Pinus clausa)* is restricted to scrub oak *(Quercus sp.)* forests, which grow on poor soils and probably receive fire only once in their lifetime (Garren 1943; Cooper 1972). Without fire sand pine would succeed to scrub oak forest.

Understory. Understory vegetation of the longleaf-slash pine type is composed principally of grasses and some shrubs. It is known east of central Alabama as the pine-wiregrass type (Hilmon and Hughes 1965). Over 90% of the herbaceous understory is made up of grasses with pineland threeawn *(Aristida stricta)* and Curtiss dropseed *(Sporobolus curtissii)* being the dominant species (Lemon 1949). Pineland threeawn alone is the dominant grass in southern Florida but is mixed in about equal proportions with Curtiss dropseed in northern Florida and southern Georgia (Hilmon and Hughes 1965). Other grasses include toothachegrass *(Ctenium aromaticum)*, Florida dropseed *(Sporobolus floridanus)*, cutover muhly *(Muhlenbergia expansa)*, buestems *(Andropogon stolonifer, A. virginicus, etc.)*, panicums *(Panicums spp.)*, and lovegrasses *(Eragrostis spp.)*. Forbs are very scattered but are generally members of the Compositae and Leguminosae families. Gallberry makes up two-thirds of shrub understory with saw-palmetto *(Serenoa repens)* being the other principal component (Fig. 15.3).

West of central Alabama, understory of longleaf-slash pine is composed of the pine-bluestem type (Fig. 15.4). Little bluestem *(Schizachyrium scoparius)*, pinehill bluestem *(S. scoparius var. divergens)*, and slender bluestem *(A. tener)* are the dominant grasses. Slender bluestem is most tolerant of fire and grazing but will gradually be replaced on open moist sites under heavy grazing by carpet grasses *(Axonopus compressus, A. affinis)* (Hilmon and Hughes 1965). Carpet grass is intolerant of shade and disappears under closing pine canopies. Other grasses contributing significant quantities of forage are panicums, paspalums *(Paspalum spp.)*, fineleaf bluestem *(A. subtenuis)*, paintbrush bluestem *(A. ternarius)*, Elliott bluestem *(A. elliottii)*, and big bluestem *(A. gerardii)* (Duvall 1962).

Grass like plants include pinehill beakrush *(Rhynchospora globularis)* and several species of *Carex* and *Cyperus*. The most common forbs are grassleaf goldaster *(Heterotheca graminifolia)* and swamp sunflower *(Helianthus angustifolius)*.

Shrubs and deciduous trees are sparse (Duvall 1962). Important species include blackjack oak *(Quercus marilandica)*, southern wax myrtle, flameleaf sumac *(Rhus copallina)*, and blackberry *(Rubus sp.)*.

EFFECTS OF FIRE

Longleaf Pine. Natural fires every two to three years will keep longleaf pine in existence for years (Chapman 1944), but today's management practices restricts its distribution and has permitted loblolly and slash pines to become the most prevalent

Fig. 15.3. Slash pine in the palmetto-gallberry fuel type near Macon, Georgia. Top photo—no burning treatment for 10 years and shrubs have become large and robust, shading much of the ground. Bottom photo—a one-year-old burn. More open space is available between the shrubs for growth of grasses and forbs. (Photos courtesy of USDA Forest Service.)

pine species (Knight and McClure 1974; Grelen 1978). Longleaf pine remains in the grass stage for the first two to seven years, depending on disease infestations and frequency of fire, and is very susceptible to fire the first year after emergence (Garren 1943). After two years the crown collar is about 1.2 cm (0.5 in.) in diameter and can tolerate fire every three years. This is an especially desirable attribute to control brown-spot disease *(Septoria acicola)* because fire will consume the leaves. A fire is

Fig. 15.4.   Pine-bluestem understory on longleaf pine site in Louisiana. (Photo courtesy of USDA Forest Service.)

the equivalent of one year's loss in growth, but without it young trees will die from brown-spot disease. Fire consumes the inoculum of brown-spot disease in the fallen leaves. Seedlings burned during the winter at intervals of three years show twice as much growth as those not exposed to fire (Siggers 1934).

Almost all growth of longleaf pine for the first five years is concentrated in the roots, giving an adequate supply of food for growth of new leaves when fire consumes the foliage. The buds are well protected from fires. At two to seven years of age, when the crown collar is about 2.5 cm (1 in.) in diameter, the tree begins to grow upward (Fowells 1965) and puts on 3 m (10 ft) of growth in 3 years (Muntz 1954). During this period of growth, longleaf pine is very susceptible to fire and should be protected (Muntz 1954; Fowells 1965). After the 3-m (10-ft) height is reached, longleaf pine is very fire-tolerant, and winter fires every three years may be resumed to control hardwoods (Sackett 1975). This tree is the slowest growing of the pines during the earlier years and has a longevity of 300 years (Garren 1943).

Pole or saw-log trees can tolerate a one-third to one-half defoliation of living crown without serious damage to growth. Annual burns can reduce basal area growth 22 to 44 percent in young longleaf pine (Garren 1943) but other studies show no reduction (Bruce 1947). Biennial burning on May 1 permitted 12-year-old trees in Louisiana to grow in height to 10.7 m (35 ft) with a dbh of 15 cm (6 in.) compared to

Fig. 15.5. A longleaf pine stand on upland site that has not been burned for 15 years. Notice how rapidly the hardwood species are over-taking the pine species. (Photo courtesy of USDA Forest Service.)

a height of 6.5 m (21.3 ft) and an average dbh of 9.6 cm (3.8 in.) for unburned trees (Grelen 1975). Both sites had similar stocking densities. Once a pine stand is well established there is no difference in timber yields between burned and unburned stands (Wahlenberg 1935; Sackett 1975). If left unburned, longleaf pine will gradually succeed to a southern mixed hardwood community (Fig. 15.5).

Pineland Threeawn-Curtiss Dropseed Understory. East of central Alabama grass yields vary from 450 kg/ha (400 lb/acre) in decadent stands of grass to 3930 kg/ha (3500 lb/acre) on recently burned areas (Fig. 15.6) (Hilmon and Hughes 1965; Lewis and Hart 1972) depending on canopy cover (Halls et al. 1952). Generally production is around 2250 kg/ha (2000 lb/acre), providing canopy cover does not exceed 25 percent (Halls et al. 1952). Maximum yields occur three years after burning (Lemon 1949) but will decline 50 percent in six to eight years if not reburned (Lewis and Hart 1972). Pineland threeawn will disappear more quickly with protection than Curtiss dropseed (Hilmon and Hughes 1965). Decadent stands of pineland threeawn are especially evident after 8 to 10 years.

For maximum forage yields, season of burning has a marked influence. Burns in March or May produced twice and four times as much herbage, respectively, 60 days after fire as ranges burned in October and November (Lewis 1964). Moreover late spring and early summer burns stimulate seed yields of native grasses more than fires in winter or early spring (Biswell and Lemon 1943).

Forage quality is generally low in the pine-wiregrass type. Protein, phosphorus, and calcium are below minimum requirements for maintenance [8 percent for dry

Fig. 15.6. A healthy stand of broomsedge bluestem in understory of longleaf pine near Tallahassee, Florida that has been burned annually for 10 years.

cows (Duncan and Epps 1958)] most of the year (Hilmon and Hughes 1965). All of these elements will increase after burning, but only slightly (Hilmon and Lewis 1962). For example, protein content was increased 2 percent by burning. Three months after burning, burned range had no advantage over unburned range (Halls et al. 1952). Protein will decrease from 10 or 13 percent to 2 or 4 percent within a five-month period (Hilmon and Lewis 1962).

Pine-Bluestem Understory.   Annual burns on pine-bluestem ranges (west of central Alabama) were shown by Wahlenberg et al. (1939) to produce twice as much herbage [5455 kg/ha (4855 lb/acre), green wt] as unburned [2490 kg/ha (2215 lb/acre), green wt] ranges, and cattle gains were 37 percent higher on burned ranges. In addition, they found 16 species of legumes on burned ranges and only 11 species on unburned ranges. On similar vegetation, Duvall (1962) reported average yields on burned plots of 3145 kg/ha (2800 lb/acre).

Herbage on unburned pine-bluestem range is generally deficient in protein after early summer and in phosphorus all year (Campbell and Cassady 1951). A combination of winter and spring fires will materially increase protein content without a loss in yield (Grelen and Epps 1967).

Forbs and Shrubs.   Many leguminous forbs increase after burning. Beggarweed *(Desmodium* sp.), partridge pea *(Cassia* sp.), *Clitoria,* and *Galactia* all increase after spring burns (Czuhai and Cushwa 1968). Wild strawberry *(Fragaria* sp.) increased after winter burning (Czuhai and Cushwa 1968).

Both major shrubs, gallberry and saw-palmetto, are well adapted to fire (Fig.

15.3). Without fire gallberry becomes large and robust, shading much of the ground (Lewis and Hart 1972). Burning at three- to four-year intervals will keep gallberry and saw-palmetto in check (Halls et al. 1964), as well as southern wax myrtle (Grelen and Enghardt 1973). Other data on shrubs is given in a subsequent section titled "Shrub Understory of Pines."

Slash Pine.   Natural slash pine in the longleaf-slash pine region was restricted to moist areas around pond margins that usually escape fires for the first five or six years (Garren 1943; Chapman 1944). However, today it has greatly increased its acreage due to commercial plantings and reduction in wildfire acreage (Knight and McClure 1974; Grelen 1978). Slash pine initiates height growth at once but is killed by the least touch of fire during the first two years (Chapman 1944). By its rapid growth and accompanying increase in bark thickness it gradually attains immunity from fire when burned carefully. Within three or four years a moderately hot fire is required to kill all trees. After 10 to 12 years of age, winter fires, which do not crown, will leave all trees practically untouched. Grazing by livestock can help prevent disastrous wildfires and undesirable prescribed burn effects for this species and other pines throughout the South. Prescribed burns may enhance height and diameter of slash pine growth as long as less than 40 percent of the crown is scorched (Johansen 1975).

Since slash pine has a shallow, widespreading root system, it does not compete well with longleaf pine on dry, upland sites. Drought years will reduce the number of seedlings and saplings. Without drought years, slash pine will invade longleaf pine sites where there has been an initial exclusion from fire for five or six years. Where slash and loblolly pine trees invade moderately moist sites, 20 to 30 percent of them become infected with fusiform rust (Shoulders 1976). Fusiform rust is least frequent on wet or dry sites (Shoulders 1976). Thus growing pines on their natural sites can help to minimize disease problems. Understory plants are mainly pine-wiregrass type.

Sand Pine.   Many sand pine stands owe their existence to wildfires (Cooper 1972). The species is most prevalent in north-central Florida and bears closed cones (Fowells 1965). Historically the trees probably grew to an age of 25 to 70 years before a fire swept through the community. This fire frequency enabled the cones on cone-bearing trees to open, release tremendous quantities of seed, and reestablish themselves (Fowells 1965). Under ideal burning conditions in spring, dense sand pine stands crown easily and are difficult to control; however, during many months of the year, stands of sand pine are an "asbestos forest" (Cooper 1972).

Unfortunately, prescribed fire in north-central Florida has not been a particularly effective tool to manage sand pine (Cooper 1972). The fuel is either explosive or does not burn. Through more experimental work prescribed burning may become an effective tool as it is in west Florida where understory fuels are light. Sand pine grows well on dry infertile sites and is commonly used to convert scrub oak sites to pine. However if two consecutive burns (double burns) occur on sand pine sites within a span of 10 years, sand pine is replaced by scrub oak (Cooper 1972).

Upper Coastal Plain

Uplands of the Coastal Plain and the Piedmont, which stretch from Maryland to east Texas (Boyce 1972), are known as the loblolly-shortleaf pine (Fig. 15.7) type with the exception of bottomland hardwoods along the Mississippi Delta Region (Garren 1943). Before the arrival of Europeans much of this area was dominated by longleaf pine where fire frequency was high (Chapman 1926). Loblolly and shortleaf pine thrive where fires occur about every 10 years (Garren 1943; Chapman 1944). Loblolly pine grows on a wide variety of soils. It grows best on fine-textured soils (Coile 1940) with poor surface drainage, a deep surface layer, and a firm subsoil (Fowells 1965; Harlow and Harrar 1968). Shortleaf pine grows on coarse-textured and drier soils that are neither highly acid nor strongly alkaline. Its best development is in the Piedmont region on sites with friable subsoils (Fowells 1965). These sites are fine sandy loams or silt loams without a distinct profile but with good internal drainage.

VEGETATION

Climax vegetation for this region as well as for the Piedmont region is the oak-hickory association (Küchler 1964). Associate species of loblolly and shortleaf pine in the seral stages include Virginia pine *(Pinus virginiana)*, eastern redcedar *(Juniperus virginiana)*, black oak *(Quercus velutina)*, southern red oak *(Quercus falcata)*, blackjack oak, post oak *(Q. stellata)*, chestnut oak *(Q. prinus)*, sweet gum, black gum *(Nyssa sylvatica)*, mockernut hickory *(Carya tomentosa)*, American elm *(Ulmus americana)*, flameleaf sumac, ash, flowering dogwood *(Cornus floridana)*, and red maple (Garren 1943; Oosting 1958). Ground cover in typical loblolly pine-hardwoods does not contain more than 3 to 8 percent grass (Lewis and Harshbarger 1976), except temporarily on old fields (Chapman 1944). A mixture of hardwood species (35 to 40% cover) generally dominate the understory (Lewis and Harshbarger 1976).

EFFECT OF FIRE

Loblolly Pine.   Understory of loblolly pine forests consists primarily of brush and leaves which are less combustible than roughs in longleaf-slash pine type (Fig. 15.7) (Chapman 1944). Nevertheless, fire plays an important role in favoring natural loblolly pine reproduction over hardwoods (Brender and Cooper 1968; McNab 1976), although it cannot be tolerated for 10 years or so when the trees are getting established (Chapman 1944). Otherwise young pines will not have had enough time to develop the heat resistance in bark or the height of crown which enables them to survive winter fires. Hence one cannot reproduce and grow loblolly pine successfully in mixture with longleaf pine. Either loblolly pine will be killed out by the fires necessary to longleaf pine survival or longleaf pine seedlings will be suppressed and die under the shade of the faster growing loblolly pine. On sites naturally adapted to loblolly pine, on which longleaf pine may have intruded because of past uncontrolled fires, the ex-

Fig. 15.7. Top photo—Loblolly and shortleaf pine with shrub understory that has not been burned for 10 years. The understory should be burned at this time to suppress hardwoods. Bottom photo—Loblolly and shortleaf pine with shrub understory that is 15 to 20 years old. Such areas cannot be burned without severe damage to the pines. Ultimately this plant community will evolve to a mixture of pine and hardwoods. (Photos courtesy of USDA Forest Service.)

clusion or control of these fires will permit growing of loblolly pine as a more profitable crop (Knight and McClure 1974).

Thinning and Establishment of Loblolly. Low intensity backfires can be used to effectively thin dense stands of loblolly pine if there is a wide range of diameters [1 to 16 cm (0.1 to 6.5 in.)] (McNab 1977). Conducting a low intensity fire [flame height 0.4 to 0.6 m (1.3 to 2.0 ft)] where fine fuel was 8760 kg/ha (3.9 tons/acre) reduced number of stems by 65 percent. No tree larger than 10 cm (4 in.) in diameter was killed. Size class distribution after burning followed a normal distribution and stocking percentage remained unchanged.

Ashes from slash burns will increase survival of one-year old loblolly pine seedlings fivefold (McNab and Ach 1977). Survival was 87 percent in ash piles versus 18 percent on control areas. Similarly, height of seedlings in ash piles was 0.45 m (1.5 ft) versus 0.21 m (0.7 ft) where there were no ashes. Minerals from ash, increased soil pH, and reduced competition from hardwoods will greatly enhance the establishment of loblolly pine seedlings (McNab 1976; McNab and Ach 1977).

Burning Frequency. After loblolly pines have become well established [an adequate number of trees in the 5- to 10-cm (2- to 4-in.) groundline diameter class], the first fire can be conducted to thin trees and set hardwoods back (McNab 1977). Thereafter burns should be conducted at least every 5 to 10 years during November through February (and sometimes into March) to keep hardwoods suppressed (Lotti 1962; Brender and Cooper 1968). Under this burning frequency, the percentage cover of herbaceous plants will be 3 to 8 percent and coverage by woody plants will be 37 to 40 percent (Lewis and Harshbarger 1976). Annual winter or biennial summer burns will increase herbaceous cover to 20 percent and decrease shrub cover at least 20 percent. However, Lewis and Harshbarger (1976) recommend burning during the winter every two or three years to maintain the best combination of forage for cattle and wildlife. Over a 10-year period prescribed burning neither increased nor decreased the growth of dominant trees (Lotti 1962).

Understory. Lewis and Harshbarger (1976) summarized the long-term effects of several fire frequencies on understory species in loblolly pine forests. Forb cover increased slightly after burning. Principal forbs that increased were composites, lespedeza, beggarweed, and partridge pea. Broomsedge bluestem *(Andropogon virginicus)* and panicum grasses were the most important grasses. Annual winter fires increased cover of blackberry and flameleaf sumac but decreased cover of gallberry, southern wax myrtle, and yellow jessamine *(Gelsemium sempervirens)*. Periodic winter burns had little effect on most species other than sweet gum, black gum, and summer grape *(Vitis aestivalis)*. Greenbriar *(Smilax glauca)* was significantly reduced by all burning treatments.

Shortleaf Pine. Shortleaf pine is more fire-tolerant than loblolly pine and grows on drier, coarse-textured, and less acid soils (Harlow and Harrar 1968). At least half the young shortleaf pine plants [up to a height of 2.4 m (8 ft)] will resprout after fire (Garren 1943). Thus if a fire occurred five years after establishment of pines, shortleaf pine will have the edge over loblolly pine. On the other hand loblolly pine is a faster-growing tree and, at least on wet sites, it will overtake shortleaf pine once fire frequencies approach a 10-year interval. Loblolly pine is a little more tolerant of shade than shortleaf pine (Harlow and Harrar 1968).

Once shortleaf pine reaches pole or saw-timber age, fires every two to three years increase yields (Somes and Moorehead 1954). Thinning without fire has no effect on growth of shortleaf pine.

Shrub Understory of Pines. Effects of fires on shrubs and trees throughout the Southeast can be summarized from a variety of sources (Garren 1943; Lay 1956; Lotti 1956, 1962; Stoddard 1931, 1962; Halls et al. 1964; Grelen 1975; Lewis and Harshbarger 1976). Those species favored include sweet gum, flameleaf sumac, French mulberry *(Callicarpa americana)*, viburnum *(Viburnum molle)*, summer grape, sweetleaf *(Symplocos tinctoria)*, wax myrtle *(Myrica pusillus)*, eastern baccharis *(Baccharis halimifolia)*, New Jersey tea ceanothus *(Ceanothus americanus)*, and American cyrilla *(Cyrilla rademiflora)*. Occasional fires are desirable to prune and stimulate fruiting for huckleberries *(Gaylussacia* spp.), and blackberries and dewberries *(Rhus* spp.) (Stoddard 1931). Fire will check the growth of saw-palmetto and oaks, but these species will maintain their vigor (Hughes 1975; Lewis and Harshbarger 1976).

Shrubs and trees that are harmed by fire include eastern redcedar, black gum, gallberry, southern wax myrtle, yellow jessamine, yaupon *(Ilex vomitoria)*, American holly, flowering dogwood, blueberries *(Vaccinium* spp.), and tree huckleberry *(Vaccinium arboreum)*. Greenbriar decreased dramatically following burns in South Carolina (Lewis and Harshbarger 1976) but increased dramatically after burning in east Texas (Lay 1956). Understory mast yield is always seriously reduced by fire (Lay 1956).

Canebrakes. Switch cane *(Arundinaria tecta)* is a rhizomatous "shrubby grass" that grows on moist to wet sites throughout the Southeast from Florida to Louisiana north to New Jersey and Oklahoma (Fernald 1950). This vegetation type used to be abundant throughout the Southeast, but a combination of fire, uncontrolled grazing, and clearing for land cultivation has eliminated it from many areas. Nevertheless about 0.8 million hectares (2 million acres) persist in the Carolinas and Virginia (Hughes 1966). Here the cane is frequently associated with pond pine (Shepherd et al. 1951) and, to some extent, loblolly pine and lowland hardwoods (Smart et al. 1960).

Switch cane is one of the most nutritious of the native forage plants growing in the eastern United States (Shepherd and Dillard 1953; Hilmon and Hughes 1965). Crude protein varies from a low of 12 to 13 percent for most of the year to a high of 15.8 percent in the spring (Smart et al. 1960). Biswell et al. (1945) reported a high of 20 percent in June, decreasing to 14 percent in September and October, and reaching a low of 12 percent in December. Digestible protein by the lignin method varies from a low in the winter of 6.2 percent to a high in the spring of 11.1 percent (Smart et al. 1960). Burning seems to have very little effect on protein content of switch cane.

Fire is essential for renovating decadent switch cane stands (Hughes 1957). Burning restores vigor, productivity, and reduces wildfire hazard for one to three years. Fire reduces switch cane productivity for the first year after burning, and it should be partially protected from grazing during the summer months immediately after burning (Hughes 1957). But after two years switch cane stands are fully recovered from burning and can withstand a 55 percent grazing use (Shepherd et al. 1951; Hughes 1957).

Maximum productivity [4500 kg/ha (2 tons/acre) annually] is reached two to four years after a fire and good yields [2250 kg/ha (1 ton/acre) or more annually] can be maintained for 10 years after a fire (Hilmon and Hughes 1965; Hughes 1966). Thereafter, production declines rapidly and fire is necessary to rejuvenate the stand (Shepherd et al. 1951; Hughes 1957). Rest from grazing will not rejuvenate the stand (Hughes 1966). Apparently the 11,240 to 15,730 kg/ha (5 to 7 tons/acre) of accumulated litter is the main deterrent to growth (Hughes 1966).

Historically, fire records dating back to the early colonial days indicate the usual fire interval in canebrakes was five to seven years (Hughes 1966). Even in poorly drained pocosins new cane growth seldom went beyond five years without a fire (Wells 1942).

Piedmont Region

Shortleaf pine makes its best growth in the Piedmont Region along with some loblolly pine, but Virginia pine and pitch pine *(Pinus rigida)* are other pines that occur in the Piedmont Region (Fowells 1965). Virginia pine is intolerant of fire and grows on dry unproductive sites (Garren 1943; Harper 1962). Pitch pine is very hardy and can grow on dry unproductive sites but grows best on moist sandy loam soils (Little 1964; Harlow and Harrar 1968). Pitch pine can resprout after fire but needs protection from fire for long periods of time to do well (Little 1964).

Mississippi Region

Bottomland hardwoods of the Mississippi Region are dominated by laurel oak, water oak, sweet gum, black gum, and red maple (Garren 1943). Slash pine and longleaf pine are intermixed with hardwoods where fire has been a factor. Fire is very damaging to hardwoods because it initiates decay in most of the trees (Kaufert 1933).

Southern Florida

VEGETATION

Southern Florida contains a mosaic of vegetation interspersed with glades of coarse grasses. The entire area known as the Everglades is underlain by marine limestones of the Pleistocene Epoch (Klukas 1972). In some areas the limestone is exposed or highly eroded but most of the area is covered with deposits of peat and calcitic or marl soil. These organic deposits vary from a few inches to more than 3 m (10 ft) in depth (Craighead 1971). Precipitation averages 149 cm (58.8 in.) with 60 to 80 percent of it falling between May and late October (Craighead 1971). Sea level, fires, and flooding all have considerable influence in determining plant communities [mangrove forests, coastal hardwood hammocks, coastal prairie, swamps, sawgrass *(Cladium jamaicensis)* sloughs, and pinelands] in this region of south Florida (Klukas 1972; Wade et al. 1980).

Mangroves occupy a nonfire habitat type along the Gulf Coast. Dominant tree species are red mangrove *(Rhizophora mangle)*, white mangrove *(Laguncularia racemosa)*, black mangrove *(Avicennia germinana)*, and buttonwood *(Conocarpus erectal)*. One of the main functions of these mangroves seems to be to absorb the coastal shock of hurricanes (Robertson 1962). Within the mangrove forests or high ground [0.3 to 0.9 m (1 to 3 ft) above sea level] are coastal prairies and hardwood hammocks. Common species in the hammocks include buttonwood, wild tamarind *(Lysiloma latisiliqua)*, Jamaica dogwood *(Piscidia piscipula)*, cabbage palmetto *(Sabal palmetto)*, live oak *(Quercus virginiana)*, bustic *(Bumelia solicifolia)*, Joewood *(Jacquinia Keyensis)*, cat claw *(Pithecellobium unguis-cati)*, baycedar *(Suriana maritima)*, white stopper *(Eugenia axillaris)*, pigeon plum *(Cocoloba diversifolia)*, randia *(Randia aculeata)*, gumbo-limbo *(Bursera simaruba)*, sea ox-eye *(Borrichia frutescens)*, poison wood *(Melopium toxiferum)*, dildo *(Acanthocerus floridanus)*, pricklypear *(Opuntia dillenii)*, stoppers *(Eugenia* spp.), strangler *(Ficus aurea)*, century plant *(Agave decipens)*, mastic *(Mastichodendron foctidissimum)*, and black ironwood *(Krugiodendron ferrcum)*. The coastal prairie areas are vegetated by grasses such as cordgrasses *(Spartina* spp.), dropseeds *(Sporobulus* spp.), and panic grass *(Panicum* spp.)

The estuarine or coastal marsh is another distinct plant community within the mangrove zone (Klukas 1972). Dominant plant species are cordgrasses, juncus *(Juncus roemerianus)*, and fringerush *(Fimbristylis castanea)* with sawgrass patches occupying areas less saline and slightly higher in elevation (Craighead 1971).

Inland from the mangroves, fresh water swamps (wet prairies) and sawgrass sloughs form the glades (Fig. 15.8). Fresh water swamps are vegetated by grasses, sedges *(Carex* spp.), and rushes *(Juncus* spp.) that are 0.9 to 1.2 m (3 to 4 ft) or less in height. Some areas are broken up by numerous islands of trees that occupy either rock outcroppings (hardwood hammocks), elevated accumulations of organic soils (bayheads or willowheads), or bedrock depressions (cypress heads).

Fresh water swamps are dissected by sawgrass sloughs. These areas occupy troughs in the limestone bedrock and are flooded more deeply for greater lengths of time than surrounding fresh water swamps (Klukas 1972). A heavy, continuous, unbroken cover of sawgrass is the dominant plant on deep organic soils with long hydroperiods in the central Everglades (Wade et al. 1980). More extensive than these pure, dense stands, however, are vast areas occupied by sawgrass with other plants such as arrowhead *(Sagittaria lancifolia)* and maidencane *(Panicum hemitomon)* (Wade et al. 1980).

Fig. 15.8. Sawgrass glades and tree islands. Burning when water levels are 0.3 m (1 ft) high in the Everglades will permit sawgrass to dominate and the tree islands, which are rich in wildlife species, to extend their boundaries into the marsh. (Photo courtesy of National Park Service.)

EFFECTS OF FIRE

Klukas (1972) presented the historical role of fires in the Everglades National Park. Fires rarely occurred in mangroves along the coasts or in hardwood hammocks that occur within the glades. Coastal marshes, however, appeared to be fire-dependent because mangroves will invade these marshes without fire. Fresh water swamps (wet prairies) and sawgrass sloughs, which form the glades, had histories of severe fires. Moreover, fires seemed to be more common in fresh water swamps than in the past. Before 1900 more of these prairies were covered with sawgrass 2 to 3 m (6 to 10 ft) high. Recent fires, especially during drought years, removed deep peat deposits on which sawgrass thrives best (Craighead 1971; Wade et al. 1980). The sloughs should be burned when water level is high to permit peat deposits to build up and to allow tree islands (Fig. 15.8) to expand their boundaries into the marsh (Robertson 1962).

   The dominant pine species in the Everglades is South Florida slash pine. This species requires fire for survival, for hardwood hammocks occupy those areas that escape frequent fires (Klukas 1972). Recently the punk tree *(Melaleuca quinquenervia)*, introduced from Australia, has become widespread around the Everglade National Park (Wade et al. 1980). It constantly threatens to move into the park and cannot be kept out by fire alone. The seed capsules are serotinous, so fire will enhance the species. To minimize spread of punk tree it is recommended that the trees be sprayed with a herbicide and followed with a burn before the seedlings reach 15 cm (6 in.) in height (Wade 1980).

Everglade National Park has an active prescribed burning program to maintain and restore fire-dependent habitats within the Park (Klukas 1972; Wade et al. 1980). Fire-dependent communities are being burned every three to six years when water levels are 0.3 m (1 ft) high, depending on the rapidity with which fuels accumulate. This is necessary to keep out serious hammock- and peat-destroying fires, which occur during very dry years. After many decades this burning program in the Everglades will allow a buildup of peat in the sawgrass glades and an extension of boundaries of tree islands (Robertson 1962).

MANAGEMENT IMPLICATIONS

Without fire the Southeast would not have pure stands of pine trees (Chapman 1932; Hebb and Clewell 1976). The Lower Coastal Plains would be dominated by the southern mixed hardwood forest (Monk 1965; Hebb and Clewell 1976), and the Upper Coastal Plains and Piedmont would be dominated by an oak-hickory-pine forest (Küchler 1964). In both cases pine would be only a small portion of the climax. Thus the long history of lightning fires, as well as those set by aboriginal and European men (Chapman 1932) and the intensive prescribed burning programs of today have enabled land managers to maintain productive stands of pine in the Southeast. Natural frequency of fire and soil type along with climate helped determine which species of pine would dominate certain areas of this region.

Fire hazard reduction is one of the principle reasons for using prescribed fire in the Southeast (Cooper 1975; Sackett 1975) because prescribed burns reduce the frequency of wildfires manytimes (Cooper 1975). For example the percentage of area burned by wildfires was 0.03 percent in roughs less than a year old, 0.14 percent in five-year old roughs, and 7.0 percent in the roughs older than five years (Davis and Cooper 1963). However, control of hardwoods, control of brown-spot disease, and establishment of a mineral seedbed for pine seedlings are equally important in fire management of pine in the Southeast.

Longleaf pine should be burned every two to three years until height growth initiates. If not burned this frequently, loblolly, shortleaf (Bruce 1947), or slash pine (Langdon and Bennett 1976) can invade longleaf pine where the site is suitable for their growth. After height growth initiates, longleaf pine should be left untreated until the trees reach a height of 3.6 m (12 ft). At this time burning should be resumed every three years to reduce surface fuels (Sackett 1975). Slash pine should not be burned for the first five years since it initiates height growth immediately. Loblolly and shortleaf pine also initiate height growth immediately but need about 10 years for the bark to get thick enough to tolerate surface fires. Then prescribed burns should be used every three years, as for the other pines, to suppress hardwoods and reduce rough (Sackett 1975). This should continue until the final saw-timber harvest.

The first commercial thinning cut is often made at 15 years of age. Pulpwood rotations are completed in 30 to 40 years, and saw-log rotations are completed in 60 years or less.

Even-aged management of pines under both the seed tree and shelterwood systems are satisfactory for all pines in the Southeast (Gaines 1951; Fowells 1965; Langdon and Bennett 1976). Generally this involves several winter burns prior to initial harvest to prepare a seedbed and control shrubs. After harvest slash may be burned as well as understory vegetation in winter or summer before seedfall. One to three years after seedlings are established, the seed trees or shelterwoods should be removed and the seedlings should be precommercially thinned to 1000 stems per acre, especially if saw-timber, plywood, or gum are primary objectives (Evans and Gruschow 1954; Langdon and Bennett 1976).

Wildlife considerations should be given for songbirds, bobwhite quail, wild turkey *(Meleagris gallapavo)*, white-tailed deer *(Odocoileus virginianus)*, and tree squirrels *(Sciurus carolinensis, S. niger)*. Hardwoods should be protected along the rivers and streams for squirrel and turkey as well as other wildlife species. Turkey need large trees for roosting, mast for winter food, as well as newly burned areas for fresh green sprouts in the spring. Bobwhite quail populations can be enhanced by leaving about 20 percent of the understory unburned each year. Shrubs will grow in these areas, and bobwhite quail will use them for cover. The areas left unburned should be changed every two to three years to keep shrub height below 0.9 to 1.5 m (3 to 5 ft).

Forage for grazing is another important resource in longleaf and slash pine forests. Moderate grazing use (45%) has no effect on survival of planted or seeded longleaf pine; however, heavy grazing use reduces density of young trees at least 20 percent (Pearson 1974). A suggested management scheme to improve forage quality and maximize cattle gains is to burn one-third of each grazing unit every year (Duvall and Whitaker 1964; Grelen and Whitaker 1973). Utilization averaged 42 percent with 78-, 31-, and 18-percent use occurring on first, second, and third year burns, which allowed for full recovery of vegetation before a reburn (Duvall and Whitaker 1964). Consequently such a system can be used to manage timber and livestock together. In canebrakes fires are necessary every 5 to 10 years to rejuvenate cane yields.

Wise and careful use of prescribed fire in the Southeast can accomplish more objectives at a cheaper cost than any of our alternate management methods (Cooper 1975). Fortunately fire is widely used as a management tool in this region.

REFERENCES

Betts, H. S. 1954. The southern pines. USDA For. Serv. American Woods Series.

Biswell, H. H., R. W. Collins, J. E. Foster, and T. S. Boggess, Jr. 1945. Native forage plants. N. C. Agric. Exp. Stn. Bull. 353. Raleigh.

Biswell, H. H., and P. C. Lemon. 1943. Effect of fire upon seed-stalk production of range grasses. *J. For.* **41:**844.

Bonninghausen, R. A. 1962. The Florida Forest Service and controlled burning. *Proc. Tall Timbers Fire Ecol. Conf.* **1:**43–60.

Boyce, S. G. 1972. Research to increase loblolly pine production. *For. Farmer* **31:**10, 11.

Brender, E. V., and R. W. Cooper. 1968. Prescribed burning in Georgia's Piedmont loblolly pine stands. *J. For.* **66:**31–36.

Bruce, D. 1947. Thirty-two years of annual burning in longleaf pine. *J. For.* **45**:809–814.

Campbell, R. S., and J. T. Cassady. 1951. Grazing values for cattle on pine forest ranges in Louisiana. La. Agric. Exp. Stn. Bull. 452. Baton Rouge.

Chapman, H. H. 1926. Factors determining natural reproduction of longleaf pine on cut-over lands in LaSalle Parish, La. Yale Univ. School of For., Bull. 16.

Chapman, H. H. 1932. Is the longleaf type a climax? *Ecology* **13**:328–334.

Chapman, H. H. 1936. Effects of fire in propagation of seedbed for longleaf pine seedlings. *J. For.* **34**:852–854.

Chapman, H. H. 1944. Fire and pines. *Amer. For.* **50**:62–64, 91–93.

Coile, T. S. 1940. Soil changes associated with loblolly pine succession on abandoned agricultural land of the Piedmont Plateau. Duke Univ. School of For., Bull. **5**:1–85.

Cooper, R. W. 1972. Fire and sand pine. In Sand Pine Symposium Proc. Panama City Beach, Florida. USDA For. Serv., Southeastern For. Exp. Stn., Asheville, N. C., and Florida Div. For. and Florida Coop. Ext. Serv., pp. 207–212.

Cooper, R. W. 1975. Prescribed burning. *J. For.* **73**:776–780.

Cost, N. D. 1968. Forest statistics for the southern Coastal Plain of South Carolina. USDA For. Serv. Res. Bull. SE-12. Southeastern For. Exp. Stn., Asheville, N.C.

Crafts, E. C. 1958. A summary of the timber resouce review. In Timber Resources for America's Future. USDA For. Res. Rep. No. 14. Washington, D.C., pp. 1–109.

Craighead, F. C., Sr. 1971. *The Trees of South Florida.* Univ. Miami Press, Miami, Fla.

Czuhai, E., and C. T. Cushwa. 1968. A resume of prescribed burnings on the Piedmont National Wildlife Refuge. USDA For. Serv. Res. Note SE-86. Southeastern For. Exp. Stn., Asheville, N.C.

Davis, L. S., and R. W. Cooper. 1963. How prescribed burning affects wildlife occurrence. *J. For.* **61**:915–917.

Dixon, M. J. 1965. A guide to fire by prescription. USDA For. Serv., Southern Reg. Atlanta, Ga.

Duncan, D. A., and E. A. Epps, Jr. 1958. Minor mineral elements and other nutrients on forest ranges in central Louisiana. La. Agric. Exp. Stn. Bull. 516. Baton Rouge.

Duvall, V. L. 1962. Burning and grazing increase herbage on slender bluestem *(Andropogon tener)* range. *J. Range Manage.* **15**:14–16.

Duvall, V. L., and L. B. Whitaker. 1964. Rotation burning: a forage management system for longleaf pine-bluestem ranges. *J. Range Manage.* **17**:222–226.

Evans, T. C., and G. F. Gruschow. 1954. A thinning study in flatwoods longleaf pine. *J. For.* **52**:9, 10.

Fernald, M. L. 1950. *Gray's Manual of Botany.* American Book Co., New York.

Fowells, H. A. 1965. Silvics of forest trees of the United States. USDA Handb. No. 271. Washington, D.C.

Gaines, E. M. 1951. A longleaf pine thinning study. *J. For.* **49**:790–792.

Garren, K. H. 1943. Effects of fire on vegetation of the southeastern United States. *Bot. Rev.* **9**:617–654.

Green, S. W. 1931. The forest that fire made. *Amer. For.* **37**:583.

Grelen, H. E. 1975. Vegetative response to twelve years of seasonal burning on a Louisiana longleaf pine site. USDA For. Serv. Res. Note SO-192. Southern For. Exp. Stn., Pineville, La.

Grelen, H. E. 1978. Forest grazing in the South. *J. Range Manage.* **31**:244–250.

Grelen, H. E., and H. G. Enghardt. 1973. Burning and thinning maintain forage in longleaf pine plantation. *J. For.* **71**:419, 420, 425.

Grelen, H. E., and E. A. Epps, Jr. 1967. Herbage responses to fire and litter removal on southern bluestem range. *J. Range Manage.* **20**:403, 404.

Grelen, H. E., and L. B. Whitaker. 1973. Prescribed burning rotations on pine-bluestem range. *J. Range Manage.* **26:**152, 153.

Halls, L. K., R. H. Hughes, R. S. Rummell, and B. L. Southwell. 1964. Forage and cattle management in the longleaf-slash pine forests. USDA Farmers' Bull. 2199. Washington, D.C.

Halls, L. K., F. E. Knox, and V. A. Lazer. 1952. Burning and grazing in coastal plain forests: A study of vegetation and cattle responses to burning frequency in longleaf-slash pine forests of Georgia. Georgia Coastal Plain Exp. Stn. Bull. 51. Tifton.

Harlow, W. M., and E. S. Harrar. 1968. *Textbook of Dendrology.* McGraw-Hill, San Francisco.

Harper, R. M. 1913. The forest resources of Alabama. *Amer. For.* **19:**657–670.

Harper, R. M. 1962. Historical notes on the relation of fire to forest. *Proc. Tall Timbers Fire Ecol. Conf.* **1:**11–29.

Hebb, E. A., and A. F. Clewell. 1976. A remnant stand of old-growth slash pine in the Florida Panhandle. *Bull. of the Torrey Bot. Club* **103:**1–9.

Henderson, J. R., and F. B. Smith. 1957. Florida and flatwoods. In Soil. Yearbook of Agriculture, USDA, Washington, D.C., pp. 595–598.

Heyward, F. 1936. Soil changes associated with forest fires in the longleaf pine region of the South. *Amer. Soil Survey Assoc. Bull.* **17:**41, 42.

Heyward, F. 1937. The effect of frequent fires of profile development of longleaf pine forest soils. *J. For.* **35:**23–27.

Heyward, F. 1938. Soil temperatures during forest fires in the longleaf pine region. *J. For.* **36:**478–491.

Heyward, F. 1939a. Some moisture relations of soil from burned and unburned longleaf pine forests. *Soil Sci.* **47:**313–324.

Heyward, F. 1939b. The relation of fire to stand composition of longleaf pine forests. *Ecology* **20:**287–304.

Hilmon, J. B., and R. H. Hughes. 1965. Forest Service research in the use of fire in livestock management in the South. *Proc. Tall Timbers Fire Ecol. Conf.* **4:**261–275.

Hilmon, J. B., and C. E. Lewis. 1962. Effect of burning on south Florida range. USDA For. Serv. Res. Paper 146. Southeastern For. Exp. Stn., Asheville, N.C.

Hughes, R. H. 1957. Response of cane to burning in the North Carolina Coastal Plain. N.C. Agric. Exp. Stn., Bull. 420. Raleigh.

Hughes, R. H. 1966. Fire ecology of canebrakes. *Proc. Tall Timbers Fire Ecol. Conf.* **5:**148–158.

Hughes, R. H. 1975. The native vegetation in south Florida related to month of burning. USDA For. Serv. Res. Note SE-222. Southeastern For. Exp. Stn., Asheville, N.C.

Johansen, R. W. 1975. Prescribed burning may enhance growth of young slash pine. *J. For.* **73:**148, 149.

Kaufert, F. H. 1933. Fire and decay in the southern bottomland hardwoods. *J. For.* **31:**64–67.

Klukas, R. W. 1972. Control burn activities in Everglades National Park. *Proc. Tall Timbers Fire Ecol. Conf.* **12:**397–425.

Knight, H. A., and J. P. McClure. 1974. Georgia's timber, 1972. USDA For. Serv. Res. Bull. SE-27. Southeastern For. Exp. Stn., Asheville, N.C.

Küchler, A. W. 1964. Potential natural vegetation of the conterminous United States. Manual to accompany the map. Am. Geogr. Soc. Special Pub. No. 36. (With map, rev. ed., 1965, 1966.)

Langdon, O. G., and F. A. Bennett. 1976. Management of natural stands of slash pine. USDA For. Serv. Res. Paper SE-147. Southeastern For. Exp. Stn., Asheville, N. C.

Lay, D. W. 1956. Effects of prescribed burning on forage and mast production in southern forests. *J. For.* **54:**582–584.

Lemon, P. C. 1949. Successional responses of herbs in the longleaf-slash pine forest after fire. *Ecology* **30:**135–145.

Lewis, C. E. 1964. Forage response to month of burning. USDA For. Serv. Res. Note SE-35. Southeastern For. Exp. Stn., Asheville, N.C.

Lewis, C. E., and T. J. Harshbarger. 1976. Shrub and herbaceous vegetation after 20 years of prescribed burning in the South Carolina Coastal Plain. *J. Range Manage.* **29**:13–18.

Lewis, C. E., and R. H. Hart. 1972. Some herbage responses to fire on pine-wiregrass range. *J. Range Manage.* **25**:209–213.

Little, E. L., S. Little, and W. T. Doolittle. 1967. Natural hybrids among pond, loblolly, and pitch pines. USDA For. Res. Paper NE-67. Northeast For. Exp. Stn., Upper Darby, Pa.

Little, S. 1964. Fire ecology and forest management in the New Jersey Pine Region. *Proc. Tall Timbers Fire Ecol. Conf.* **3**:34–59.

Lotti, T. 1956. Eliminating understory hardwoods with summer prescribed fires in coastal plains loblolly pine stands. *J. For.* **54**:191, 192.

Lotti, T. 1962. The use of prescribed fire in the silviculture of loblolly pine. *Proc. Tall Timbers Fire Ecol. Conf.* **1**:109–120.

Lovvorn, R. L. 1948. Grasslands in the South: a wide and versatile empire. In Grass. Yearbook of Agriculture, USDA. Washington, D.C., pp. 455–462.

McNab, W. H. 1976. Prescribed burning and direct-seeding old clearcuts in the Piedmont. USDA For. Serv. Res. Note SE-229. Southeastern For. Exp. Stn., Asheville, N.C.

McNab, W. H. 1977. An overcrowded loblolly pine stand thinned with fire. *South. J. Appl. For.* **1**(1):24–26.

McNab, W. H., and E. E. Ach. 1977. Slash burning increases survival and growth of planted loblolly pine in the Piedmont. *Tree Plant. Notes* **28**(2):22–24.

Mobley, H. E., R. S. Jackson, W. E. Balmer, W. E. Ruziska, and W. A. Hough. 1973. A guide for prescribed fire in southern forests. USDA For. Serv., Southern Reg. Atlanta, Ga. (Rev. 1978.)

Monk, C. D. 1965. Southern mixed hardwood forest of northcentral Florida. *Ecol. Monogr.* **35**:335–354.

Muntz, H. H. 1954. How to grow longleaf pine. USDA Farmers' Bull. 2061. Washington, D.C.

Oosting, H. J. 1958. *The Study of Plant Communities.* Freeman, San Francisco.

Pearson, H. A. 1974. Utilization of a forest grassland in southern United States. In Sectional Papers: "Grassland Utilization." Proc. 12th International Grassland Cong. Moscow.

Pearson, R. W., and L. E. Ensminger. 1957. Southeastern uplands. In Soil. Yearbook of Agriculture, USDA Washington, D.C., pp. 579–594.

Riebold, R. J. 1971. The early burning of wildfires and prescribed burning. In Prescribed Burning Symp. Proc. USDA For. Serv. Southeastern For. Exp. Stn., Asheville, N.C., pp. 11–20.

Robertson, W. B., Jr. 1962. Fire and vegetation in the Everglades. *Proc. Tall Timbers Fire Ecol. Conf.* **1**:67–80.

Sackett, S. S. 1975. Scheduling prescribed burns for hazard reduction in the Southeast. *J. For.* **73**:143–147.

Shepherd, W. O., and E. V. Dillard. 1953. Best grazing rates for beef production on cane range. N.C. Agric. Exp. Stn. Bull. 384. Raleigh.

Shepherd, W. O., E. V. Dillard, and H. L. Lucas. 1951. Grazing and fire influences on pond pine forests. N.C. Agric. Exp. Stn. Tech. Bull. 97. Raleigh.

Shoulders, E. 1976. Site characteristics influence relative performance of loblolly and slash pine. USDA For. Res. Paper SO-115. Southern For. Exp. Stn., Pineville, La.

Siggers, P. V. 1934. Observations on the influence of fire on the brown spot needle blight on longleaf pine seedlings. *J. For.* **32**:556–562.

Smart, W. W. G., Jr., G. Matrone, W. O. Shepherd, R. Hughes, and F. E. Knox. 1960. Comparative consumption and digestibility of cane forage. N.C. Agric. Exp. Stn. Tech. Bull. 140. Raleigh.

Somes, H. A., and G. R. Moorehead. 1954. Do thinning and prescribed burning affect the growth of shortleaf pine? USDA For. Exp. Stn. Res. Note 34. Northeast For. Exp. Stn., Upper Darby, Pa.

Stoddard, H. L., Sr. 1931. *The Bobwhite Quail: Its Habits, Preservation, and Increase.* Scribner's, New York.

Stoddard, H. L., Sr. 1962. Use of fire in pine forests and game lands of the deep Southeast. *Proc. Tall Timbers Fire Ecol. Conf.* **1:**31–42.

U. S. Department of Interior. 1970. *Geological Survey. The National Atlas of the United States of America.* Washington, D. C.

Wade, D. D. 1980. Some Melaleuca-fire relationships including recommendations for homesite protection. (In press). In Melaleuca Symp., Sept. 23, 24, Ft. Myers, Florida.

Wade, D. D., J. Ewel, and R. Hofstetter. 1980. Fire in South florida ecosystems. USDA For. Serv. Gen. Tech. Rep. SE-17. Southeastern For. Exp. Stn., Asheville, N.C.

Wahlenberg, W. G. 1935. Effect of fire and grazing on soil properties and the natural reproduction of longleaf pine. *J. For.* **33:**331–338.

Wahlenberg, W. G., S. W. Greene, and H. R. Reed. 1939. Effects of fire and cattle grazing on longleaf pine lands as studied at McNeill, Mississippi. USDA Tech. Bull. 683. Washington, D. C.

Welch, R. L. 1968. Forest statistics for the northern Coastal Plain of South Carolina. USDA For. Serv. Res. Bull. SE-10. Southeastern For. Exp. Stn., Asheville, N.C.

Wells, B. W. 1942. Ecological problems of the southeastern United States coastal plain. *Bot. Rev.* **8:**533–561.

CHAPTER

16

PRESCRIBED BURNING

Prescribed burning is both a science and an art requiring a background in weather, fire behavior, fuels, and plant ecology along with the courage to conduct burns, good judgment, and experience to integrate all aspects of weather and fire behavior to achieve planned objectives safely and effectively. Experience in fighting wildfires does not necessarily qualify one to conduct prescribed burns, because setting fires is the opposite of controlling fires. People with experience in fighting fires are often too cautious and waste time trying to burn when conditions are too moist. Others are careless because of a lack of experience and fail to post personnel and equipment in strategic locations. Personnel should be carefully selected and trained for prescribed burning work. Moreover trainees should be exposed to fire behavior that ranges from conditions that are too wet to too dry. There is no substitute for experience, common sense, and reason, which are needed in applying available prescriptions.

Prescriptions vary from one locality to another and there is more prescription data for some types of vegetation than others. Therefore, prescriptions are simply guides. Often they have limitations and may not fit a particular situation. Costs will vary depending on precautions taken, fuel type, and experience of personnel. However, as fire is used more extensively in a particular fuel type, costs will go down.

Learning to conduct prescribed burns cannot be done just by reading instructions. One must learn by conducting many small fires. Once the basics of controlled burning have been learned, the novice fireman can read or listen to more experienced burners and improve technique, increase efficiency of effort, safety, and desired results.

There are many judgments to be made on prescribed burns and generally there are unforeseen circumstances peculiar to each burn. Thus the fire boss should be experienced, for experience is the best teacher. We learn by doing, and by doing, we learn rapidly.

Wildfires often are frightening but prescribed burns are usually less frightening because they should be conducted under moderate weather conditions with prepared firelines. Prescribed burning is, however, dangerous to the inexperienced. It is very dangerous to personnel who are not fully experienced, for they can easily become overconfident and, inadvertently, let fires get away. Escape fires are common one to four days after a burn if fire crews do not do a good job of "mopping up" the day after a burn.

Prescription models are helpful in the planning phases of a burn and their use is recommended. However, these modeling exercises do not protect you against unusual situations. For example, a hill on one side of the burn area that might cause unusual winds; a canyon on the lee side that may aid in the formation of an intense firewhirl which will throw firebrands at greater distances than normal; unusual fuel densities that can create intense firewhirls; the possibility of a nighttime low-level jet of wind; or volatile fuel material. One must gain experience in fire behavior before becoming a fire boss in charge of a prescribed burning operation.

Experienced people consider the intangibles as well as weather and fuel conditions when planning a prescribed burn. Generally they use two to four critical variables (e.g., wind speed, relative humidity, quantity of fine fuel, fine fuel moisture, duff moisture, fuel load, and ambient temperature) before making a decision as to whether to burn a specific fuel type. In addition, a host of "experience factors" are involved.

The secret to all prescribed burning is to let the weather work for you. When all environmental factors are right the job is easy. With the proper weather, burns in most fuel types can be conducted safely with a crew of 6 to 10 people, 2 pickups, 1 or 2 pumpers, 1 dozer (or extra pumper), 2 weather kits, 5 drip torches, an adequate quantity of diesel-gas fuel (70:30% mixture), and 4 two-way radios.

In an unfamiliar fuel type, an experienced fire manager should supplement past training by burning a few small test plots based on the best prescribed burning data available. Experienced personnel will need one or two seasons of experience with a new fuel type before they feel comfortable using prescribed fire in a new area. Those beginning with no experience may need four or five seasons of experience, particularly in high-volatile fuels.

For maximum safety always prepare adequate firelines, especially where headfires are being used to burn volatile brush fuels [e.g., chaparral or pinyon-juniper (*Pinus-Juniperus* spp.)]. Burn out firelines that will more than adequately hold the headfire and prevent spot fires outside the fireline under desired burning conditions. The extra expense to burn out wide firelines [150 m (500 ft)] is minimal and will more than offset the cost of extra pumpers and the risk of fire getting away because of a less than adequate fireline.

Prescribed fires should always be conducted on a management-unit basis. A small burned area within a large field or wild ungulate winter range will be preferentially grazed for several years. Burning portions of large fields often makes prescribed burning look like the worst possible management practice. The problem is not with the fire but with inadequate planning. Adequate soil moisture [7.5 to 10 cm (3 to 4 in.) of precipitation] should also be a prerequisite for most prescribed burns, especially when burning grasslands in the spring.

FUEL CONSIDERATIONS

Fuel Load

A minimum of 670 to 1120 kg/ha (600 to 1000 lb/acre) of fine fuel is necessary to conduct prescribed burns in grasslands (Wink and Wright 1973; Beardall and Sylvester 1976), although wildfires will carry with as little as 340 kg/ha (300 lb/acre) of fine fuel [less than 3 mm (0.12 in.) diameter]. Grasslands containing shrubs or dozed material have similar requirements for minimum fine fuel. However the closer shrubs grow together or the closer that dozed piles of debris are deposited adjacent to each other on the ground, the lower the quantity of fine fuel needed to carry the fire (Fig. 16.1) (Wink and Wright 1973; Britton and Ralphs 1979). Moreover the more fine fuel or rough that is present, the more flexibility there is with prescriptions.

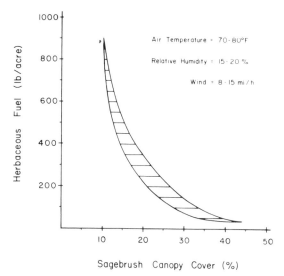

Fig. 16.1. Hypothetical relationship of sagebrush canopy cover and herbaceous fuel loading for successful fall burning. (From Britton and Ralphs 1979.)

Chaparral communities will not burn well in the absence of forbs and grasses when the shrubs are vigorous and green (Cable 1975). These fuels need to be continuous with 30 percent or more dead material before they will burn easily. Live big sagebrush *(Artemisia tridentata)* must have a canopy cover of at least 20 percent for fire to carry well and live juniper *(Juniperus* spp.) must have a canopy cover of 35 to 40 percent to carry well.

In forests it is important to know the quantity and moisture content of duff, needles and grass, 1.2-cm (0.5-in.) fuel, 2.5-cm (1.0 in.) fuel, and 7.5-cm (3.0 in.) fuel before initiating a burn. Fine fuels and duff are most important in underburning programs and the 1.2-cm (0.5-in.) fuels are most important in broadcast slash burning because they need to be well distributed for a continuous burn.

Volatility

Vegetation for prescribed burning is classed as one of two basic fuel types: low-volatile or high-volatile [fuels containing either extractives such as waxes, oils, terpenes, and fats (Countryman 1967; Philpot 1969)]. Grasses and hardwoods are low-volatile fuels, whereas chaparral, conifers, dead aspen *(Populus tremuloides)*, and dead juniper are high-volatile fuels. Sagebrush, oaks *(Quercus* spp.), rough beneath southeastern pines, live aspen, and slash are moderately volatile.

Low-volatile fuels such as grasses are relatively safe to burn, whereas high-volatile fuels are explosive and create firebrand problems. They can be burned safely, however, but wide firelines [up to 150 m (500 ft)] that are burned out using various techniques and a thorough knowledge of weather and fire behavior are necessary (Green 1970; Bunting and Wright 1974).

Firebrands

Firebrands can be a problem when burning high-volatile fuels (Bunting and Wright 1974). Usually the brands ignite low density fuels that have low heat capacity, such as rotten wood or dung (Fig. 16.2). After these materials are ignited, they will ignite surrounding fuels—a spot fire (Fig. 16.3). Occasionally the brands will fall into tight crevices such as between bark and wood or in a pile of matted leaves and start fires after a lapse of hours or days.

It is not possible to totally eliminate the potential of spot fires (Fig. 16.3) and still accomplish prescribed burning objectives. The best precaution is to prepare adequate firelines (instructions are in sections that follow) and use wind speed, relative humidity, air temperature, and fuel moisture as guides for spot-fire danger and the amount of patrolling that may be necessary.

Volatile fuels should be burned with a pumper and a dozer on standby. One dozer is worth 15 to 20 men in such fuels. When these precautions are used, spot fires do not get away. This technique has been tested under very hazardous burning conditions [fine fuel moisture 1%, relative humidity 11%, temperature 27°C (80°F), and wind speed 24 km/h (15 mi/h)]. Where dung and rotten wood are scarce, the risk of

Fig. 16.2. Cow dung is a low density fuel that is easily ignited by glowing firebrands.

Fig. 16.3. Spot fire in fescue grassland that started from glowing dead aspen bark.

spot fires is greatly reduced. Firebrand problems are generally minimized when burning out firelines if windspeed is less than 10 km/h (6 mi/h) relative humidity is above 40 percent, and ambient temperature is below 16°C (60°F).

Fuel Moisture

In the absence of precipitation, fine fuel moisture is closely related to relative humidity (Countryman 1964; Countryman 1971; Mobley et al. 1973); thus relative humidity has a direct effect on fine fuel moisture. The most important effect of fuel moisture may be described as a smothering process in which water vapor coming out of the fuels dilutes the oxygen in surrounding fuels (Davis 1959). This effect is especially important in getting fires started.

The threshold moisture at which fine fuels will or will not burn in sunlight is 30 to 40 percent, depending on type of fuel and overstory species. Many researchers (e.g., Mobley et al. 1973) use a figure of 30 percent. Below 20 percent fine fuel moisture has relatively little effect on prescription objectives in comparison to wind speed (Britton and Wright 1971). The preferred range of fine fuel moisture for prescribed burns is from 7 to 20 percent (Mobley et al. 1973).

Surface fuel moisture of woody fuels as indicated by relative humidity seems to be more important than total fuel moisture. In a dozed Ashe juniper *(Juniperus ashei)* community containing 2250 kg/ha (2000 lb/acre) of grass, one particular fire burned all the grass and left most of the piles of dead juniper. Relative humidity was 66 percent, fine fuel moisture was 10 percent, and there had been no precipitation for six months. Surface fuel moisture of the woody fuels is our only explanation for this unique behavior of fire.

Other threshold fine fuel moisture contents, are five percent, 7 to 8 percent, and 11 percent. Below 5 percent fine fuel moisture (relative humidity less than 20%) spot fires are certain, whereas spot fires are rare when fine fuel moisture is above 11 percent (relative humidity 65%). The 7- to 8-percent fine fuel moisture corresponds to a relative humidity of 40 percent, which is the minimum relative humidity at which firebrands usually cease to be a problem in dry grass (Wright 1974; Green 1977). Fine fuel moisture can be measured directly (Norum and Fischer 1980; Sackett 1980).

Following a rain, fine fuels such as grass reach 80 percent of their equilibrium moisture content with ambient weather within 1 hr (Britton et al. 1973). However, limbwood that is 5 cm (2 in.) in diameter may require up to four days at a constant relative humidity and temperature to reach equilibrium, and logs may require weeks or even months (Countryman 1971).

Ten-hour timelag fuel moisture [based on moisture in 1.2-cm (0.5-in.) diameter pine dowels] is a good indicator of moisture for fuels in the 1.2-cm (0.5-in.) size class and is often a good indicator of moisture in heavy fuels (Lancaster 1970; Fosberg 1977; Deeming et al. 1978). Prescribed fires will carry well when the 10-hr timelag fuel moisture is between 6 and 15 percent (Beaufait 1966). However, 1.2-cm (0.5-in.) fuels burn very quickly at 6 percent moisture, and they essentially stop burning at 15 percent. In slash an ideal moisture content of 1.2-cm (0.5-in.) fuels may be 7 to 8 percent and 10 to 12 percent for 2.5-cm (1.0-in.) fuels. Heavy fuels will burn up to a moisture content of 17 percent, but will not carry a fire.

Moisture content of fuels adjacent to a planned burn is extremely important. If it exceeds 30 to 40 percent in duff and fine fuels or 15 percent in 10-hr timelag fuels, there is minimal danger from firebrands (Bunting and Wright 1974).

WEATHER CONSIDERATIONS

Optimum weather conditions necessary for prescribed burning vary widely depending on fuel types and objectives of the burn. For example, where there is ample fine fuel [2.2 to 3.0 metric tons/ha (6.0 to 8.0 tons/acre)], such as in prairie marshes or roughs in southeastern forests, a temperature range of $-7°$ to $10°C$ ($20°$ to $50°F$) and a relative humidity of 30 to 50 percent are desired for prescribed burning (Mobley et al. 1973). By contrast where fine fuels are light and brush needs to be removed, as in sagebrush-grass vegetation, the preferred burning prescription is a temperature range of $24°$ to $32°C$ ($75°$ to $90°F$) and a relative humidity of 15 to 25 percent (Ralphs et al. 1976). Thus there is no universal optimum weather for prescribed burns. Prescriptions must be developed for each vegetation type, depending on the objectives and quantity of fine fuel (Martin and Dell 1978). Even with prescriptions there are times when they must be adjusted. The experienced fire boss will know how much modification of the prescription can be done safely under field weather and site conditions (Wright 1974).

Relative Humidity

A relative humidity of 40 percent is a threshold value (Britton and Wright 1971; Lindenmuth and Davis 1973). Below this value fine fuels burn easily with about the same intensity until the relative humidity drops below 20 percent. As the relative humidity creeps above 40 percent the rate of spread slows significantly (Lindenmuth and Davis 1973), standing woody material is difficult to ignite (Britton and Wright 1971), and danger from firebrands is noticeably low (Green 1977).

At a relative humidity of 50 percent glowing firebrands rarely start fires and glowing firebrands do not ignite material adjacent to burns when the relative humidity is above 55 percent. Below a relative humidity of 20 percent fire causes fine fuels to burn loudly and the danger from firebrands is always present. Thus, except for special situations, burns should seldom be conducted when the relative humidity is below 25 percent, unless winds are less than 10 km/h (6 mi/h) and temperatures are below 4°C (40°F). Another exception is west of the Rocky Mountains where a relative humidity below 20 percent is required for burning pinyon-juniper and sagebrush-grass. When the relative humidity exceeds 60 percent fires burn very spottily unless the fuel bed is at least 10 cm (4 in.) deep and has dry leaves or needles in the lower portion of the fuel bed.

Temperature

Ambient temperature plays a more critical role in fire behavior than one might think. As shown by Bunting and Wright (1974), danger from firebrands is low if ambient temperatures are below 15°C (60°F) but increases exponentially if ambient temperature is above 15°C (60°F) (Fig. 16.4). A threshold temperature of 19°C (67°F) is supported in the rate-of-spread model for Arizona chaparral by Lindenmuth and Davis (1973) and by practitioners in the interior west (Great Basin and adjacent regions) who say that prescribed fires are difficult to start in brush if the air temperature is below 21°C (70°F) (Stinson 1978).

Except for the dry interior west, a temperature of 27°C (80°F) is considered the upper limit for safe burning of volatile fuels unless the relative humidity is greater than 40 percent and the wind speed is less than 16 km/h (10 mi/h). In the Great Basin and surrounding regions where the relative humidity is usually below 30 percent, prescribed burns are often conducted with air temperatures as high as 32° to 35°C (90° to 95°F).

When temperature is below 4°C (40°F), firebrands will not ignite dung, although rotten wood will ignite down to 0°C (32°F). Below 0°C (32°F) grass will not support a fire unless the fuel is dense. Under these conditions piles of debris can be burned without risk from flaming firebrands, but if the piles are surrounded by grass, one must be on guard for hold-over fires in days to follow. Smoldering roots, partially covered logs, peat, and leaf mold are frequently a problem if wind speeds increase or the relative humidity drops on days following the burn.

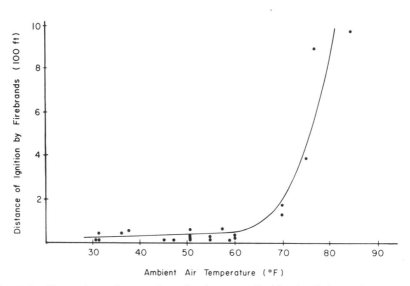

Fig. 16.4.   The maximum distance of spot fires from prescribed fires in relation to air temperature. Spot fires that started when temperatures were below 16°C (60°F) were caused primarily by flaming firebrands. (From Bunting and Wright 1974.)

Wind Speed

Wind speed [height 2 m (7 ft)] affects the burning rate of fuel directly by influencing the rate of oxygen supply to burning fuel (Davis 1959). Also strong wind speeds increase the rate of fire spread by tilting the flames forward so that unburned fuels receive energy by radiation and convection at an increased rate (Countryman 1976, 1978). These two mechanisms are especially important in causing smaller fires to build their intensity. As wind speed increases, however, it has a cooling effect and retards fuel moisture losses (Britton et al. 1973).

A wide range of wind speeds are used for controlled burns and to achieve most objectives some wind is preferred. The threshold value for igniting and burning standing debris is 13 km/h (8 mi/h) (Britton and Wright 1971). Thus to remove dead hardwood material or top-kill shrubs, a wind speed in excess of 13 km/h (8mi/h) but not exceeding 24 km/h (15 mi/h) is preferred to achieve the desired result (Wright 1974). This wind speed is necessary for one to two hours after ignition to be assured of consumption of low-volatile wood. High-volatile fuels ignite and are consumed easily regardless of wind (Wink and Wright 1973), although the preferred wind speed for burning most volatile fuels is 13 to 24 km/h (8 to 15 mi/h) (Wright 1974). The threshold wind speed for igniting mature or decadent chaparral is 5 km/h (3 mi/h), although about 8 to 13 km/h (5 to 8 mi/h) is usually best for running fire through chaparral (L. M. Green, pers. comm.). When winds exceed 32 km/h (20 mi/h), tumbleweeds, chunks of rolling debris, and firebrands can suddenly become serious problems.

Proximity of natural low pressure centers to fire and canyons will affect wind movements. For example wind will move from bodies of water to land during the day, but then move from the land to water at night (Schroeder and Buck 1970). Similarly, winds move up canyons during the warm part of the day and down canyons at night. Different colors of the landscape cause differential heating patterns which influence the ''light and variable'' wind patterns that are often forecast on calm days. This is why burning on a calm day can be risky. A steady wind is preferred for prescribed burning.

Nocturnal low-level jets are also of concern when burning at night. Many times at night sudden increases in wind speeds last for several hours. These jets usually set in shortly before sunset when cooling begins and last until midnight. They are most common at 460 to 760 m (1500 to 2500 ft) in valley bottoms and at 305 to 1070 m (1000 to 3500 ft) on level terrain (Davis 1959).

Weather Forecasts

A good fire weather forecaster, experienced in giving ''spot weather forecasts,'' is very important in the success of a prescribed burning program. A good forecaster can save many unnecessary trips to the field. Fewer trips will be made to the burn area when weather and fuel conditions are not suitable. This is particularly the case when high winds or winds from the wrong direction are forecast. A forecaster will also alert the fire boss to the passage of weather fronts or unusual weather conditions. Many times large tracts have been burned safely and then had firebrand problems the next day after passage of a front with 50- to 80-km/h (30- to 50-mi/h) winds. A good job of mopping up after a fire will minimize problems the day after a fire.

Fire weather forecasters can save 30 to 50 percent of costs in travel expenses per year. With good information from a fire weather forecaster, crews may be 95 percent certain that they can burn when they get to the field. When dependent on local forecasts, one can only be 50 percent certain of being able to burn when one gets to the field.

IGNITION

Ignition of woody materials depends on fuel size (diameter), moisture content (Fons 1950), density (Countryman 1975), arrangement, bark characteristics, and volatility. Dry, low-density fuels (dung and rotten wood) are easy to ignite. As fuel size increases it takes longer for the material to ignite, but bark characteristics and decay or wood borer activity in the wood can aid ignition. For example hardwood with wood borer activity is twice as easy to ignite at 480°C (900°F) than wood with no borer activity (Burton et al. 1972) because the wood with wood borer activity has a lower heat capacity (Countryman 1975). This characteristic is unimportant when temperatures are in excess of 590°C (1100°F) because the surface of the wood is being heated faster than the heat can be conducted toward the center of the log.

Logs laying on the soil surface are easier to keep ignited than wood that is perched in the air. The reason for this is that as heat radiates to the soil and reradiates back to the log, the log will stay ignited easier than if the heat radiates in all directions with none of it reflected back. Once air-dry volatile wood such as juniper is lit, it will continue to burn regardless of the relative humidity, wind speed, or ambient temperature. By contrast hardwood will go out easily if the wind drops below 13 km/h (8 mi/h) or the humidity increases dramatically, although dead oak is less likely to go out than dead honey mesquite *(Prosopis glandulosa* var. *glandulosa).*

TOPOGRAPHY

Slope

In addition to weather and fuel characteristics, topography affects fire behavior (Southwest Interagency Fire Council 1968). A fire spreading up a steep slope resembles a fire spreading before a strong wind (Davis 1959). Based on many individual fire reports from California (Southwest Interagency Fire Council 1968), researchers showed the relative rate of spread for different slopes (Table 16.1). A fire burning up a slope of +20 to +39 percent will spread twice as fast as a fire on level terrain. Nevertheless under the influence of a brisk wind, fire spread is slowed when a fire's intensity in the center of the burn becomes high enough to produce a strong indraft opposite the direction of firespread.

Table 16.1. Relative Rate of Forward Spread of Flame Front (No Allowance for Spotting or Roll) in Relation to Percentage Slope

Slope (%)	Rate of Spread	
	(ft/min)	(m/min)
Fires Burning Down Slope		
−70 to −40	1.0	0.3
−39 to −20	1.5	0.5
−19 to − 5	2.5	0.8
Fires Burning Upslope		
− 4 to + 4	5.0	1.5
+ 5 to +19	7.5	2.3
+20 to +39	10.0	3.0
+40 to +70	22.5	6.8

Data from Southwest Interagency Fire Council (1968).

Firewhirls

Topography can affect the formation of firewhirls. If a fire burns across a ridge, firewhirls will usually develop on the lee side of the ridge. They will pick up large chunks of material because of internal winds up to 485 km/h (300 mi/h) (Countryman 1971) and create a potential for spot fires at distances greater than normal from the leading edge of the fire. Thus burn with ridges, not across them.

Firewhirls can also develop on very still days on flat terrain by lighting the entire perimeter of an area (Haines and Updike 1971). In west Texas a whirl formed on level grassland when winds were light and variable that was 12 to 15 m (40 to 50 ft) in diameter at the base and lasted for 10 min. However it was in the center of the fire and did not move. Burning headfires into backfires will also cause whirls. A 15-m (50-ft) streak of fire crossed a 6-m (20-ft) fireline on one occasion when a headfire burned into a backfire. Corners of burns, heavy fuel loads on forest sites, edges of roads, and curves in topography are other potential sites for firewhirls.

FIRELINES

Fireline widths vary considerably depending on the type of fire and objectives for burning in specific vegetation types. For burning in southeastern pine communities, slash, and litter in grasslands a fireline width of 2 to 6 m (5 to 20 ft) is adequate. In sagebrush-grass communities (moderately volatile fuel type) in Utah, a 75-m (250-ft) fireline is usually preferred. In high-volatile fuels, however, such as those in west Texas where heavy grass is a continuous cover between dead piles of juniper, and in California chaparral a 150-m (500-ft) fireline is desired. Depending on slope and aspect wide firelines are not always necessary. Firelines on ridgetops can be narrower, especially if a south slope is being burned and the north slope on the opposite side of the ridge is more moist.

One point of grave concern is to never mix woody material with soil along firelines, for this is a prime source of "holdover" fire (Beaufait 1966; Schimke and Green 1970). **Always plow firelines away from the area to be burned.** Fires can break out as long as 8 to 12 weeks after pile burning has been completed. Such fires start from sparks that blow from dozed piles of soil and wood material. The risk of holdover fires increases where woody material has been mixed with soils having high levels of organic matter.

Firelines are usually dozed or plowed because these techniques are versatile in a wide variety of terrain and vegetation. However more attention should be given to the preparation of safe firelines without the use of dozers where possible (Davis 1976) to minimize the effect on aesthetics. In cheatgrass the wetline technique has proven to be successful (Martin et al. 1977) and is equally useful in other light to moderate grass fuels on relatively smooth terrain (Dubé 1977). A sprinkler wetline system has been successfully used on broadcast burns in lodgepole pine *(Pinus contorta)* slash (Quintilio 1972). Fire retardants and the use of chemicals are possible alternatives for firelines. Building firelines with liquid explosives (Dell and Ward 1970) is used widely in Idaho and Montana. Shredded lines are used in tallgrass prairies where soils are wet at the time of burning. In general, dozed or plowed lines allow the greatest flexibility in planning and conducting a burn on most terrain over a wide variety of vegetation types, but natural firebreaks and other alternatives should be used when possible.

AIR QUALITY AND SMOKE MANAGEMENT

Airborne particulates are the primary pollutant of wildfires and prescribed burns (Komarek 1970; Dieterich 1971) and account for 23.7 percent of all particulates emitted into the atmosphere (Martin et al. 1977). Their most objectionable feature is their effect on visibility. However, burning under favorable dispersal conditions minimizes the visibility problem (Perovich 1977). Their effect is short-lived. The large particles settle out rapidly, but small particles may remain suspended for several days (Martin et al. 1977). The effect of small particles on humans may not be as innocuous as we once thought but the effects remain unclear (Sandberg et al. 1979).

Hydrocarbons are the second most important combustion product from fires, but few, if any, appear in the combustion of wood products that are important in photochemical reactions (Fritchen et al. 1970). Prescribed burns and wildfires account for about 6.9 percent of all hydrocarbons emitted into the atmosphere (Martin et al. 1977).

Carbon monoxide is given off in substantial quantities [25 kg/metric ton (60 lb/ton)] in forest fuels (Martin et al. 1977) but it seems to oxidize quite readily (Fritchen et al. 1970) and does not pose a threat to people, plants, or animals (Dieterich 1971). Sulfur is almost absent in woody fuels and nitrogen oxides are not formed, as their formation generally requires a temperature higher than that generated by burning wood (Sandberg and Martin 1975).

The most objectionable feature about smoke is that it looks bad and can obstruct visibility. To minimize these undesirable effects burning should be done after the morning inversion has broken and before the evening inversion forms, when mixing depth and wind is most favorable for smoke dispersal (Nikleva 1972). Adhering to such a time frame is not always convenient but it should be observed around populated areas. Moreover, burns should be conducted in as short a time as possible. People are willing to tolerate smoke for a few hours but not for several days. Once the public starts to react to smoke more restrictions are usually placed on prescribed burning. This ultimately limits the use of fire. Our interests in the use of prescribed fire are best served by adhering to state pollution guidelines and burning when the conditions for dispersal of smoke are optimum and the wind direction will take smoke away from settled areas.

FIRING TECHNIQUES

Headfires, backfires, strip-headfires, flank fires, center ignition, and area ignition are all useful ignition methods (Fig. 16.5) (Davis 1959; Dixon 1965; Southwest Interagency Fire Council 1968; Sando and Dobbs 1970; Mobley et al. 1973). Since these methods have been thoroughly discussed in the above references, they will only be briefly discussed here, although their use will be more fully illustrated in sections that follow.

Headfires (fires that move with the wind) (Fig. 16.6) are most effective for killing shrubs and trees (Fahnestock and Hare 1964; Gartner and Thompson 1972; Sackett 1975) and in getting an effective burndown of standing dead trees (Britton and Wright 1971). They are also effective in burning low quantities of fine fuel [670 to 1125 kg/ha (600 to 1000 lb/acre)] to efficiently clean up debris and brush (Heirman and Wright 1973; Wink and Wright 1973). Backfires (fires that back into the wind) work well when the fuel exceeds several thousand kilograms per hectare; when you wish to maintain good control, particularly in high-volatile fuels; when you wish to reduce heat damage to overstory conifers (Biswell et al. 1973; Mobley et al. 1973); and when the weather is more risky than is desirable.

Strip-headfires and flank fires are variations in types of ignition techniques used to control fire intensities. They are usually used when backfires move too slowly but a headfire would be undesirable or too dangerous. Area ignition (Fenner et al. 1955; Schimke et al. 1969) is used to set the entire area on fire at once and cause a fire to move toward the middle. Center ignition (Beaufait 1966) is similar to area ignition although the center is lit first and the intensity of the fire increases more slowly over time than area ignition. These latter two ignition techniques usually result in very in-

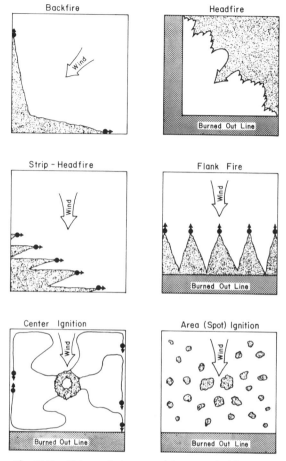

Fig. 16.5. Firing techniques used for prescribed burning.

Fig. 16.6. Fires that move with the wind are called headfires.

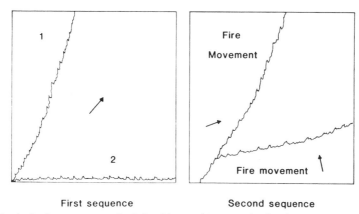

First sequence   Second sequence

Fig. 16.7.   In the first sequence a fire is lit with a southwest wind. After the two sides have burned for a while, the fire from both sides begin moving toward each other (sequence two), even though the prevailing wind is from the southwest.

tense fires and occasional firewhirls. Center ignition and sometimes area ignition are generally used for slash burning (clear-cut and no overstory) when winds are light. Their primary value is to create intense fires that suck winds into the center of the fire and minimize danger from firebrands.

The way that a fire is ignited affects fire behavior. Fires in one location can be used to draw fire from another even though the prevailing wind may be blowing against the latter. The diagrams in Fig. 16.7 illustrate this point. Also canyons can aid ignition patterns. Fires move more rapidly up a canyon than adjacent areas and this intense burning, which creates a displacement of air, will draw fire from adjacent areas.

PRESCRIPTIONS FOR LOW-VOLATILE FUELS

Detailed prescriptions for conducting burns in low-volatile grassland fuels have been given by Wright (1974) and Wright and Bailey (1980). Depending on the objectives and the quantity of fine fuel, a wide variety of prescriptions can be used. Experience is the best teacher. Generally, firebrands (glowing embers) are not a problem in grasslands with low-volatile shrubs and trees, but chunks of glowing debris will easily roll on the ground when wind gusts above 32 km/h (20 mi/h). Thus, depending whether the objective of a prescribed burn is to remove litter, burn debris, or top-kill shrubs and trees, the desired fireline width will vary with quantity of fine fuel in adjacent pastures and the wind speed and relative humidity needed to accomplish objectives. Firelines in low-volatile fuels may range from the width of a cow trail to 30 m (100 ft). Firelines greater than 3 m (10 ft) are usually burned out between two dozed lines using special prescriptions.

Fig. 16.8.   Fire in the semidesert grass-shrub type (top photo) is useful to kill burroweed (predominant subshrub in foreground), cactus, and young velvet mesquite. Thick stands of sacaton (bottom photo) are frequently burned to remove litter and increase utilization. (Top photo courtesy of USDA Forest Service.)

Semidesert Grass-Shrub

Use of prescribed burning in southern desert grasslands (Fig. 16.8) is controversial, but it as been beneficial to kill burrowweed *(Aplopappus tenuisectus)*, to kill 50 percent of various cactus *(Opuntia* spp.) species, and to kill 20 percent of young velvet mesquite *(Prosopis glandulosa* var. *velutina)* and some other shrubs (Cable 1967). These uses of fire have been restricted to southeastern Arizona. In New Mexico there is no clear-cut need for prescribed fire, although it is used occasionally in tobosagrass *(Hilaria mutica)* and sacaton *(Sporobolus airoides, S. wrighti)* (Fig. 16.8, bottom) flats to remove litter and increase utilization by livestock.

Prescription guides have not been developed for the semidesert grass-shrub type, although Cable (1967) and Dwyer (1972) have reported the conditions under which they burned. Cable (1967) conducted burns on areas with 340 to 785 kg/ha (300 to 600 lb/acre) of fine fuel when air temperature was 25°C (77°F), relative humidity was 15 to 18 percent, and wind speed was 5 to 10 km/h (3 to 6 mi/h). In heavier fuel [3190 to 4010 kg/ha (2840 to 3570 lb/acre)] Dwyer (1972) burned when air temperature was 29° to 36°C (85° to 96°F), relative humidity was 5 to 20 percent and wind speed was slight.

Our recommendation is that no burning should be attempted unless there is at least 670 kg/ha (600 lb/acre) of fine fuel. Where a good stand of burroweed is present to help carry the fire (Fig. 16.8, top) one can get by with 225 to 335 kg/ha (200 to 300 lb/acre) of fine fuel (Cable 1967). About 240 ha (600 acres) is a reasonable unit to burn and firelines should be prepared as shown in Fig. 16.9.

In light fuels, doze a fireline 3.0 to 3.7 m (10 to 12 ft) wide around the area to be burned (Fig. 16.9). Strip-headfire (Fig. 16.10) a 30-m (100-ft) strip on the leeward sides of the planned burn during evening or morning hours in May or June when weather conditions are approximately as follows: air temperature 21°C (70°F), relative humidity 15 to 50 percent, and wind speed less than 13 km/h (8 mi/h). At the 30-m (100-ft) point, the fire could either be stopped with another dozed line or put out with a pumper, depending on the quantity of fuel. On the same day headfire the re-

Fireplan for Shortgrass Prairie

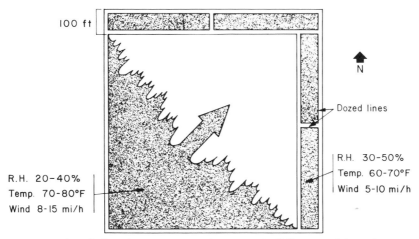

Fig. 16.9.   To burn fuels in shortgrass prairie, doze a 3-m (10-ft) line around the entire area to be burned and around the area that you plan to burnout for a fireline, as shown in the figure. Strip-headfire the fireline when relative humidity is 30 to 50 percent and wind speed is less than 16 km/h (10 mi/h). Headfire the remainder of the pasture when relative humidity is 20 to 40 percent, air temperature is 21° to 27°C (70° to 80°F), and wind speed is 13 to 24 km/h (8 to 15 mi/h).

mainder of the area when air temperature is 21° to 27°C (70° to 80°F), relative humidity is 15 to 40 percent, and wind speed is 13 to 24 km/h (8 to 15 mi/h). Burning is usually done in early June.

In heavy fuels, such as tobosagrass or sacaton, a backfire may be adequate to accomplish objectives. In such cases wind speed is not needed to carry the fire and temperature up to 36°C (96°F) and relative humidity as low as 5 percent can be tolerated. However where fuel is light [less than 1125 kg/ha (1000 lb/acre) of fine fuel] a wind speed in excess of 13 km/h (8 mi/h) will be necessary for the fire to carry through the grass.

Shortgrass Prairie

Generally, fire is not used in the shortgrass prairie because it seldom enhances the growth of grasses (Wright 1978). However, it is a beneficial tool to remove chained mesquite and to kill cactus species. Most burning is done during February and March, before the warm-season grasses begin to grow.

If you wish to clean up chained debris in buffalograss *(Buchloe dactyloides)* that has 2250 kg/ha (2000 lb/acre) of fine fuel, a 30-m (100-ft) fireline should be burned out on the north and east sides (Fig. 16.9) when the relative humidity is 30 to 50 percent and wind speed is 8 to 16 km/h (5 to 10 mi/h). Air temperature is usually 16° to 24° C (60° to 75°F). Strip-headfire the fireline (Fig. 16.10). The rest of the pasture

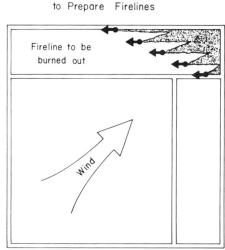

Fig. 16.10.   The strip-headfire technique usually involves the combination of a backfire (lead man) and several staggered strip-headfires. The men are staggered so that the fire will not over-run anyone. Also the line of the second man may only be 3 to 6 m (10 to 20 ft) from the dozed line, whereas the men will usually be spaced progressively further apart [e.g., 10, 25, 50 m (33, 82, 164 ft)]. This is a very common technique to burn firelines in grasslands, tallgrass prairie, slash, and other vegetation types.

should be burned with a headfire when the relative humidity is 20 to 40 percent, wind is out of the southwest (direction of prevailing dry winds) at 13 to 24 km/h (8 to 15 mi/h), and air temperature is 21° to 27°C (70° to 80°F).

Mixed Grass Prairie

Fire is desirable in mesquite-tobosagrass communities to remove excess litter; increase yields of tobosagrass; increase palatability of tobosagrass; reduce mesquite canopy to acceptable levels; top-kill mesquite and leave the stems in a state where wood borers will attack and aerate them to such an extent that they will be easily consumed by a reburn; kill 40 to 80 percent of three species of cactus (Bunting et al. 1980); and kill undesirable annual weeds such as common broomweed *(Xanthocephalum dracunculoides)* (Wright 1972).

If dead mesquite is standing and you wish to burn it down (Fig. 16.11) you need a minimum of 3370 kg/ha (3000 lb/acre) of fine fuel, a relative humidity below 40 percent, and wind speed in excess of 13 km/h (8 mi/h). Firelines are prepared by burning a 30-m (100-ft) line on the north and east sides when the relative humidity is 50 to 60 percent and the wind speed is less than 13 km/h (8 mi/h) (Fig. 16.12). Then the rest of the pasture is headfired when the relative humidity is 25 to 40 percent, wind speed is 13 to 24 km/h (8 to 15 mi/h), and air temperature is 21° to 27°C (70° to 80°F). Burns in the southern mixed prairie are conducted in late February and March.

The primary dangers in burning mixed prairie grasslands come from tumbleweeds and firewhirls. Tumbleweeds will ignite and tumble, leaving flames in their path. Firewhirls develop where wind shears occur such as when a headfire runs into a backfire, or a fire goes upslope into a wind. Huge firewhirls can develop when wind speeds are light and variable. For these reasons burns should be conducted with a steady wind and one should never burn into backfires, unless there is a 90-m (300-ft) fireline. Fires should be planned to move with the ridges, not across them.

Fire is generally not used in the mixed prairie of the central and northern Great Plains. In most mesic areas adjacent to the tallgrass prairie on the east and fescue prairie to the north, prescribed fire has a place. It can be used to enhance wildlife habitat; reduce litter buildup in selected communities promoting better livestock distribution; and top-kill shrubs, particulary snowberry *(Symphoricarpos* spp.). Prescription guides have not been developed but they should resemble others used for grasslands where brush is a problem.

Tallgrass Prairie

In the tallgrass prairie (Fig. 16.13), particularly the Flint Hills region, many ranchers have burned for years and know how to do it. Launchbaugh and Owensby (1978) have written a bulletin that contains burning guidelines for this vegetation type. Except for some areas that contain eastern redcedar *(Juniperus virginiana)*, firebrands are not a problem. The primary goal on most burns is to remove excess litter which may be 3370 to 4495 kg/ha (3000 to 4000 lb/acre). Therefore most burns are con-

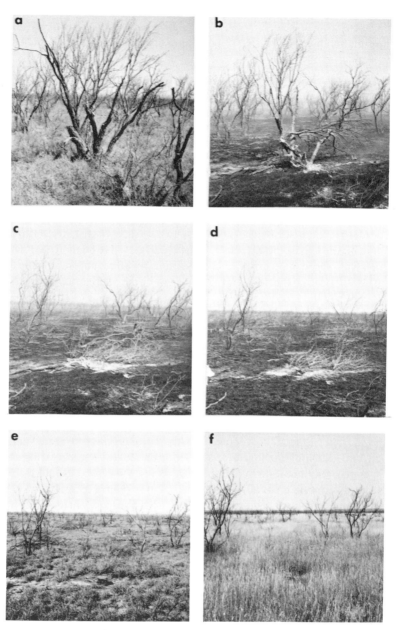

Fig. 16.11. A sequence of photos in a mesquite-tobosa community before burning until three months after burning. (a) Plot before burning on March 20, 1969—4573 kg/ha (4070 lb/acre) of fine fuel, wind speed 31 km/h (19 mi/h), air temperature 21°C (70°F), and relative humidity 22 percent; (b) 20 min after ignition; (c) 60 min after ignition; (d) 85-percent burndown one day after ignition; (e) grasses growing well one month after burn; (f) yields 3146 kg/ha (2800 lb/acre) four months after burning compared with 1236 kg/ha (1100 lb/acre) on control. Mortality of mesquite was 25 percent.

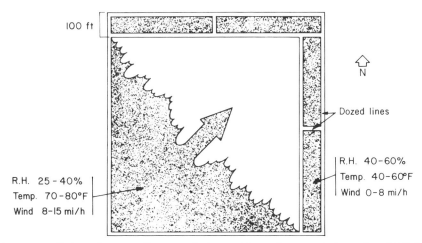

Fig. 16.12. To burn down standing mesquite (sprayed in previous years), doze a 3-m (10-ft) line around the entire area to be burned and within the main area as shown. Strip-headfire the firelines on the downwind sides.

Fig. 16.13. Tallgrass prairie in central Missouri.

Fire Plan for Tallgrass Prairie

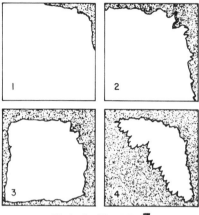

Wind 5 - 20 mi/h ↗

Fig. 16.14. In tallgrass prairie, natural fire-breaks, including roads, trails and fenceline cowpaths, are used to the extent possible. In some cases a wetline may be put down with a sprayer where there is no natural break. A backfire is started on the downwind side (1) and lit simultaneously in each direction on the downwind sides (2). After the backfire has burned 15 to 30 m (50 to 100 ft) on the lee sides the remainder of the area is lit (3) and the rest of the area is burned with a headfire (4). Wind speeds may vary from 8 to 32 km/h (5 to 20 mi/h). (From Launchbaugh and Owensby 1978.)

ducted when the relative humidity is 50 to 60 percent, minimizing any risk from potential firebrands.

Launchbaugh and Owensby (1978) begin burning on the leeward sides of a pasture and let the fire back into the pasture (Fig. 16.14). Normally, firing begins along a road or cow trail with two pumpers patrolling the fire and laying down a wetline when necessary. Firelines are rarely dozed or plowed in tallgrass prairie. After the fire has backed up 30 to 45 m (100 to 150 ft), they headfire the rest of the pasture.

In Missouri, Clair Kucera (pers. comm.) uses a similar procedure to burn prairie land, but mows the fireline [5 to 7 m (16 to 23 ft)] with a shredder before burning (Fig. 16.15). In this area the soils are 60 percent clay and usually have standing water when burned. In the Black Hills of South Dakota, Gartner and Thompson (1972) have used prescribed burning to kill ponderosa pine *(Pinus ponderosa)* seedlings and saplings invading tallgrass prairie.

Headfires are set in tallgrass prairie with 8 to 16 km/h (5 to 10 mi/h) wind speeds if improvement of forage quality is the only concern. Where brush control is important (Fig. 16.16) Launchbaugh and Owensby (1978) recommend burning with average wind speeds that vary from 8 to 24 km/h (5 to 15 mi/h) and when relative humidity is about 40 percent. Ambient temperature is not as important but generally varies from 16° to 24°C (60° to 75°F).

Fescue Prairie

Burning is not commonly practiced in this fuel type in the northern plains and adjacent Rocky Mountain foothills. Fire was caused by Indians and lightning prior to settlement (Nelson and England 1971). Brush is invading at an alarming rate wherever rough fescue is a dominant in the northern Great Plains, in the foothills, and in the interior plateau of British Columbia (Maini 1960; Bird 1961; Johnston and Smoliak 1968; Bailey and Wroe 1974; Scheffler 1976; Strang and Parminter 1980). The

Fig. 16.15. Shredded area (left) used as a fire break (right) in tallgrass prairie in central Missouri.

Fig. 16.16. Where eastern redcedar is present in tallgrass prairie, Launchbaugh and Owensby (1978) recommend resting the pasture until it has at least 3371 to 4494 kg/ha (3000 to 4000 lb/acre) of fine fuel. Backfires on the lee sides should burn 30- to 60-m (100- to 200-ft) before a headfire with a 24- to 32-km/h (15- to 20-mi/h) wind speed is set to maximize kill of eastern redcedar.

greatest potential use of prescribed fire is to top-kill shrubs and trees invading grasslands (Bailey 1978). Ranchers are quickly becoming aware of the possible uses of fire but they are generally reluctant to use it. Other uses of prescribed burning include increased nutritive values and/or palatability of forage and browse; reduced litter buildup and improved livestock distribution; increased accessibility of forage to livestock; and restored parklike appearance of the landscape.

Cool-season grasses predominate and their forage production is generally not enhanced the first year after burning. There are indications that spring or fall burning may promote higher production the second and third growing season after fire (Sinton 1980; Willms et al. 1980). When possible, burning should be done when there will be no loss in forage production. Anderson and Bailey (1978) found no change in forage production for 1971 after either a fall or a spring burn but Sinton (1980) found that burning in 1978 and later than three weeks after spring snowmelt reduced forage production. Prolonged annual early spring burning of this grassland reduced herbage production by 30 percent (Anderson and Bailey 1980).

When headfires are to be used, a 30-m (100-ft) fireline should be prepared on leeward edges. Burns can be conducted in spring or fall when the relative humidity is 40 to 65 percent, temperatures are 4° to 24°C (40° to 75°F) and winds are 3 to 19 km/h (2 to 12 mi/h). Backfiring can be started along a trail or road. Allow the fire to back into the grassland if wide firelines are objectionable and if the law permits [Alberta requires that a 10-m (30-ft) bare area encircle the area to be burned]. Under these circumstances, backfiring should only be done with a relative humidity of 50 to 60 percent and winds of 3 to 13 km/h (2 to 8 mi/h). Once the fire has backed up 30 m (100 ft) on the leeside, or 100 m (300 ft) if brush is abundant, the remainder of the area can be ignited using a headfire. Only experienced people should use the procedure, provided they have two or more pumpers and a crew of six or more experienced fire suppression people on hand. Brush up to 2.5 cm (1.0 in.) in diameter at flame height will be top-killed by a backfire while most brush up to 5 cm (2 in.) in diameter will be killed by a headfire. A minimum of 1100 kg/ha (1000 lb/acre) of fine fuel is required for continuous fire coverage. Resting a range for a year followed by early spring burning will result in a higher percentage of top-killed brush because of the higher fuel loading of 2200 to 4400 kg/ha (2000 to 4000 lb/acre). The fire will burn more intensely and kill brush up to 8 cm (3 in.) in diameter.

Interspersed in fescue grassland are patches of shrubs. The most common are western snowberry *(Symphoricarpos occidentalis)*, silverberry *(Elaeagnus commutata)* and serviceberry *(Amelanchier alnifolia)*. In spring before leaf flush, snowberry is a very flammable type that can release many firebrands. It will ignite at 50 percent relative humidity or lower and will shower firebrands for 100 m (300 ft) in a 16- to 24-km/h (10- to 15-mi/h) wind. To burn fescue grassland with minimum risk when there are snowberry patches conduct the fire under the following conditions: 40 to 50 percent relative humidity; 4° to 24°C (40° to 75°F) temperature; 3 to 18 km/h (2 to 12 mi/h) wind; burn into a 100-m (300-ft) fireline.

Marshes

Marshes are usually burned to reduce the density of common reed *(Phragmites communis)*, create new feeding areas, and create more edge effect for nesting areas. In the north-central Nebraska Sandhills, Schlichtemeier (1967) found that thick, dense stands of marsh could easily be burned. Excellent results were achieved after

Table 16.2. Weather Conditions During Four Burns in
a Common Reed Marsh in North-Central Nebraska

Burn Number	Date	Relative Humidity (%)	Temperature		Wind	
			(°C)	(°F)	(km/h)	(mi/h)
1	2/23/66	72	−1	30	24	15
2	2/24/66	60	3	37	11	7
3	1/18/67	67	−8	17	37	23
4	1/19/67	52	6	42	27	17

Data from Schlichtemeier (1967).

ice was 23 to 30 cm (9 to 12 in.) thick and 5 to 10 cm (2 to 4 in.) of snow covered the surrounding range. Four burns were conducted in January and February under the conditions shown in Table 16.2. Combustibility of vegetation varied only slightly on all four burns. Density of common reed and bulrush (*Scirpus* sp.) decreased 60 to 85 percent. Schlichtemeier (1967) burned successfully in freezing weather because marshes have large quantities of dry material in thick layers partially protected from the weather. Burning under the cold, icy conditions in this study prevented ignition of organic soils. Light or moderate grass fuels will not burn under similar conditions.

Wet meadow vegetation containing 7100 kg/ha (6300 lb/acre) of fine fuel should be burned in nonfreezing weather. Air temperatures of 10° to 18°C (50° to 65°F), relative humidity of 30 to 40 percent, and 0 to 3 km/h (0 to 2 mi/h) of wind are desirable burning conditions (Britton et al. 1980).

Open Stands of Pinyon-Juniper with Grass Understory

Fire is effective in this vegetation type (Fig. 16.17) to kill pinyon and juniper trees that are less than 1.2 m (4 ft) tall (Jameson 1962; Arnold et al. 1964; Dwyer and Pieper 1967). At least 670 kg/ha (600 lb/acre) of fine fuel is needed to carry a fire. Trees greater than 1.2 m (4 ft) tall will not be killed unless they have accumulations of tumbleweeds at the base of the trees (Jameson 1962).

A fireline 3.0 to 3.7 m (10 to 12 ft) wide should be dozed around the entire area to be burned. About 182 ha (450 acres) is a good-sized unit to burn (Davis 1976). Use a strip-headfire (Fig. 16.10) on the leeward sides of the planned burn during evening or morning hours in the spring to burn out a 30-m (100-ft) strip (Fig. 16.12). Where fire continues to back up beyond 30 m (100 ft) put it out with a pumper. Headfire the remainder of the area when air temperature is 21° to 23°C (70° to 74°F), relative humidity is 20 to 40 percent, and wind speed is 16 to 32 km/h (10 to 20 mi/h) (Jameson 1962; Dwyer and Pieper 1967).

Fig. 16.17. Open stand of pinyon-juniper with grass understory in eastern New Mexico.

PRESCRIPTIONS FOR MODERATELY VOLATILE FUELS

Big Sagebrush and Grass

The primary value of burning sagebrush-grass communities (Fig. 16.18) is to reduce cover of big sagebrush, increase forbs and eventually grass, and to increase accessibility of forage to livestock. Prescribed burning is not necessary until cover of big sagebrush exceeds 11 to 20 percent (Pechanec et al. 1954; Ralphs et al. 1976). Antelope bitterbrush *(Purshia tridentata)* is very susceptible to fire, and rabbitbrush *(Chrysothamnus* spp.) and horsebrush *(Tetradymia canescens)* respond vigorously to fire. Thus these factors need to be considered before planning a burn.

Big sagebrush is usually difficult to burn in the spring unless there is at least 670 kg/ha (600 lb/acre) of herbaceous fuel (Beardall and Sylvester 1976), although Britton and Ralphs (1979) hypothesize that minimum herbaceous fuel is a function of big sagebrush canopy cover (Fig. 16.1). For a range of 10 to 50 percent canopy cover, Britton and Ralphs indicate that the quantity of fine fuel necessary to carry fires in big sagebrush communities in the fall could vary from 1000 kg/ha (890 lb/acre) down to 50 kg/ha (44 lb/acre). However, closed stands, such as shown in Fig. 16.19, may need pretreatment with mechanical or chemical methods if the canopy cover is less than 30 percent. Plowing and seeding is another alternative.

After the decision to burn has been made, be sure that there is adequate fine fuel on the ground to carry the fire and try to avoid burning immediately after heavy seed crops, because big sagebrush can reoccupy the burned area rapidly during years with good moisture (Pechanec et al. 1954; Johnson and Payne 1968). Early spring or fall (September, October) burns may be the most preferable (Blackburn and Bruner

Fig. 16.18. Big sagebrush-grass community in southern Idaho. (Photo courtesy of University of Idaho.)

Fig. 16.19. Closed stands of sagebrush (no grass understory) may need mechanical pretreatment before they can be burned. (Photo courtesy of USDA Forest Service.)

1975; Beardall and Sylvester 1976). Soils should be moist down to 45 cm (18 in.) before burning in the spring, but in the fall soil moisture is not critical for most plant species.

Based on burning prescriptions in sagebrush-grass vegetation by Pechanec et al. (1954), Davis (1976), Ralphs et al. (1976), and Britton and Ralphs (1979), a 3.0- to 3.7-m (10- to 12-ft) fireline should be dozed around the entire area to be burned [pref-

Fire Plan for Sagebrush-Grass

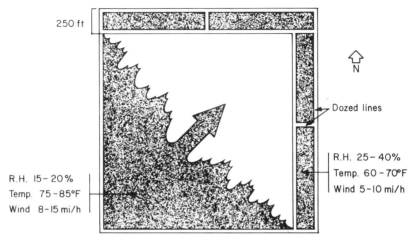

250 ft

N

Dozed lines

R.H. 25-40%
Temp. 60-70°F
Wind 5-10 mi/h

R.H. 15-20%
Temp. 75-85°F
Wind 8-15 mi/h

Fig. 16.20. Prescribed burning plan for sagebrush-grass communities. Doze a 3 m (10 ft) line around the entire area to be burned and as indicated on the above figure. Strip-headfire the fireline on the lee sides during the morning hours and during early afternoon set the main headfire. Most burns are conducted in the fall, but spring fires are equally satisfactory (maybe better if fine fuel is adequate and the weather is not too cold).

erably about 180 ha (450 acres) according to Davis (1976)]. Then strip headfire (Fig. 16.10) a 75-m (250-ft) line on the leeward sides during the morning hours when relative humidity is 25 to 40 percent and wind speed is 8 to 16 km/h (5 to 10 mi/h). If the fire backs up in heavy fuel beyond the 75-m (250-ft) strip, use a pumper or another fireline to stop the fire. About 2:00 p.m. headfire the remaining area when air temperature is 24° to 29°C (75° to 85°F), relative humidity is 15 to 20 percent, and wind speed is 13 to 24 km/h (8 to 15 mi/h) (Fig. 16.20).

A wetline may be used in lieu of a plowed line, particularly in cheatgrass *(Bromus tectorum)* (Martin et al. 1977). This technique involves wetting a line and letting the fire back away from it, or burning between two wet lines.

Big Sagebrush and Low Sagebrush

Where dense stands of big sagebrush are mixed with low sagebrush *(Artemisia arbuscula)*, no firelines need to be prepared because fires will rarely carry in low sagebrush even during a hot day with wind speeds up to 40 km/h (25 mi/h) (Beardall and Sylvester 1976). Snow banks are equally effective as natural barriers in big sagebrush. As a consequence, spring burns can be conducted on warm days with steady winds wherever it is convenient.

Beardall and Sylvester (1976) suggest burning in early spring when relative humidity is below 60 percent, wind speed is above 13 km/h (8 mi/h), and there is more than 670 kg/ha (600 lb/acre) of fine fuel. Late summer burning has not been tested in areas dominated by low sagebrush because of the potential to harm sensitive grasses such as medium to large Idaho fescue *(Festuca idahoensis)* plants.

Sand Shinnery Oak with Little Bluestem (Western Oklahoma)

Where there is a good stand of little bluestem *(Schizachyrium scoparium)* growing in sand shinnery oak *(Quercus havardii)*, fire is often desirable to reduce litter of the little bluestem and accumulations of oak leaves on the ground. Sand shinnery oak is a vigorous sprouter and may increase in density after the fire, but total grass yields also increase about 340 kg/ha (300 lb/acre) following the first growing season after burning.

Burning techniques have not been developed for sand shinnery oak plant communities (McIlvain and Armstrong 1966), but these communities have been burned experimentally in Oklahoma where little bluestem is the primary understory. Fine fuel generally amounts to 2250 to 4500 kg/ha (2000 to 4000 lb/acre). Oak leaves are good firebrands and spotting may occur 15 to 20 m (50 to 65 ft) ahead of fire fronts in sand shinnery oak. Therefore we recommend prescriptions for burning out the firelines and the headfire that are essentially identical to those shown in Fig. 16.12. The primary difference would be to burn out a fireline 75-m (250-ft) wide, using a combination of backfires and strip-headfires.

Southeast Pine

Prescribed burning was pioneered and went through intensive criticism for 30 years in the Southeast before it was accepted as a management practice. Reasons for burning are to control brown spot disease in longleaf pine *(Pinus palustris)* while it is in the grass stage; reduce excess fuels; prepare mineral seedbeds for germination of new pine seedlings; control understory hardwoods; and improve wildlife habitat (Mobley et al. 1973).

Pine trees are subclimax to a variety of hardwood species in the southeastern United States. Thus the rough in the understory (Fig. 16.21) is a composite of grasses, young shrubs and deciduous trees, pine needles, and occasionally stands of saw palmetto *(Sereonoa repens)* and gallberry *(Ilex glabra)*. Ground fuels will accumulate to 27 to 38 metric tons/ha (10 to 14 tons/acre) (Taylor and Wendel 1954; Fahnestock and Hare 1964; Cooper 1971) in a period of 10 to 15 years (Bruce 1951). Bruce found that in open pine stands fuels averaged from 6.8 to 13.8 metric tons/ha (2.5 to 5.1 tons/acre) in one to five years whereas under dense pine stands they averaged from 18.5 to 26.4 metric tons/ha (6.8 to 9.7 tons/acre) over the same period of time. So in this region a lot of fine fuel accumulates in a very short time. The fuel is moderately volatile; however, even with such large quantities on the ground, the fuel can be burned effectively and safely with low temperatures, high humidities, and low wind speeds (Dixon 1965; Mobley et al. 1973).

Dixon (1965), Mobley et al. (1973), and Sackett (1975) have given detailed fire

Fig. 16.21. Rough in understory of longleaf pine. (Photo courtesy of USDA Forest Service.)

prescriptions for burning in the Southeast (Fig. 16.22). Weather and fuel data should be as follows: wind speed 8 to 29 km/h (5 to 18 mi/h) in openings at a 6-m (20-ft) high station, which is equivalent to 3 to 16 km/h (2 to 10 mi/h) in a pine stand; air temperature $-7°$ to $10°C$ ($20°$ to $50°F$); relative humidity 30 to 50 percent, although under special conditions a range of 20 to 60 percent can be used (Mobley et al. 1973); and moisture content of upper litter is between 20 and 25 percent (Sackett 1975). Wind is necessary to limit the heat that rises into the crowns of trees from backfires and to enhance consumption of rough (Sackett 1975). Fires are usually conducted in winter and spring as backfires in heavy roughs (the fire backs into the wind) 24 to 48 h after passage of a cold front that has delivered 1.2 to 2.5 cm (0.5 to 1.0 in.) of rain. If the humidity becomes too high or the wind gets too low for a good backfire, strip-headfires are used to speed up the burning. On many areas the goal is to use more headfires because they cover more area in a shorter time and top-kill more shrubs. Headfires can be used in one- to two-year-old rough, but care must be used in three-year-old rough (Sackett 1975). Headfires can easily damage crop trees in four-year-old roughs.

Firelines about 1.5 m (5 ft) wide are plowed 20 to 40 m (65 to 130 ft) apart in the area to be burned (Mobley et al. 1973). Total acreage planned for burning in one day should be limited to no more than 405 ha (1000 acres) (Dixon 1965). On the day of the burn, begin backfiring in mid-morning and continue to set fire until 5 p.m.

Burns are generally conducted every two to five years during the winter to control brown spot disease and hardwoods. A three-year burning interval is optimum for minimizing wildfire damage (Sackett 1975). Once the trees begin to put on height, fire should be excluded from longleaf pines until they are 1.8 to 2.4 m (6 to 8 ft) high

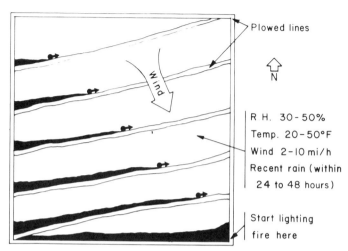

Fig. 16.22.   In heavy rough plow a 2-m (6-ft) line around the entire area to be burned (use roads when appropriate) and plow lines at 20-m (66-ft) intervals throughout the unit to be burned. (From Mobley et al. 1973.)

and from loblolly *(Pinus taeda)* and slash *(P. elliottii)* pines until they are 4.6 m (15 ft) or more in height (Dixon 1965). Fire should not scorch more than one-third to one-half of the tree crown, which is equivalent to the loss of one year's growth. Just before seed fall prior to a harvest burning should be conducted annually for three or four consecutive falls to prepare a good seedbed, although today most areas are planted with seedlings of superior stock.

Where excess flammable fuels have accumulated, several burns will be necessary to reduce fuel accumulations. The first burn in such fuels should be conducted 2 to 3 h after a rain with a temperature of 4°C (40°F) or less.

Spot fires from firebrands are generally not a problem when burning forests in the Southeast, primarily because fires are conducted with low temperatures and high humidities. Moreover, burning every two to five years will keep low-density fuels (rotten wood and cow chips), which are the primary ignition sources for firebrands, to a minimum.

Ponderosa Pine

Objectives of prescribed burning in ponderosa pine stands have been outlined by Biswell et al. (1973) as follows: reduce the volume of dead, highly flammable fuels; thin dense thickets of pine saplings and pole-size stands; create a mineral seedbed in openings for germination of new seeds; raise crown level on crop trees; keep shade-tolerant trees out of the understory and thus destroy the ladders that carry surface fires into the crowns of the larger trees; and keep pine needles compacted as closely as possible to the soil surface by removing understory shrubs and other debris perched in the air.

HEAVY FUELS

With logging slash or excess debris (Fig. 16.23) a cautious burning prescription is necessary (Biswell 1963; Buck 1971) and it may take more than one burn to get a 51- to 86-percent reduction of heavy fuels (Weaver 1957; Biswell 1963). Following the first fall rains, however, good consumption can be achieved with one burn if the large logs are dry inside (Martin et al. 1979). A second burn may be needed to clean up intermediate-sized woody debris and other surface fuels. Burning is generally done in November after grasses have cured but can be done through winter and spring if weather permits (Biswell 1963). Some people prefer spring burning because they often have more days to burn. At this time of the year heavy fuels may be soaked and several burns may be needed to clean up all debris (flashy fuels first and then the larger material).

Five to eight centimeters (2 to 3 in.) of early fall precipitation is desirable before prescribed burns are planned in heavy fuels. This will thoroughly soak the litter, leaving the large logs dry inside. After one to three days the fine fuels are dry and ready for burning. A 1- to 3-m (3- to 10-ft) fireline around the area to be burned is adequate. Temperatures can range from a minimum of 4°C (40°F) to an absolute maxi-

Fig. 16.23. Excess debris in a stand of ponderosa-white fir/pinegrass that will take a cautious burning prescription and three consecutive burns to reduce the heavy fuels 51 to 86 percent. (Photo courtesy of USDA Forest Service.)

mum of 16°C (60°F). Relative humidity should be between 20 and 40 percent and fuel stick moisture should not exceed 15 percent. Wind speeds in open areas may be from 8 to 24 km/h (5 to 15 mi/h) (steady and established). Fine fuels and heavy logs generally burn with this prescription following the first fall rains.

These are general guidelines and good judgment must be used at all times. For example, if you use 16°C (60°F) and 20 percent relative humidity, you will get excessive scorch on live trees, although ponderosa pine will live with an 80- to 85-percent scorch (Dieterich 1979). With 4°C (40°F) and 40 percent relative humidity the fuels will not burn. Also if the 10-h timelag moisture is below 15 percent, but not below 7 or 8 percent, burning will probably be safe if the relative humdity is above 40 percent but not above 60 percent.

Biswell (1963) cites a test fire when heavy fuel in untreated areas was dry to the mineral soil. Relative humidity was 64 percent, air temperature was 16°C (60°F), and the fire burned fiercely. This example stresses the importance of fuel moisture in heavy, moderately volatile fuels. Under similar fuel moisture and weather conditions in west Texas a grass fire would not ignite dead piles of juniper very easily. When the piles ignited they did not burn fiercely and firebrands were rarely seen. In Biswell's fire he probably had fuels that were draped with dead needles since he often refers to these conditions in heavy fuels. This could have been a reason for the difference in fire behavior.

Starting at the top of a ridge or high elevations, backing fires should be the primary method of burning (Fig. 16.24). Strip headfires may be used as necessary to control fire intensity. Rate of spread will be 0.3 m/min (1 ft/min), depending on the degree of slope. The steeper the slope, the slower the movement of the fire and the greater the degree of control (Biswell 1958; Buck 1971). Test burns on 0.04-ha (0.1-acre) plots can be helpful in making decisions to burn if one is in doubt. Remember that patience is one of the most important attributes of a prescribed burner in these types of fuels. Without patience the job will be improperly done and unnecessary damage to the overstory may result.

Rotten logs in an area will burn for prolonged periods of time and can be troublesome if they are too close to the fireline. Such fuels are ideal sources of material upon which firebrands can ignite (Bunting and Wright 1974) and, although seemingly saturated with water, they smoulder hotly (Gordon 1967). When mopping up, such fuels should be placed in bare areas at least 15 m (50 ft) within the edge of the burn so that they will burn up without causing a delayed outbreak of fire as the surrounding fuels become drier.

LIGHT FUELS

Once the heavy accumulation of debris has been eliminated from a stand of ponderosa pine, the light fuels (Fig. 16.25) can be burned every five to seven years (Biswell et al. 1973) and can be burned under a wide range of weather conditions (Wagle and Eakle 1979). Following a 1.2-cm (0.5-in.) rain, air temperatures should not exceed 24°C (75°F) (Biswell and Schultz 1956), the relative humidity should not

Fire Plan for Burning Understory
of Ponderosa Pine

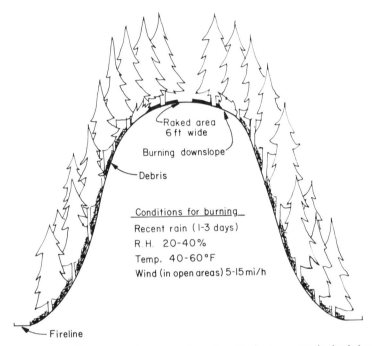

Fig. 16.24. Prescribed burns in ponderosa pine. Sometimes the fire is permitted to back down both sides of a ridge simultaneously. A narrow line is cleared at the top of the ridge (raked) and then the fire is lit and allowed to back down the ridge. Recent rains (4 to 5 days before burning) will keep duff moisture high, but needles and small limbs will be dry enough to burn. Occasionally strip-headfires may be used to speed up the burning, especially when conditions become too cool for backfires to move down the ridge.

be below 20 percent (Biswell et al. 1955), and maximum wind speed in open areas should not exceed 24 km/h (15 mi/h) (Buck 1971). A wind speed of 8 to 24 km/h (5 to 15 mi/h) in open areas is comparable to 3 to 11 km/h (2 to 7 mi/h) in stands of pine (Dixon 1965). Ten-hour timelag fuel moisture should be 8 to 12 percent. Burning may begin as early as October 1.

Optimum burning conditions in light fuels would be 25 to 40 percent relative humidity, 16° to 21°C (60° to 70°F) air temperature, 6.5 to 8.0 km/h (4 to 5 mi/h) of wind speed in the stand, and a 10-h timelag fuel moisture of 7 to 10 percent. Experience will show that much flexibility can be built into a prescription.

Unless fire is to be used as a nonselective thinning tool, fires should be kept out of stands of young trees until they are 3.7 to 4.6 m (12 to 15 ft) tall. Generally, this is easily done because of the lack of pine needles in areas with smaller trees (Biswell

Fig. 16.25.   Light fuels in ponderosa pine, particularly in the Southwest, should be burned at least every five to seven years.

1972). With low intensity fires, trees 1.8 to 2.4 m (6 to 8 ft) high may be burned without severe scorching (Gartner and Thompson 1972), but trees 1 m (3 to 4 ft) tall will be killed (Biswell 1972).

THINNING

Prescriptions for thinning ponderosa pine stands are still in the developmental stages. Biswell (1972) indicates that if you wait until young trees reach a height of 3.7 to 4.6 m (12 to 15 ft), gentle surface fires from the moderate amount of needles under them will kill the small suppressed trees. This is the beginning of the thinning process. As the trees grow and additional burning is done on the area the thinning process continues and no new trees become established in the understory. When blow-ups (hot spots) occur in old decadent areas, openings occur and new even-aged groups of reproduction become established—a process that occurs continually. Weaver (1967) mentions that this process may be crude, but it is cheaper than the current thinning technique.

For overstocked stands (Fig. 16.26) there may not be an adequate prescription (Gordon 1967) without paying a severe price for 20 to 40 years. However research in this area may provide great dividends. Practitioners in Oregon are trying to thin the trees by dozing excess trees. Then the debris is crushed to a thickness of 0.5 m (1.5 ft) and burned. Some burns have been successful and others have not. More research will be needed before the method can be recommended.

Fig. 16.26.   Overstocked stands of ponderosa pine need considerable modification with mechanical or hand-cutting techniques before fire can be introduced as a management tool. (Photo courtesy of USDA Forest Service.)

Douglas Fir-Western Larch

Understory burning in Douglas fir *(Pseudotsuga menziesii)* and western larch *(Larix occidentalis)* (Fig. 16.27) in Montana has been done safely to remove excess fuel accumulations without harm to thick-bark tree species (Norum 1975, 1976). Surface fires with a flame height of 3 m (9 ft) resulted in a scorch height of 5 m (17 ft) and did not kill any trees greater than 13 cm (5 in.) dbh.

Based on research by Norum (1975) and Swanson (1974, 1976) at the University of Washington, the desired prescription for underburning in Douglas fir, where fuel loading ranges from 15 to 200 metric ton/ha (5.5 to 74 ton/acre), is air temperature between 13° and 24°C (55° and 75°F) with a preferred air temperature of about 16°C (60°F), relative humidity of 40 to 55 percent, wind speed within the stand between 3 to 6 km/h (2 to 4 mi/h), and a fine fuel moisture content between 9 and 18 percent. Preferably burns should be conducted three to five days after the last rain and usually with strip headfires between September and October. Using these guidelines a fire removed about 71 percent of the duff [70 metric ton/ha (26 ton/acre)] and about 47 percent of the total down and dead fuel [49 metric ton/ha (18 ton/acre)].

Fig. 16.27.   Douglas fir-Larch-Ponderosa pine communities can be burned to remove excess fuel accumulations in the understory.

Red and White Pine

Natural stands of red *(Pinus resinosa)* and white pine *(P. strobus)* benefit from fire of a certain periodicity and intensity (Van Wagner and Methven 1977). The reason for this lies in the regeneration requirements of these species. Optimum conditions for regeneration as explained by Van Wagner and Methven (1977) include the following:

1   A seedbed either bared to mineral soil or with its duff cover substantially reduced.
2   Relative freedom from competition by shrubs and understory trees of undesired species.
3   A live overhead seed source.
4   Considerable opening in the overhead canopy.

Prescribed burning is recommended by Van Wagner and Methven (1977) in two steps before any cutting in a stand of red or white pine. The first burn in an untouched stand should be conducted in the spring before flushing. The second burn in the following year can be initiated after the leaves are out. Then the first cut should be taken in five years or so, preferably after a good seed year. Stands as young as 60 years can be burned with great care, but trees of 80 years or more will sustain little damage.
     The recommended prescription for burning the understory of red and white pine is as follows (Van Wagner and Methven 1977): Fine Fuel Moisture Code, 90–95; Initial

Spread Index, 8–16; Buildup Index, <52; Fire Weather Index, 12–24. Flame height with these conditions is generally 1 m (3 ft) when the burns are conducted during May and June. The fine fuel moisture code of 90 to 95 relates to a litter moisture content of 6 to 11 percent (Van Wagner 1974) and a duff moisture content less than 40 percent (Van Wagner 1972). These moisture contents are extremely important to burn through raw humus and achieve 30 percent bared mineral soil for germination of pine seeds (Van Wagner 1972; Chrosciewicz 1974). Weather usually involves light winds [<10 km/h (<6 mi/h)] and low relative humidity (20 to 50%) (Van Wagner 1963). Dozed firelines are generally 3 m (10 ft) wide with some backfiring to widen lines when necessary.

Slash

Prescribed burning generally involves the removal of slash (Fig. 16.28) for planting or seeding following logging operations although hazard reduction, range improvement for livestock and wildlife, cover type conversion, and insect and disease control are other uses for fire (Beaufait 1966). Following interviews with 62 fire specialists throughout the intermountain region, Beaufait gave several general guidelines for slash burning. The most widely accepted weather and fuel conditions were relative humidity 20 to 50 percent; wind speed 0 to 16 km/h (0 to 10 mi/h), but preferably wind speeds that are less than 10 km/h (<6 mi/h) (Beaufait 1966); and a 10-h timelag fuel moisture of 6 to 15 percent. There was no consensus on air temperature among the people interviewed, although Perala (1974) says that it should be above 18°C

Fig. 16.28.   Logging slash must be removed by some means for planting or seeding that follows logging operations. Fire is the most effective means by which to accomplish this and other objectives at once. (Photo courtesy of USDA Forest Service.)

(64°F) to burn aspen slash. Slash should be cured for one to two summer months. A fireline width of 3.6 to 6 m (12 to 20 ft) was considered the most desirable.

The preferred ignition technique by most men with experience was the strip-headfire technique, starting at the top of a block or downwind side (Beaufait 1966). Center ignition is used by some people, and occasionally area ignition (Fig. 16.5). Tankers, spotters with radios, and dozers, if considered necessary, are strategically located. Crews of 10 to 12 men or more are used to burn blocks of slash that range from 40 to 160 ha (100 to 400 acres). When vegetation around the slash blocks is green at the time of controlled burning, the probability of spot fires is reduced as long as ignition of the block continues. Continuous lighting enables the fire to draw wind into the fire from the surrounding area and minimize the chance for spot fires. Most burning is done in the fall, with some burning in late spring. In the boreal forest summer fires are recommended for jack pine *(Pinus banksiana)* (Chrosciewicz 1978) and spruce-fir *(Picea-Abies* spp.) (Randall 1966; Kiil 1969) slash where the slash fire hazard is to be reduced and seeding or planting conditions are to be enhanced through removal of parts of the duff layer. Randall's prescription for burning spruce slash was within the desired range of parameters found by Beaufait (1966) and Van Wagner and Methven (1977).

Most practitioners prefer to burn after only a small amount of rain. They get better cleanup for planting and seedbed preparation and have fewer holdover fires. Also the areas are cheaper to burn. Most holdover fires can be avoided by not mixing wood with soil along the firelines.

Normally burning should begin in the late afternoon when temperature has peaked for the day and wind is steady at maximum speeds. Then, as burning continues into the evening, temperature and wind speed will drop, relative humidity will rise, all of which minimize control problems.

Live Aspen Forest

Aspen is widely distributed in North America but prescribed burning research has been restricted to the aspen parkland, boreal forest, and Intermountain Region.

The live aspen forest has been described as the ''asbestos forest'' in the Rocky Mountains because it does not burn as readily during summer conflagrations as adjacent coniferous forest types. Aspen is often a scrubby, noncommercial tree on millions of acres in the aspen parkland and in parts of the adjacent boreal forest. The most common method of disposal of this forest for agriculture (a practice which is becoming very common) is to doze the trees, pile and burn the windrows, plow the land, and seed to cereals or perennial forages.

Prescribed burning can be used to convert aspen forests to grass-shrublands. The ash created by the fire is a favorable seedbed for forages (Bailey 1977, 1978). Woody suckers and forage after burning are productive (Bailey 1978) and are desirable for livestock and wildlife food. After four years, however, the aspen saplings will be out of reach of all ruminants but moose *(Alces alces).*

The live aspen forest requires relative humidities below 35 percent for successful

Fig. 16.29.   Live, burning aspen forest. Flames will usually be 1 m (3 ft) high.

burning (Fig. 16.29). It will burn in the period between snowmelt and leaf flush. Occasionally autumn burning can be done after leaf fall. Usually there are about two weeks of favorable spring weather every second or third year. Acceptable burning conditions are 15 to 30 percent relative humidity; temperature 18° to 27°C (65° to 80°F); 6- to 24-km/h (4- to 15-mi/h) wind speed; at least 14 days since the snow melted in adjacent grassland or three drying days after the last rain. Surface duff layer moisture must be less than 20 percent.

A 150-m (500-ft) fireline is prepared on the leeside and a 30-m (100-ft) fireline on the remaining sides. Strip-headfire ignition is the preferred ignition procedure. Firebrands will be a problem if there is much wind shift. Two pumpers and a six-man crew should patrol downwind from the leeside. Firebrands have caused spot fires 60 m (200 ft) downwind from the source in a 13- to 16-km/h (8- to 10-mi/h) wind. Spot fires will go greater distances in winds above 13 km/h (10 mi/h).

If the forest to be burned is in the aspen parkland where groves of trees are interspersed amongst fescue grassland (Fig. 16.30), a two-stage burning scheme can be used (Fig. 16.31). A week after snowmelt, the grassland can be burned when the relative humidity is 45 to 60%, temperature 4° to 24°C (40° to 75°F) and wind is 3 to 19 km/h (2 to 12 mi/h) using the live aspen forest as a fireline (Fig. 16.32). Two weeks later, when the forest duff has dried, the aspen groves can be burned as described above. The burned grassland will then act as a fireline.

Fig. 16.30. Aspen parkland vegetation contains a mixture of aspen, snowberry, and grasses.

Fire Plan for Aspen Parkland

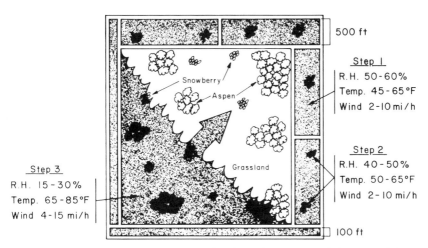

Fig. 16.31. In preparation for a fire in the aspen parkland, doze a 3-m (10-ft) line around the entire area to be burned and between aspen groves as shown in the figure. Burn out the grass in the fireline after it has had one day to dry. Burn out patches of snowberry in fireline after they have had four days to dry. Lastly burn out the main portion of the burn after 14 days of drying weather when relative humidity is below 30%, air temperature is above 18°C (64°F), and wind speed is 13 to 26 km/h (8 to 16 mi/h).

427

Fig. 16.32. Firelines can be prepared around live aspen by burning the grass around the aspen clones within 10 days after snowmelt. During this time the aspen understory is too wet to burn. The aspen can be burned about 14 to 28 days after snow melts out of the forest.

Dead Aspen Forest

Since the live aspen forest can only be burned in the northern Great Plains about one year out of two or three, and since risky weather conditions are required, it may be desirable to kill the aspen forest first with a herbicide or some other means. An aspen forest that has been dead for at least two years can be burned under more moderate weather conditions. This is very desirable since bark of dead aspen trees is a danger-

Fig. 16.33. Dead aspen flame heights can quickly reach 10 m (30 ft) and give off dangerous firebrands. This particular fire quickly subsided with flame lengths of 3 m (10 ft) when it ran into green aspen (left background).

ous firebrand (Fig. 16.3). Burning should not be conducted when the relative humidity is less than 35 percent because of the hazard from firebrands. The recommended weather conditions for spring burning are 35 to 50 percent relative humidity, temperature 4° to 24°C (40° to 75°F), wind of 3 to 19 km/h (2 to 12 mi/h), 10 days of drying weather after spring snow melts or two drying days since the last measurable rain. A 130-m (400-ft) fireline should be prepared on the leeside of the burn and 30-m (100-ft) on the remaining sides. Strip-headfiring is the desirable ignition procedure if the person lighting can walk around the perimeter of the groves. In a continuous forest, perimeter ignition is the acceptable procedure. It is too dangerous for the ignition crew to attempt to light strip-headfires through a dead aspen forest. Walking is difficult and flame lengths of 10 m (30 ft) are realized within 2 min of ignition (Fig. 16.33). When a dead forest is totally surrounded by live aspen forest a 3-m (10-ft) fireline can be constructed around the perimeter of the dead forest. Weather conditions as described above are acceptable except that relative humidity should be 40 to 50 percent.

PRESCRIPTIONS FOR HIGH–VOLATILE FUELS

High-volatile fuels require more preparation before burning than other fuel types. This is particularly true for California chaparral and dozed pinyon-juniper. Firelines on the leeward sides should be about 150 m (500 ft) wide and the trees and shrubs in the firelines should preferably be crushed, chained, or dozed before being burned. If a third or more of the chaparral brush is dead, crushing is not necessary, but crushed brush is always safer and easier to burn than standing brush. The firelines with a 3-m (10-ft) dozed strip on each side need to be burned when relative humidity is 40 to 60 percent, air temperature is 4° to 16°C (40° to 60°F), and wind speed is less than 16 km/h (10 mi/h). When air temperature drops to 4°C (40°F) and relative humidity rises to 60 percent ignition is difficult if not impossible unless the 10-h fuels are dry inside. Then the larger fuels burn and dry out the smaller fuels, unless the fuel is discontinuous. As long as the relative humdity does not drop below 40 percent and wind speeds are light, there will be little risk from volatile firebrands (Bunting and Wright 1974; Green 1977).

After the firelines have been prepared crushed strips in brush or grass-shrubs will be helpful to ignite the headfire in 20- to 30-year-old stands. Decadent stands will burn without crushed strips. For headfires, air temperatures of 21° to 27°C (70° to 80°F), relative humidity of 25 to 40 percent, and wind speeds of 13 to 24 km/h (8 to 15 mi/h) are recommended, provided you have a 150-m (500-ft) fireline to burn into.

California Chaparral

Prescribed burning in this high-volatile fuel (8 to 11 percent ether extractives, Countryman 1967; Philpot 1969) is done to reduce wildfire frequency, increase plant productivity (particularly shrubs), improve wildlife habitat, or create a good seedbed for converting brushland to grassland planting. Prescribed burning is usually preferred over other methods of conversion because it is cheaper, does not disturb the soil, leaves a clean seedbed, and temporarily raises the level of soil nutrients for better establishment and growth of a new vegetative cover (Bentley 1967).

Chaparral (Fig. 16.34) can be burned under quite a variety of weather conditions depending on the characteristics of adjacent fuels and the ease with which firelines can be prepared. Some preparation of brush fuels either by crushing or with desiccants is desired before planning a late fall, early winter, or possibly spring prescribed burn (Bentley 1967; Green 1970; Bentley et al. 1971). Crushed strips of chaparral around the edge of the burn and at various intervals throughout the burn (Fig. 16.35) greatly add to the success of the burn. Without such preparation, burning of chaparral in Arizona must be done under hazardous conditions (Baldwin 1968), probably in the range of 38°C (100°F), less than 10 percent relative humidity, and 24- to 32-km/h (15- to 20-mi/h) wind speeds, which are undesirable. Green (1981) recommends 10 prescriptions for burning California chaparral depending on age of stand, proportion of dead fuel, and objectives (Table 16.3).

Green (1970) burned a 1.5-ha (4-acre) plot in May with a relative humidity of 28 percent, air temperature of 27°C (80°F), and a wind speed of 16 to 19 km/h (10 to 12 mi/h). Occasional gusts of wind to 32 km/h (20 mi/h) caused spotting across a 90-m (300-ft) fireline. From 95 to 100 percent of all dead and crushed fuel was consumed, whereas 40, 45, 75, and 85 percent of the green standing scrub oak *(Quercus dumosa)*, western mountain-mahogany *(Cercocarpus betuloides)*, manzanita *(Arctostaphylos glauca* and *A. glandulosa)*, and chamise *(Adenostoma fasciculatum)* were consumed, respectively. In chamise plants about 25 percent of the material is dead

Fig. 16.34. California chamise chaparral, a highly volatile fuel. (Photo courtesy of USDA Forest Service.)

Fire Plan for California Chaparral

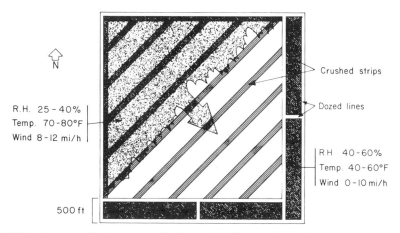

Fig. 16.35. In preparation for a prescribed burn in California chaparral, doze a 3-m (10-ft) line around the entire area to be burned and around the lee fireline as indicated in the figure. Next, crush a 150-m (500-ft) strip on the leeward sides of the area that you plan to burn for the fireline and crush 3-m (10-ft) strips at 20-m (66-ft) intervals throughout the burn. The crushed areas should be prepared in late June or July (2 months before burning). Begin burning firelines with a strip-headfire (Fig. 16.10) in late August or September, preferably during late afternoon or evening when weather is appropriate. Burn in 0.8-km (0.5-mi) strips so that the burning can be stopped when it is too hazardous or will not carry. After the fireline has been prepared, burn the entire area with a headfire under the prescribed conditions indicated on the figure. If chaparral contains 30 percent or more dead material, dozed strips throughout main area are not necessary before lighting headfire.

(Countryman and Philpot 1970), which probably acounts for some of the higher cleanup compared to other genera.

Based on the experience of Bentley (1967), Green (1970, 1977, 1981), Bentley et al. (1971), and Schimke and Green (1970) in burning California chaparral and other experiences in high-volatile fuels, a general prescription as shown in Figure 16.35 is recommended for high-volatile fuels. With this prescription fires can be conducted safely with crews of 6 to 10 men, although the 150-m (500-ft) fireline on the lee sides would have to be prepared on a different day than the main headfire. Burning would be most effective in the fall, although it can be done in the spring with the aid of desiccants (Green 1970). A crew of six men can usually burn 2 to 3 km (1.2 to 1.8 mi) of fireline in a 4 to 5-hr period. The crushed borders and strips throughout the burn should be prepared two months before burning (late spring or early summer). The main headfire should be set in September.

In central California firelines can be prepared in woodland grass vegetation that often borders chaparral. Layout of the burn, topography, adjacent vegetation, method of lighting as well as weather, fuel moisture, and one's experience should determine how and when to burn. Every situation is different and each of them will dictate a different prescription (all of which can be safe) to reach the desired objectives.

Table 16.3. Aligning Brush Stands, Fuel Reduction Objective, and Burning Prescription[a]

Age of Stand (Years)	Proportion of Fuel That is Dead (%)	Fuel Reduction Objective (by prescription numbers)		
		75% or Greater	50 to 75%	Less Than 50%
7 to 20	5 to 20	1[b]	1	1
20 to 35	20 to 30	2	3	4
35 or more	30 to 45	5	6	7
50 or more	45 to 65	8	9	10

	Prescription Number					
	2	4	5	7	8	10
Windspeed, mi/h	10	6	8	3	6	0
Relative Humidity, %	17	28	25	40	35	70
Air temperature, °F	90	65	85	60	75	40
Fuel stick moisture, %	5	8	7	10	9	16
Live fuel moisture, %	60	65	65	75	65	85

From Green (1981). Reproduced by permission of Lisle R. Green and the USDA Forest Service.
[a]To arrive at an approximate prescription, select the proportion of fuel that is dead (upper left) and the burn objective (upper right), then go to the corresponding prescription number at top of lower table. The prescription will be within a small range of that specified in the column below. Prescription No. 3 would fall between 2 and 4, 6 between 5 and 7, and so on.
[b]Prescription No. 1—Crush, or spray with herbicide. This brush is too green for successful prescribed burning. It may also be too flexible for successful crushing. Prescription Nos. 2 and 3 will also need treatment if the dead fuel is in the lower part of the range.

Trained, experienced personnel will be the key to success because the prescriptions for safety and success can vary so widely.

Arizona Chaparral

Arizona chaparral is virtually impossible to burn until the stands are about 20 years old (Cable 1975). Even after this age the chaparral grows in various sized clumps and is generally not as continuous as stands of California chaparral. As a consequence Arizona chaparral does not need to be burned as frequently as California chaparral to improve wildlife habitat and water yields. Lindenmuth and Davis (1973) have done an intensive study in Arizona to show when fires will spread from clump to clump and when they will not. Their study showed that old stands of chaparral were easier to

Fire Plan for Arizona Chaparral

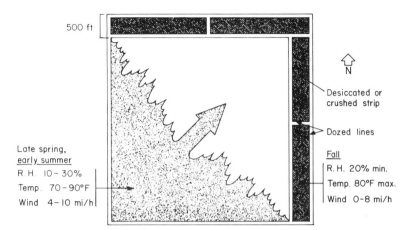

Fig. 16.36. In preparation for a prescribed burn in Arizona chaparral, doze a 3-m (10-ft) line around the entire area to be burned and around the fireline as indicated in the figure. Then either desiccate or crush a 150-m (500-ft) strip on the leeward sides of the area. The desiccated areas should be prepared several weeks before fall burning. If the firelines are to be crushed instead of desiccated this should be done in August. Burn the firelines using the strip-headfire technique in September or October. After the fireline has been prepared, wait until the following May or June to set the main headfire. For further refinements on ROS, see Lindenmuth and Davis (1973) and Davis and Dietrich (1975). The ROS should be less than 20 for green fuel when burning out the fireline and it should be above 20 when setting the main headfire.

burn in late spring and summer than in the fall because the foliar phosphorus content is highest during dry springs. If, however, 7.5 cm (3 in.) of precipitation has been received during the winter, then late spring, early summer and fall burns will behave very similarly. A field workbook prepared by Davis and Dieterich (1975), prepared from research by Lindenmuth and Davis (1973), can be very helpful to practitioners in preparing prescribed burns.

A second point of major concern is that green chaparral will not burn very easily until air temperature is at least 20°C (68°F). As temperature rises above 24°C (75°F) it has relatively little effect on rate of spread (ROS). Wind has a major effect and relative humidity is usually low and only has a minor effect.

Lindenmuth and Davis (1973) have observed short-range spotting on a wildfire with a rate of spread of 13.7 m/min (45 ft/min), a hazardous ROS. Thus spotting (spot fires) does not appear to be near the problem in Arizona chaparral as it is in California chaparral.

The suggested fire plan in Fig. 16.36 takes advantage of the least dangerous fall burning season to prepare firelines and the highest rate of spread season to burn the major portion of the area in late May and June. After some experience one may find that narrower firelines 75 m (250 ft) are adequate. However, it is just about as easy to prepare a wide fireline as a narrow one and fewer people are needed to patrol the main headfire when it is set on a later date.

Dozed Juniper in Mixed Prairie (Texas)

This fuel type (Fig. 16.37) is burned with a prescription similar to that for California chaparral because it is a high volatile fuel that gives off firebrands which ignite cow dung easily. Generally dozed or chained areas are not burned until three to five years after treatment. This gives the grass a chance to recover and time for most leaves to fall off the dead trees. The primary objective of burning is to remove dead piles and to kill young juniper trees.

A 150-m (500-ft) fireline is burned on the north and east sides of a pasture (Fig. 16.38). Dead piles of brush (4 to 5 years old) are burned out of this line in May or early June when the grass is green. Later (January or February), where little bluestem is the primary fine fuel, the 150-m (500-ft) firelines are burned out, using the prescription as shown in Fig. 16.38. Where buffalograss occurs, a higher wind speed and lower relative humidity are required.

After the fireline has been burned out, the rest of the area is burned with a southwest wind. After a rain, wait at last two or three days to burn.

For safety, avoid burning backfires into headfires and avoid burning across ridges. Firewhirls can easily develop under these situations. In highly volatile fuels such as juniper, burn into heavily graze pastures, when possible, to minimize risk. In this fuel type, one should have at least two seasons of burning experience before assuming responsibility for conducting a burn.

Fig. 16.37. Dozed juniper in central Texas, a mixture of dead piles of juniper, young juniper trees, and grass.

Fire Plan for Dozed Juniper
(Mixed Prairie)

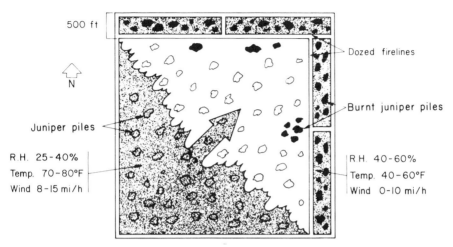

500 ft

N

Juniper piles

R.H. 25-40%
Temp. 70-80°F
Wind 8-15 mi/h

Dozed firelines

Burnt juniper piles

R.H. 40-60%
Temp. 40-60°F
Wind 0-10 mi/h

Fig. 16.38. When the grass is green, juniper piles in the 150-m (500-ft) strip (black splotches) on the downwind sides (north and east) are burned with wind velocities less than 16 km/h (10 mi/h) and relative humidity above 45 percent. Eight months later, when grass is dormant, the grass in the 150-m (500-ft) strip is burned (strip-headfire technique) when the wind speed is less than 16 km/h (10 mi/h) and relative humidity is between 40 and 60 percent. Lower relative humidities may be used if the grass fuel is less than 2247 kg/ha (2000 lb/acre). All large concentrations of piles are back-fired on the downwind sides of main area to be burned and then the entire area is burned into the prepared firelines with a wind speed of 13 to 24 km/h (8 to 15 mi/h) and a relative humidity of 25 to 40 percent.

Chained Juniper in Mixed Prairie (Texas)

Chained juniper is less hazardous to burn than piled juniper because the dry juniper fuel is closer to the ground and burning embers are not likely to travel more than 75 m (250 ft), although we use the 150-m (500-ft) fireline in this fuel type for safety. Thus the dead trees in the fireline can be burned at the same time as the grass. Follow the procedure as outlined in Fig. 16.39.

Closed Stands (No Grass Understory) of Pinyon and Juniper

Thick stands of pinyon and juniper eliminate the understory vegetation (Fig. 16.40) which reduces forage for wildlife and livestock (Bruner and Klebenow 1979). Pre-scribed burning is an inexpensive tool to restore and maintain desirable mixtures of shrubs and grasses in these communities. Where wood utilization is not a feasible alernative fire will be used in more pinyon-juniper communities in the future than it has been in the past, once the techniques for burning have been refined. Rest from

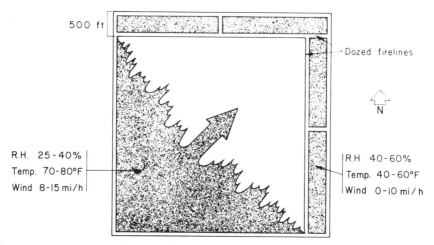

Fig. 16.39. In preparation for a prescribed burn in chained juniper, doze a 3-m (10-ft) line around the entire area to be burned and in interior areas of burn as shown in the figure. Burn out the firelines in January and February. In March, light the main headfire.

Fig. 16.40. Closed stand of pinyon and juniper (no fine fuel in the understory).

grazing will not restore closed stands of pinyon-juniper back to the original vegetation with grasses and shrubs (Stinson 1978).

Mixtures of pinyon and juniper with 740 or more standing green trees/ha (300 or more trees/acre) in 35- to 45-cm (14- to 18-in.) rainfall areas can be burned on hot, windy days if prepared properly (Truesdell 1969; Blackburn and Bruner 1975). However closed stands of juniper with a canopy cover less than 30 percent are almost impossible to burn (Blackburn and Bruner 1975) except under extreme burning conditions and would probably require wind speeds in excess of 55 km/h (35 mi/h) to carry a fire. As the proportion of pinyon to juniper increases and density increases, the stands are easier to burn (Hester 1952; Truesdell 1969; Blackburn and Bruner 1975). A minimum cover of 35 to 40 percent appears necessary to sustain a continuous crown fire.

MIXTURE OF PINYON AND JUNIPER

Where firebrands are not a problem on a hot day (i.e., burning into the Grand Canyon or on top of a mesa), dense stands of pinyon and juniper can be prepared for burning in March or April by clearing a strip 6 to 15 m (20 to 50 ft) wide every 0.4 km (0.25 mi) and pushing the windrows against green trees on the windward side. The dozed trees should cure for 60 to 90 days. Then, in late June or early July, burning should be conducted when air temperatures vary from 27° to 35°C (80° to 95°F), relative humidity is less than 10 percent, and wind speeds exceed 13 km/h (8 mi/h) (Truesdell 1969; Blackburn and Bruner 1975).

If you cannot burn into a safe area, chaining may be used as a method to construct firelines (Fig. 16.41) (Davis 1976). The necessary width is not known, but a 150-m (500-ft) line should be adequate where there is essentially no grass in the understory. The lines could be chained in the winter and burned under moderate conditions in the spring or early summer (depending on location) when surrounding vegetation is green (Davis 1976). When there is no grass and limited big sagebrush in the understory, green pinyon and juniper are very difficult to burn in the spring and can serve as an adequate fireline. Occasionally some areas with canopy cover over 50 percent will burn for short distances, but not with light and variable winds.

After the firelines have been burned out, then the windrows (also prepared in winter) on the upwind side can be lit on a dry, hot, windy day in June, or July. The heat from the windrows should carry the fire into the green trees and the wind then carries the fire in the crowns of the trees between windrows. Without windrows, the fire will go out and not carry through the stand.

MIXTURE OF PINYON–JUNIPER AND SAGEBRUSH

Mixtures of big sagebrush and pinyon-juniper (Fig. 16.42) are common throughout the Great Basin. Dense patches of pinyon-juniper and big sagebrush that vary in size from 2 to 24 ha (5 to 60 acres) can easily be burned without preparing firelines (Bruner and Klebenow 1979).

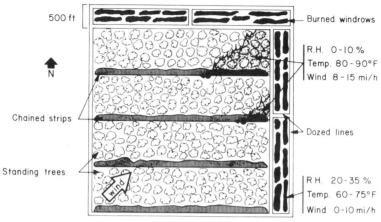

Fig. 16.41. In preparation for a prescribed burn in mixtures of pinyon and juniper, doze a 3-m (10-ft) line around the entire area to be burned and around fireline as indicated in the figure. Chain or crush the strips of juniper in the fireline and at 0.4 km (0.25 mi) intervals throughout the main area to be burned during the winter (60 to 90 days before planning to burn). Burn the fireline in spring or early summer under the conditions listed in the figure. After the fireline has been burned, start on the downwind side and light each chained strip. When lighting the strips in the main burn, the combination of air temperature (°F), wind (mi/h), and shrub and tree cover (%) should exceed 125.

Fig. 16.42. Mixture of sagebrush and pinyon-juniper.

438

Klebenow and Bruner (1976) have found that mixtures of pinyon-juniper and big sagebrush with a total shrub and tree cover of 45 to 60 percent can be burned when spring air temperatures vary from 16° to 24°C (60° to 75°F), relative humidity is below 25 percent, and maximum wind speeds vary from 8 to 40 km/h (5 to 25 mi/h). They recommend using an index to determine whether to burn (Bruner and Klebenow 1979) equal to the sum of maximum wind (mi/h), shrub and tree cover (%), and air temperature (°F).

If the index is 110 or higher, a fire will carry and will kill large pinyon and juniper trees. If the index is above 130, it may be too dangerous to burn unless you are burning into no-fuel areas. Further testing on a large scale of this technique is desired because retorching of trees is necessary until the index is over 125. Thus it is desirable for burning out thick patches of pinyon and juniper, but not large areas.

Where it is desired to burn large areas of pinyon-juniper and big sagebrush it would be preferable to prepare a fireline as shown in Fig. 16.41 and burn the remaining area with a headfire, when the index of Bruner and Klebenow (1979) is above 125.

JUNIPER

In pure stands of juniper, a more acceptable method of killing trees in closed stands is to chain, burn, and seed (Aro 1971; Stinson 1978). Two or three months after chaining juniper can be burned with little risk when the wind is blowing into an untreated closed stand or into a recently treated area with little fine fuel. Burn the lee sides (Fig. 16.43) of the area 75 m (250 ft) during the morning hours when the fire potential in this type is low (Stinson 1978). Then burn the remainder of the area under the conditions shown in Fig. 16.43 (Stinson 1978). Firebrands are not a serious problem because there is very little fine fuel on the soil surface.

It takes large crews to do this kind of burning because much of the material is in piles or windrows. Advantages of burning chained areas are to remove the debris and young trees which provide an ideal microenvronment for the establishment of pinyon and juniper seedlings (Meagher 1943) and to provide a good seedbed for grasses. Moreover, green stands of juniper without pinyon are very difficult to burn unless they are chained (Blackburn and Bruner 1975).

PLANNING A PRESCRIBED BURN

Fire planning can be a simple matter once the objectives are known and if the prescription for achieving the objectives has been tried and proven reliable. The first step is to visit the area that someone wants to burn and look at the fuel type. Is it high-volatile fuel (e.g., juniper) or a low-volatile fuel (e.g., grass and mesquite)? Assuming the fuel is highly volatile, has the area been chained or dozed? One must be extremely careful on dozed sites but there is more leeway on chained sites. On forested

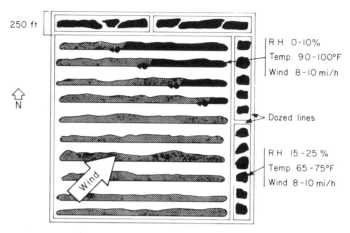

Fire Plan for Chained Juniper
(no grass understory)

250 ft

N

Wind

R.H. 0-10%
Temp. 90-100°F
Wind 8-10 mi/h

Dozed lines

R.H. 15-25 %
Temp. 65-75°F
Wind 8-10 mi/h

Fig. 16.43. In preparation for a prescribed burn in chained juniper, burn out firelines during the morning hours when fire danger is low. Then in the afternoon burn the remaining rows of debris as indicated in the diagram. If grass is present in understory of adjacent fuels, widen firelines on leeward side to 150 m (500 ft.).

areas fuel moisture in all size classes of fuel must be measured and looked at closely on the area to be burned as well as in adjacent areas not to be burned.

The next step is to look at the lay of the land and decide how it should be burned. Note direction of prevalent winds. Plan the burn so that wind will move parallel to the ridges to minimize the risk of forming firewhirls.

Once we have looked the area over and decided how we want to burn it, we draw a map with all roads and proposed firelines. On the map we draw in major ridges, hills and so on, and write prescriptions for backfires and the main headfire. After thinking the fireplan over for a few days we go back to the proposed burn area and flag all the lines that need to be dozed. We draw a map and go over the fireplan and dozed lines with the rancher or person to be in charge of the fire. Plans are made for someone to doze the firelines and to be sure that the operator gets precise instructions.

All of this work may take place three or four months before a planned burn. Meanwhile we encourage the person in charge of the land to let the neighbors, local fire departments, and local news media know what is going on. Public relations work is very important. Also we encourage the land owner or manager to graze pastures heavily on the lee side of the fire to deplete the quantity of fine fuels.

One month before the main fire is to take place, we begin to look for the appropriate weather conditions to burn out firelines. For example in chained juniper grasslands in central Texas a person may start burning firelines when maximum air temperature for the day is about 16°C (60°F), minimum relative humidity is 40 percent, and maximum wind speeds are less than 16 km/h (10 mi/h). We watch the local weather every day until the right day seems to be near. Then we call the U.S. Weather Bureau or the nearest weather station that gives fire weather forecasts and ask for the person in charge of giving fire weather forecasts. This should be done the evening before an anticipated burn.

If the preliminary forecast for the next day looks good, we alert all members of our fire crew that we will probably burn the next day and give them the approximate time that we plan to leave. Before leaving, we check the local weather and we check with our weather forecaster again. If the forecast is still good, we pass out the final word that we will be burning.

A good rule to remember in this regard is that for every 11°C (20°F) increase in temperature, the relative humidity will drop 50 percent. Thus, if you get up in the morning and the high predicted for the day is 16°C (60°F) you can quickly measure the relative humidity and current temperature to estimate the low relative humidity for the day. Meanwhile we gather up a crew of 6 to 12 people, four two-way radios, six shovels, six swatters or backpack spray pumps, two pickups (one with a 100-gal slip-on pumper), five drip torches, 110 liters (30 gal) of fuel (70:30% diesel-gas mixture), two belt-weather kits, and boxes of matches. A D-7 dozer is on standby.

When we arrive at the burning site, we check wind direction to see where we should start burning. If weather conditions are on the ''high safe'' side, we begin burning in areas with the lightest fuels and save the heavier fuels for later. Burning of firelines is usually done using the strip-headfire technique mentioned earlier, with two to three people with suppression equipment spaced behind the lead torch and a pumper following them. At least four people should have two-way radios.

If air temperature is 16°C (60°F), relative humidity is 40 percent, and wind speed is 16 km/h (10 mi/h) when burning firelines in chained juniper, we should proceed cautiously. However, after the relative humidity rises above 50 percent, we issue all drip torches and tell everyone to move as fast as they can. Only a pumper crew needs to patrol the firelines. When air temperature drops to 4°C (40°F) and relative humidity rises to 60 percent, all personnel may have to quit burning because only the heaviest stands of grass will burn. The next day we check for smoldering debris and move it 15 m (50 ft) inside the fireline. The weather conditions here are simply an example for one fuel type and not intended to be used as anything other than an illustration.

The above procedures are repeated until all firelines are burned out. This usually involves three trips for an 810-ha (2000-acre) burn.

After all firelines have been burned, we start looking for suitable weather for the headfire: air temperature 21° to 27°C (70° to 80°F), relative humidity 25 to 40 percent, and wind speed 13 to 24 km/h (8 to 15 mi/h) in chained juniper. Again we follow local weather until a suitable day is approaching and then we ask the local forecaster for a fire weather forecast. Before leaving to burn the next day, we get an updated forecast.

If everything looks good, we call the land owner with instructions to notify neighbors and the fire department again. Despite all of this telephone work, the owner's phone will still ring off the hook when burning begins, but the word eventually gets out. Be sure to tell the fire department not to come out unless you personally tell them that you are in trouble. After the burn is over, plan to spend the next day mopping up to be sure that nothing is burning within 15 m (50 ft) of the firelines. A dozer is highly desired for this work.

Lastly, do not tell people how to burn a piece of property over the telephone. Your information based on many years of experience does not mean much to most people, and they will most likely burn on the day that fits into their schedule regardless of the weather. On the other hand, when people ask for help, do what you can to help them. Otherwise many of them will proceed without help and burn several thousand hectares of their neighbors land and many kilometers of fence, as has been witnessed more than once.

REFERENCES

Anderson, H. G., and A. W. Bailey. 1980. Effects of annual burning on grassland in the aspen parkland of east-central Alberta. *Can J. Bot.* **58:**985–996.

Anderson, M. L., and A. W. Bailey. 1978. Prescribed burning of a *Festuca-Stipa* grassland. *J. Range Manage.* **31:**446–449.

Arnold, J. F., D. A. Jameson, and E. H. Reid. 1964. The pinyon-juniper type of Arizona: Effect of grazing, fire, and tree control. USDA Prod. Res. Rep. No. 84. Washington, D.C.

Aro, R. S. 1971. Evaluation of pinyon-juniper conversion to grassland. *J. Range Manage.* **24:**188–197.

Bailey, A. W. 1977. Prescribed burning as an important tool for Canadian rangelands. In S. B. R. Peters and A. W. Bailey (eds.) *Range Improvement in Alberta: A Literature Review.* Univ. Alberta, Edmonton, pp. 361–381.

Bailey, A. W. 1978. Use of fire to manage grasslands of the Great Plains: Northern Great Plains and adjacent forests. In Proc. of the First Int. Rangeland Congr., Denver, Colo., pp. 691–693.

Bailey, A. W., and R. A. Wroe. 1974. Aspen invasion in a portion of the Alberta parklands. *J. Range Manage.* **27:**263–266.

Baldwin, J. J. 1968. Chaparral conversion on the Tonto National Forest. *Proc. Tall Timbers Fire Ecol. Conf.* **8:**202–218.

Beardall, L. E., and V. E. Sylvester. 1976. Spring burning for removal of sagebrush competition in Nevada. *Proc. Tall Timbers Fire Ecol. Conf.* **14:**539–547.

Beaufait, W. R. 1966. Prescribed fire planning in the Intermountain West. USDA For. Serv. Res. Paper INT-26. Intermt. For. and Range Exp. Stn., Ogden, Utah.

Bentley, J. R. 1967. Conversion of chaparral areas to grassland: techniques used in California. USDA Handbook No. 328. Washington, D.C.

Bentley, J. R., C. E. Conrad, and H. E. Schimke. 1971. Burning trials in shrubby vegetation desiccated with herbicides. USDA For. Serv. Res. Note PSW-241. Pac. Southwest For. and Range Exp. Stn., Berkeley, Calif.

Bird, R. D. 1961. Ecology of the aspen parkland of western Canada. Research Branch. Can. Dept. Agric. Pub. 1066. Res. Br., Ottawa, Ont.

Biswell, H. H. 1958. Prescribed burning in Georgia and California compared. *J. Range Manage.* **11:**273–297.

Biswell, H. H. 1963. Research in wildland fire ecology in California. *Proc. Tall Timbers Fire Ecol. Conf.* **2:**62–97.

Biswell, H. H. 1972. Fire ecology in ponderosa pine-grassland. *Proc. Tall Timbers Fire Ecol. Conf.* **12:**69–96.

Biswell, H. H., H. R. Kallander, R. Komarek, R. J. Vogl, and H. Weaver. 1973. Ponderosa fire management: A task force evaluation of controlled burning in ponderosa pine forests in central Arizona. Misc. Pub. No. 2, Tall Timbers Res. Stn., Tallahassee, Fla.

Biswell, H. H., and A. M. Schultz. 1956. Removal of tinder in ponderosa. *Calif. Agric.* **10**(2):6, 7. Berkeley.

Biswell, H. H., A. M. Schultz, and J. L. Launchbaugh. 1955. Brush control in ponderosa pine. *Calif. Agric.* **9**(1):3, 14. Berkeley.

Blackburn, W. H., and A. D. Bruner. 1975. Use of fire in manipulation of the pinyon-juniper ecosystem. In *The Pinyon-Juniper Ecosystem: A Symposium.* Utah State Univ., Logan, pp. 91–96.

Britton, C. M., J. E. Cornely, and F. A. Sneva. 1980. Burning, haying, grazing, and non-use of flood meadow vegetation. Ore. Agric. Exp. Stn. Spec. Rep. 586. Oregon State Univ., Corvallis.

Britton, C. M., C. M. Countryman, H. A. Wright, and A. G. Walvekar. 1973. The effect of humidity, air temperature, and wind speed on fine fuel moisture content. *Fire Technol.* **9**:46–55.

Britton, C. M., and M. H. Ralphs. 1979. Use of fire as a management tool in sagebrush ecosystems. In *The Sagebrush Ecosystem: A Symposium.* Utah State Univ., Logan, pp. 101–109.

Britton, C. M., and H. A. Wright. 1971. Correlation of weather and fuel variables to mesquite damage by fire. *J. Range Manage.* **23**:136–141.

Bruce, D. 1951. Fuel weights on the Osceola National Forest. *USDA For. Serv. Fire Cont. Notes* **12**(3):20–23.

Bruner, A. D., and D. A. Klebenow. 1979. Predicting success of prescribed fires in pinyon-juniper woodland in Nevada. USDA For. Serv. Res. Paper INT-219. Intermt. For. and Range Exp. Stn., Ogden, Utah.

Buck, B. 1971. Prescribed burning on the Apache National Forest. In *Symposium on Planning for Fire Management.* Southwest Interagency Fire Counc. Tucson, Ariz., pp. 165–172.

Bunting, S. C., and H. A. Wright. 1974. Ignition capabilities on nonflaming firebrands. *J. For.* **72**:646–649.

Bunting, S. C., H. A. Wright, and L. F. Neuenschwander. 1980. Long-term effects of fire on cactus in the southern mixed prairie of Texas. *J. Range Manage.* **33**:85–88.

Burton, C. E., W. M. Portnoy, and H. A. Wright. 1972. Borer activity and mesquite ignition parameters. *Proc. South. Weed Sci. Soc.* **25**:303–313.

Cable, D. R. 1967. Fire effects on semidesert grasses and shrubs. *J. Range Manage.* **20**:170–176.

Cable, D. R. 1975. Range management in the chaparral type and its ecological basis: The status of our knowledge. USDA For. Serv. Res. Paper RM-155. Rocky Mtn. For. and Range Exp. Stn., Fort Collins, Colo.

Chrosciewicz, Z. 1974. Evaluation of fire-produced seedbeds for jack pine regeneration in central Ontario. *Can. J. For. Res.* **4**:455–457.

Chrosciewicz, Z. 1978. Large-scale operational burns for slash disposal and conifer reproduction in central Saskatchewan. Can. For. Serv. Inf. Rep. Nor-X-201. North. For. Res. Cent., Edmonton, Alberta.

Cooper, R. W. 1971. The pros and cons of prescribed burning in the South. *For. Farmer* **31**(2):10–12, 39, 40.

Countryman, C. M. 1964. Mass fires and fire behavior. USDA For. Serv. Res. Pap. PSW-19. Pac. Southwest For. and Range Exp. Stn., Berkeley, Calif.

Countryman, C. M. 1967. Thermal characteristics of pinyon pine and juniper fuels used in experimental fires. In Proc. Tripartite Tech. Cooperation Program Panel N-3 (Thermal Radiation) Mass Fire Symp., DASA 1949, DASA Information and Analysis Cent. Special Rep. 59, pp. 309–319.

Countryman, C. M. 1971. Fire whirls . . . why, when, and where. USDA For. Serv., Pac. Southwest For. and Range Exp. Stn., Berkeley, Calif.

Countryman, C. M. 1975. The nature of heat: Heat—Its role in wildland fire. Part 1. USDA For. Serv., Pac. Southwest For. and Range Exp. Stn., Berkeley, Calif.

Countryman, C. M. 1976. Radiation and wildland fire: Heat—Its role in wildland fire. Part 5. USDA For. Serv., Pac. Southwest For. and Range Exp. Stn., Berkeley, Calif.

Countryman, C. M. 1978. Radiation: Heat—Its role in wildland fire. Part 4. USDA For. Serv., Pac. Southwest For. and Range Exp. Stn., Berkeley, Calif.

Countryman, C. M., and C. Philpot. 1970. Physical characteristics of chamise as a wildland fuel. USDA For. Serv. Res. Paper PSW-66. Pac. Southwest For. and Range Exp. Stn., Berkeley, Calif.

Davis, J. R., and J. H. Dieterich. 1976. Predicting rate of spread (ROS) in Arizona oak chaparral: Field workbook. USDA For. Serv. Gen. Tech. Rep. RM-24. Rocky Mtn. For. and Range Exp. Stn., Fort Collins, Colo.

Davis, K. P. 1959. *Forest Fire: Control and Use.* McGraw-Hill, New York.

Davis, W. F. 1976. Planning and constructing firebreaks for prescribed burning within the Intermountain range ecosystem. In *Use of Prescribed Burning In Western Woodland and Range Ecosystems: A Symposium.* Utah State Univ., Logan, pp. 65–68.

Deeming, J. E., R. E. Burgan, and J. D. Cohen. 1978. The national fire-danger rating system—1978. USDA For. Serv. Gen. Tech. Rep. INT-39. Intermt. For. and Range Exp. Stn., Ogden, Utah.

Dell, J. D., and F. R. Ward. 1970. Building firelines with liquid explosive. USDA For. Serv. Res. Note PSW-200. Pac. Southwest For. and Range Exp. Stn., Berkeley, Calif.

Dieterich, J. H. 1971. Air-quality aspects of prescribed burning. In Proc. Prescribed Burning Symp., Charleston, S. C. USDA For. Serv. Southeastern For. Exp. Stn., Asheville, N.C., pp. 139–151.

Dieterich, J. H. 1979. Recovery potential of fire-damaged southwestern ponderosa pine. USDA For. Serv. Res. Note RM-379. Rocky Mtn. For. and Range Exp. Stn., Fort Collins, Colo.

Dixon, M. J. 1965. A guide to fire by prescription. USDA For. Serv., Southern Reg. Atlanta, Ga.

Dubé, D. 1977. Prescribed burning in Jasper National Park. For. Rep. **5**(2):4, 5. Can. For. Serv., North. For. Res. Cent., Edmonton, Alberta.

Dwyer, D. D. 1972. Burning and nitrogen fertilization of tobosagrass. New Mexico State Univ., Agric. Exp. Stn. Bull. 595. Las Cruces.

Dwyer, D. D., and R. D. Pieper. 1967. Fire effects on blue grama-pinyon-juniper rangeland in New Mexico. *J. Range Manage.* **20:**359–362.

Fahnestock, G. R., and R. C. Hare. 1964. Heating of tree trunks in surface fires. *J. For.* **62:**799–805.

Fenner, R. L., R. K. Arnold, and C. C. Buck. 1955. Area ignition for brush burning. USDA For. Serv. Tech. Paper No. 10. Pac. Southwest For. and Range Exp. Stn., Berkeley, Calif.

Fons, W. L. 1950. Heating and ignition of small wood cylinders. *Ind. and Eng. Chem.* **42:**2130–2133.

Fosberg, M. A. 1977. Forecasting the 10-hour timelag fuel moisture. USDA For. Serv. Res. Paper RM-187. Rocky Mtn. For. and Range Exp. Stn., Fort Collins, Colo.

Fritchen, L. J., H. Bovee, K. Buettner, R. Charlson, L. Monteith, S. Pickford, and J. Murphy. 1970. Slash fire atmospheric pollution. USDA For. Serv. Res. Paper PNW-97. Pac. Northwest For. and Range Exp. Stn., Portland, Ore.

Gartner, F. R., and W. W. Thompson. 1972. Fire in the Black Hills forest-grass ecotone. *Proc. Tall Timbers Fire Ecol. Conf.* **12:**37–68.

Gordon, D. T. 1967. Prescribed burning in the interior ponderosa pine type of northern California—a preliminary study. USDA For. Serv. Res. Paper PSW-45. Pac. Southwest For. and Range Exp. Stn., Berkeley, Calif.

Green, L. R. 1970. An experimental prescribed burn to reduce fuel hazard in chaparral. USDA For. Serv. Res. Note PSW-216. Pac. Southwest For. and Range Exp. Stn., Berkeley, Calif.

Green, L. R. 1977. Fuel breaks and other fuel modification for wildland fire control. USDA Handb. No. 499. Washington, D.C.

Green, L. R. 1981. Burning by prescription in chaparral. USDA For. Serv. Gen. Tech. Rep. PSW-51. Pac. Southwest For. and Range Exp. Stn., Berkeley, Calif.

Haines, D. A., and G. Updike. 1971. Fire whirlwind formation over flat terrain. USDA For. Serv. Res. Paper NC-71. North Cent. For. Exp. Stn., St. Paul, Minn.

Heirman, A. L., and H. A. Wright. 1973. Fire in medium fuels of west Texas. *J. Range Manage.* **26**:331–335.

Hester, D. A. 1952. The piñon-juniper fuel type can really burn. *Fire Contr. Notes* **13**(1):26–29.

Jameson, D. A. 1962. Effects of burning on a galleta-black grama range invaded by juniper. *Ecology* **43**:760–763.

Johnson, J. R., and G. F. Payne. 1968. Sagebrush re-invasion as affected by some environmental influences. *J. Range Manage.* **21**:209–212.

Johnston, A., and S. Smoliak. 1968. Reclaiming brushland in southwestern Alberta. *J. Range Manage.* **21**:404–406.

Kiil, A. D. 1969. Fuel consumption by a prescribed burn in spruce-fir logging slash in Alberta. *For. Chron.* **45**:100–102.

Klebenow, D. A., and A. D. Bruner. 1976. Determining factors necessary for prescribed burning. In *Use of Prescribed Burning in Western Woodland and Range Ecosystems: A Symposium. Utah State Univ.*, Logan, pp. 69–74.

Komarek, E. V., Sr. 1970. Controlled burning and air pollution: An ecological review. *Proc. Tall Timbers Fire Ecol. Conf.* **10**:141–173.

Lancaster, J. W. 1970. Timelag useful in fire danger rating. *Fire Contr. Notes* **31**(3):6–8, 10.

Launchbaugh, J. L., and C. E. Owensby. 1978. Kansas rangelands: Their management based on a half century of research. Kansas Agric. Exp. Stn. Bull. 622. Manhattan.

Lindenmuth, A. W., Jr., and J. R. Davis. 1973. Predicting fire spread in Arizona oak chaparral. USDA For. Serv. Res. Paper RM-101. Rocky Mtn. For. and Range Exp. Stn., Fort Collins, Colo.

Maini, J. S. 1960. Invasion of grassland by *Populus tremuloides* in the Northern Great Plains. Ph.D. Diss. Univ. Saskatchewan, Saskatoon.

Martin, R. E., H. E. Anderson, W. D. Boyer, J. H. Dieterich, S. N. Hirsch, V. J. Johnson, and W. H. McNab. 1979. Effects of fire on fuels: A state-of-knowledge review. USDA For. Serv. Gen. Tech. Rep. WO-13. Washington, D.C.

Martin, R. E., R. W. Cooper, A. B. Crow, J. A. Cuming, and C. B. Phillips. 1977. Report of task force on prescribed burning. *J. For.* **75**:297–301.

Martin, R. E., and J. D. Dell. 1978. Planning for prescribed burning in the Inland Northwest. USDA For. Serv. Tech. Rep. PNW-76. Pac. Northwest For. and Range Exp. Stn., Portland, Ore.

McIlvain, E. H., and C. G. Armstrong. 1966. A summary of fire and forage research on shinnery oak rangelands. *Proc. Tall Timbers Fire Ecol. Conf.* **5**:127–129.

Meagher, G. S. 1943. Reaction of piñon and juniper seedlings to artificial shade and supplemental watering. *J. For.* **41**:480–482.

Mobley, H. E., R. S. Jackson, W. E. Balmer, W. E. Ruziska, and W. A. Hough. 1973. A guide for prescribed fire in southern forests. USDA For. Serv., Southern Reg. Atlanta, Ga. (Rev. 1978.)

Nelson, J. G., and R. E. England. 1971. Some comments on the causes and effects of fire in the northern grasslands area of Canada and the nearby United States. ca. 1750–1900. *Can. Geogr.* **15**:295–306.

Nikleva, S. 1972. The air pollution potential of slash burning in southwestern B. C. *For. Chron.* **48**:187–189.

Norum, R. A. 1975. Characteristics and effects of understory fires in western larch/Douglas-fir stands. Ph.D. Diss. Univ. Montana, Missoula.

Norum, R. A. 1976. Fire intensity-fuel reduction relationships associated with understory burning in larch/Douglas-fir stands. *Proc. Tall Timbers Fire Ecol. Conf.* **14:**559–572.

Norum, R. A., and W. C. Fischer. 1980. Determining the moisture content of some dead forest fuels using a microwave oven. USDA For. Serv. Res. Note INT-277. Intermt. For. and Range Exp. Stn., Ogden, Utah.

Pechanec, J. F., G. Stewart, and J. P. Blaisdell. 1954. Sagebrush burning—good and bad. USDA Farmers' Bull. 1948. Washington, D.C.

Perala, D. A. 1974. Prescribed burning in an aspen-mixed hardwood forest. *Can. J. For. Res.* **4:**222–228.

Perovich, J. M. 1977. Facing up to smoke management. *Southern Lumberman.* February:8, 9.

Philpot, C. W. 1969. Seasonal changes in heat content and ether extractive content of chamise. USDA For. Serv. Res. Paper INT-61. Intermt. For. and Range Exp. Stn., Ogden, Utah.

Quintilio, D. 1972. Fire spread and impact in lodgepole pine slash. M.S. Thesis. Univ. Montana, Missoula.

Ralphs, M., D. Schen, and F. E. Busby. 1976. General considerations necessary in planning a prescribed burn. In *Use of Prescribed Burning in Western Woodland and Range Ecosystems: A Symposium.* Utah State Univ., Logan, pp. 49–53.

Randall, A. G. 1966. A prescribed burn following a clearcut in the spruce type. Maine Agric. Exp. Stn., Misc. Pub. 675. Univ. Maine, Orono.

Sackett, S. S. 1975. Scheduling prescribed burns for hazard reduction in the Southeast. *J. For.* **73:** 143–147.

Sackett, S. S. 1980. An instrument for rapid, accurate, determination of fuel moisture content. *Fire Manage. Notes* **41**(2): 17, 18.

Sandberg, D. V., and R. E. Martin. 1975. Particle sizes in slash fire smoke. USDA For. Serv. Res. Paper PNW-199. Pac. Northwest For. and Range Exp. Stn., Portland, Ore.

Sandberg, D. V., J. M. Pierovich, D. G. Fox, and E. W. Ross. 1979. Effects of fire on air: A state-of-knowledge review. USDA For. Serv. Gen. Tech. Rep. WO-9. Washington, D.C.

Sando, R. W., and R. C. Dobbs. 1970. Planning for prescribed burning in Manitoba and Saskatchewan. Liaison and Serv. Note MS-L-9. Can. Dept. Fish. and For., Winnipeg, Manitoba.

Scheffler, E. J. 1976. Aspen forest vegetation in a portion of the east central Alberta parklands. M.S. Thesis. Univ. Alberta, Edmonton.

Schimke, H. E., J. D. Dell, and F. R. Ward. 1969. Electrical ignition for prescribed burning. USDA For. Serv. Pac. Southwest For. and Range Exp. Stn., Berkeley, Calif.

Schimke, H. E., and L. R. Green. 1970. Prescribed fire for maintaining fuel-breaks in the central Sierra Nevada. USDA For. Serv. Pac. Southwest For. and Range Exp. Stn., Berkeley, Calif.

Schlichtemeier, G. 1967. Marsh burning for waterfowl. *Proc. Tall Timbers Fire Ecol. Conf.* **6:**40–46.

Schroeder, M. J., and C. C. Buck. 1970. Fire weather. USDA Handb. 360. Washington, D.C.

Sinton, H. M. 1980. Effect of seasonal burning and mowing on *Festuca hallii* Loom. *(Festuca scabrella* Torr.). M.S. Thesis. Univ. Alberta, Edmonton.

Southwest Interagency Fire Council. 1968. Guide to prescribed fire in the Southwest. Watershed Manage. Dept., Univ. Arizona, Tucson.

Stinson, K. J. 1978. Range and wildlife habitat improvement through prescribed burning in New Mexico. *Soc. Range Manage. Abstr.* **31:**23.

Strang, R. M., and J. V. Parminter. 1980. Conifer encroachment on the Chilcotin grasslands of British Columbia. *For. Chron.* **56:**13–18.

Swanson, R. J. 1974. Prescribed underburning for wildfire hazard abatement in second-growth stands of west-side Douglas-fir *[Pseudotsuga menziesii (Mirb.) Franco]. M.S. Thesis. Univ. Washington, Seattle.*

Swanson, R. J. 1976. Hazard abatement by prescribed underburning west-side Douglas-fir. *Proc. Tall Timbers Fire Ecol. Conf.* **15:**235–238.

Taylor, D. F., and G. W. Wendel. 1964. Stamper tract prescribed burn. USDA For. Serv. Res. Paper SE-14. Southeastern For. Exp. Stn., Asheville, N.C.

Truesdell, P. S. 1969. Postulates of the prescribed burning program of the Bureau of Indian Affairs. *Proc. Tall Timbers Fire Ecol. Conf.* **9:**235–240.

Van Wagner, C. E. 1963. Prescribed burning experiments: red and white pine. Dept. For. Pub. No. 1020. For. Res. Br., Ottawa, Ont.

Van Wagner, C. E. 1972. Duff consumption by fire in eastern pine stands. *Can. J. For. Res.* **2:**34–39.

Van Wagner, C. E. 1974. Structure of the Canadian Forest Fire Weather Index. Dept. Environ., Can. For. Pub. 1333. Ottawa, Ont.

Van Wagner, C. E., and I. R. Methven. 1977. Prescribed fire for site preparation in white and red pine. In Proc. Symp. sponsored by Ont. Min. Nat. Res. and Can. For. Serv., Chalk River, Ont., pp. 95–100.

Wagle, R. F., and T. W. Eakle. 1979. A controlled burn reduces the impact of a subsequent wildfire in a ponderosa pine vegetation type. *For. Sci.* **25:**123–129.

Weaver, H. 1957. Effects of prescribed burning in ponderosa pine. *J. For.* **55:**133–138.

Weaver, H. 1967. Fire and its relationship to ponderosa pine. *Proc. Tall Timbers Fire Ecol. Conf.* **7:**127–149.

Willms, W., A. W. Bailey, and A. McLean. 1980. The effects of clipping or burning on the morphological characteristics of bluebunch wheatgrass plants. *Can. J. Bot.* **58:**2309–2312.

Wink, R. L., and H. A. Wright. 1973. Effects of fire on an Ashe juniper community. *J. Range Manage.* **26:**326–329.

Wright, H. A. 1972. Fire as a tool to manage tobosa grasslands. *Proc. Tall Timbers Fire Ecol. Conf.* **12:**153–167.

Wright, H. A. 1974. Range Burning. *J. Range Manage.* **27:**5–11.

Wright, H. A. 1978. Use of fire to manage grasslands of the Great Plains: Central and Southern Great Plains. In *Proc. of the First Int. Rangeland Congr.*, Denver, Colo., pp. 694–696.

Wright, H. A., and A. W. Bailey. 1980. Fire ecology and prescribed burning in the Great Plains—A research review. USDA For. Serv. Gen. Tech. Rep. INT-77. Intermt. For. and Range Exp. Stn., Ogden, Utah.

SUBJECT INDEX

PLANT INDEX

ANIMAL INDEX